中 外 物 理 学 精 品 书 系
本 书 出 版 得 到 " 国 家 出 版 基 金 " 资 助

国家出版基金项目
NATIONAL PUBLICATION FOUNDATION

中外物理学精品书系

引 进 系 列 · 7 3

Atoms and Molecules Interacting with Light:
Atomic Physics for the Laser Era

原子及分子与光的相互作用
——激光时代的原子物理

（影印版）

〔荷〕彼得·范德施特拉滕（Peter van der Straten）
〔美〕哈罗德·梅特卡夫（Harold Metcalf） 著

北京大学出版社
PEKING UNIVERSITY PRESS

著作权合同登记号　图字：01-2020-3099

图书在版编目(CIP)数据

原子及分子与光的相互作用：激光时代的原子物理＝Atoms and Molecules Interacting with Light: Atomic Physics for the Laser Era：英文/(荷)彼得·范德施特拉滕，(美)哈罗德·梅特卡夫著. —影印本. —北京：北京大学出版社，2020.8
（中外物理学精品书系）
ISBN 978-7-301-31407-4

Ⅰ. ①原… Ⅱ. ①彼… ②哈… Ⅲ. ①原子物理学-研究-英文 Ⅳ. ①O562

中国版本图书馆 CIP 数据核字(2020)第 115866 号

Atoms and Molecules Interacting with Light: Atomic Physics for the Laser Era (ISBN-13: 9781107009790) by P. van der Straten and H. Metcalf, first published by Cambridge University Press 2016.
All rights reserved.

This reprint edition for the People's Republic of China is published by arrangement with the Press Syndicate of the University of Cambridge, Cambridge, United Kingdom.
© Cambridge University Press & Peking University Press 2020.

This book is in copyright. No reproduction of any part may take place without the written permission of Cambridge University Press and Peking University Press.

This edition is for sale in the People's Republic of China (excluding Hong Kong SAR, Macau SAR and Taiwan Province) only.
此版本仅限在中华人民共和国(不包括香港、澳门特别行政区及台湾地区)销售。

Copies of this book sold without a Cambridge University Press sticker on the cover are unauthorized and illegal.
本书封面贴有 Cambridge University Press 防伪标签，无标签者不得销售。

书　　　名	Atoms and Molecules Interacting with Light: Atomic Physics for the Laser Era（原子及分子与光的相互作用——激光时代的原子物理）（影印版）
著作责任者	〔荷〕彼得·范德施特拉滕（Peter van der Straten） 〔美〕哈罗德·梅特卡夫（Harold Metcalf） 著
责任编辑	刘　啸
标准书号	ISBN 978-7-301-31407-4
出版发行	北京大学出版社
地　　　址	北京市海淀区成府路 205 号　100871
网　　　址	http://www.pup.cn　新浪微博：@北京大学出版社
电子信箱	zpup@pup.cn
电　　　话	邮购部 010-62752015　发行部 010-62750672　编辑部 010-62754271
印　刷　者	北京中科印刷有限公司
经　销　者	新华书店
	730 毫米×980 毫米　16 开本　33.5 印张　630 千字
	2020 年 8 月第 1 版　2020 年 8 月第 1 次印刷
定　　　价	116.00 元

未经许可，不得以任何方式复制或抄袭本书之部分或全部内容。
版权所有，侵权必究
举报电话：010-62752024　电子信箱：fd@pup.pku.edu.cn
图书如有印装质量问题，请与出版部联系，电话：010-62756370

"中外物理学精品书系"
(二期)
编 委 会

主　任：王恩哥
副主任：夏建白
编　委：(按姓氏笔画排序,标*号者为执行编委)

丁　洪	王力军	王孝群	王　牧	王雪华
王鼎盛	石　兢	田光善	冯世平	邢定钰
朱邦芬	朱　星	向　涛	刘　川*	汤　超
许宁生	许京军	李茂枝	李建新	李新征*
李儒新	吴　飙	汪卫华	张立新	张振宇
张　酣*	张富春	陈志坚*	武向平	林海青
欧阳钟灿	罗民兴	周月梅*	郑春开	赵光达
钟建新	聂玉昕	徐仁新*	徐红星	郭　卫
资　剑	龚新高	龚旗煌	崔　田	阎守胜
谢心澄	解士杰	解思深	樊铁栓*	潘建伟

秘　书：陈小红

序　言

物理学是研究物质、能量以及它们之间相互作用的科学。她不仅是化学、生命、材料、信息、能源和环境等相关学科的基础，同时还与许多新兴学科和交叉学科的前沿紧密相关。在科技发展日新月异和国际竞争日趋激烈的今天，物理学不再囿于基础科学和技术应用研究的范畴，而是在国家发展与人类进步的历史进程中发挥着越来越关键的作用。

我们欣喜地看到，改革开放四十年来，随着中国政治、经济、科技、教育等各项事业的蓬勃发展，我国物理学取得了跨越式的进步，成长出一批具有国际影响力的学者，做出了很多为世界所瞩目的研究成果。今日的中国物理，正在经历一个历史上少有的黄金时代。

在我国物理学科快速发展的背景下，近年来物理学相关书籍也呈现百花齐放的良好态势，在知识传承、学术交流、人才培养等方面发挥着无可替代的作用。然而从另一方面看，尽管国内各出版社相继推出了一些质量很高的物理教材和图书，但系统总结物理学各门类知识和发展，深入浅出地介绍其与现代科学技术之间的渊源，并针对不同层次的读者提供有价值的学习和研究参考，仍是我国科学传播与出版领域面临的一个富有挑战性的课题。

为积极推动我国物理学研究、加快相关学科的建设与发展，特别是集中展现近年来中国物理学者的研究水平和成果，北京大学出版社在国家出版基金的支持下于2009年推出了"中外物理学精品书系"，并于2018年启动了书系的二期项目，试图对以上难题进行大胆的探索。书系编委会集结了数十位来自内地和香港顶尖高校及科研院所的知名学者。他们都是目前各领域十分活跃的知名专家，从而确保了整套丛书的权威性和前瞻性。

这套书系内容丰富、涵盖面广、可读性强，其中既有对我国物理学发展的梳理和总结，也有对国际物理学前沿的全面展示。可以说，"中外物理学精品书系"力图完整呈现近现代世界和中国物理科学发展的全貌，是一套目前国内为数不多的兼具学术价值和阅读乐趣的经典物理丛书。

"中外物理学精品书系"的另一个突出特点是，在把西方物理的精华要义"请进来"的同时，也将我国近现代物理的优秀成果"送出去"。物理学在世界范围内的重要性不言而喻。引进和翻译世界物理的经典著作和前沿动态，可以满足当前国内物理教学和科研工作的迫切需求。与此同时，我国的物理学研究数十年来取得了长足发展，一大批具有较高学术价值的著作相继问世。这套丛书首次成规模地将中国物理学者的优秀论著以英文版的形式直接推向国际相关研究的主流领域，使世界对中国物理学的过去和现状有更多、更深入的了解，不仅充分展示出中国物理学研究和积累的"硬实力"，也向世界主动传播我国科技文化领域不断创新发展的"软实力"，对全面提升中国科学教育领域的国际形象起到一定的促进作用。

习近平总书记在 2018 年两院院士大会开幕会上的讲话强调，"中国要强盛、要复兴，就一定要大力发展科学技术，努力成为世界主要科学中心和创新高地"。中国未来的发展在于创新，而基础研究正是一切创新的根本和源泉。我相信，在第一期的基础上，第二期"中外物理学精品书系"会努力做得更好，不仅可以使所有热爱和研究物理学的人们从中获取思想的启迪、智力的挑战和阅读的乐趣，也将进一步推动其他相关基础科学更好更快地发展，为我国的科技创新和社会进步做出应有的贡献。

"中外物理学精品书系"编委会主任
中国科学院院士，北京大学教授
王恩哥
2018 年 7 月于燕园

ATOMS AND MOLECULES INTERACTING WITH LIGHT

Atomic Physics for the Laser Era

PETER VAN DER STRATEN
Utrecht University

HAROLD METCALF
Stony Brook University

Contents

Preface		page xi
Part I	**Atom–light interaction**	1
1	**The classical physics pathway**	3
1.1	Introduction	3
1.2	Damped harmonic oscillator	4
1.3	The damped driven oscillator	6
1.4	The Bohr model	7
1.5	de Broglie waves	9
Appendix 1.A	Damping force on an accelerating charge	10
Appendix 1.B	Hanle effect	11
Appendix 1.C	Optical tweezers	12
2	**Interaction of two-level atoms and light**	16
2.1	Introduction	16
2.2	Quantum mechanical view of driven optical transitions	16
2.3	Rabi oscillations	17
2.4	The dressed atom picture	21
2.5	The Bloch vector and Bloch sphere	23
Appendix 2.A	Pauli matrices for motion of the Bloch vector	25
Appendix 2.B	The Ramsey method	26
Appendix 2.C	Echoes and interferometry	30
Appendix 2.D	Adiabatic rapid passage	34
Appendix 2.E	Superposition and entanglement	36
3	**The atom–light interaction**	40
3.1	Introduction	40
3.2	The three primary approximations	40

3.3	Light fields of finite spectral width	44
3.4	Oscillator strength	46
3.5	Selection rules	47
Appendix 3.A	Proof of the oscillator strength theorem	50
Appendix 3.B	Electromagnetic fields	50
Appendix 3.C	The dipole approximation	55
Appendix 3.D	Time resolved fluorescence from multilevel atoms	56

4 "Forbidden" transitions — 64

4.1	Introduction	64
4.2	Extending the electric dipole approximation	65
4.3	Extending the perturbation approximation	70
Appendix 4.A	Higher-order approximations	78

5 Spontaneous emission — 80

5.1	Introduction	80
5.2	Einstein A- and B-coefficients	80
5.3	Discussion of this semi-classical description	83
5.4	The Wigner–Weisskopf model	84
Appendix 5.A	The quantum mechanical harmonic oscillator	87
Appendix 5.B	Field quantization	89
Appendix 5.C	Alternative theories to QED	91

6 The density matrix — 93

6.1	Introduction	93
6.2	Basic concepts	93
6.3	The optical Bloch equations	96
6.4	Power broadening and saturation	98
Appendix 6.A	The Liouville–von Neumann equation	101

Part II Internal structure — 105

7 The hydrogen atom — 107

7.1	Introduction	107
7.2	The Hamiltonian of hydrogen	108
7.3	Solving the angular part	109
7.4	Solving the radial part	110
7.5	The scale of atoms	117
7.6	Optical transitions in hydrogen	118
Appendix 7.A	Center-of-mass motion	122
Appendix 7.B	Coordinate systems	123

	Appendix 7.C	Commuting operators	124
	Appendix 7.D	Matrix elements of the radial wavefunctions	125

8 Fine structure — 131

8.1	Introduction	131
8.2	The relativistic mass term	132
8.3	The fine-structure "spin–orbit" term	133
8.4	The Darwin term	138
8.5	Summary of fine structure	138
8.6	The Dirac equation	138
8.7	The Lamb shift	140
Appendix 8.A	The Sommerfeld fine-structure constant	142
Appendix 8.B	Measurements of the fine structure	144

9 Effects of the nucleus — 149

9.1	Introduction	149
9.2	Motion, size, and shape of the nucleus	149
9.3	Nuclear magnetism – hyperfine structure	152
Appendix 9.A	Interacting magnetic dipoles	157
Appendix 9.B	Hyperfine structure for two spin-$1/2$ particles	160
Appendix 9.C	The hydrogen maser	161

10 The alkali-metal atoms — 164

10.1	Introduction	165
10.2	Quantum defect theory	166
10.3	Non-penetrating orbits	168
10.4	Model potentials	170
10.5	Optical transitions in alkali-metal atoms	171
Appendix 10.A	Quantum defects for the alkalis	175
Appendix 10.B	Numerov method	176

11 Atoms in magnetic fields — 181

11.1	Introduction	181
11.2	The Hamiltonian for the Zeeman effect	182
11.3	Zeeman shifts in the presence of the spin–orbit interaction	183
Appendix 11.A	The ground state of atomic hydrogen	188
Appendix 11.B	Positronium	190
Appendix 11.C	The non-crossing theorem	192
Appendix 11.D	Passage through an anti-crossing: Landau–Zener transitions	194

12 Atoms in electric fields — 198
- 12.1 Introduction — 198
- 12.2 Electric field shifts in spherical coordinates — 199
- 12.3 Electric field shifts in parabolic coordinates — 202
- 12.4 Summary — 205

13 Rydberg atoms — 208
- 13.1 Introduction — 208
- 13.2 The Bohr model and quantum defects again — 209
- 13.3 Rydberg atoms in external fields — 211
- 13.4 Experimental description — 218
- 13.5 Some results of Rydberg spectroscopy — 219

14 The helium atom — 227
- 14.1 Introduction — 227
- 14.2 Symmetry — 227
- 14.3 The Hamiltonian for helium — 230
- 14.4 Variational methods — 233
- 14.5 Doubly excited states — 237
- Appendix 14.A Variational calculations — 239
- Appendix 14.B Detail on the variational calculations of the ground state — 239

15 The periodic system of the elements — 244
- 15.1 The independent particle model — 245
- 15.2 The Pauli symmetrization principle — 247
- 15.3 The "Aufbau" principle — 248
- 15.4 Coupling of many-electron atoms — 249
- 15.5 Hund's rules — 254
- 15.6 Hartree–Fock model — 256
- 15.7 The Periodic Table — 258
- Appendix 15.A Paramagnetism — 261
- Appendix 15.B The color of gold — 264

16 Molecules — 272
- 16.1 Introduction — 272
- 16.2 A heuristic description — 274
- 16.3 Quantum description of nuclear motion — 276
- 16.4 Bonding in molecules — 282
- 16.5 Electronic states of molecules — 286
- 16.6 Optical transitions in molecules — 290
- Appendix 16.A Morse potential — 299

17	**Binding in the hydrogen molecule**		303
	17.1 The hydrogen molecular ion		303
	17.2 The molecular orbital approach to H_2		306
	17.3 The valence bond approach to H_2		311
	17.4 Improving the methods		312
	17.5 Nature of the H_2 bond		314
	Appendix 17.A	Confocal elliptical coordinates	316
	Appendix 17.B	One-electron, two-center integrals	317
	Appendix 17.C	Electron–electron interaction in molecular hydrogen	318
18	**Ultra-cold chemistry**		321
	18.1 Introduction		321
	18.2 Long-range molecular potentials		322
	18.3 LeRoy–Bernstein method		328
	18.4 Scattering theory		331
	18.5 The scattering length		334
	18.6 Feshbach molecules		337
Part III	**Applications**		345
19	**Optical forces and laser cooling**		347
	19.1 Two kinds of optical forces		347
	19.2 Low-intensity laser light pressure		348
	19.3 Atomic beam slowing and collimation		352
	19.4 Optical molasses		353
	19.5 Temperature limits		355
	19.6 Experiments in three-dimensional optical molasses		356
	19.7 Cooling below the Doppler temperature		359
20	**Confinement of neutral atoms**		367
	20.1 Dipole force optical traps		367
	20.2 Magnetic traps		370
	20.3 Magneto-optical traps		373
	20.4 Optical lattices		376
21	**Bose–Einstein condensation**		382
	21.1 Introduction		382
	21.2 The road to BEC		384
	21.3 Quantum statistics		385
	21.4 Mean-field description of the condensate		388
	21.5 Interference of two condensates		391
	21.6 Quantum hydrodynamics		394

21.7	The superfluid–Mott insulator transition	400
Appendix 21.A	Distribution functions	405
Appendix 21.B	Density of states	410

22 Cold molecules — 413
22.1 Slowing, cooling, and trapping molecules — 413
22.2 Stark slowing of molecules — 415
22.3 Buffer gas cooling — 418
22.4 Binding cold atoms into molecules — 420
22.5 A case study: photo-association spectroscopy — 426

23 Three-level systems — 433
23.1 Introduction — 433
23.2 The spontaneous and stimulated Raman effects — 435
23.3 Coherent population trapping — 436
23.4 Autler–Townes and EIT — 438
23.5 Stimulated rapid adiabatic passage — 440
23.6 Slow light — 442
23.7 Observations and measurements — 444
Appendix 23.A General case for $\delta_1 \neq \delta_2$ — 445

24 Fundamental physics — 448
24.1 Precision measurements and QED — 449
24.2 Variation of the constants — 452
24.3 Exotic atoms and antimatter — 455
24.4 Bell inequalities — 459
24.5 Parity violation and the anapole moment — 461
24.6 Measuring zero — 463

Part IV Appendix — 465

Appendix A Notation and definitions — 467

Appendix B Units and notation — 471

Appendix C Angular momentum in quantum mechanics — 473

Appendix D Transition strengths — 479

References — 490
Index — 508

Preface

By any measure, atomic physics is among the fastest-growing, most dynamic, and best-recognized areas of physics. The student attendance at conferences devoted to this subject area has burgeoned, and the number of new university tenure track positions in atomic physics is disproportionately high. Recently there have been several documents from national and international science-oriented agencies extolling the growth and importance of atomic physics, with statements such as "Light influences our lives today in ways we could never have imagined a few decades ago," and "AMO science (not only) provides the basis for new technology, it is also a source of the intellectual capital on which science and technology depend for growth and development." The General Assembly of the United Nations and UNESCO declared 2015 as the "International Year of Light", further underscoring its importance to the world. Applications are not restricted to further exploration of our specialized field, but have expanded to include substantial impact on other areas of physics such as condensed matter, quantum information, thermodynamics, and fluid mechanics. For these and other reasons, we have decided to provide an introductory text appropriate to the emerging laser era in atomic, molecular, and optical science.

This book is intended for multiple purposes. First and foremost, we are experimentalists, so the material is presented in an intuitive and very physical tone. Many ideas are developed from the classical physics perspective rather than from mathematical formalism. We try to connect the concepts with measurements where it makes sense to do so, and to motivate each topic by the observations that produced the information about it.

Our intent is to use it as a text for a course in atomic physics. It requires a knowledge of quantum mechanics at the level of the well-known textbooks by Griffiths or Liboff, and of elementary electricity and magnetism. Thus it is suitable for an advanced undergraduate course or a beginning graduate course (certainly for a course that serves both populations together). In addition, we have tried to write in a sufficiently familiar style that a student can read and understand the material even without the benefit of a course. That is, the material is presented in a descriptive mode to maximize understanding. Some applications and detailed calculations relevant to a particular chapter are provided as appendices to the specific chapter.

Atomic physics underwent a renaissance with the advent of tunable lasers in the 1970s, and we have taken this revolution to heart. Modern atomic physics is intimately coupled to the interaction of atoms with laser light, and we have chosen to emphasize this aspect. Since almost all students in physics or chemistry have some minimal notions about atomic structure before they undertake a course devoted to this subject, this book begins with several chapters on transitions of two-level systems driven by a single frequency of light. We have chosen this approach because it is most appropriate for the frontiers of research in atomic physics at present. Most of the books currently available emphasize atomic structure first, and the interaction with light is treated as a secondary topic. Thus this book is distinct because its initial approach is based on the interaction of atoms with light as opposed to the elements of atomic structure.

The text begins with a discussion of classical physics as it relates to atomic physics because there are so many striking similarities. The second chapter has a discussion of two-level systems with appendices on Ramsey spectroscopy, adiabatic rapid passage, entanglement, and other topics. Then we generalize to multilevel systems with selection rules, the usual radiative approximations, and a discussion of electromagnetic fields from various sources. In the next chapter we relax these approximations to include M1/E2 transitions and multiphoton processes. The following chapters introduce spontaneous emission, the density matrix, field quantization, and several related topics.

These discussions of electromagnetic radiation at the start are not to suggest that atomic structure is neglected. Beginning with the hydrogen atom in Chapter 7, the conventional topics are discussed in some detail. After chapters on the fine and hyperfine structure of hydrogen including measurements of the Lamb shift and fine structure, there are chapters on helium and heavier atoms, followed by a treatment of external fields (Zeeman and Stark effects). What is important is that they are interspersed with material that is rarely found in textbooks, such as intuitive description of quantum defects. We have found that students rarely understand why the dominant transitions of many atoms are between two states of the same principal quantum number: the Bohr energy formula is often considered sacrosanct. This second part of the book continues with a discussion of the structure of Rydberg atoms, helium, and heavier atoms. These are followed by a chapter on molecular structure and a second one on the paradigm of molecular physics, H_2.

The third part of the book has separate chapters on various applications of the first two parts. These include but are not limited to laser cooling, trapping, BEC, applications to fundamental physics such as parity violation, exotic atoms, and three-level systems. References are made to the earlier sections of the book where the underlying science has been introduced. Since the development of these applications are contemporary, there are many more references to the current literature than to the standard textbooks that discuss the topics of the first two parts. These are quite likely to be outdated by further progress, whereas the standard topics in Parts I and II will stand the test of time.

Some of the material in Parts I and II has a small overlap, including a few figures, with the early chapters of our previous work, *"Laser Cooling and Trapping"* [1], where the background needed to understand laser cooling and trapping is summarized. However, this book provides more depth and more complete description of the atomic physics needed for present-day research in this field. By contrast, Chapters 19 and 20 here present a summary of laser cooling and trapping, but much more detailed discussion can be found in Ref. [1]. Students are encouraged to consult that book for further studies.

The chapters are complemented by exercises of two types. Some of these exercises are simply mathematical calculations to derive something in the text that is dismissed with "it can be shown that...". Others are extensions of the text where the students are asked to find something new based on a different approach than that taken in the text. Since most of Part III addresses current progress, it does not lend itself as well to exercises as do Parts I and II, so it contains only around 10% of the approximately 130 exercises in the book.

Although there are many systems of units in use, this book is restricted to SI units. Atomic units are very convenient and are summarized in an appendix, but SI units enable numerical evaluation of formulas needed for laboratory work where experimental parameters must be chosen. We think that the price of the frequent appearance of $4\pi\varepsilon_0$ and other constants is worth it.

Part I
Atom–light interaction

1
The classical physics pathway

1.1 Introduction

While it might appear that the nineteenth-century physics presented in this chapter has no place in a topic as quantum mechanically oriented as atomic physics, this is simply not the case. The purpose of this first chapter is to try to convince the readers that the material learned in their early years of studying physics is not disjoint from modern topics, but in fact can provide a foundation for their further understanding. Other examples will occur later.

The reader may be a bit surprised to see that the results of the classical description turn out to be so close to the quantum mechanical results. Where there are significant disagreements between classical and quantum descriptions, it is necessary to explain why these occur. The correspondence principle requires that in the limit of large quantum numbers the complete quantum mechanical description must agree with the classical one. It is always better to need fewer of these explanations and to be able to treat physics as a coherent whole instead of a collection of topics.

The consequences of classical physics are intuitively familiar, and this is often most helpful in gaining deeper understanding. In fact, the approach taken in very many of the chapters in this book is to start with a purely classical description of a topic, and then transport it directly into the Schrödinger equation. Radiation from an accelerated charge, as treated here in the harmonic case, provides an ideal setting for such pedagogy. Even though classical mechanics is used in place of quantum mechanics, for the case of harmonic motion, the results are quite compatible.

The early history of atomic physics contains many topics that are well worth reading about, especially the experimental facilities. For example, there were no electric motors, wire was not readily available, and making even a crude vacuum was an experimental tour de force. Very many textbooks have discussions of these topics, and so they will not be presented here. But the reader is warned that certain oft-repeated modern descriptions are simply wrong.

Perhaps the most important notion to be gained from this chapter is that the classical description of a harmonically bound electron provides quite accurate results about the interaction between atoms and light. Of course, the utility of this description is limited because the classical harmonic oscillator has no internal structure, is infinitely deep, and has a continuous energy spectrum. Atoms do not radiate or interact with waves that have too low a frequency, and they will ionize if exposed to too high a frequency. Still, this chapter will show that the classical description gives very good answers for damping rate of the excited state, the cross-section for light scattering, and the emitted spectrum of the light.

1.2 Damped harmonic oscillator

1.2.1 Introduction

The fact that atoms survive quite violent collisions surprisingly effectively suggests that disturbances to the electrons' motion are countered by restoring forces, and the lowest-order term in an expansion of such a force is linear. Thus $\vec{F} = -k(\vec{r} - \vec{r}_0)$, where $\vec{r}_0$ corresponds to a stable orbit about an infinitely massive nucleus. In this approximation, the equation of motion in one dimension is $\ddot{x} + \omega_0^2 x = 0$, where $\omega_0^2 = k/m$.

An electron in such an orbit radiates because it is an accelerating charge, and the radiation field is calculated in the appendix to this chapter. The energy flux (W/m^2) of the radiated field is given by the Poynting vector, $\vec{S} = \vec{\mathcal{E}} \times \vec{\mathcal{H}}$. Conservation of energy requires that the radiated energy be lost from the motion of the oscillator, and the result is equivalent to a damping force $\vec{F} = -(e^2 \omega_0^2/6\pi\varepsilon_0 c^3)\vec{v} \equiv -\beta \vec{v}$ (also shown in App. 1.A).

The equation of motion for a damped harmonic oscillator is $\ddot{x} + \gamma \dot{x} + \omega_0^2 x = 0$ with solution

$$x = x_0 e^{-\gamma t/2} \cos(\omega_0' t + \phi), \quad \omega_0' = \sqrt{\omega_0^2 - \frac{\gamma^2}{4}}, \quad \gamma = \frac{e^2 \omega_0^2}{6\pi\varepsilon_0 m c^3}, \quad (1.1)$$

where $\gamma \equiv \beta/m$ comes from the damping coefficient following Eq. (1.17) in App. 1.A. This solution is plotted in Fig. 1.1a.

For optical transitions (visible light) $\gamma \approx 5 \times 10^8$ rad/s. This value of γ corresponds to a lifetime $\tau \equiv 1/\gamma \approx 2$ ns and is typical of the value of atomic excited state lifetimes. It is most surprising that such a completely classical consideration results in an answer that matches laboratory measurements. Furthermore, for $\gamma \ll \omega_0$ the frequency shift from Eq. (1.1) is approximately $\gamma^2/8\omega_0 \sim 2$ rad/s, which is negligibly small compared with $\omega_0 \sim 10^{16}$ rad/s and even compared with $\gamma \sim 10^9$ rad/s, so it is not even apparent in Fig. 1.1b. In the following ω_0' will be replaced by ω_0.

1.2 Damped harmonic oscillator

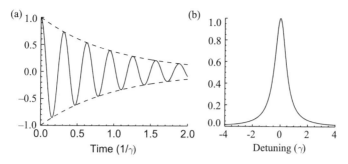

Figure 1.1 (a) Amplitude of a damped harmonic oscillator in the Lorentz model, where the motion is damped because of the acceleration of the electron. (b) Spectrum of the emitted radiation, where the width is inversely proportional to the lifetime.

The quantity $Q \equiv \omega_0/\gamma = 6\pi\varepsilon_0 mc^3/e^2\omega_0$ is the number of cycles during one damping time τ, which is called the quality factor. For optical transitions $Q \approx 2 \times 10^7$. The high Q domain extends all the way up to frequencies $3mc^2/2\hbar$ which includes the ultraviolet, X-ray, and gamma-ray regions, and ends where $E \sim 50\,\mathrm{MeV}$.

1.2.2 Spectrum of emitted radiation

The spectrum of the emitted light can be found by calculating the Fourier transform of the emitted radiation. The emitted field for $t > 0$ is

$$\vec{\mathcal{E}}_{\mathrm{rad}}(t) \equiv \vec{\mathcal{E}}_{\mathrm{rad}}^{(0)} e^{-\gamma t/2} \cos(\omega_0 t + \phi) \qquad (1.2)$$

where $\vec{\mathcal{E}}_{\mathrm{rad}}^{(0)} \equiv e\omega_0^2 x_0/4\pi\varepsilon_0 c^2 r$ using Eq. (1.13) in Appendix 1.A and $\ddot{x} = -\omega_0^2 x$ for harmonic motion. The resulting Fourier transform is

$$\begin{aligned}\vec{\mathcal{E}}_{\mathrm{rad}}(\omega) &= \frac{\vec{\mathcal{E}}_{\mathrm{rad}}^{(0)}}{2\pi} \int_0^\infty e^{-\gamma t/2} \cos(\omega_0 t + \phi) e^{-i\omega t} dt \\ &= \frac{\vec{\mathcal{E}}_{\mathrm{rad}}^{(0)}}{2\pi} \int_0^\infty e^{-\gamma t/2} \frac{1}{2}\left[e^{i(\omega_0 t+\phi)} + e^{-i(\omega_0 t+\phi)}\right] e^{-i\omega t} dt \\ &= \frac{\vec{\mathcal{E}}_{\mathrm{rad}}^{(0)}}{4\pi} \left[\frac{e^{i\phi}}{\gamma/2 + i(\omega - \omega_0)} + \frac{e^{-i\phi}}{\gamma/2 + i(\omega + \omega_0)}\right]. \end{aligned} \qquad (1.3)$$

Since $\omega_0 \gg \gamma$, the second term in Eq. (1.3) can be dropped, and the energy spectrum in the vicinity of $\omega \sim \omega_0$ is described by the Lorentzian

$$\left|\vec{\mathcal{E}}_{\mathrm{rad}}(\omega)\right|^2 = \frac{1}{16\pi^2}\left|\vec{\mathcal{E}}_{\mathrm{rad}}^{(0)}\right|^2\left[\frac{1}{\gamma^2/4 + (\omega - \omega_0)^2}\right], \qquad (1.4)$$

which is plotted in Fig. 1.1b.

1.3 The damped driven oscillator

1.3.1 Radiated power

For a driven oscillating charge, the amplitude will be constant after the initial transients die out, so the radiation will be emitted at a constant rate. The frequency eventually settles to the driving frequency (not the natural frequency). Since the oscillator is constantly radiating energy, it can only be that the energy is taken from the driving field. Conservation of energy allows the calculation of the rate of re-radiation starting with the work done by the driving force $\int \vec{F} \cdot \vec{v}$ averaged over one cycle of the oscillation.

For a driving field $\vec{\mathcal{E}} = \vec{\mathcal{E}}_0 \cos(\omega t)$ the resulting motion is $x = x_0 \cos(\omega t + \phi)$ with ω the driving frequency. The velocity $\dot{x}$ is calculated from the derivative of this motion, and for such a driving force

$$x_0 = \frac{e|\vec{\mathcal{E}}_0|/m}{\sqrt{(\omega_0^2 - \omega^2)^2 + \gamma^2 \omega^2}} \quad \text{and} \quad \tan \phi = \frac{\gamma \omega}{\omega^2 - \omega_0^2}. \tag{1.5}$$

The average scattered power must equal the work done by the driving force, and can be found by integrating over one period $\Delta t = 2\pi/\omega$:

$$\begin{aligned} \overline{P}_{\text{rad}} &= \frac{1}{\Delta t} \int_t^{t+\Delta t} \vec{F} \cdot \vec{v} \, dt \\ &= \frac{\omega e |\vec{\mathcal{E}}_0| x_0}{\Delta t} \int_t^{t+\Delta t} \sin(\omega t + \phi) \cos(\omega t) \, dt \\ &= m\omega^2 x_0^2 (\gamma/2) = \frac{e^2 \mathcal{E}^2}{2m\gamma}, \end{aligned} \tag{1.6}$$

when Eq. (1.5) is evaluated at the resonance condition $\omega = \omega_0$.

1.3.2 Scattering of radiation

By conservation of energy, the rate of work done by the force exerted on the electron by the driving field must equal the radiated power. But knowing the fields allows calculation of the intensity, which is the power per unit area of the incoming light, not its total power. In order to relate scattered power to this incident intensity, there is need for an "effective area" or cross-section σ for the atom that characterizes its effective size for light scattering so that the scattered light power equals the incident light intensity times σ. Using Eq. (1.6) gives

$$\sigma = \frac{e^2 |\vec{\mathcal{E}}_0|^2 / 2m\gamma}{\varepsilon_0 c |\vec{\mathcal{E}}_0|^2 / 2} = \frac{e^2}{\varepsilon_0 m c \gamma} = \frac{6\pi c^2}{\omega_0^2} = \frac{3\lambda^2}{2\pi}. \tag{1.7}$$

where the last steps come from using Eq. (1.1) for γ. Note that the incident light intensity has equal contributions from both the electric and magnetic fields, $\varepsilon_0 c \mathcal{E}^2/2 + c\mathcal{B}^2/2\mu_0$, and each of these is proportional to $\cos^2 \omega t$, whose time average is $1/2$. Thus the intensity is $\varepsilon_0 c \mathcal{E}_0^2/2$.

This is a most interesting and curious result. The resonant cross-section depends only on the wavelength of the incident light. It is completely independent of the type of atom and the properties of the electron. Not only that, it is huge. It is 10,000 times larger than the size of the atom itself.

1.4 The Bohr model

1.4.1 Introduction

In the nineteenth century Balmer had already noticed certain regularities in spectrum of electrical discharges of hydrogen gas. He was interested in number theory and its relation to natural phenomena, and studied the measured values of the wavelengths. He found that the wavelengths of the hydrogen spectrum satisfied

$$\nu_n = \frac{c}{\lambda_n} \propto \left(1 - \frac{4}{n^2}\right), \tag{1.8}$$

where $n \geq 3$ is an integer. In studies that were unrelated at the time, Planck had solved the long-standing problem of the spectrum of light from hot objects in 1900. He did this by assuming, with no physical basis whatsoever, that the light was emitted from oscillators in energy packets that had to be discrete, and the discretization was in terms of the light frequency ν. His now-famous formula is $E = h\nu$ where the constant $h = 6.626 \times 10^{-34}$ Js now bears his name.

At that time, the prevailing theory of atomic structure was Thomson's "plum pudding" model: tiny negative particles embedded in a uniform sea of positive charge that had the atomic size. Thomson's idea was replaced after Rutherford's experiments in 1908 that could only be explained by the "solar system" model still currently in use. However, the "solar system" model conflicted directly with Maxwell's electromagnetic theory that underlies Sec. 1.2. It requires that the orbiting electrons would necessarily emit radiation, because of their acceleration, until the atom collapsed.

Finally, in 1913 Bohr proposed his model. He supposed the "solar system" model but with two additional postulates that were outside the realm of classical physics. One was that the electrons in certain special discrete orbits do not emit classical radiation, and the other was that when they make a transition between discrete orbits whose energy differs by ΔE, they emit radiation according to Planck's formula $\Delta E = h\nu$. A more detailed description is in Sec. 1.4.2. He found that found the measured wavelengths agreed with his calculations to very high precision. In many texts it is asserted that Bohr quantized the orbital angular momentum, but this is simply not true [2, 3].

1.4.2 Energy levels

In his 1913 paper [2] (see also Ref. [3]) Bohr addressed the problem posed by the Rutherford model with two postulates. First, in spite of the tenets of classical mechanics and electrodynamics, there were only discrete allowed orbits of the Kepler motion of electrons in the Coulomb field of a massive nucleus, and these were stable against radiation. Second, transitions between these allowed orbits conserved energy by emitting or absorbing radiation whose frequency is given by the Planck condition $E = h\nu$. This was a radical choice because classical electrodynamics requires that the radiation be at the harmonic frequencies of the orbital motion, but classical electrodynamics had to be discarded for the quantum theory anyway because of the postulate of orbital stability.

Newtonian mechanics for a circular orbit in the Coulomb field requires

$$F = \frac{mv^2}{r} = \frac{e^2}{4\pi\varepsilon_0 r^2}. \tag{1.9}$$

The total energy of such an orbit is given by the sum of kinetic and potential energy terms, and Eq. (1.9) leads to

$$E = T + V = \frac{mv^2}{2} - \frac{e^2}{4\pi\varepsilon_0 r} = \frac{1}{4\pi\varepsilon_0}\left[\frac{e^2}{2r} - \frac{e^2}{r}\right] = -\frac{e^2}{8\pi\varepsilon_0 r}. \tag{1.10}$$

The ratio of kinetic to potential energy is $1/2$, which is an example of the virial theorem. The total energy is negative because these are the bound states of the atoms.

Bohr adapted Balmer's formula for the measured frequencies in the spectrum of atomic hydrogen and the Ritz combination principle to find that the energies of these discrete orbits were given by $E_n = -R_\infty/n^2$ where n is an integer and

$$R_\infty \equiv \frac{e^2}{8\pi\varepsilon_0 a_0} \tag{1.11}$$

from Eq. (1.10). The length a_0 is characteristic of atomic dimensions. He combined this formula for E_n with Eq. (1.10) to find that the size of these fixed orbits was $r_n = n^2 e^2/8\pi\varepsilon_0 R_\infty \equiv n^2 a_0$.

He then reasoned that orbits with very high values of n should correspond to classical motion (the embryo of his later correspondence principle) so that the frequency of the emitted radiation should be the same as the orbital frequency ω_{rot}. Equation (1.9) gives $\omega_{rot} = v/r = \sqrt{e^2/(4\pi\varepsilon_0 mr^3)}$ and the energy separation between adjacent high n states is $\Delta E = -(2R_\infty/n^3)\Delta n$ with $\Delta n = 1$. Then setting

1.5 de Broglie waves

$\hbar\omega_{\text{rot}} = 2R_\infty/n^3$ and using $r_n = n^2 a_0$ gives $a_0 = 4\pi\varepsilon_0\hbar^2/me^2 = 0.529 \times 10^{-10}$ m, so the total energy is

$$E = -\frac{\hbar^2}{2ma_0^2 n^2} \equiv -\frac{R_\infty}{n^2}. \qquad (1.12)$$

The Rydberg constant $R_\infty \approx 13.6$ eV sets the scale for all energies of atomic phenomena. These results are simpler to reach by quantizing orbital angular momentum as $mvr = n\hbar$ but, in spite of statements in many textbooks, this was *not* the path Bohr followed.

1.5 de Broglie waves

In 1923 de Broglie [4] suggested matter waves and claimed that they must undergo constructive interference in atomic orbits in order to have the stationary states of the Bohr model. The length of the orbit must be an integral number of de Broglie wavelengths as shown in Fig. 1.2, and since their wavelength is $\lambda = h/p$, the constructive interference condition is $2\pi r = n\lambda$. This condition is the same as quantizing the angular momentum of the orbit yielding $mvr = n\hbar$ and many authors claim that this is the derivation used by Bohr. It is not. In fact, it is not even sensible because the orbital angular momentum quantum number ℓ is quite different from the principal quantum number n.

Sometime later Debye asked the question "What is the wave equation for a de Broglie matter wave?" There are wave equations for acoustic, water, and electromagnetic waves, so there should be one for matter waves. Later, Schrödinger proposed the one that bears his name, and it became the basis of modern quantum mechanics because of its success. The first major accomplishment was the

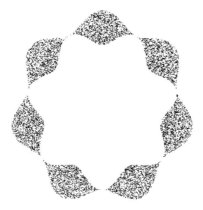

Figure 1.2 Artist sketch of the standing de Broglie waves in a circular orbit of an H-atom with seven nodes.

calculation of Bohr's formula for the energy levels of the hydrogen atom starting from its time-independent form.

Appendices

1.A Damping force on an accelerating charge

An electron bound to a nucleus in a Kepler-like orbit can be represented by linear harmonic oscillators. Since it is an accelerating charge it should radiate an electric field as illustrated in Fig. 1.3 (see Ref. [5], section 14.2). The electric and magnetic fields at position $\vec{r}$ ($\hat{r}$ is a unit vector in the $\vec{r}$ direction) are given by

$$\vec{\mathcal{E}} = \frac{e}{4\pi\varepsilon_0 c^2}\left[\frac{\hat{r}\times(\hat{r}\times\ddot{\vec{x}})}{r}\right]_{t-r/c}, \quad \vec{\mathcal{H}} = \sqrt{\frac{\varepsilon_0}{\mu_0}}(\hat{r}\times\vec{\mathcal{E}}). \tag{1.13}$$

Note that $\vec{\mathcal{E}} \perp \hat{r}$ and also that $\vec{\mathcal{E}} \not\perp \ddot{\vec{x}}$.

The energy of the radiated field given by the Poynting vector $\vec{S} = \vec{\mathcal{E}}\times\vec{\mathcal{H}}$ must be lost from the oscillating charge, and this loss implies a "radiative" force that does work on the charge. To calculate this force, start with the rate of energy loss, which is the same as the rate of radiated energy. Begin by calculating $\vec{S}$ from the fields:

$$\begin{aligned}\vec{S} &= \sqrt{\frac{\varepsilon_0}{\mu_0}}\,|\vec{\mathcal{E}}|^2\,\hat{r} = \frac{e^2}{16\pi^2\varepsilon_0^2 c^4}\frac{|\hat{r}\times\ddot{\vec{x}}|^2}{r^2}\sqrt{\frac{\varepsilon_0}{\mu_0}}\,\hat{r}\\ &= \frac{e^2}{4\pi\varepsilon_0}\frac{1}{4\pi r^2}\frac{(\ddot{x}\sin\theta)^2}{c^3}\,\hat{r},\end{aligned} \tag{1.14}$$

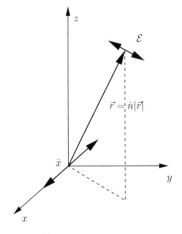

Figure 1.3 The electric field $\vec{\mathcal{E}}$ at location $\vec{r}$ from a charge accelerating along the x-axis at the origin with acceleration $\ddot{x}$.

where θ is the angle between $\hat{r}$ and $\ddot{\vec{x}}$, and $|\hat{r}\times(\hat{r}\times\ddot{\vec{x}})|$ simplifies to $|\hat{r}\times\ddot{\vec{x}}|$ because $\hat{r} \perp (\hat{r}\times\ddot{\vec{x}})$. The total radiated power at a given instant of the oscillator is then the integral of $\vec{S}$ over a closed surface. For a sphere, the area element $dA = (r d\theta)(r \sin\theta d\phi) = r^2 \sin\theta d\theta d\phi$ so the total radiated power at any instant is

$$\oiint \vec{S} \cdot d\vec{A} = \frac{e^2 \ddot{x}^2}{16\pi^2 \varepsilon_0 c^3} \int_0^{2\pi} d\phi \int_0^{\pi} d\theta \sin^3\theta = \frac{e^2 \ddot{x}^2}{6\pi\varepsilon_0 c^3}, \qquad (1.15)$$

where the integral over ϕ yields 2π and the integral over θ becomes $4/3$. Since $\ddot{x}$ is time-dependent, it is necessary to average Eq. (1.15) over one cycle. Then

$$\langle \ddot{x}^2 \rangle = \frac{1}{\Delta t} \int_t^{t+\Delta t} \ddot{x}^2 dt' = \frac{1}{\Delta t} \ddot{x}\dot{x}\Big|_t^{t+\Delta t} - \frac{1}{\Delta t} \int_t^{t+\Delta t} \dddot{x}\dot{x} dt', \qquad (1.16)$$

where $\Delta t = 2\pi/\omega_0$. Because $\ddot{x}\dot{x}$ is periodic, the average $\langle \ddot{x}\dot{x} \rangle = 0$. The energy lost by the oscillator is the total radiated power and is given by

$$P = \left\langle \oiint \vec{S} \cdot d\vec{A} \right\rangle = -\frac{e^2}{6\pi\varepsilon_0 c^3} \frac{1}{\Delta t} \int_t^{t+\Delta t} \dddot{x}\dot{x} dt' = -\frac{1}{\Delta t} \int_t^{t+\Delta t} \vec{F}\cdot\vec{v} dt' \qquad (1.17)$$

where the $(-)$ sign arises because it is an energy loss. Using $\dot{\vec{x}} = \vec{v}$, and identifying the terms in Eq. (1.17), leads to $\vec{F} = -(e^2\omega_0^2/6\pi\varepsilon_0 c^3)\vec{v}$ since $\dddot{x} = -\omega_0^2 \dot{x}$ for harmonic motion. Notice that $\vec{F} \propto -\vec{v}$, as is the usual case for a damping force. Such a dependence is found in very many common physical systems, for example air resistance at low Reynold's number.

1.B Hanle effect

Suppose a harmonically bound electron is oscillating along the y-axis, for example excited by the electric field of linearly polarized light incident along the x-axis and polarized in the y-direction as shown in Fig. 1.4. The doughnut-shaped radiation pattern expected from Eq. (1.13) is shown at the origin. It clearly shows that there is no power radiated in the y-direction because that is the direction of the acceleration, and also that the light radiated in any direction in the x–z plane is linearly polarized parallel to the y-axis.

Now suppose a magnetic field $\vec{B} = \hat{z}B$ is applied in the z-direction. Since there is a velocity v_y there is a force F_x causing the motion to deviate from the y-axis. In fact, the Lorentz force $e(\vec{v} \times \vec{B})$ combined with the harmonic force produces a complicated motion in the x–y plane and the result is radiation emitted in all directions. A characteristic frequency of this motion is $\omega_L = v/r = e|\vec{B}|/m$. However, the oscillator decays at a rate given by Eq. (1.1) and so the instantaneous y-directed power P_y is calculated from its electric field using $|\vec{\mathcal{E}}|^2 = e^{-\gamma t}\sin^2\omega_L t$ and its average value is given by

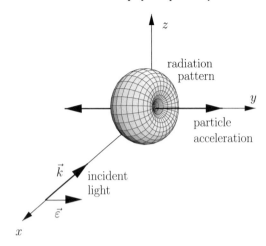

Figure 1.4 Schematic diagram of the Hanle effect showing the incident light with the motion it produces, and the radiated light field at $\vec{B} = 0$.

$$\overline{P}_y \propto \int_0^\infty e^{-\gamma t} \sin^2 \omega_L t \, dt = \frac{4\omega_L^2}{\gamma^2 + \omega_L^2}. \tag{1.18}$$

For $\vec{B} = 0$ there is no light radiated in the y-direction, and as $\vec{B} \to \infty$ the radiated power increases to some maximum value determined by the proportionality constant implied in Eq. (1.18). Somewhere between those extremes there is a field value $\mathcal{B}_{1/2}$ where $\gamma = \omega_L$ so the radiated power reaches $1/2$ of that maximum value, and by measuring this value one can find $\gamma = (e/m)\mathcal{B}_{1/2}$. For γ given by Eq. (1.1) and $\omega \sim 5 \times 10^{15}$/s as for typical visible light frequencies, characteristic values of $\mathcal{B}_{1/2}$ are $\sim 1/4$ mT = few Gauss.

Moreover, the component of x-polarized light emitted in the z-direction has the same dependence as in Eq. (1.18), so an x-oriented linear polarizer on the z-axis can also be used to extract γ. Finally, y-polarized light emitted in the z-direction must decrease as the x-polarized light increases, and so two detectors in the $\pm z$-direction with orthogonal polarizers provide complementary signals that can be subtracted to remove many kinds of spurious experimental effects. Therefore the easy measurement of a modest magnetic field allows the determination of a nanosecond decay rate, and it is made possible by using the oscillator itself as a clock with a ticking rate of several MHz.

1.C Optical tweezers

In the optical frequency range, the interaction of light with macroscopic objects mimics that of atoms, although quantum mechanics is rarely used to describe such

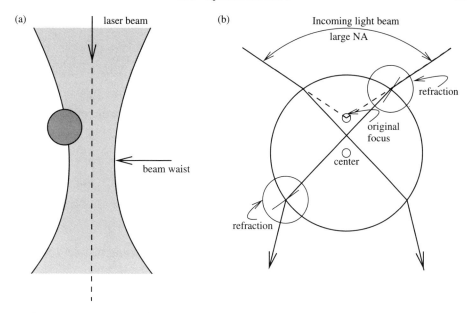

Figure 1.5 (a) Using a focused laser beam, a macroscopic object can be attracted to the center of the beam causing it to be trapped. (b) Ray diagram for a beam focused by a high numerical aperture (NA) lens onto a transparent spherical object. The outermost rays of a focused cone of light enter from the top at a large angle because of the high NA. They were originally directed above the sphere's center so the refraction bends them downward. They exit the sphere as shown and are again refracted downward. Since the wavelength above and below the sphere is the same, the outgoing light has a larger downward momentum than the incident light. Thus there is an upward force on the sphere that can be made strong enough to support it against gravity. A similar description can be made for the lateral force illustrated in (a).

objects. One important example is optical tweezers, a scheme whereby objects as small as 1 μm or as large as 100 μm (10^6 times the volume) can be manipulated with focused laser beams. A schematic diagram of a typical setup is shown in Fig. 1.5a. Usually, macroscopic objects such as plastic spheres attached to DNA strands or entire bacteria have their optical resonances in the ultraviolet, so visible light is always below the resonance. In this case, the particles are drawn into the most intense region of a light field, and simply steering the beam around allows manipulation without physical contact. Moreover, the optical system used to focus the light is also readily used to observe the manipulation simultaneously. Atoms can be similarly trapped in focused light tuned just below their resonances, and such atom traps are widely used. Because of the internal structure of atoms, light tuned just above the resonance is easily produced and can repel them from the intense field region, allowing different kinds of atomic manipulation. The similarities and differences are made clear by the classical description of the interaction of radiation and matter.

Exercises

1.1 A classical model of a one-electron atom, like the Thomson model, consists of a positive charge of amount $+e$ uniformly distributed throughout the volume of a sphere, radius R, together with a point electron of charge $-e$ which is free to move within the sphere. Show that the electron will oscillate about its equilibrium position with simple harmonic motion and find the frequency of oscillation, $\omega_0/2\pi$. If the atomic radius R is equal to the radius of the first Bohr orbit, $a_0 = 4\pi\varepsilon_0\hbar^2/me^2$, show that $\hbar\omega = e^2/4\pi\varepsilon_0 a_0 = 2R_\infty$.

1.2 Evaluate numerically the formula for γ in Eq. (1.1) using visible light. Find the lifetime $\tau \equiv 1/\gamma$.

1.3 Show that the quality factor $Q = 6\pi\varepsilon_0 mc^3/e^2\omega_0$ for $n = 1$.

1.4 To detect distant SO_2 in the atmosphere, laser light is aimed through it and the resonantly scattered light is detected. What is the mean free path of the $\lambda = 300$ nm light if the background concentration is only 1 part per billion (10^9)?

1.5 Use the formula for the Coulomb force $F = e^2/4\pi\varepsilon_0 r^2$ and the de Broglie condition $2\pi r = n\lambda$ to derive the Bohr energy condition $E = -\hbar^2/2ma_0^2 n^2$.

1.6 For large n show that the orbital frequency ω_{rot} satisfies $\omega_{\text{rot}} = \omega_n - \omega_{n-1}$, if $r_n = n^2 a_0$. Use this result along with the classical condition for a circular orbit in a Coulomb field to find the energy levels of hydrogen.

1.7 Bohr's theory of the atomic structure pre-dated de Broglie's wave hypothesis. But if Bohr had known of this hypothesis he might have chosen to assume that the orbits of the stationary states are defined by the condition that only integral multiples of h/p "fit" into the orbital circumference. Show that this is the same as assuming that the angular momentum is quantized in units of $\hbar$, and that it would lead to the correct energies that he derived using the correspondence principle.

1.8 Using the Bohr formula and an approximate value of the Rydberg constant, $R_\infty \cong 109,737$ cm$^{-1} \cong 13.6$ eV, construct energy, level diagrams for $n = 1$–10 for atomic hydrogen and singly ionized helium. Choose the same scale and zero of energy for both.

1.9 Using the infinite nuclear mass approximation in the Bohr model, obtain the wavelengths of the first four lines of the Lyman, Balmer, and Paschen series for H, He$^+$, Li^{2+}, and C^{5+}.

1.10 Positronium is a bound state of a positron (anti-electron) and an electron, that is, a hydrogen atom with the proton replaced by a positron. What are the Bohr energies of positronium?

1.11 Calculate the energy E_r with which a hydrogen atom recoils when making a transition from the $n = 4$ level to the $n = 1$ level, and show that E_r is negligible in comparison with the energy difference between the two levels.

1.12 When combined with the de Broglie hypothesis $mv = p = h/\lambda$, the usual expression for kinetic energy leads to $E = \frac{1}{2}mv^2 = hv/2$ for $v = \lambda v$. Discuss this conflict with the Planck formula.

1.13 Consider an electron with momentum $\vec{p}$ in the Coulomb field of a proton. The total energy is

$$E = \frac{p^2}{2m} - \frac{e^2}{4\pi\varepsilon_0 r},$$

where r is the distance of the electron from the proton. Assuming that the uncertainty Δr of the radial coordinate is $\Delta r = r$ and that $\Delta p = p$, use Heisenberg's uncertainty principle to obtain estimates of the size and of the energy of the hydrogen atom in the ground state.

1.14 Figure 1.5b is a ray diagram of vertical confinement in optical tweezers. Light from a large numerical aperture (NA) lens is focused tightly into a dielectric sphere that has dropped below the focal point. Entering light is refracted toward the surface normal as shown at the upper right, and exiting light away from the surface normal as shown at the lower left. The resulting light has more $\hat{z}$-directed momentum (downward) than the original beam so that there is a net upward force on the sphere, raising it toward the original focus. Draw a similar ray diagram for the force on a sphere that has been displaced laterally. Explain why both cases require a lens with large NA, such as a microscope objective.

2
Interaction of two-level atoms and light

2.1 Introduction

The use of light has been a very powerful tool to study the internal structure of atoms. Since absorption of light takes place only when the frequency of the light is nearly resonant with an atomic transition, and since the frequency of laser light can be controlled to a high degree, detailed information can be obtained this way. However, the careful analysis of the interaction between atoms and light is complicated and merits a study of its own. This chapter starts with the analysis of the simplest case, namely when only two atomic levels are important. The study of the internal structure of more complicated atoms is the subject of the second part of this book.

The next few chapters address transitions in atoms under the influence of an applied field, typically a laser beam. The topic is related to the driven oscillator of Chap. 1, hence the title of the next section. The notion of spontaneous emission or natural decay is deferred until Chap. 5. The quantitative interaction between the light and atoms is not specified in this chapter, but simply labeled as $\hbar\Omega$. A more detailed discussion is given in Chap. 3.

2.2 Quantum mechanical view of driven optical transitions

The possible states of a free atom are determined by the atomic Hamiltonian $\mathcal{H}_0$ whose stationary eigenfunctions ϕ_n have eigenenergies $E_n \equiv \hbar\omega_n$. Specifically, $\mathcal{H}_0 \phi_n = E_n \phi_n$. Such atomic structure is the subject of the second part of this book. Shining light on the atoms adds time-dependent terms to the Hamiltonian, denoted by $\mathcal{H}'(t)$, and the consequence of such radiation is that these stationary eigenstates are mixed. Since the eigenfunctions ϕ_n form a complete set, the wavefunction of the atom can always be expressed as

$$\Psi(t) = \sum_n c_n(t) \phi_n e^{-i\omega_n t}, \qquad (2.1)$$

Usually the $\mathcal{H}'(t)$ term does not commute with the atomic Hamiltonian $\mathcal{H}_0$ so they do not share the same set of eigenfunctions (see App. 7.C). Instead, $\mathcal{H}'(t)$ makes the coefficients $c_n(t)$ time-dependent and thereby changes the occupation probabilities $|c_n(t)|^2$. The light is described by a classical electromagnetic field whose effects are contained in the interaction Hamiltonian $\mathcal{H}'(t)$.

The time-dependent Schrödinger equation is

$$\mathcal{H}\Psi(t) = i\hbar \frac{\partial \Psi(t)}{\partial t}, \qquad (2.2)$$

where the Hamiltonian $\mathcal{H} = \mathcal{H}_0 + \mathcal{H}'(t)$. The task of solving the time-dependent Schrödinger equation has now been transformed to finding the solutions for the coefficients $c_n(t)$ in Eq. (2.1) by substituting it into Eq. (2.2). After the time differentiation there are equal sums on each side that drop out and the remaining two sums are next multiplied by ϕ_j and integrated over all space. The orthonormality of the eigenfunctions ϕ_n then yields

$$\dot{c}_j(t) = \frac{1}{i\hbar} \sum_n c_n(t) \mathcal{H}'_{jn}(t) e^{i\omega_{jn} t}, \qquad (2.3)$$

where the matrix elements $\mathcal{H}'_{jn}(t) \equiv \langle \phi_j | \mathcal{H}'(t) | \phi_n \rangle$ and the frequency $\omega_{jn} \equiv \omega_j - \omega_n$. Equation (2.3) is exactly equivalent to the Schrödinger equation. It has no approximations and no new information.

The task of solving Eq. (2.3) is still large, since the evolution of one coefficient $c_j(t)$ depends on all other coefficients. Therefore analytical solutions can be obtained only in certain cases.

2.3 Rabi oscillations

2.3.1 Introduction

The usual textbook approach to solving Eq. (2.3) uses perturbation theory, an approximation of very limited utility in this modern era of laser spectroscopy. Similar limits were recognized by Rabi as early as 1937 for magnetic resonance experiments (usually in the radio frequency range between Zeeman or hyperfine states) where atoms could be completely transferred from one state to another.

An alternate path is to avoid the perturbation approximation entirely but to note that in the case of very narrow band excitation whose frequency is very close to atomic resonance, only two states are connected by the radiation field. Since some approximation is required to solve Eq. (2.3), the choice is to restrict the sum to only the two terms associated with the pair of connected states, and to write the sums explicitly. The resulting coupled differential equations can be solved directly.

The notation can be clarified with the replacement $\phi_n \to |g\rangle$ for the ground state and $\phi_j \to |e\rangle$ for the excited state in Eq. (2.3), and restricting the Hamiltonian to the electric dipole approximation, discussed later in Sec. 3.2.1, so that $\mathcal{H}'_{eg}(t)$ has only off-diagonal terms. Then the "sums" for $n = e, g$ on the right-hand side of Eq. (2.3) have only one term, $\mathcal{H}'_{eg}(t) = \mathcal{H}'_{ge}{}^*(t)$, so they become

$$i\hbar \frac{dc_g(t)}{dt} = c_e(t)\mathcal{H}'_{ge}(t)\, e^{-i\omega_{eg}t} \tag{2.4a}$$

and

$$i\hbar \frac{dc_e(t)}{dt} = c_g(t)\mathcal{H}'_{eg}(t)\, e^{i\omega_{eg}t}. \tag{2.4b}$$

For single frequency radiation constituting a traveling wave, the coupling matrix element is $\mathcal{H}'_{eg}(t) = +\hbar\Omega \cos(kz - \omega t)$, where $|\Omega|$ is called the Rabi frequency. For the continuous wave (cw) field discussed here there is no loss of generality if Ω is chosen to be real, but there are many cases with two or more light fields (see Apps. 2.B and 2.C). There may be a relative phase shift between them, and it will be treated by choosing Ω to be complex. In Sec. 3.2.1 the electric dipole approximation will be invoked to define $\Omega \equiv -\vec{\mu}_{eg} \cdot \vec{\mathcal{E}}_0/\hbar$ where $\vec{\mathcal{E}}_0$ is the electric field strength of the light, and $\vec{\mu}_{eg}$ is the electric dipole moment connecting the ground and excited state, but for now, Ω is just a complex coupling constant.

2.3.2 The rotating wave approximation and rotating frame transformation

There is no loss of generality by assuming that the atom is at $z = 0$ and writing $\mathcal{H}'_{eg}(t) = \hbar\Omega \cos \omega t = \hbar\Omega(e^{i\omega t} + e^{-i\omega t})/2$. In this two-level atom case, Eqs. (2.4) can be solved directly instead of doing a time integration as in Sec. 3.2.3, but still the rotating wave approximation (RWA) introduced there is necessary to proceed with the solution. There are terms in Eqs. (2.4) that oscillate at

$$\delta \equiv \omega - \omega_{eg} \tag{2.5}$$

and others that oscillate at $\omega + \omega_{eg} \gg \delta$. In the first case, the two-level atom is driven close to its resonance frequency ω_{eg} so that the oscillation is slow, whereas in the second case it is driven at approximately twice its resonance frequency producing small rapid oscillations. Thus the coupling of the first term is more effective than the second term so it can be dropped. Doing so is called the rotating wave approximation and is equivalent to dropping the second term in Eq. (1.3).

The solutions of Eqs. (2.4) are simplified by making a slightly different choice for the phase of the c_j coefficients because the incident radiation has a single known

2.3 Rabi oscillations

frequency ω. This is done by replacing $c_e(t)$ using $\tilde{c}_e(t) \equiv c_e(t)e^{i\delta t}$. It is an algebraic equivalent to the usual textbook rotating frame transformation [6]. The substitution ensures that the temporal evolution of $\tilde{c}_e(t)$ is the same as that of the field so that the large optical frequencies $\omega \pm \delta$ and ω_{eg} can now drop out of Eqs. (2.4). The name of the transformation is appropriate because Ω can be considered as a vector that is rotating at the optical frequency ω. Note that the rotating frame transformation is exact, as is the algebraic equivalent used here, and is completely different from the RWA discussed above, although the names are similar.

With both the RWA and the rotating frame transformation, the oscillatory terms drop out and Eqs. (2.4) become

$$\frac{dc_g(t)}{dt} = -i\frac{\Omega^*}{2}\tilde{c}_e(t) \tag{2.6a}$$

and

$$\frac{d\tilde{c}_e(t)}{dt} = -i\frac{\Omega}{2}c_g(t) + i\delta\tilde{c}_e(t). \tag{2.6b}$$

2.3.3 Dynamical solutions

The two Eqs. (2.6) can be uncoupled by differentiating the first one and substituting for $\tilde{c}_e(t)$ in the second one to find

$$\frac{d^2 c_g(t)}{dt^2} - i\delta\frac{dc_g(t)}{dt} + \frac{|\Omega|^2}{4}c_g(t) = 0 \tag{2.7a}$$

and conversely

$$\frac{d^2 \tilde{c}_e(t)}{dt^2} - i\delta\frac{d\tilde{c}_e(t)}{dt} + \frac{|\Omega|^2}{4}\tilde{c}_e(t) = 0. \tag{2.7b}$$

Since the field-free wavefunctions $|g\rangle$ and $|e\rangle$ are not eigenfunctions of $\mathcal{H}$, it is not surprising that the solution of Eqs. (2.7) for the initial conditions $c_g(0) = 1$ and $c_e(0) = 0$ are time-dependent. The steady-state solutions are discussed later, but these initial conditions yield

$$c_g(t) = \left(\cos\frac{\Omega' t}{2} - i\frac{\delta}{\Omega'}\sin\frac{\Omega' t}{2}\right)e^{+i\delta t/2} \tag{2.8a}$$

and

$$\tilde{c}_e(t) = c_e(t)e^{i\delta t} = -i\frac{\Omega}{\Omega'}\sin\frac{\Omega' t}{2}e^{+i\delta t/2}, \tag{2.8b}$$

where

$$\Omega' \equiv \sqrt{|\Omega|^2 + \delta^2}. \tag{2.8c}$$

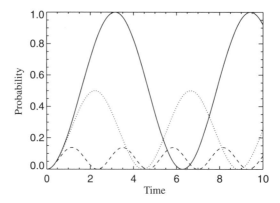

Figure 2.1 Plot of the probability $|c_e(t)|^2$ for an atom to be in the excited state for $\delta = 0$ (solid line), $\delta = |\Omega|$ (dotted line), and $\delta = 2.5|\Omega|$ (dashed line). Time is in units of $1/|\Omega|$. (Figure from Ref. [1].)

Figure 2.1 shows that the probability for finding the atom in the excited state $|e\rangle$, $|c_e(t)|^2$, oscillates at frequency Ω', and that increasing the detuning $|\delta|$ increases the frequency of the oscillation while decreasing its amplitude. Of course, the sum of the probabilities $|c_e|^2 + |c_g|^2 = 1$. The segment of the oscillation associated with the transition from the excited state down to the ground state corresponds exactly to stimulated emission, and the result here illustrates clearly why the Einstein coefficients B_{ge} and B_{eg} are equal (see Chap. 5). When $\sin^2(\Omega' t/2)$ is between its extreme values, the system may be driven toward either ground or excited state depending on the relative phase between the optical electric field $\mathcal{E}_0$ and the oscillations of $\Psi(t)$.

It seems clear that applying a pulse of frequency $\omega = \omega_{eg}$ ($\delta = 0$ so $\Omega' = |\Omega|$), and duration $t_\pi = \pi/|\Omega|$, inverts the population. This is called a π-pulse. A pulse of half the duration of a π-pulse (a $\pi/2$-pulse) produces an equal superposition of the ground and excited states, whose relative phase evolves after the pulse ends because the two states have different energies.

2.3.4 Eigenvalues and eigenfunctions

In the presence of the off-diagonal Hamiltonian matrix elements of the operator $\mathcal{H}'(t)$, the energies $E_{e,g}$ that are the eigenvalues of $\mathcal{H}_0$ are no longer the eigenvalues of the full Hamiltonian. The energy levels are shifted by an amount that depends on $|\Omega|$ and vanishes for $\Omega = 0$. The solution for the eigenenergies of the linear, homogeneous Eqs. (2.6) can be found by diagonalizing the matrix

$$\mathcal{H}' = \frac{\hbar}{2}\begin{pmatrix} -2\delta & \Omega \\ \Omega^* & 0 \end{pmatrix}, \tag{2.9}$$

where the order of the states in the matrix is e, g. The eigenvalues of Eq. (2.9) are

$$E_{1,2} = \frac{\hbar}{2}(-\delta \pm \Omega'), \quad (2.10)$$

The light mixes the states by an amount expressed in terms of a mixing angle θ given by $\tan(2\theta) \equiv -|\Omega|/\delta$ with $0 \leq \theta \leq \pi$, so that each state is mixed with a component of ground and excited state. The eigenstates corresponding to $E_{1,2}$ are called the "dressed states" of the atom [7] and are given by

$$|1\rangle = \begin{pmatrix} \cos\theta \\ \sin\theta\, e^{-i\varphi} \end{pmatrix} \quad \text{and} \quad |2\rangle = \begin{pmatrix} -\sin\theta \\ \cos\theta\, e^{-i\varphi} \end{pmatrix}, \quad (2.11)$$

where $\varphi = \arg(\Omega)$ takes into account that Ω can be complex. Thus the identification of the eigenstates with $|e\rangle$ and $|g\rangle$ is ambiguous because they are linear combinations of both ground and excited states as shown by Eqs. (2.11).

In the limit where $|\Omega| \ll |\delta|$ and $\delta > 0$, the mixing angle becomes $\theta \approx \pi/2$ and state $|1\rangle$ is predominantly the ground state, whereas state $|2\rangle$ is predominantly the excited state. In the same limit but $\delta < 0$, the angle $\theta \approx 0$, and now state $|2\rangle$ is predominantly the ground state and state $|1\rangle$ the excited state. In both cases the resulting energies are shifted by $E_g \approx \hbar|\Omega|^2/(4\delta)$ and $E_e \approx -\hbar|\Omega|^2/(4\delta)$. Since the light intensity is proportional to $|\Omega|^2$, $E_{e,g}$ is appropriately called the light shift. In the opposite limit $|\Omega| \gg |\delta|$, the solutions give $E_1 = \hbar|\Omega|/2$ and $E_2 = -\hbar|\Omega|/2$, where both states are in an equal superposition of the ground and excited state.

In a standing wave, the light shifts of these atomic dressed states vary from zero at the nodes to a maximum at the antinodes. The spatially oscillating energies found from Eq. (2.10) are not sinusoidal, except in the limit of $|\delta| \gg |\Omega|$. This is apparent because these oscillatory terms will always be dominated by δ^2 in the vicinity of a node. Thus, for any value of $|\Omega| \gg |\delta|$, the expansion of Eq. (2.10) in a standing wave as $E \sim |\hbar\Omega \cos kz/2|$ will eventually fail near a node.

Because the eigenstates in the presence of the field are mixtures of the ground and excited states there are several possible transitions as shown in Fig. 2.2. This results in a spectrum that has three peaks corresponding to these transitions. There is a clear semi-classical view of this phenomenon that arises because the probability of the atom being in the excited state oscillates. Since it can only radiate when it is in the excited state, the field emitted looks like an amplitude-modulated sine wave as shown in Fig. 2.3a whose spectrum is exactly that shown in Fig. 2.3b, namely a carrier and two sidebands. This spectrum is commonly called the Mollow triplet [8].

2.4 The dressed atom picture

Absorption of light by an atom can only conserve energy if the energy of the electromagnetic field is correspondingly reduced. But the Hamiltonian of Eq. (2.2)

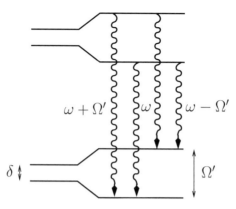

Figure 2.2 Energy levels of the dressed states for the case of $\delta \neq 0$. Note that the splitting of each state in two leads to four different possibilities for the transitions, for which two coincide at a frequency of ω. (Figure from Ref. [1].)

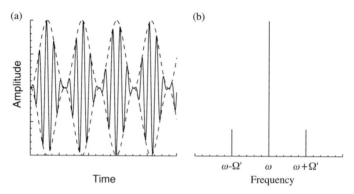

Figure 2.3 The classical counterpart of the Mollow triplet. Since the atom can only radiate if it is in the excited state, the emitted light is amplitude-modulated (part a) leading to three frequencies (part b): one carrier and two side peaks.

does not have the energy of the field itself included, so the subsequent description is lacking such information. To include the interaction with the field in the Hamiltonian, consider operators that raise ($a^\dagger$) or lower (a) the energy of the light field by $\hbar\omega$ since atomic energy levels are separated by approximately this amount. Then the interaction operator that describes these transitions is

$$\mathcal{H}_{\text{int}} = \hbar\Omega \left(a^\dagger |g\rangle \langle e| + a |e\rangle \langle g| \right). \tag{2.12}$$

Clearly it preserves energy conservation because it changes $|g\rangle$ to $|e\rangle$ while lowering the field energy with the operator a and conversely. The operator $\mathcal{H}_{\text{int}}$ results from making the RWA on the fully quantized Hamiltonian, and is called the Jaynes–Cummings model [9]. Its eigenstates are often used interchangeably with the

2.5 The Bloch vector and Bloch sphere

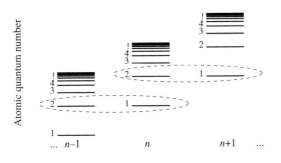

Figure 2.4 The energy level diagram for the atom plus field Hamiltonian. In each vertical column there is the familiar level scheme of a typical atom, but the columns are vertically displaced by the addition of $\hbar\omega$ per column. The nearly degenerate pairs are indicated. In the presence of the coupling interaction, each of these pairs is a mixture of ground and excited states, so each can decay by spontaneous emission. (Figure from Ref. [1].)

dressed states of Eqs. (2.11) [7]. Of course, $\mathcal{H}_{\text{int}}$ needs to operate on something, so the optical field must also be included in the wavefunction. In the quantized field description, it is characterized by a total energy $(n + 1/2)\hbar\omega$ where n is an integer not to be confused with the principal quantum number of atomic states.

The energy level diagram of the combined atom plus field system consists of the ordinary atomic energies repeated for each value of n and vertically displaced by $\hbar\omega$ each time, as shown schematically in Fig. 2.4. Attention is focused on the two atomic states coupled by the laser light that form closely spaced pairs of one excited state and one ground state separated by $\hbar\delta$, as shown in Fig. 2.4. They are each mixtures of the ground and excited states, found by diagonalizing the Hamiltonian of the combined atom plus field system.

The interaction between the atom and the field embodied in $\mathcal{H}_{\text{int}}$ couples the ground and excited states that form each of these pairs through the off-diagonal matrix elements $\mathcal{H}'_{ge}(t)$. This splits the energy levels further apart to $\hbar\Omega'$ as given in Eq. (2.10). Note that Ω' is independent of the sign of δ, and the shift $\hbar(\Omega' - |\delta|)/2$ is the light shift of each dressed state.

2.5 The Bloch vector and Bloch sphere

Because the overall phase of the wavefunction has no physical meaning, there are really only three free parameters in the solutions given in Eqs. (2.8) for the complex coefficients $c_j(t)$. In a classic paper, Feynman, Vernon, and Hellwarth [10] combined the real and imaginary parts of the coefficients $c_j(t)$ to form the three real parameters, denoted commonly as

$$u \equiv 2\text{Re}(c_g \tilde{c}_e^*) \qquad v \equiv 2\text{Im}(c_g \tilde{c}_e^*), \quad \text{and} \quad w \equiv |\tilde{c}_e|^2 - |c_g|^2. \qquad (2.13)$$

The equations of motion (Eq. (2.6)) can be used to calculate the time dependence of the parameters u, v, and w, and one finds

$$\frac{du}{dt} = +\delta v - \Omega_i w \qquad \frac{dv}{dt} = -\delta u - \Omega_r w \quad \text{and} \quad \frac{dw}{dt} = +\Omega_i u + \Omega_r v, \qquad (2.14)$$

with $\Omega \equiv \Omega_r + i\Omega_i$. The result bears a striking resemblance to a vector cross product, and so the notation can be made more compact by defining two artificial vector quantities $\vec{\Omega} = (\Omega_r, -\Omega_i, -\delta)$ and $\vec{R} = (u, v, w)$. Then the evolution equation for $\vec{R}$ becomes

$$\frac{d\vec{R}}{dt} = \vec{\Omega} \times \vec{R}. \qquad (2.15)$$

An algebraically different but physically equivalent way to arrive at this result uses the well-known Pauli matrices to represent the Hamiltonian, and is discussed in App. 2.A.

The vector $\vec{R}$ is called the Bloch vector after Felix Bloch. Notice that the time derivative of $\vec{R}$ is always perpendicular to $\vec{R}$. This means that the magnitude $|\vec{R}|$ is a constant, which is unity as seen from its components in Eq. (2.13). The notion of $\vec{\Omega}$ causing the precession of $\vec{R}$ according to Eq. (2.15) is clearest in a reference frame where $\vec{\Omega}$ is stationary, so most textbooks suggest viewing the dynamics in the rotating frame. Either way, the path taken after the rotating frame transformation led to the readily solved time-independent Hamiltonian matrix of Eq. (2.9) and the dressed atom picture of Sec. 2.4. It will become clear that the dynamic solutions can lead to much more than the Rabi oscillations of Fig. 2.1.

The artificial vector $\vec{R}$ therefore moves on the surface of a unit sphere called the Bloch sphere, as shown in Fig. 2.5. The south (north) poles of this sphere correspond to the ground (excited) states of the atom, and equatorial plane corresponds to equal superpositions with various phases. If $\vec{R}$ is in the equatorial plane, $w = 0$ so $c_1 c_1^* = c_2 c_2^*$ and the eigenstates $\Psi_\pm$ are equal mixtures of excited and ground states. This mixture can be written as $\Psi_+ = \frac{1}{2}\sqrt{2}(|g\rangle + e^{i\chi}|e\rangle)$ and $\Psi_- = \frac{1}{2}\sqrt{2}(|g\rangle - e^{i\chi}|e\rangle)$ where χ is a phase that depends on the details of the state preparation.

For the case of $\delta = 0$, $\vec{\Omega}$ is in the equatorial plane and an atom starting in state $|g\rangle$ executes polar orbits. The probability of finding it in state $|e\rangle$ oscillates, and is exactly the solid curve of Fig. 2.1. If $\delta \neq 0$, such orbits do not reach the north pole because $\vec{\Omega}$ is off the equator, as plotted in the other two curves of Fig. 2.1.

In the discussion of the dynamical solutions of the Schrödinger equation just above Eqs. (2.8), specific initial conditions were chosen. In general, the response of an atom initially in any superposition of ground and excited states depends strongly

2.A Pauli matrices for motion of the Bloch vector

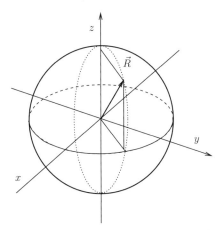

Figure 2.5 Graphical representation of the Bloch vector $\vec{R}$ on the Bloch sphere. (Figure from Ref. [1].)

on the initial components of $\vec{R}$ and thus on the process that produced the superposition. For example, if the initial state is $|e\rangle$ the interaction with radiation tuned to resonance ($\delta = 0$) causes stimulated emission. If the initial state is an equal superposition of ground and excited states ($\vec{R}$ on the equator), the response to on-resonance light depends on the initial preparation. Then the system could be driven to either the ground or the excited state depending on the relative phase of the superposition.

Section 2.3 shows that a light field tuned near atomic resonance causes the populations of the ground and excited states to oscillate at the Rabi frequency, and by contrast, Eqs. (2.11) give time-independent expressions for the eigenstates (θ is constant). This apparent contradiction can be reconciled by examination of the Bloch sphere. The eigenstates on the Bloch sphere given by Eqs. (2.11) are parallel and antiparallel to the torque vector $\vec{\Omega}$ so that $\vec{\Omega} \times \vec{R} = 0$. Thus these particular states do not evolve, and are the stationary states of the Hamiltonian. An interesting special case is $\vec{R}$ on the equator after a $\pi/2$-pulse, followed by exposure to light having $\delta = 0$ but with a phase shift from the $\pi/2$-pulse such that $\vec{\Omega}$ is parallel or antiparallel to $\vec{R}$ so that $\vec{\Omega} \times \vec{R} = 0$.

Appendices

2.A Pauli matrices for motion of the Bloch vector

Most of the discussion in Secs. 2.4 and 2.5 has been directed toward the two-level atom, a simple quantum system driven by a classical field. This has been shown to have quite complicated response, and the subsequent appendices in this chapter show even further richness. It is therefore useful to have an alternative calculational scheme for describing some of the phenomena, and the well-known Pauli matrices can provide this.

All Hermitian operators on the two-dimensional Hilbert space of a two-state system can be represented by combinations of the identity matrix I and the Pauli matrices, because they span the space. The Pauli matrices are

$$\sigma_x = \begin{pmatrix} 0 & 1 \\ 1 & 0 \end{pmatrix} \quad \sigma_y = \begin{pmatrix} 0 & -i \\ i & 0 \end{pmatrix} \quad \sigma_z = \begin{pmatrix} 1 & 0 \\ 0 & -1 \end{pmatrix}. \tag{2.16}$$

Thus the Hamiltonian is $\mathcal{H} = \mathcal{H}_0 + \mathcal{H}'$ and $\mathcal{H}_0 = E_0 I + 1/2\hbar\omega_{eg}\sigma_z$ where E_0 allows for an offset of the zero of energy and $E_e - E_g = \hbar\omega_{eg}$. The atom–light coupling becomes $\mathcal{H}' = +\hbar(\Omega_r\sigma_x - \Omega_i\sigma_y)$. The total Hamiltonian is therefore

$$\mathcal{H} = E_0 I + 1/2\hbar\omega_{eg}\sigma_z + 1/2\hbar(\Omega_r\sigma_x - \Omega_i\sigma_y). \tag{2.17}$$

These Pauli matrices are particularly convenient because their commutation relations satisfy $[\sigma_i, \sigma_j] = 2i\epsilon_{ijk}\sigma_k$. Also, their expectation values satisfy the usual equation of motion: $\langle[\sigma_j, \mathcal{H}]\rangle = i\hbar\langle\dot\sigma_j\rangle$. These expectation values can be calculated using the basis wavefunction $\Psi(t) = c_g(t)\phi_g + c_e(t)\phi_e$, and then

$$\langle\sigma_x\rangle = \langle\Psi|\sigma_x|\Psi\rangle = (c_e^* \ c_g^*)\begin{pmatrix} 0 & 1 \\ 1 & 0 \end{pmatrix}\begin{pmatrix} c_e \\ c_g \end{pmatrix} = c_g^*c_e + c_e^*c_g = u \tag{2.18}$$

and likewise for the other components of the artificial vector $\vec{R}$ of Sec. 2.5. Thus $\langle\sigma_j\rangle = \langle\Psi|\sigma_j|\Psi\rangle = R_j$.

For the Hamiltonian after the rotating frame transformation of Eq. (2.9), $\mathcal{H} = -1/2\hbar\delta(I + \sigma_z) + 1/2\hbar(\Omega_r\sigma_x - \Omega_i\sigma_y)$. Using the relation for $\langle\dot\sigma_j\rangle$ above readily produces the equation of motion $d\vec{R}/dt = \vec{\Omega} \times \vec{R}$.

2.B The Ramsey method

The motivation for this section emerges from a consideration of precision measurements of the frequency ω_{eg}. Clearly the Fourier transform limit $\Delta\omega\Delta t > 1$ imposed by the interaction time Δt determines the precision $\Delta\omega$ of any measurement. Moreover, the motion of atoms defines the maximum value of Δt simply because of the size limit of the apparatus. Even for ponderous Cs atoms whose thermal velocity is typically only a few hundred m/s, a meter-size vacuum system limits $\Delta\omega/2\pi$ to $\sim 10^2$ Hz. This is the width of the signal, but of course, a high signal/noise ratio would allow resolution to a tiny fraction of the width. Longer beamlines could be built, but applying radiation and/or a homogeneous magnetic field over such a long region is a difficult task. In 1950, Ramsey described a method whereby such requirements were needed only near the start and end regions of a considerably longer path [11].

2.B The Ramsey method

Section 2.3.3 describes the Rabi oscillations between state $|g\rangle$ and $|e\rangle$, and near its end describes how interrupting on-resonance exciting light after time $t = \pi/\Omega$ leaves the atom in state $|e\rangle$ by a "π-pulse", where Ω is assumed to be real in this section without loss of generality (see Eqs. (2.8) and Fig. 2.1). This is called a π-pulse because the Bloch vector $\vec{R}$ is rotated by π. If the pulse duration is $\pi/2\Omega$ it is called a $\pi/2$-pulse, and clearly a π-pulse is equivalent to two sequential $\pi/2$-pulses. However, if the two $\pi/2$-pulses are separated in time so that the atom is in the dark between them, a totally different situation applies. To discuss the new phenomena that arise, it is necessary to consider the dynamical description.

The solutions for the coefficients $c(t)$ in Eqs. (2.8) lead to a curious observation about Fig. 2.1. The solid curve (for $\delta = 0$ so $\vec{\Omega} = (\Omega, 0, 0)$ and $\Omega > 0$) passes through $|c_e(t)|^2 = 1/2$ (where $w = 0$) at times $\pi/2\Omega$ and $3\pi/2\Omega$, where the atom is in an equal superposition of state $|g\rangle$ and $|e\rangle$. However, the evolution of the wavefunction is very different following these two instants. For $t = \pi/2\Omega$ the state evolves toward $|e\rangle$ whereas for $t = 3\pi/2\Omega$ the state evolves toward $|g\rangle$.

The explanation can be found by evaluating v of Eq. (2.13) at these two times (where $u = 0 = w$) as shown by the solid lines in Fig. 2.6a each leading from the south pole to points A and B. Thus $v = +1$ for $t = \pi/2\Omega$ at point A but $v = -1$ for $t = 3\pi/2\Omega$ at point B. Therefore the cross product $\vec{\Omega} \times \vec{R}$ rotates $\vec{R}$ in accordance with Eq. (2.15) in the same sense, upward from point A to the north pole when

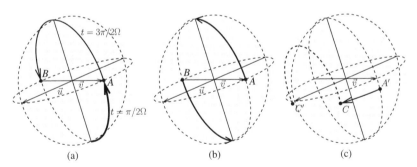

Figure 2.6 Part (a) shows the motion of the Bloch vector $\vec{R}$ starting from the ground state $|g\rangle$ (south pole) under the influence of pulses of two different durations, $\pi/2\Omega$ and $3\pi/2\Omega$ for detuning $\delta = 0$ (solid lines ending at A and B respectively). Part (b) shows what happens when a second pulse of duration $\pi/2\Omega$ is applied: $\vec{R}$ can go to either pole (solid lines from A and B) depending on the duration of the first pulse (now shown as dotted lines). Part (c) shows the motion of $\vec{R}$ when a pulse of duration $\pi/2\Omega$ is applied with $\delta \neq 0$ bringing $\vec{R}$ to A, then some time elapses, and then a similar pulse is applied. Since $\delta \neq 0$, $\vec{R}$ moves parallel to the equator during the dark period between the pulses. It is clear that $\vec{R}$ undergoes a more complicated motion and can be driven to arbitrary points on the Bloch sphere.

$\vec{R} = (0, 1, 0)$, and downward from point B to the south pole when $\vec{R} = (0, -1, 0)$, as shown by the solid lines in Fig. 2.6b. Even though $\vec{R}$ may have started on the equator in both cases of $|c_e(t)|^2 = 1/2$ in Fig. 2.1, it is in diametrically opposite places on the Bloch sphere so that $\vec{\Omega}$ causes precession toward opposite poles.

Now suppose that the light field is interrupted so that the atom is exposed to two $\pi/2$-pulses separated by a time T much larger than each pulse duration $\pi/2\Omega$. The initial conditions are $\vec{R} = (0, 0, -1)$ and $\vec{\Omega} = (\Omega, 0, 0)$. For $\delta = 0$, $\vec{\Omega} \perp \vec{R}$ is preserved during the first pulse that brings $\vec{R}$ to the equator at point A as shown in Fig. 2.6a, and then the light goes off. To see what happens to the atom in the dark, note that in the dark $\vec{\Omega} = (0, 0, 0)$, there is no torque and the Bloch vector does not precess so that during the second pulse the atom is driven to the north pole.

It is instructive to consider how the detuning can be described when the atoms are not interacting with the light. In the dark the superposition of the ground and excited state evolves with a phase $e^{-i\omega_{eg}t}$ given by the difference between the frequencies of the two atomic states. At the same time the light field evolves with a phase $e^{-i\omega t}$ and it is the difference between these two phases, namely $\Delta\phi$, that affects the atoms during the second pulse. One of the requirements is that the light in the two pulses is phase stable so that the atoms do not experience an additional phase shift during the second pulse. When $\delta \neq 0$ and $-\delta T \equiv \Delta\phi = \pi$, the Bloch vector is changed by just the right amount so that the second $\pi/2$-pulse rotates $\vec{R}$ toward the south pole just as for the case of $t = 3\pi/2\Omega$ in Fig. 2.6b. For larger δ so that $\Delta\phi = 2\pi$, the second pulse again brings $\vec{R}$ toward the north pole and this produces the familiar oscillations of the Ramsey signals (see Fig. 2.7).

Alternatively, the Bloch picture of Sec. 2.5 can be used to describe what happens to the atoms in the dark for $\delta \neq 0$. The vector $\vec{\Omega} = (0, 0, -\delta)$ is along the z-axis and this according to Eq. (2.15) causes $\vec{R}$ to precess in a horizontal plane at a rate $-\delta$. If $\Omega \gg |\delta|$ the first $\pi/2$-pulse brings the Bloch vector very close to the equator, $\vec{R}$ rotates in the dark in the equatorial plane and acquires a phase $\Delta\phi = -\delta T$. This phase is fictitious, since the atoms do not interact with the light. However, when the second $\pi/2$-pulse interacts with the atoms, its action depends on $\Delta\phi$ and causes $\vec{R}$ to rotate to the north pole, south pole, or anything in between. The signal thus oscillates between zero and the maximum for $\delta = 0$.

For $\delta \neq 0$ and $\Omega \sim |\delta|$ the first $\pi/2$-pulse does not bring $\vec{R}$ to point A, but to point A' as shown in Fig. 2.6c. During the dark interval between pulses $\vec{R}$ evolves depending on the detuning δ bringing $\vec{R}$ to point C in Fig. 2.6c. The second pulse rotates $\vec{R}$ about $\vec{\Omega}$ by an angle different from $\pi/2$ just as the first one did, leaving it at point C' on the sphere. Note that for $\delta \neq 0$ the first pulse does not bring the atoms to the equator so the second pulse is not likely to leave them at either pole; thus the amplitude of the oscillations is damped.

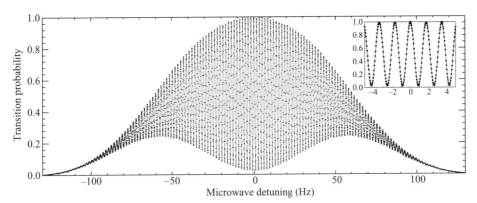

Figure 2.7 A plot of the measured signal of the Ramsey fringe pattern from a fountain clock at the PTB laboratory in Germany. The dots are the data points and the solid line is to guide the eye. It shows the oscillations that result from using two separated fields with a delay time T between them to drive a transition in the Cs atom. The inset is an enlargement of the central part showing resolution of well below 1 Hz at a frequency of 9.1 GHz. The horizontal scale shows that the full width of each "peak" corresponds to a detuning of $\sim \pi/T$, about half the width of the peak expected for a signal that was driven continuously for a time T. (Figure from Ref. [12].)

Precisely where $\vec{R}$ lands in each of these cases now depends on $\Delta \phi$. Different atoms may have different values of $\Delta \phi$ because of Doppler shifts, different transit times between two laser beams, etc., so that there is some change of w for $\Delta \phi$ not ideal as shown in Fig. 2.6c. Also for $\delta \neq 0$, $\vec{\Omega}$ is not perpendicular to $\vec{R}$ at the start, and the angle between $\vec{R}$ and $\vec{\Omega}$ is not preserved in the dark so that the action of a second $\pi/2$-pulse rotates $\vec{R}$ to some point other than the appropriate pole, depending on the values of the parameters.

For this separated pulse technique, a plot of w or $|c_e(t)|^2$ vs. δ has a peak at $\delta = 0$ and reaches a minimum as shown in Fig. 2.7 where the parameters have combined to put $\vec{R}$ on the opposite pole of the sphere. Subsequent measurement of w, e.g. by measurement of fluorescence, oscillates as ω is swept to vary δ, and the largest peak is at $\delta = 0$, with other maxima reduced by inhomogeneous effects resulting from different $\Delta \phi$-values. Moreover, the width of the central maximum is determined by T and is narrower than that of the usual resonance curve for continuous light with the same duration. Of course, T can be made long with suitable arrangements, and is not as limited by field inhomogeneities.

The earliest application of this Ramsey method, for which N. Ramsey received the 1989 Nobel Prize, was for magnetic resonance at microwave frequencies on the hyperfine sublevels of ground-state atoms, notably hydrogen in the hydrogen maser (see Sec. 9.3.3) and Cs in the first atomic clocks. The Cs clocks used a 9.1

30   *Interaction of two-level atoms and light*

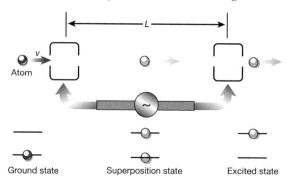

Figure 2.8 A schematic diagram of the molecular beam setup for a clock based on the Ramsey separated oscillatory field method. Atoms traveling through this system are exposed to a microwave field in zone 1, then remain in the "dark" as they fly toward zone 2, and then are exposed to a microwave field with its phase preserved from zone 1. Phase preservation is implemented by splitting the microwave input as shown, and carefully adjusting both arms to equal lengths. (Figure from Ref. [13].)

GHz microwave field at two points a few meters apart (see Fig. 2.8) to drive the Caesium hyperfine ground state transition, and became so accurate that the second was eventually defined in terms of this frequency by international agreement. Caesium clocks have been the time standard of the world for decades.

The latest version of Cs atomic clocks begins with a sample of laser-cooled atoms that are launched vertically through a microwave field that puts $\vec{R}$ on the equator, and then they fall back through the same field for the second pulse. Because of the laser cooling, the velocity distribution is very narrow resulting in uniform excitation and a distribution of T-values that is narrow. Such "fountain clocks" (see Fig. 2.9) have become commonplace, but still work on the 9.1 GHz splitting of the ground state of Cs.

Optical clocks using Ramsey fringes produced by transitions to electronically excited states have been constructed in many laboratories, and will soon replace microwaves as the international time standard. They use optical frequency comb synthesizers at optical frequencies that have much higher precision simply because of their higher frequencies. Invention of these frequency comb synthesizers received the Nobel Prize in 2005.

2.C Echoes and interferometry

When $\vec{R}$ is not at one of the poles of the Bloch sphere, perhaps removed by the first $\pi/2$-pulse of the Ramsey method as at point A in Fig. 2.6a where $\vec{R} = (0, 1, 0)$, the atom is in a superposition of states $|g\rangle$ and $|e\rangle$, a circumstance that has no

2.C Echoes and interferometry

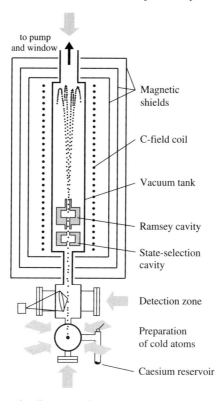

Figure 2.9 A schematic diagram of an atomic fountain clock. Atoms are laser-cooled in a magneto-optical trap at the base, and them launched upward by a "walking wave" that results from counterpropagating laser beams of slightly different frequencies. They pass through a probe laser beam and a microwave cavity on the way up, and also on the way down. The two sequential passes through the microwave field constitute the separated oscillatory fields for the clock transition, and phase coherence is assured because it is the same field: arm length differences do not exist. The first passage through the laser probe beam prepares the atomic ground state, and the second passage measures it after two separated exposures to the microwaves. (Figure from Ref. [12].)

analog in classical physics and is discussed further in App. 2.E below. For now, the question of $\Delta\phi$ being different for different atoms needs to be discussed further. Although the internal frequency ω_{eg} is the same for all free atoms, atoms embedded in a material or otherwise perturbed, e.g. by collisions, may have shifted energy levels. The precession about the polar axis in the dark could rotate each $\vec{R}_j$ by $\delta_j T/2$ relative to $\vec{\Omega}$ during time $T/2$. The range of their shifted frequencies can be sufficiently large that during the interval $T/2$ their evolutions result in dispersal on the Bloch sphere so that at this later time they are spread out as shown at points A and B in Fig. 2.10a.

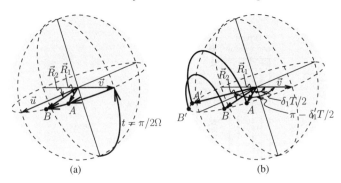

Figure 2.10 Part (a) shows the evolution of different atoms starting from $\vec{R} = (0, 1, 0)$ on the equator to points A and B possibly because of different environments. The one ending at A has rotated through an angle $\delta_1 T/2$ as shown in part (b). Then a π-pulse moves them from A to A' and B to B' respectively. For the $\vec{R}$-vector at A' the angle from $\vec{R} = (0, 1, 0)$ is now $\pi - \delta_1 T/2$ so that a subsequent evolution that rotates it by another $\delta_1 T/2$ brings it to $\vec{R} = (0, -1, 0)$ (not shown for clarity).

Consider what happens if a π-pulse is applied to such a sample of atoms at $t = T/2$, midway between the two $\pi/2$-pulses discussed in App. 2.B. Independent of the position of the various vectors $\vec{R}_j$ of different atoms after the evolution time $T/2$, the π-pulse inverts their component perpendicular to $\vec{\Omega}$ because the vectors $\vec{R}_j$ are rotated by π about $\vec{\Omega}$, bringing them to A' and B' as shown in Fig. 2.10b. That is, the π-pulse now changes that phase angle to $\pi - \delta_j T/2$ as shown in Fig. 2.10b. During the second interval of $T/2$ the $\vec{R}_1$-vector will precess by the same amount, going from point A' by the same angle $\delta_1 T/2$ as before, bringing it to $\vec{R} = (0, -1, 0)$. The $\vec{R}_2$-vector at point B' will precess by $\delta_2 T/2$, also bringing it to $\vec{R} = (0, -1, 0)$. Thus the evolution during the subsequent $T/2$ brings all the vectors $\vec{R}_j$ precisely to $\vec{R} = (0, -1, 0)$.

Of course, the excited-state part of the superposition created by the first $\pi/2$-pulse can decay, but the radiation emitted by different atoms at time t has different phases, depending on $\delta_j t$, because the phase of the emitted radiation depends on the relative phase of state $|g\rangle$ and $|e\rangle$. Clearly the radiated power is greatest when these phases are all the same, and this condition is achieved when the various vectors $\vec{R}_j$ are all brought together at $\vec{R} = (0, -1, 0)$. The sample can now radiate efficiently producing a pulse at time T called a "photon echo" (see Ref. [14]). Alternatively, a second $\pi/2$-pulse could produce a strong stimulated emission for systems whose upper state radiates slowly (such as spins).

As with many coherent optical phenomena, including the Ramsey method described in App. 2.B, the notion of echoes first arose in connection with magnetic resonance (see Ref. [15]). Such "spin echoes" are vitally important in material studies, and especially in medical MRI.

2.C Echoes and interferometry

When an atom absorbs light, its energy $\hbar\omega$ goes into the excitation of the atom, its angular momentum $\hbar$ goes into the orbital motion of the electrons (recall $\Delta\ell = \pm 1$), but its linear momentum $\hbar\omega/c = \hbar k$ can go only into the overall translational motion of the atom. If an atom undergoes an absorption-stimulated emission sequence from counterpropagating beams, it could end up back in its initial internal state, but its momentum state may be changed by 0, $\pm 2\hbar k$, and these three possibilities are indeed distinct states. If such an event has a transition probability of $1/2$, the sequence can be described as a $\pi/2$-pulse. The resulting superposition has no component of state $|e\rangle$ so there is no subsequent spontaneous emission, but it is indeed a superposition of three corresponding momentum states differing by $2\hbar k$. The three components of the superposition have different velocities and therefore travel different paths. It is completely equivalent to say that the atomic wavefunction has been diffracted by the periodic standing-wave light field.

Sometimes the up–down sequence connects two sublevels of the atomic ground state, for example different hyperfine states (see Chap. 9). Then the slightly different frequencies can be exploited to determine that $\Delta p = 0$ is not allowed, and either $\Delta p = +1$ or $\Delta p = -1$ can be selected.

A subsequent π-pulse at time $T/2$ later again leaves atoms in state $|g\rangle$ (or one of its hyperfine sublevels) but can exchange these linear momenta so that some of the momentum components can later overlap at time T. A final $\pi/2$-pulse can have significant probability to recombine the two components of the superposition to the same momentum state, but since the components have traveled on different paths, the resulting phase differences can determine the nature of the recombination. The final momentum state depends on this phase shift, and the process comprises an atom interferometer.

A common scheme is schematically illustrated in Fig. 2.11. A beam of atoms could enter from the left, or the horizontal axis could simply represent time for a cloud of stationary atoms. They are subject to a light pulse from source L1 and either a reflected beam (standing wave) or a slightly different frequency beam (imagine the mirror to be replaced by a second source). Some time elapses while each atomic wavefunction is in a superposition state, and then the π-pulse from L2 inverts the populations, still leaving the system in a superposition state. Finally the $\pi/2$-pulse from L3 collapses the superposition into one or the other final states, depending on the relative phase shifts since the interaction with L1.

There are many geometrical variations of such interferometers, including the widely used "Ramsey–Bordé" scheme, but they all depend upon the superposition state created when $\vec{R}$ is moved away from a pole of the Bloch sphere by the first two-step transition. That is, each atom must have a non-zero probability of traveling each pathway, rather than some atoms going on one and some going on the other. The latter case would lead to determinable path choices and destroy the interference pattern just as in the more familiar optical case.

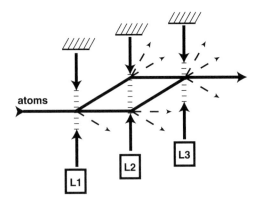

Figure 2.11 Schematic diagram of the interaction of atoms in an interferometer with three light pulses delivered by L1–L3. The solid lines could represent actual atomic trajectories or simply a horizontal time axis. The consequences of each light pulse are discussed in the text.

In the cases considered here it is assumed that the atoms start at the south pole and are excited by a $\pi/2$-pulse, but of course, there can be variations of this as well. In all these cases, the initial and final events are analogous to the beam splitters in optical interferometers, but the optical element in these cases is a light field and the object being "split" is matter, exactly complementary to the optical case. Thus the name "atom optics" seems perfectly appropriate. The phenomenon of echoes corresponds to the white light central fringe of optical interferometry.

2.D Adiabatic rapid passage

The motion of $\vec{R}$ on the Bloch sphere allows a particularly graphic interpretation of a phenomenon called adiabatic rapid passage (ARP). If the frequency of the applied field is swept through resonance, an atom initially in the ground state is left in the excited state (and *vice versa*) with very high probability. At the beginning of the frequency sweep, $|\delta|$ is chosen to be very much larger than $|\Omega|$ so that $\vec{\Omega}$ is nearly polar as shown in Fig. 2.12a. For an atom initially in state $|g\rangle$, $\vec{R}$ executes small, rapid orbits near the south pole, and these evolve toward the equator as $|\delta|$ sweeps toward 0 because the precession axis approaches the equatorial plane as shown in Fig. 2.12b. At $\delta = 0$, $\vec{R}$ executes circles in a vertical plane because $\vec{\Omega}$ is now in the equatorial plane. As the frequency sweep continues, the continually rotating $\vec{\Omega}$ now moves the center of the orbit on the surface of the sphere toward the north pole. Near the end of the sweep, $\vec{R}$ executes small, rapid orbits near the north pole as shown in Fig. 2.12c, and at the end of the sweep, $\vec{R}$ is left at the north pole, and the atom is left in the excited state. Another most attractive view in terms of the dressed states is given below.

2.D Adiabatic rapid passage

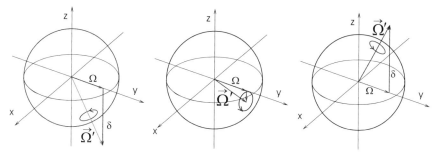

Figure 2.12 The evolution of the Bloch vector $\vec{R}$ during ARP. (a) It begins with small precessions near the south pole because the atom starts in the ground state and the large detuning makes $\vec{\Omega}$ pass near the poles. (b) As the detuning approaches zero the precession of $\vec{R}$ becomes large circles centered on a point near the equator as shown in the center picture. (c) Finally the detuning is very large in the opposite sense to its beginning, resulting in small circular precession near the north pole. Thus the atom is left in the excited state. In all three pictures, the $\hat{x}$-component of $\vec{\Omega}$ is chosen to be zero, and the $\hat{y}$-component is constant as shown. The detuning is represented by the $\vec{z}$-vector pointing downward at the start (a) and upward at the end (c). (Figures adapted from Ref. [1].)

The name *adiabatic rapid passage* may seem a bit enigmatic: how can something be both adiabatic and rapid? During the process of raising $\vec{R}$ from the south to the north pole of the Bloch sphere, there is always some component of the excited state present, with a corresponding probability of spontaneous decay. Thus this coherent excitation process can succeed only if it occurs in a time short compared with the natural lifetime of the excited state $|e\rangle$, so it must be fast. Needless to say, it must also be slow enough for the precessing Bloch vector $\vec{R}$ to follow the evolving axis of $\vec{\Omega}$ adiabatically. Thus there are boundaries determined by the atomic parameters on the rate of sweeping the detuning $d\delta/dt$. In practice, these limits can be satisfied with ordinary lasers and atoms, but it takes some effort.

Compared with the π-pulse method of inverting atomic populations discussed in Sec. 2.3, ARP is very much more effective and robust. It does not depend on precise control of δ that can be compromised by experimental conditions such as Doppler shifts or laser frequency drifts, and it is not critically dependent on light intensity either. As long as the intensity is high enough to satisfy the adiabaticity condition above, the efficiency of inversion by ARP is not affected.

The dressed atom picture provides an excellent way to envision ARP. The energy separation between a pair of coupled levels in this picture is simply $\hbar\Omega'$ as given by Eq. (2.10). An important aspect of the dressed atom picture for the present purposes is the energy ordering of the eigenstates. In the low-intensity limit the upper eigenstate approaches the ground state $|g\rangle$ and the lower one approaches

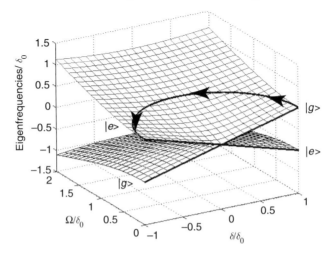

Figure 2.13 A 3D plot of the dressed-state energy levels given by Eq. (2.10). Atoms starting in state $|g\rangle$ at the right-hand edge can be completely transferred to state $|e\rangle$ by following the path indicated by the heavy line. Care must be taken to avoid a non-adiabatic transition to the opposite energy sheet.

the excited state $|e\rangle$ for the case of $\delta > 0$, but the reverse is true for $\delta < 0$. This difference is given by Eq. (2.10), and the plot in Fig. 2.13 shows these limits near the $\Omega = 0$ plane, and also that these bare ground and excited states $|g\rangle$ and $|e\rangle$ are otherwise mixed on two eigenenergy sheets away from the low-intensity limit (see Ref. [16]).

The process of ARP in this view involves a synchronized sweep of both the amplitude and frequency of the light so that the energy of the system follows a trajectory similar to that of the heavy line in Fig. 2.13. As long as travel along this trajectory is slow enough to avoid a non-adiabatic transition to the complementary eigenenergy sheet, the population will be completely inverted. Travel on the lower energy sheet can be similarly adiabatic.

2.E Superposition and entanglement

Perhaps the most important distinction between quantum and classical mechanics is the possibility of superposition of states. Schrödinger went so far as to say it is "the heart of the matter". Quantum systems in such states are not divided between two possibilities with certain probabilities like classical objects, but instead are in a superposition state that has no classical analog. Only multiple measurements that can distinguish among the possibilities can determine the probabilities of finding each state, but these are not sensitive to the phase of the superposition: some information is always lost in measurements.

When an atom is subject to a $\pi/2$-pulse, $\vec{R}$ is driven to the equator of the Bloch sphere and it is in a superposition of state $|g\rangle$ and $|e\rangle$. The phase is determined at first by $\vec{\Omega}$, but it can evolve in the dark by precession about the polar axis. For $\delta \neq 0$, $\vec{R}$ lands somewhere else on the Bloch sphere and evolves differently. What is important is that there are *two* independent parameters, e.g. latitude and longitude on the sphere, and a measurement can determine only one of them while the other is projected out and its information is lost. This property is at the core of quantum information studies, particularly the quantum non-cloning theorem.

There is a very special kind of superposition called an entangled state. Consider the case of the momentum change associated with the absorption of light produced by a $\pi/2$-pulse discussed above. The total wavefunction $\Psi(t)$ must include the atomic motion, and it is correlated with the internal state. It can be written as $\Psi(t) = \alpha(t)|g\rangle|\vec{p}\rangle + \beta(t)|e\rangle|\vec{p}+\hbar\vec{k}\rangle$, where $\alpha(t)$ and $\beta(t)$ are time-dependent complex numbers satisfying $|\alpha|^2 + |\beta|^2 = 1$. It is an entangled state because there are no components of the wavefunction $|e\rangle|\vec{p}\rangle$ or $|g\rangle|\vec{p}+\hbar\vec{k}\rangle$ – that is, $\Psi(t)$ cannot be written as the product of two sums, but only as the sum of products. It is analogous to a polarizing beam splitter for light: the direction is correlated with the polarization.

One can imagine very many different kinds of entanglement. The necessary condition is the presence of two independent properties that can be correlated by some kind of interaction. It is even possible to entangle two separate particles by their different motions and orientations, and these can form the basis elements for quantum computing. Although the language was very different, such ideas originated from the famous "EPR" paper of 1935 (Ref. [17]) that considered the annihilation of an electron–positron pair into gamma-rays with opposite $\vec{k}$-vectors correlated with their helicities. The literature abounds with other implementations of quantum entanglement for quantum computing, cryptography, communication, and quantum teleportation.

Exercises

2.1 Show that the transformation for the expansion coefficients $c'_g(t) = c_g(t)$ and $c'_e(t) = c_e(t)e^{+i\delta t}$ is equivalent to the transformation to the rotating frame given in Ramsey [18], page 147, Allen & Eberly [6], page 41 ff, or Mandel & Wolf [19], page 753.

2.2 When the transformation of Exercise 2.1 is done, the Hamiltonian matrix has the form of Eq. (2.9). Draw an energy level diagram showing the energies of the states for $\Omega = 0$, calculate the eigenvalues for arbitrary Ω (see Eq. (2.10)), and then show the shifts that depend on these off-diagonal terms on your energy level diagram (light shifts). Calculate the shift in the limit of $|\delta| \gg |\Omega|$ and $|\delta| \ll |\Omega|$.

2.3 In this exercise the solutions of Eq. (2.8) are further considered.
(a) Derive Eqs. (2.7) from Eqs. (2.6).
(b) Show that the solutions of Eq. (2.7) are given by Eq. (2.8).
(c) Show that $|c_g(t)|^2 + |c_e(t)|^2 = 1$ for the solutions given by Eq. (2.8).
(d) Argue that the solutions imply that the Einstein B-coefficients for stimulated emission and absorption are equal.

2.4 Explain why the eigenenergies of Eqs. (2.6) can be found by diagonalizing the matrix of Eq. (2.9).

2.5 Show that the eigenvectors of the matrix in Eq. (2.9) can be written as

$$|1\rangle = \cos\theta |e\rangle + \sin\theta |g\rangle \quad \text{and} \quad |2\rangle = -\sin\theta |e\rangle + \cos\theta |g\rangle$$

and show that the mixing angle θ satisfies $\tan 2\theta = |\Omega|/\delta$.

2.6 The state of a two-level atom in a radiation field is described by the Bloch vector.
(a) Show that the three components of the Bloch vector $\vec{R}$ of Eq. (2.13) are real.
(b) Show that $R_j = \langle \sigma_j \rangle$, where σ is the Pauli matrix as defined in Eq. (2.16).
(c) Show that the equation of motion for the Bloch vector is $d\vec{R}/dt = \vec{\Omega} \times \vec{R}$.

2.7 Show that the Pauli matrices of Eq. (2.16) satisfy the relations $[\sigma_x, \sigma_y] = 2i\sigma_z$ and cyclic permutations, and also $\sigma_x^2 = \sigma_y^2 = \sigma_z^2 = I$, where I is the identity matrix.

2.8 Since the equation of motion for the Bloch vector is $d\vec{R}/dt = \vec{\Omega} \times \vec{R}$, it means that $\vec{R}$ is time-dependent in the presence of an optical field. The same problem yields solutions in the dressed state picture that are time-independent eigenfunctions as given above in terms of the mixing angle θ. Explain how the same problem can yield two solutions, one time-dependent and the other stationary.

2.9 Use the symbolic manipulation program of your choice to solve the coupled differential equations Eq. (2.14) (or Eq. (2.15)) for the case of adiabatic rapid passage (ARP). Show that a frequency sweep produces nearly perfect inversion for a sweep rate ω_m slow enough to satisfy $\omega_m \ll \delta_0$, where the optical frequency ω is swept from below atomic resonance ω_{eg} to above it by an amount δ_0. That is, the frequency sweep covers $\omega - \delta_0$ to $\omega + \delta_0$ and the Rabi frequency $|\Omega|$ is kept constant at approximately the same as δ_0. Use a functional dependence of the sweep of your choice – you might want to compare a few such as cosine, linear, etc. Finally, show that the sweep suggested by Fig. 2.13 where Ω is synchronously swept with the optical frequency to form a pulse results in an even more robust inversion of the atomic state on the Bloch sphere. Explain why this is so.

2.10 Appendix 2.B describes the method of separated oscillatory fields for precision measurements and atomic clocks. Using Eq. (2.1) as suggested, calculate the evolution of $\Psi(t)$ during the dark interval between the pulses, and show that there is no change in $|c_e(t)|^2$ as a result. This means that the Bloch vector $\vec{R}$ remains on the equator of the Bloch sphere. However, the wavefunction does change during the dark interval. Explain what happens to it and how anything can happen in the dark.

3
The atom–light interaction

3.1 Introduction

This chapter begins with three very common, first-order approximations used to solve the Schrödinger equation for an atom exposed to nearly resonant, monochromatic light. These are the electric dipole approximation, the perturbation expansion, and then a reprise of the RWA discussed in Sec. 2.3.2. In Chap. 4 two of them will be extended to higher order, but here the analysis is for the case that the light acts as a small perturbation on the atom. Thus the rate of absorption of nearly resonant light in the limit of weak-intensity light will be examined. Here the restriction of only two atomic levels will be relaxed, resulting in a more general description.

3.2 The three primary approximations

In Chap. 2 the Schrödinger equation was manipulated without approximation to the form of Eq. (2.3)

$$\dot{c}_j(t) = \frac{1}{i\hbar} \sum_k c_k(t) \mathcal{H}'_{jk}(t) e^{i\omega_{jk}t}, \tag{3.1}$$

where $\mathcal{H}'_{jk}(t)$ is the matrix element for the incident radiation in the basis of the field-free atomic states ϕ_k. This equation cannot be solved without approximation, and the choice there was to truncate the sum to two terms. Here a different and more conventional choice is made.

3.2.1 Electric dipole approximation

To begin, it is necessary to provide an expression for the operator $\mathcal{H}'(t)$. It was introduced in Eq. (2.3) and later simply replaced by $\hbar\Omega \cos\omega t$ in Eqs. (2.6), but now it is necessary to evaluate it to actually calculate the matrix elements $\mathcal{H}'_{jk}(t)$.

3.2 The three primary approximations

The general expression for the Hamiltonian of an atom plus a field requires the conservation of the total momentum of atom plus field, so the field momentum $e\vec{A}$ is included in the Hamiltonian as

$$\mathcal{H}(t) = \frac{1}{2m}\left(\vec{p} + e\vec{A}\right)^2 + V. \tag{3.2}$$

The squared term can be expanded remembering that the vector potential $\vec{A}$ and the momentum $\vec{p} = i\hbar\vec{\nabla}$ are operators that may not necessarily commute. However, they do commute in the Coulomb gauge where $\vec{\nabla}\cdot\vec{A} = 0$ so that $\vec{p}\cdot\vec{A} + \vec{A}\cdot\vec{p} = 2\vec{A}\cdot\vec{p}$.

The $\mathcal{A}^2$ term of Eq. (3.2) is generally much smaller than the $\vec{A}\cdot\vec{p}$ term, as can be seen by calculating the ratio of their magnitudes. Here $p \approx \alpha mc$ can be estimated from the momentum of the electron in the ground state of hydrogen (see Sec. 7.5) and the ratio becomes

$$\frac{e\mathcal{A}}{p} = \frac{e\mathcal{E}/\omega}{m\alpha c} = \frac{e\mathcal{E}a_0}{\hbar\omega}. \tag{3.3}$$

Here $|\vec{A}| = |\vec{\mathcal{E}}|/\omega$ is used for monochromatic light because $\vec{\mathcal{E}} = -\partial\vec{A}/\partial t$. The numerator here is essentially $\hbar|\Omega|$ as shown below, and typical values of $|\Omega|$ are $\sim 10^8$ Hz for light beams of moderate intensity. This is to be compared to $\omega \sim 10^{15}$ Hz. Thus the $|\vec{A}|^2$ term only becomes important for atoms in highly excited states where the energy level spacing and hence ω can be very small, or in very intense light beams. Thus the Hamiltonian of Eq. (3.2) becomes

$$\mathcal{H}(t) = \frac{p^2}{2m} + V + \frac{e\vec{A}(t)\cdot\vec{p}}{m} \equiv \mathcal{H}_0 + \mathcal{H}'(t). \tag{3.4}$$

where the explicit oscillatory time dependence of $\vec{A}(t)$ has been indicated.

The dependence of the coefficients $c_j(t)$ on the radiation field is in the form of the matrix elements $\mathcal{H}'_{jk}(t)$ as in Eq. (3.1), and the wavefunctions ϕ_k in these integrals vanish exponentially on the distance scale of the Bohr radius a_0. Since the vector potential can be written as $\vec{A}(t) = \tfrac{1}{2}\vec{A}_0\left[\exp[i(\vec{k}\cdot\vec{r} - \omega t + \phi)] + c.c.\right]$ for an atom at the origin with an arbitrary phase ϕ, the expansion of its spatial exponential $e^{i\vec{k}\cdot\vec{r}} = 1 + i\vec{k}\cdot\vec{r} + O(\vec{k}\cdot\vec{r})^2$ can be truncated to its lowest-order term of unity because $|\vec{k}\cdot\vec{r}| \approx 2\pi a_0/\lambda \sim \alpha \ll 1$ for the optical case. This is called the electric dipole approximation, for which an alternative description can be found in App. 3.C. For now write $\vec{A}_0 \equiv \hat{\varepsilon}\mathcal{A}_0$ to find:

$$(m/e)\mathcal{H}'_{jk}(t) = \langle j|\vec{A}(t)\cdot\vec{p}|k\rangle \approx \mathcal{A}_0\cos(\omega t + \phi)\hat{\varepsilon}\cdot\langle j|\vec{p}|k\rangle. \tag{3.5}$$

It is useful to write the matrix element $\langle j|\vec{p}|k\rangle$ in another form using the Heisenberg equation of motion

$$\vec{p} = m\frac{d\vec{r}}{dt} = \frac{im}{\hbar}[\mathcal{H}_0, \vec{r}]. \tag{3.6}$$

Then the matrix element becomes

$$\langle j|\vec{p}|k\rangle = \frac{im}{\hbar}\langle j|[\mathcal{H}_0, \vec{r}\,]|k\rangle = \frac{im}{\hbar}(E_j - E_k)\langle j|\vec{r}|k\rangle = \frac{-im\omega_{jk}}{e}\vec{\mu}_{jk}, \quad (3.7)$$

where $\mathcal{H}_0$ operates to the left on $\langle j|$ and to the right on $|k\rangle$ in the commutator, and $\vec{\mu}_{jk}$ is the matrix element of the dipole operator $-e\vec{r}$. Thus the electric dipole approximation gives

$$\mathcal{H}'_{jk}(t) = -i\omega_{jk}\mathcal{A}_0 \cos(\omega t + \phi)\hat{\varepsilon} \cdot \vec{\mu}_{jk} \equiv +\hbar\Omega_{jk}\cos(\omega t) \quad (3.8)$$

where for monochromatic light $\vec{\mathcal{E}}(t) = -d\vec{\mathcal{A}}/dt = \vec{\mathcal{E}}_0 \cos\omega t$ for $\phi = -\pi/2$ has been used, and

$$\Omega_{jk} \equiv -\frac{\mathcal{E}_0 \hat{\varepsilon} \cdot \vec{\mu}_{jk}}{\hbar}. \quad (3.9)$$

The frequency $|\Omega_{jk}|$ is a measure of the interaction strength introduced in Sec. 2.2 and is called the Rabi frequency. The quantity $\hat{\varepsilon} \cdot \vec{\mu}$ is the projection of the dipole moment operator on the electric field of the light, hence the name of the approximation.

Thus it is clear that the transitions arising in this lowest-order approximation derive from the electric field of the light (see also App. 3.C). This electric dipole interaction is the same as that obtained from a classical treatment of the interaction of a dipole with an electromagnetic field.

3.2.2 Perturbation approximation

Now it remains to solve the Schrödinger Eq. (3.1), beginning with the case where the coupling term $\mathcal{H}'(t)$ is weak. In practical terms this means that if the system is initially in state ϕ_i ($c_i(0) = 1$, $c_j(0) = 0$ for $j \neq i$), then $|c_{j\neq i}(t)| \ll 1$ for all time so that only one term contributes to the sum in Eq. (3.1). The difference between this and Eqs. (2.4) is that here there are many equations, one for each value of j on the left-hand side, whereas there are only two coefficients in Eqs. (2.4) and they are written out explicitly.

For the case where the atom is initially in state ϕ_i, $c_i(t) \approx 1$ so it does not change significantly with time in this approximation. The nearly constant $c_i(t)$ can come out of the integral and Equation (3.1) is now integrated, so after a finite interaction time τ the amplitudes $c_j(\tau)$ for all states with $j \neq i$ are given by

$$c_j(\tau) = \frac{1}{i\hbar}\int_0^\tau dt\, \mathcal{H}'_{ji}(t) e^{i\omega_{ji}t}, \quad (3.10)$$

and the matrix elements $\mathcal{H}'_{ji}(t)$ are given in Eq. (3.8). Note that the approximation of a weak coupling assumes that either the elements $\mathcal{H}'_{ji}(t)$ are small or the interaction time τ is small.

3.2 The three primary approximations

The evolution of the atomic wavefunction as given by these coefficients can now be found by substituting Eq. (3.8) into Eq. (3.10), where the electric field is given in terms of a monochromatic wave with frequency ω:

$$\vec{\mathcal{E}}(t) = \mathcal{E}_0 \hat{\varepsilon} \cos(\omega t) \equiv \frac{\mathcal{E}_0 \hat{\varepsilon}}{2} \left(e^{i\omega t} + e^{-i\omega t} \right). \tag{3.11}$$

Here the last substitution is made because it simplifies the integral in Eq. (3.10). This integration can now be carried out, resulting in

$$c_j(\tau) = -\frac{\Omega_{ji}}{2} \left[\frac{e^{i(\omega_{ji}+\omega)\tau} - 1}{\omega_{ji} + \omega} + \frac{e^{i(\omega_{ji}-\omega)\tau} - 1}{\omega_{ji} - \omega} \right]. \tag{3.12}$$

where Ω_{ji} is defined by Eq. (3.8).

3.2.3 The rotating wave approximation revisited

An important special case arises when the denominator of one of the two terms between brackets in Eq. (3.12) becomes close to zero. For instance, if $\omega_{ji} \approx \omega$ the denominator of the second term becomes small and the light efficiently couples the state i to the state j, where the energy of state j is higher than that of state i by an energy $\hbar \omega_{ji}$. This is the case of absorption, since the atom absorbs energy from the field to go to a higher state (see Fig. 3.1). Since in that case $(\omega_{ji}+\omega) \gg |\omega_{ji}-\omega|$, the first term can be neglected and only the second term of Eq. (3.12) is retained. This is referred to as the rotating wave approximation (RWA), discussed in Sec. 2.3.2, since the only retained term of $\vec{\mathcal{E}}(t)$ in Eq. (3.11) is the one that oscillates (rotates) with nearly the same frequency as the difference frequency of the atomic states. Corrections to the RWA are called the Bloch–Siegert shift.

The probability of finding the atom in state j is then given by

$$P_j(\tau) = |c_j(\tau)|^2 = |\Omega_{ji}|^2 \frac{\sin^2(\delta\tau/2)}{\delta^2}, \tag{3.13}$$

where $\delta \equiv \omega - \omega_{ji}$ and the fraction on the right side is the familiar sinc-function. Note that the requirement that the interaction is weak is fulfilled as long as

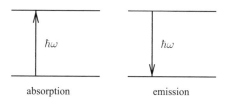

Figure 3.1 Absorption (left) and emission (right) of radiation by the atom, where the light is resonant with the transition frequency $(E_e - E_g)/\hbar$.

$P_j(\tau) \ll 1$ and can be realized by keeping either the interaction strength $|\Omega_{ji}|^2$ or the interaction time τ small.

So far only the second term in Eq. (3.12) has been considered. However, in the case that $-\omega_{ji} = \omega_{ij} \approx \omega$, the first term in brackets in Eq. (3.12) is no longer negligible and the same discussion as above applies, where the role of the two terms is reversed. Since the frequency of the light ω is always taken to be positive, the energy of the final state i is lower than the energy of the initial state j by an energy $\hbar\omega_{ij}$ and the transition process is called stimulated emission, since the atom loses energy to the field to go to a lower state (see Fig. 3.1). The probability $P_j(\tau)$ is unchanged in Eq. (3.13) when ω_{ji} is replaced by ω_{ij}, showing that the process of absorption and stimulated emission are of equal strength, just as discussed by Einstein in his famous 1917 paper [20] (see Sec. 5.2).

3.3 Light fields of finite spectral width

3.3.1 Averaging over the spectral width

Figure 3.2 shows the excitation probability of Eq. (3.13) after a finite interaction time τ vs. detuning between the light and the atoms. The probability is peaked around $\omega = \omega_{ji}$ and becomes rapidly smaller for larger detuning $\delta = \omega - \omega_{ji}$. For the excitation with a light field that has a finite spectral width, this probability has to be averaged over all the frequencies that compose the spectral intensity profile of the light source $S(\omega)$ (see App. 3.B.2), so Eq. (3.13) is replaced by

$$\overline{P_j(\tau)} = \frac{2|\hat{\varepsilon} \cdot \vec{\mu}_{ji}|^2}{\varepsilon_0 \hbar^2 c} \int_0^\infty S(\omega) \frac{\sin^2(\delta\tau/2)}{\delta^2} d\omega, \qquad (3.14)$$

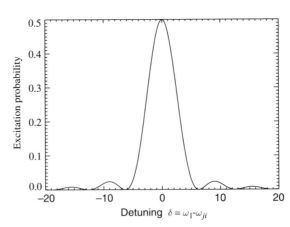

Figure 3.2 Excitation probability as a function of the difference between the light frequency ω and the transition frequency ω_{ji} in units of $1/\tau$, where τ is the interaction time.

3.3 Light fields of finite spectral width

where $\hat{\varepsilon}$ is the polarization of the light. This is not truly an approximation in the sense of those in Sec. 3.2 because it can be made arbitrarily accurate for narrow atomic transitions. Although the polarization plays an essential role in the coupling between atoms and light, it couples to the internal state of the atom, which has not been discussed so far. The discussion about the effects of the polarization will be deferred to Sec. 3.5.

The width of the excitation probability $1/\tau$ has to be compared with a width of the spectral intensity of the light source, which is for a typical continuous laser of the order of 1 MHz. If the excitation time is longer than about 10 μs, the spectral profile $S(\omega)$ of the light is much broader than the width of the excitation probability and $S(\omega)$ can be treated as a constant so it comes out of the integral and can be evaluated at the atomic resonance frequency ω_{ji}. Then integrating the sinc-function gives

$$\overline{P_j(\tau)} = \frac{\pi |\hat{\varepsilon} \cdot \vec{\mu}_{ji}|^2 S(\omega_{ji})\tau}{\varepsilon_0 \hbar^2 c}. \quad (3.15)$$

It is important to remember that the perturbation approximation used to derive this result must always be satisfied, namely $|c_j|^2 \ll 1$ at all times, and this puts an upper limit on the light intensity. The time-independent transition rate between the initial state i and final state j can be written as

$$R_{ij} = \frac{d\overline{P_j(\tau)}}{d\tau} = \frac{\pi |\hat{\varepsilon} \cdot \vec{\mu}_{ji}|^2 S(\omega_{ji})}{\varepsilon_0 \hbar^2 c} = \frac{4\pi^2 \alpha |\hat{\varepsilon} \cdot \vec{\mu}_{ji}|^2 S(\omega_{ji})}{e^2 \hbar}, \quad (3.16)$$

This result in Eq. (3.16) can also be obtained by directly applying Fermi's Golden Rule [21], but the present derivation clearly brings out the role of the three approximations involved in the calculation of the excitation process. First, there is the dipole approximation to calculate the effective Hamiltonian for the dominant interaction between atoms and light, then the perturbation approximation to limit the effects of the infinite summation in Eq. (3.1), and finally the RWA is required. Note that the transition rate is now time-independent because of the integration over the spectral profile of the exciting light and the choice that its width is larger than the atomic excitation profile. In the case of sharp transitions or short excitation times, Eq. (3.16) is no longer valid.

3.3.2 Scattering cross-section calculation again

As discussed earlier in Sec. 1.3.2, the relation between the incident and scattered light must involve some sort of effective size or cross-section for the atoms. The intensity of the light is related to $|\vec{\mathcal{E}}|^2$ whereas the atomic calculations result in the scattering rate. The first step is to recognize that the absorption rate multiplied by the transition's energy change is the rate of absorption of energy, and, by conservation of energy, must be the same as the emission or scattering rate. As before, the scattered light power is equal to the incident light intensity multiplied by the cross-section σ, and then Eq. (3.16) gives

$$\sigma = \frac{4\pi^2 \alpha |\vec{\mu}_{ji}|^2 S(\omega_{ji})}{\hbar e^2} \times \hbar\omega_{ji} \times \frac{1}{S(\omega_{ji})\Delta\omega_{ji}} = \frac{2\pi^2 \alpha \hbar}{m\Delta\omega_{ji}}. \tag{3.17}$$

Here it is estimated that $|(\hat{\varepsilon} \cdot \vec{r})_{ji}| \approx a_0/2$, that $mc^2\alpha^2/2 = \hbar\omega$, and that the incident intensity can be written as $S(\omega_{ji})\Delta\omega_{ji}$ (see App. 3.B.2). Figure 1.1b suggests that an appropriate spectral width $\Delta\omega_{ji}$ would be proportional to the damping rate, so for a Lorentzian spectral profile the appropriate choice is $\Delta\omega_{ji} = \pi\gamma/2 = \pi\hbar\omega_0^2\alpha/3mc^2$ (see App. 3.B.2). Then $\sigma = 3\lambda^2/2\pi$, the same as the classical value given in Eq. (1.7).

3.4 Oscillator strength

An interesting and useful dimensionless quantity to consider is the oscillator strength f_{kj} defined by

$$f_{kj} \equiv \frac{2m\omega_{kj}}{\hbar} |\hat{\varepsilon} \cdot \vec{r}_{kj}|^2 \tag{3.18}$$

because it satisfies

$$\sum_k f_{kj} = 1 \tag{3.19}$$

and the proof is given in App. 3.A. Some care is needed in applying this very useful theorem. If the state $|j\rangle$ is not the atomic ground state, Eq. (3.19) is still true, of course, but since the summation is over the complete set of wavefunctions $|k\rangle$, some of the frequencies ω_{kj} will be negative. Thus the other terms can sum to more than unity, or even some individual terms can exceed unity.

The oscillator strength is a useful quantity in cases where fine and/or hyperfine structure is to be resolved. For instance, in the case of the alkali-metal atoms the oscillator strength for transitions from the ground S state to the first excited P state is nearly one. However, the excited state is split into two fine-structure states labeled $P_{1/2}$ and $P_{3/2}$ (see Chap. 8). Including the fine-structure interaction in the calculation shows that for sodium the oscillator strength of 0.9238 of the 3S→3P transition is split into 0.3079 for the 3S→3P$_{1/2}$ transition and 0.6159 for the 3S→3P$_{3/2}$ transition. Similarly, including the hyperfine interaction in the calculation, leading to a further splitting into several hyperfine states of both the ground and excited states, leads to the oscillator strengths for the different levels, as shown in Tab. 3.1. Note that the sum of oscillator strength of all transitions is 0.9238,

3.5 Selection rules

Initial state $3S_{1/2}(F=1)$		Initial state $3S_{1/2}(F=2)$	
Final state	f_{eg}	Final state	f_{eg}
$3P_{1/2}(F=1)$	0.0256	$3P_{1/2}(F=1)$	0.0770
$3P_{1/2}(F=2)$	0.1283	$3P_{1/2}(F=2)$	0.0770
$3P_{3/2}(F=0)$	0.0513	$3P_{3/2}(F=0)$	$\equiv 0$
$3P_{3/2}(F=1)$	0.1283	$3P_{3/2}(F=1)$	0.0154
$3P_{3/2}(F=2)$	0.1283	$3P_{3/2}(F=2)$	0.0770
$3P_{3/2}(F=3)$	$\equiv 0$	$3P_{3/2}(F=3)$	0.2156

Table 3.1 Oscillator strength for the transition of the ground states of sodium to the first excited states. F = total angular momentum.

which is again the oscillator strength for the S→P transition in the absence of both fine and hyperfine effects.

3.5 Selection rules

3.5.1 What are selection rules?

To evaluate the expressions in the previous sections it is necessary to calculate $\hat{\varepsilon} \cdot \vec{\mu}_{ji}$, and this depends on the wavefunction of the initial state i and the final state j. In the case of two-state transitions considered up to now, there is only one such term to be calculated, but its numerical value still depends on the details of the states involved. However, there can also be multiple states and thus a requirement for multiple calculations.

Such calculations can be simplified or avoided if there is information that some of these quantities vanish, and identification of such zeros is called "selection rules". For example, $\vec{\mu}_{ji}$ vanishes identically if $|i\rangle$ and $|j\rangle$ have the same parity, and this is indeed the case for the two-state transition between the hyperfine levels of ground-state atomic hydrogen that is so important in cosmology, medical MRI, and a host of other things. Understanding such transitions requires the extension of the approximations made earlier in Sec. 3.2, as done in Sec. 4.2.

3.5.2 Selection rules for electric dipole transitions

The electric field $\vec{\mathcal{E}}$ or the unit vector $\hat{\varepsilon}$ of Eq. (3.11) can in general be decomposed in a basis set of three orthogonal vectors. It is natural to select the three Cartesian directions ($\hat{x}$, $\hat{y}$, and $\hat{z}$) for this purpose. In that case one would need to evaluate $\hat{x} \cdot \vec{\mu}_{ji}$, $\hat{y} \cdot \vec{\mu}_{ji}$, and $\hat{z} \cdot \vec{\mu}_{ji}$, and this turns out to be complicated given the eigenfunctions $|i\rangle$ and $|j\rangle$. A better choice for the basis vectors $\hat{\varepsilon}_q$ turns out to be

$$\hat{\varepsilon}_{-1} = \frac{(\hat{x} - i\hat{y})}{\sqrt{2}} \qquad \hat{\varepsilon}_0 = \hat{z} \qquad \hat{\varepsilon}_{+1} = -\frac{(\hat{x} + i\hat{y})}{\sqrt{2}}. \qquad (3.20)$$

It is easy to show that these three vectors are normalized and are mutually orthogonal. The polarization $\hat{\varepsilon}_0$ refers to light linearly polarized in the z-direction (π-light) and since light has a transverse nature the propagation direction $\vec{k}$ is in the x, y-plane. The polarizations $\hat{\varepsilon}_{\pm 1}$ refer to circularly polarized light with -1 for left-handed (σ^--light) and $+1$ for right-handed (σ^+-light), and here the $\vec{k}$ vector is in the z-direction. In principle, only two basis vectors are required if one uses the fact that $\vec{k} \perp \hat{\varepsilon}_q$, but the choice of three vectors makes the discussion more general.

The matrix element that determines the coupling between two states by the field in the electric dipole approximation is given by $\langle j|\hat{\varepsilon}_q \cdot \vec{\mu}|i\rangle$. Using Eq. (3.20) and the definitions of the spherical harmonics $Y_{\ell m}(\theta, \phi)$ in Cartesian coordinates that can be readily found from Tab. 7.1, this matrix element can be written

$$\hat{\varepsilon}_q \cdot \vec{r} = \sqrt{\frac{4\pi}{3}} r Y_{1q}(\theta, \phi). \tag{3.21}$$

In Chap. 7 it is shown that the angular dependence of the eigenfunctions of the electronic states of atoms are given by $Y_{\ell m}(\theta, \phi)$, where ℓ is the quantum number for the angular momentum of the electron and m its projection along the z-direction. The angular part of the matrix element is thus given by

$$\mathcal{A}^q_{\ell' m', \ell m} = \sqrt{\frac{4\pi}{3}} \int \sin\theta \, d\theta \, d\phi \, Y^*_{\ell' m'}(\theta, \phi) \, Y_{1q}(\theta, \phi) \, Y_{\ell m}(\theta, \phi). \tag{3.22}$$

The orthogonality of the spherical harmonics cannot be used for the evaluation of such integrals because it does not apply to products of three $Y_{\ell,m}$s. However, a product of two of them can be replaced by a sum of two of them using Eqs. (C.21) in the end-of-book appendix. Then the orthogonality of the spherical harmonics can be used to determine certain selection rules, i.e. special cases where the integral in Eq. (3.22) vanishes (see Sec. 7.6.3).

Since the rank of the Y_{1q} is necessarily 1 as indicated by the subscript, Eqs. (C.21) require that electric dipole transitions must occur between states whose ℓ-values differ by ± 1, hence the selection rule about opposite parity. Moreover, for linearly polarized light ($q = 0$), Eq. (C.21b) requires the m-values of the coupled states to be the same (called π-transitions), and for circularly polarized light ($q = \pm 1$), Eq. (C.21a) requires the m-values of the coupled states to differ by ± 1 (called $\sigma^\pm$-transitions).

These dipole selection rules, $\Delta\ell = \pm 1$ and $\Delta m = \pm 1, 0$ determine which transitions are allowed. Those that are forbidden in the electric dipole approximation may be allowed in higher order approximations, as discussed in Sec. 4.2. It is most important to recognize that these selection rules depend on the polarization of the light and not on its propagation direction.

3.5.3 Experimental application of dipole selection rules

The atomic physics literature describing experiments that exploit these selection rules is enormous. One example is the quantum beats and level crossing spectroscopy discussed in App. 3.D. Perhaps the most famous and well-known application is the optical pumping experiments that earned the Nobel Prize for Alfred Kastler in 1966. Their description requires a knowledge of the sublevel structure of atoms, a topic discussed in Part II of this book, but whose basic features are familiar to advanced undergraduates in the physical sciences.

Consider an atomic transition between two electronic states of total angular momentum j_g and j_e in the most common situation, namely $j_e = j_g + 1$. An energy level diagram is shown in Fig. 3.3 for the case of $j_g = 1$ and σ^+ light. Atoms are excited along paths indicated by straight lines, and can decay along wavy lines. Thus after a few cycles, all atoms are trapped in the cycling transition at the right edge of Fig. 3.3. After that, all spontaneous emission must be circularly polarized because it must correspond to a $\Delta m = -1$ transition, so a detector behind a circular analyzer of opposite sign will detect no light. Only if atoms are moved from the ground state $m_j = +1$ to $m_j = 0$ can they be excited to the state $m_j = 1$ which can decay with linearly polarized light to the ground state $m_j = +1$. If a magnetic field is applied so that the energy levels are split by the Zeeman effect, then radio frequency (rf) transitions tuned to the Zeeman splitting will populate states such as $m_j = 0$ and -1, and the spontaneously emitted light is not completely circularly polarized. Therefore a photocell can detect the rf transitions.

In a sense, the atoms serve as amplifiers from rf to optical frequencies, because an rf transition is reflected in the fluorescent light. It is easy to imagine many different optical and level schemes that exploit the dipole selection rules to trap

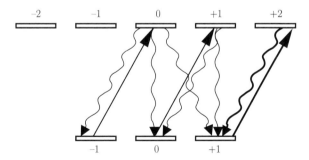

Figure 3.3 The straight lines indicate electric dipole excitations with circularly polarized light corresponding to σ^+ transitions. Thus they connect ground states of m_j with excited states of $m_j + 1$. The wavy lines indicate possible decay paths of electric dipole transitions. It is clear that after a few cycles, all atoms are pumped into the cycling transition indicated by the heavy lines because excitation between states of equal m_j is forbidden.

atoms in a single sublevel, or in a cycling transition between two sublevels as shown in Fig. 3.3.

Appendices

3.A Proof of the oscillator strength theorem

It is easiest to prove this theorem using one of the Cartesian coordinates and later combine the results for all three. The proof shown below starts by choosing the polarization vector $\hat{\varepsilon}$ in Eq. (3.18) along the x-axis, dividing it into halves and then writing out the two separate terms of the square, and using Eq. (3.7) as $\langle k|x|j\rangle = -i \langle k|p_x|j\rangle /m\omega_{kj}$. Now extract the unit operator $\sum_k |k\rangle \langle k|$ twice from the appropriate places and arrive at the result in the last step of Eq. (3.23):

$$\sum_k |\langle k|x|j\rangle|^2 = \frac{1}{2} \sum_k \langle j|x|k\rangle \langle k|x|j\rangle + \langle j|x|k\rangle \langle k|x|j\rangle$$

$$= \frac{i}{2m\omega_{kj}} \sum_k \langle j|p_x|k\rangle \langle k|x|j\rangle - \langle j|x|k\rangle \langle k|p_x|j\rangle$$

$$= \frac{i}{2m\omega_{kj}} \langle j|p_x x - x p_x|j\rangle$$

$$= \frac{i}{2m\omega_{kj}} \langle j|[p_x, x]|j\rangle = \frac{\hbar}{2m\omega_{kj}}. \qquad (3.23)$$

Here the minus sign in the second line arises because $\omega_{kj} = -\omega_{jk}$. Thus the theorem of Eq. (3.19) is proven.

3.B Electromagnetic fields

There are many ways to describe the electromagnetic fields of light, depending on the problem to be addressed. Of course, each one must be correct within its own limitations as well as be appropriate for the questions to be answered. One can generally make a distinction between non-monochromatic light, usually called classical light, and monochromatic light, usually from lasers. In the first case the light derives from a gaseous discharge or an arc source, where various conditions of the emitters of the light cause it to be spectrally broadened to a width much larger than the linewidth of the atomic transitions of interest. In that case it is appropriate to describe the light by its spectral intensity $S(\omega)$, which is non-negligible for a large range of frequencies ω. In the second case, the linewidth of the light is much smaller than the linewidth of the transition induced in the atom, and as such one can describe the light by its strength at a single frequency ω. Finally, there are situations where the quantum description of light fields is appropriate, but that description is deferred to Chap. 5 and the details are left to other textbooks devoted to quantum optics.

Classical light is usually described as a diffuse set of fields, even though it can be fashioned into beams with suitable lenses. But laser light is produced from its source in a single (or few) spatial modes and the propagation of such beams is amenable to special treatment. These spatially coherent Gaussian laser beams have special properties that need to be understood. Of course, classical light beams must also obey these equations, but typically there are so very many spatial and temporal modes present that using Gaussian beam optics is impractical.

3.B.1 Laser light

Lasers produce very nearly monochromatic light whose properties are simply described by Maxwell's equations. This appendix discusses the basic features of laser light that are needed for the understanding of its interaction with atoms. It is described in terms of a field $\vec{\mathcal{A}}$, called the vector potential, that is used in the Hamiltonian of the atom–light interaction.

Maxwell's equations for electromagnetic fields in SI units are

$$\vec{\nabla} \cdot \vec{\mathcal{E}} = \rho/\varepsilon_0 \qquad \vec{\nabla} \times \vec{\mathcal{E}} = -\frac{\partial \vec{\mathcal{B}}}{\partial t}$$

$$\vec{\nabla} \cdot \vec{\mathcal{B}} = 0 \qquad \vec{\nabla} \times \vec{\mathcal{B}} = \frac{1}{c^2}\frac{\partial \vec{\mathcal{E}}}{\partial t} + \mu_0 \vec{\mathcal{J}}, \tag{3.24}$$

where the charge density ρ and the current density $\vec{\mathcal{J}}$ satisfy $\vec{\nabla} \cdot \vec{\mathcal{J}} = -d\rho/dt$. There are four equations that describe the fields, and in two of them the electric field $\vec{\mathcal{E}}$ and the magnetic field $\vec{\mathcal{B}}$ are coupled. Since the divergence of the magnetic field is zero, it can be written in terms in terms of the vector potential $\vec{\mathcal{A}}$ as

$$\vec{\mathcal{B}} = \vec{\nabla} \times \vec{\mathcal{A}}. \tag{3.25}$$

In order to satisfy the second Maxwell equation, the electric field can be written as

$$\vec{\mathcal{E}} = -\frac{\partial \vec{\mathcal{A}}}{\partial t} - \vec{\nabla}\Phi, \tag{3.26}$$

where Φ is a scalar potential. The two remaining Maxwell equations can now be expressed in terms of the scalar and vector potentials.

The scalar and vector potentials are not uniquely defined by the relations above. There can be a "gauge transformation" that transforms these potentials by the gauge field Λ as

$$\vec{\mathcal{A}}' = \vec{\mathcal{A}} - \vec{\nabla}\Lambda, \qquad \Phi' = \Phi + \frac{\partial \Lambda}{\partial t}. \tag{3.27}$$

It can easily be seen that $\vec{\mathcal{E}}$ and $\vec{\mathcal{B}}$ are unchanged under this transformation. For electromagnetic wave fields it is customary to use the Coulomb gauge where Λ is chosen so that $\vec{\nabla} \cdot \vec{\mathcal{A}} = 0$ and the wave equations become

$$\frac{1}{c^2}\frac{\partial^2 \vec{\mathcal{A}}}{\partial t^2} - \nabla^2 \vec{\mathcal{A}} + \frac{1}{c^2}\vec{\nabla}\frac{\partial \Phi}{\partial t} = \mu_0 \vec{\mathcal{J}}, \qquad \nabla^2 \Phi = -\rho/\varepsilon_0. \tag{3.28}$$

Since the scalar potential Φ derives only from the charges, as if they are not moving, this is called the Coulomb gauge. Note that in this gauge, the inhomogeneous wave equation for $\vec{\mathcal{A}}$ depends on $\vec{\mathcal{J}}$ and Φ.

For the interaction of a single atom with an electromagnetic field in vacuum, there are no charges or currents present. Then $\Phi = 0 = \vec{\mathcal{J}}$ and there remains only the homogeneous equation for the vector potential. For monochromatic laser light at frequency ω, the time-dependent term of Eq. (3.28) becomes $-\omega^2 \vec{\mathcal{A}}/c^2$ resulting in the Helmholtz equation

$$\nabla^2 \vec{\mathcal{A}} + k^2 \vec{\mathcal{A}} = 0 \tag{3.29}$$

since $\omega = ck$. One of the solutions of the Helmholtz equation is a set of infinite plane waves travelling in the $\vec{k}$-direction

$$\vec{\mathcal{A}}(\vec{r},t) = \frac{\mathcal{A}_0 \hat{\varepsilon}}{2} \exp\left[i(\vec{k}\cdot\vec{r} - \omega t + \phi)\right] + \text{c.c.} = \mathcal{A}_0 \hat{\varepsilon} \cos(\vec{k}\cdot\vec{r} - \omega t + \phi), \tag{3.30}$$

where $\mathcal{A}_0$ is the amplitude of the field, $\hat{\varepsilon}$ is the polarization, $\vec{k}$ is the wavevector whose magnitude is $2\pi/\lambda$, $\omega = c|\vec{k}|$ is the frequency of the field, and ϕ an arbitrary phase. Note that $\vec{\mathcal{A}}$ is real and that in the Coulomb gauge $\vec{k}\cdot\hat{\varepsilon} = 0$, so the polarization is perpendicular to the direction of propagation. Then $\vec{\mathcal{E}}$ and $\vec{\mathcal{B}}$ can be written as

$$\vec{\mathcal{E}}(\vec{r},t) = \frac{i\omega \mathcal{A}_0 \hat{\varepsilon}}{2} \exp\left[i(\vec{k}\cdot\vec{r} - \omega t + \phi)\right] + \text{c.c.} \tag{3.31a}$$

$$= \mathcal{E}_0 \hat{\varepsilon} \cos\left(\vec{k}\cdot\vec{r} - \omega t + \phi + \frac{\pi}{2}\right)$$

and

$$\vec{\mathcal{B}}(\vec{r},t) = \frac{i\mathcal{A}_0(\vec{k}\times\hat{\varepsilon})}{2} \exp\left[i(\vec{k}\cdot\vec{r} - \omega t)\right] + \text{c.c.} \tag{3.31b}$$

$$= \mathcal{B}_0(\vec{k}\times\hat{\varepsilon}) \cos\left(\vec{k}\cdot\vec{r} - \omega t + \phi + \frac{\pi}{2}\right)$$

so that both $\vec{\mathcal{E}}$ and $\vec{\mathcal{B}}$ are real, have a phase difference of $\pi/2$ from $\vec{\mathcal{A}}$, and $|\vec{\mathcal{B}}| = |\vec{\mathcal{E}}|/c$. The energy flow is given by the Poynting vector $\vec{\mathcal{S}}$ as

$$\vec{\mathcal{S}} = \vec{\mathcal{E}} \times \vec{\mathcal{H}} \equiv \vec{\mathcal{E}} \times \vec{\mathcal{B}}/\mu_0 \tag{3.32}$$

so that the time-averaged light intensity is

$$I \equiv |\vec{S}|_{avg} = 1/2\, \varepsilon_0 c \mathcal{E}_0^2, \qquad (3.33)$$

where $\mathcal{E}_0 = \omega \mathcal{A}_0$.

To find a solution that corresponds to a beam it is necessary to construct a functional form that resembles observed beams, substitute it into Eq. (3.29), and see what constraints arise. A suitable solution can be found using the slowly varying envelope approximation that chooses the fractional change of the beam's transverse profile over a wavelength distance to be small compared with unity, and similarly for its derivative. For a beam along $\hat{z}$ this requires that

$$\lambda \left|\frac{\partial \mathcal{E}_0}{\partial z}\right| \ll |\mathcal{E}_0| \text{ and } \lambda \left|\frac{\partial^2 \mathcal{E}_0}{\partial z^2}\right| \ll \left|\frac{\partial \mathcal{E}_0}{\partial z}\right|. \qquad (3.34)$$

The resulting solution is a beam whose cross-section is Gaussian given by

$$\vec{\mathcal{E}}(r) = \mathcal{E}_0\, \hat{\varepsilon}\, \exp[-(\rho/w(z))^2] \qquad (3.35)$$

where $\rho \equiv \sqrt{x^2 + y^2}$ and $w(z) = w_0 \sqrt{1 + (z/z_0)^2}$. The Rayleigh length $z_0 \equiv \pi w_0^2/\lambda$ and the beam waist w_0 both depend on how the beam was formed. The intensity distribution is therefore

$$I(\rho) = 1/2\, \varepsilon_0\, c\, \mathcal{E}_0^2 \left[\frac{w_0}{w(z)}\right]^2 \exp[-2(\rho/w(z))^2]. \qquad (3.36)$$

For a beam of total power P the peak intensity at the waist is $P/\pi w_0^2$. Such Gaussian laser beams are always diverging or converging because of diffraction and always have a minimum waist w_0 at $z = 0$. But for $w_0 \sim$ few mm, a beam of $\lambda = 633$ nm (helium–neon laser) expands to $\sqrt{2}\, w_0$ over a distance of $z_0 \sim$ several tens of meters, thus appearing quite nearly collimated. The half-angular divergence is $\theta = \lambda/\pi w_0$ and is smaller than the $1.22\lambda/(2w_0)$ from diffraction by a circular aperture. The detailed behavior of Gaussian laser beam propagation can be found in many textbooks [22].

3.B.2 Light from classical sources

Any measurement of the intensity of the incident radiation as used in Eq. (1.7) or Eq. (3.14) necessarily has limited spectral resolution, and therefore measures the total intensity in some spectral bandwidth $\Delta\omega$. The measured intensity is divided by this experimental bandwidth to give the spectral intensity $\mathcal{S}(\omega)$. In order to relate the electric field of a single frequency of radiation as used earlier in this chapter to the spectral energy density or spectral intensity as used, for example, in the description of the Planck law of black body radiation, it is necessary to consider this experimental question. In the single-frequency case, the experimentally measured total intensity is the Poynting vector $\vec{S} = \vec{\mathcal{E}} \times \vec{\mathcal{H}}$.

There are similar definitions for the spectral energy density $U(\omega) = S(\omega)/c$, and this can sometimes lead to confusion with the actual electrical energy density in a single-frequency radiation field, $\varepsilon_0 |\vec{\mathcal{E}}|^2/2$. One can always do a calculation using $U(\omega)$ or $S(\omega)$, but comparison with experiment will require multiplication by some bandwidth.

It is also important to note that a thermal radiation bath contains all possible polarizations whereas the calculation that leads to Eq. (3.16) and those above it are based on a single polarization. Thus $1/3$ of the energy density in a thermal radiation bath corresponds to a polarized beam of light, so the correct substitution is therefore $U(\omega)\Delta\omega/3$. Of course, this expression has to be summed over all polarizations.

In the case of non-monochromatic light, the radiation cannot be described by a single frequency. However, the Fourier theorem allows the radiation to be described by superposition of fields given by

$$\vec{\mathcal{A}}(t) = \int \hat{\varepsilon} \mathcal{A}_0(\omega) e^{i(\vec{k}\cdot\vec{r} - \omega t + \phi(\omega))} d\omega + c.c., \tag{3.37}$$

where the phase $\phi(\omega)$ is explicitly a function of ω. The relations between the vector potential $\vec{\mathcal{A}}$, the electric field $\vec{\mathcal{E}}$, and the magnetic field $\vec{\mathcal{B}}$ given in Eq. (3.25) and Eq. (3.26) for a single frequency now hold for each frequency ω in the integrand. In particular, the amplitudes of the electric and magnetic fields are given by $|\vec{\mathcal{E}}(\omega)| = \omega \mathcal{A}_0(\omega)$ and $|\vec{\mathcal{B}}(\omega)| = |\vec{k} \times \hat{\varepsilon}| \mathcal{A}_0(\omega)$, respectively.

The intensity of the radiation can be found by integrating the spectral intensity profile $S(\omega)$ over ω, or

$$I = \int S(\omega) \, d\omega \equiv \frac{1}{2} \varepsilon_0 c \int |\vec{\mathcal{E}}(\omega)|^2 \, d\omega, \tag{3.38}$$

where Eq. (3.33) has been used for the right-hand side. When the spectral profile of the light is much larger than the absorption width, it is appropriate to evaluate $S(\omega)$ at at the center frequency ω_0 and then define a spectral width $\Delta\omega$ so that multiplying it by the peak value of $S(\omega)$ results in the total intensity. That is,

$$I = \int S(\omega) \, d\omega \equiv S(\omega_0) \Delta\omega. \tag{3.39}$$

For a Gaussian profile one finds $\Delta\omega = \sqrt{2\pi}\sigma$, where σ is the half width at $1/\sqrt{e}$ maximum, and for a Lorentzian profile $\Delta\omega = \pi\gamma/2$, where γ is the full width at half maximum.

3.C The dipole approximation

In Sec. 3.2.1 the lowest-order interaction between the optical field and atoms was found to arise from the electric field of the light acting on the light-induced atomic dipole moment. The derivation there uses the Heisenberg picture that transforms the interaction $\vec{\mathcal{A}} \cdot \vec{p}$ into the electric dipole interaction $\vec{\mu} \cdot \vec{\mathcal{E}}$ to write Eq. (3.7). This appendix presents the classical view of how an electric field interacts with the charges of a neutral system such as an atom, and that the interaction arises from the induced dipole moment. It has two parts: first it takes the lowest orders of the electrodynamic interaction, thereby making the long-wavelength expansion, and second, it makes a unitary transformation to separate and clarify the roles of the field and particle momenta.

The Hamiltonian of a cloud of particles j with charges q_j in an electromagnetic field is given by

$$\mathcal{H} = \sum_j \frac{1}{2m_j} \left[\vec{p}_j - q_j \vec{\mathcal{A}}(\vec{r}_j, t) \right]^2 + V_c + \sum_j q_j \Phi(\vec{r}_j, t), \tag{3.40}$$

where $\vec{\mathcal{A}}$ is again the vector potential, Φ is the scalar potential, and V_c is the Coulomb interaction between the charges.

The dipole approximation now uses the long-wavelength expansion and assumes that the distances between the charges are small compared with the spatial variation of the field. Let $\vec{R}$ represent the position of the center of mass of the system so that the location of each charge is $(\vec{r}_j - \vec{R})$ with respect to the center. Then the exponent $e^{i\vec{k} \cdot \vec{r}_j}$ in Eq. (3.30) becomes

$$e^{i\vec{k} \cdot \vec{r}_j} = e^{i\vec{k} \cdot \vec{R}} \left(1 + i\vec{k} \cdot (\vec{r}_j - \vec{R}) - \frac{1}{2} \left[\vec{k} \cdot (\vec{r}_j - \vec{R}) \right]^2 + \cdots \right). \tag{3.41}$$

For clouds of atomic size the terms $\vec{k} \cdot (\vec{r}_j - \vec{R})$ are $\ll 1$, since all the terms $(\vec{r}_j - \vec{R})$ are of order of the Bohr radius $a_0 \sim 0.5$ nm, and for an optical wavelength $\lambda \sim 500$ nm the wavevector $|\vec{k}| = 2\pi/\lambda \ll 1/a_0$. In the lowest-order approximation $\vec{\mathcal{A}}(\vec{r}_j, t)$ can therefore be replaced by $\vec{\mathcal{A}}(0, t)$ (this constitutes the long-wavelength approximation and puts the center of mass at the origin) or

$$\begin{aligned} \vec{\mathcal{A}}(\vec{r}_j, t) &\approx \vec{\mathcal{A}}_0(0, t) \\ \Phi(\vec{r}_j, t) &\approx \Phi(0, t) + \vec{r}_j \cdot \vec{\nabla} \Phi(0, t). \end{aligned} \tag{3.42}$$

Note that the electric field depends on the gradient of the scalar potential and thus the scalar potential has to be expanded to one order higher compared with the vector potential. Substitution of these expressions into the Hamiltonian of Eq. (3.40) yields

$$\mathcal{H} = \sum_j \frac{1}{2m_j} \left(\vec{p}_j - q_j \vec{\mathcal{A}}(0,t) \right)^2 + V_c + \vec{\mu} \cdot \vec{\nabla}\Phi(0,t), \tag{3.43}$$

where the electric dipole moment is defined by $\vec{\mu} = \sum_j q_j \vec{r}_j$.

In Eq. (3.43) the momentum of the particles $\vec{p}_j$ is "dressed" by the field term $q_j \vec{\mathcal{A}}(0,t)$, and this leads to various ambiguities for the identification of the "real" momentum of the particles. However, the unitary transformation $T(t) = \exp(-iV(t)/\hbar)$, with careful choice of the transformation term $V(t)$, removes this ambiguity ($V(t)$ not to be confused with V_c). In this particular case of the dipole approximation, choosing $V(t) = \vec{\mu} \cdot \vec{\mathcal{A}}(0,t)$ leads to

$$\mathcal{H}' = \sum_j \frac{\vec{p}_j^{\,2}}{2m_j} + V_c - \vec{\mu} \cdot \vec{\mathcal{E}}(0,t). \tag{3.44}$$

after considerable algebra [23]. With the momentum of the vector field separated from the momentum of the particle, these are now disentangled so that the last term of the Hamiltonian can be identified as the interaction term. It is called the electric dipole interaction, since it is the interaction of the dipole moment of the charges with the electric field component of the light.

It is important to remember that the symmetry of atomic wavefunctions precludes any permanent electric dipole moment. The $\vec{\mu}$ used here must be induced by the electric field of the radiation, as defined in Eq. (3.7).

3.D Time resolved fluorescence from multilevel atoms

3.D.1 Introduction

This section describes two closely related spectroscopic techniques that arise from the multiple excited-state levels of atoms. Their development in the 1960s had a profound impact on precision optical spectroscopy because the measurements were not compromised by the Doppler broadening that characterized other optical spectroscopy experiments. In those days, before the advent of laser spectroscopy, Doppler widths dominated the spectra of measured transitions, and only when the resonance frequencies were low, e.g. for rf or microwave transitions, was the spectral resolution limited by factors other than Doppler broadening

In the newer coherence experiments described here, the observed quantities resulted from the superpositions produced in the excited states of atoms and molecules, and these coherences were manifest in intensity or polarization changes of the fluorescence. Optical frequencies were not measured, and the beat frequencies from internal atomic coherences that were measured were so low that they had negligibly small Doppler shifts. Thus the atoms themselves provided the time base for the measurements.

3.D.2 Time resolved excited-state spectroscopy

Consider an atom initially in one of its ground states $|g\rangle$ that is excited by a short laser pulse at $t = 0$ to a superposition of its excited states $|e\rangle$. The wavefunction of the atom at time t can be written as [24]

$$\Psi(t) = c_g |g\rangle + \sum_e c_e |e\rangle e^{-(\gamma/2 + i\omega_e)t}, \qquad (3.45)$$

where $\hbar\omega_e$ is the energy of the state $|e\rangle$, and $\tau \equiv 1/\gamma$ is its lifetime against spontaneous decay. The excitation pulse is weak, and therefore this expression is for the perturbative regime where $|c_e| \ll 1$ and $|c_g| \approx 1$. Since the atom is excited by a laser pulse, the coefficients $c_e \propto \langle e | \vec{\mu} \cdot \vec{\mathcal{E}} | g \rangle \equiv \Omega_{eg}$, where $\vec{\mu} \cdot \vec{\mathcal{E}}$ represents the operator for the electric dipole transition between the ground state $|g\rangle$ and the excited states $|e\rangle$. Thus the quantum numbers of the states $|e\rangle$, such as j, ℓ, and m, are determined by the quantum numbers of $|g\rangle$ and the polarization of the exciting laser light according to the electric dipole selection rules. The more general case allows for multiple ground states $|g\rangle$, and this means summing the results over them to obtain the final result.

When the state $\Psi(t)$ of Eq. (3.45) decays by spontaneous emission, it can do so to many of the various ground states $|g'\rangle$ of the atom, but the trajectory from $|g\rangle$ to each final state $|g'\rangle$ is an independent event. Although these must be summed over to find the total fluorescence, only one term will be considered here. The emitted light intensity $I_{g,g'}(t)$ from the initial state $|g\rangle$ to the final state $|g'\rangle$ is proportional to the squares of the relevant decay matrix element so that

$$I_{g',g}(t) \propto \left| \langle g' | \vec{\mu} \cdot \vec{\mathcal{E}}_\omega | \Psi(t) \rangle \right|^2 \qquad (3.46)$$

where the matrix elements $\langle g' | \vec{\mu} \cdot \vec{\mathcal{E}}_\omega | \Psi(t) \rangle$ are for the electric dipole transitions of spontaneous decay (see Sec. 5.4). The $|g\rangle$ part of $\Psi(t)$ (see Eq. (3.45)) does not contribute to Eq. (3.46) because electric dipole transitions cannot connect two ground state sublevels of the same ℓ such as g and g'. Then Eq. (3.46) can be expanded to find

$$I_{g'g}(t) \propto \sum_e \langle g' | \vec{\mu} \cdot \vec{\mathcal{E}}_\omega | e \rangle \Omega_{eg} e^{-(\gamma/2 + i\omega_e)t} \sum_{e'} \Omega^*_{ge'} \langle e' | \vec{\mu} \cdot \vec{\mathcal{E}}_\omega | g' \rangle e^{-(\gamma/2 - i\omega_{e'})t}$$

$$\propto e^{-\gamma t} \sum_{ee'} \Omega^S_{g'e} \Omega_{eg} \Omega^*_{ge'} \Omega^{S*}_{e'g'} e^{i(\omega_{e'} - \omega_e)t} \qquad (3.47)$$

where Ω^S_{eg} for spontaneous emission processes is defined similarly to Ω_{eg}, and the double summation over e and e' includes all pairs of intermediate excited states.

The interesting part of the time dependence here arises not from the exponential decay of the fluorescent light as expected, but from the modulation of this light at the frequencies $\omega_{e'} - \omega_e$. There is a modulation only at frequencies for which $|e\rangle$ and $|e'\rangle$ can each be excited from the same ground state $|g\rangle$ by the same light field. These oscillations are named "quantum beats" and enable a technique called quantum beat spectroscopy. The modulation frequencies are sufficiently low that their Doppler shifts are negligible. Note that the case where $|e\rangle = |e'\rangle$ produces only an exponential decay with no modulation.

In general the excitation may take place from various initial states $|g\rangle$ and spontaneous emission may go to various final states $|g'\rangle$, Eq. (3.47) has to be summed over $|g\rangle$ and $|g'\rangle$, but this does not lead to modulations proportional to the energy difference in the ground states. The quantum beat signal can be interpreted as an analogy to the interference fringes of Young's double slit experiment. That is, the atoms have been excited to a superposition state so one cannot say which excited state $|e\rangle$ the atoms have populated so each intermediate state $|e\rangle$ acts as a "slit" in the interference experiment. The atom can pass through both of them simultaneously.

Consider an excited atomic state with total angular momentum j having $2j + 1$ components of different m_j. In a modest magnetic field $\mathcal{B}$, these are Zeeman split by $E_Z \equiv \hbar\omega_L = g_j m_j \mu_B \mathcal{B}$, where g_j is the Landé g-factor and μ_B is the Bohr magneton. Excitation of these m_j states by π, σ^+, or σ^- light cannot couple a particular ground-state sublevel to two different excited states as discussed in Sec. 3.5.2, but linearly polarized light with electric field perpendicular to $\mathcal{B}$ can do so because it is a superposition of σ^+ and σ^-. Thus there will be quantum beats at frequency $2\omega_Z = 2g_j m_j \mu_B \mathcal{B}/\hbar$ from decay to the original ground state, where the factor of 2 arises because the excited state m_j values must differ by 2. There will also be unmodulated exponential decay of each of the m_j states to various ground states resulting in a large exponential background. A plot of the time dependence of the fluorescence for pulsed excitation is shown in Fig. 3.4 (for cw excitation, the origin of time is different for each atom, and so the oscillations all have different phases and their sum over all atoms vanishes). Plotting the oscillation frequency vs. $\mathcal{B}$ allows extraction of the g-factor, and fitting the exponential allows extraction of the lifetime, both from a single experiment.

Quantum beats can give rise to systematic errors in precision lifetime measurements. Often atoms in atomic beams are excited by passage through a resonant light beam, perpendicular to their travel to eliminate Doppler shifts, so the excitation is essentially pulsed (collisional excitation of fast beams, for example in beam–foil experiments, is similar). Very small residual magnetic fields then alter the exponential decay, causing errors in its measurement and hence in the measured lifetime.

An interesting case arises when the excited state angular momentum is the sum of two contributions, say spin and orbit, resulting in multiple values of j (or F if there is also a nuclear spin). Consider the P states ($\ell = 1$) of atomic hydrogen or the

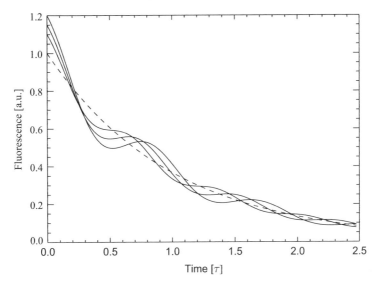

Figure 3.4 Plot of the time dependence of the fluorescence calculated in Eq. (3.47) showing oscillations at various frequency differences ($\omega_{e'} - \omega_e$). The unmodulated curve is just the exponential decay ($\omega_{e'} = \omega_e$).

alkalis. The angular momenta $s = 1/2$ and $\ell = 1$ combine to make the two values $j = 1/2, 3/2$, and each of these has Zeeman sublevels. The two states of different j are widely split by the fine structure E_{SO} (see Sec. 8.3) so that the quantum beat frequency between their sublevels is generally too high to resolve.

An applied magnetic field can shift the energy levels sufficiently so that quantum beats can be observed between states of different j (or F if there is a nuclear spin). Because $j = 3/2$ lies higher, and the energy of $(j, m_j) = (3/2, -3/2)$ decreases with increasing $\mathcal{B}$ since $m_j < 0$ (see Fig. 3.5), it can come very close to the $(1/2, 1/2)$ state whose energy increases with $\mathcal{B}$, thereby reducing their quantum beat frequency to an observable range. In the vicinity of their crossing as circled in Fig. 3.5 (actually, very weak anti-crossing) the frequency vanishes, and then changes sign (phase) on the higher field side. Thus a plot of the quantum beat frequency vs. $\mathcal{B}$ crosses zero at the point where their energies would be degenerate, and a knowledge of the Zeeman effect allows the extraction of E_{SO}. Such measurements can be extremely accurate, and are a type of level crossing spectroscopy (see below). The Fourier widths of the oscillation signals have extremely small Doppler broadening, but are dominated by the exponential decay, and hence the resolution is determined by the natural lifetime of the excited state.

Optical spectroscopy dominated by atomic natural widths was a breakthrough in precision at the time this technique was developed. It enabled sub-Doppler measurement of excited-state fine and hyperfine structure and provided detailed knowledge of atomic wavefunctions and of nuclear properties.

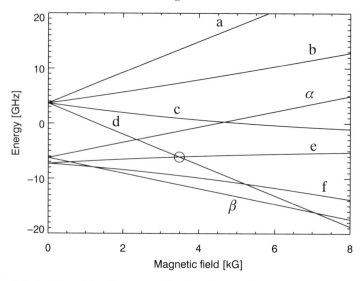

Figure 3.5 A plot of the Zeeman effect in the $n = 2$ states of hydrogen. The energy separation between the $P_{3/2}(m = -3/2)$ and $P_{1/2}(m = 1/2)$ states vanishes at a point near $B \approx 3.5$ kG enabling the observation of low-frequency quantum beats. (Figure adapted from Ref. [25].)

Another variation can also be found from Eq. (3.47). Suppose the fluorescent light is detected only in a short time interval Δt_1 that begins at t_1 after the pulsed excitation ends. Examination of Fig. 3.4 shows that the field dependence of such a signal would have oscillations as shown in Fig. 3.6 (Ref. [26]). If t_1 were larger than the natural lifetime, the width of the signals is narrower that the natural width. Such sub-natural width spectroscopy comes at the expense of signal-to-noise ratio because of the exponential decay, but it may have advantages that depend on the noise sources and on the causes of systematic errors.

3.D.3 The continuous light case

The excited state coherences that underlie quantum beat spectroscopy are also manifest under detection of the total signal. Integrating Eq. (3.47) over time using

$$\int_0^\infty dt \, e^{-(\gamma - i(\omega_{e'} - \omega_e))t} = \frac{\gamma + i(\omega_{e'} - \omega_e)}{\gamma^2 + (\omega_{e'} - \omega_e)^2} \tag{3.48}$$

leads to

$$I_{g'g} \propto 2 \sum_{ee'}{}' \left| \Omega_{g'e}^S \Omega_{eg} \Omega_{g,e'}^* \Omega_{e'g'}^{S*} \right| \left[\frac{\gamma \cos\phi - (\omega_{e'} - \omega_e)\sin\phi}{\gamma^2 + (\omega_{e'} - \omega_e)^2} \right], \tag{3.49}$$

where ϕ is the sum of the phases of all Ωs and the prime on the summation indicates summing only once over each pair of states (e, e').

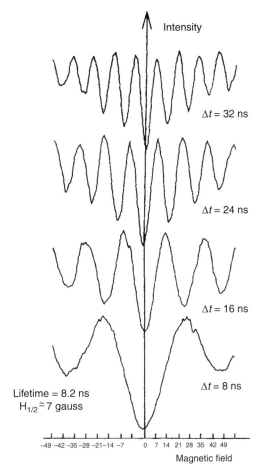

Figure 3.6 A plot of the measured fluorescence intensity from the $^1P_1$ state of Ba ($\tau \approx 8.3$ ns) in a short interval delayed by various times Δt. The calculated fluorescence shown in a parallel figure in Ref. [26] shows excellent agreement. (Figure from Ref. [26].)

For the special geometry described above, where the light is linearly polarized perpendicular to $\mathcal{B}$ (superposition of σ^+ and σ^-) and the fluorescence is observed in the direction perpendicular to both $\mathcal{B}$ and the propagation direction, $\phi = 0$ and the fluorescence shows a Lorentzian change centered at $\omega_{e'} - \omega_e = 0$. The signal width in $\mathcal{B}$-field units is given by the $\mathcal{B}$-dependence of the energy levels (slope) embodied in the $(\omega_{e'} - \omega_e)$-terms, and their natural width. The width in frequency is much smaller than the Doppler width of the fluorescence itself. Measurements using such coherence signals of atomic fine and hyperfine structure, as well as molecular rotational structure, are called level crossing spectroscopy, and permit very high accuracy and precision [27]. The name is not exactly correct because the levels do not actually cross, but the anti-crossing is usually too small to be observable.

Note that the case of $\omega_{e'} - \omega_e = 0$ also occurs among the Zeeman sublevels at $\mathcal{B} = 0$. This zero-field level crossing is precisely the Hanle effect that has been described in App. 1.B. The description here provides yet another example where the classical and quantum descriptions of a phenomenon give the same result.

Exercises

3.1 In order to include the electromagnetic field in the Hamiltonian the canonical momentum is written as $\vec{p} + e\vec{\mathcal{A}}$. Thus the Hamiltonian has a term in $|\vec{\mathcal{A}}|^2$, which is neglected. Show that this neglect is justifiable. Under what conditions does it become questionable?

3.2 Show that the Schrödinger equation $i\hbar \partial \psi/\partial t = \mathcal{H}\psi$ of a hydrogenic atom in an electromagnetic field can be transformed to the form

$$i\hbar \frac{\partial \phi}{\partial t} = \left(-\frac{p^2}{2m} + V - e\vec{\mathcal{E}}\cdot\vec{r}\right)\phi,$$

using the dipole approximation, the relation $\vec{\mathcal{E}} = -\partial \vec{\mathcal{A}}/\partial t$ (see Eq. (3.26)) and

$$\phi = \psi \exp\left(\frac{ie}{\hbar}\vec{\mathcal{A}}\cdot\vec{r}\right).$$

3.3 The interaction between radiation and matter is often described by the dipole approximation.

(a) Give a short description of the dipole approximation. Why is it a good approximation for the interaction of visible light with atoms.

(b) What are the selection rules for n, ℓ, and m_ℓ for transitions between two levels of a one-electron atom?

A hydrogen atom is in the following superposition of eigenfunctions at $t = 0$:

$$\Psi(\vec{r}, 0) = \tfrac{1}{7}\left(3\psi_{100}(\vec{r}) - 6\psi_{310}(\vec{r}) + 2\psi_{311}(\vec{r})\right).$$

(c) Calculate the expectation value for the energy of the state at $t = 0$. What is the time dependence of the energy?

At $t = 0$ the atom is irradiated for a short period τ with right-handed circular polarized light that is resonant with the $n = 2 \to n = 3$ transition of hydrogen.

(d) Which states are populated at $t = \tau$? Has the expectation value of the energy at $t = \tau$ increased or decreased with respect to $t = 0$? Explain what causes the change of energy.

3.4 Show that the formula for the transition rate R_{ij} of Eq. (3.16) indeed has the dimension of a rate. Consider a transition to a state whose natural width is γ. What intensity would be required to make the rate $R_{ij} = \gamma$?

3.5 The proof of the oscillator strength theorem (App. 3.A) can be done using some commutation relations. Start by proving

$$[r, \mathcal{H}_0] = (1/2m)([r, p]p + p[r, p]) = (i\hbar/m)p$$

and use it to evaluate $[[r, \mathcal{H}_0], r]$. Next use the unit operator $\sum_k |k\rangle\langle k| = 1$ appropriately to show that the ground-state expectation value

$$\langle j|[[r, \mathcal{H}_0], r]|j\rangle = \langle j|2r\mathcal{H}_0 r|j\rangle - \langle j|\mathcal{H}_0 r^2 + r^2 \mathcal{H}_0|j\rangle = \sum_k 2|r_{kj}|^2 \hbar \omega_{kj}.$$

Combine these to show that

$$\sum_k 2m\omega_{kj}|\vec{r}_{kj}|^2/\hbar = 1.$$

4
"Forbidden" transitions

4.1 Introduction

The selection rules discussed in Sec. 3.5 forbid many kinds of transitions that are frequently observed or exploited. It must be remembered that these "rules" are based on the approximations of Sec. 3.2, but of course, those can be extended. Such expansions to higher orders of approximation may seem to be little more than a tedious exercise, but this chapter will show that the results are extremely important, because they cause qualitatively different kinds of transitions. Since they involve both the electric and magnetic components of the light, their atomic dipole moments will be identified with the superscript $\mathcal{E}$ and $\mathcal{B}$ for the remainder of this chapter. The magnetic dipole transitions, called M1, discussed in Sec. 4.2.1 that arise from extension of the electric dipole approximation encompass all of magnetic resonance, including atomic clocks, forced evaporative cooling in Bose–Einstein Condensation (BEC), MRI, and astrophysical processes such as generation of the all-important 21 cm radio frequency line. The electric quadrupole transitions, called E2, discussed in Sec. 4.2.2 describe myriad stellar spectral features as well as the vast majority of nuclear γ-ray transitions.

Even with these extensions there are many cases where the relaxation of the electric dipole (E1) selection rules to include M1 and E2 transitions as in Sec. 4.2 still do not account for observed processes. Thus other approximations have to be extended so that their limitations can be relaxed as well. Extension of the perturbation approximation (see Sec. 4.3.1) provides an analytical description of most of non-linear optics, including second harmonic generation, four wave mixing, Raman spectroscopy, and the fabled decay of the hydrogen 2S state.

A semi-classical description of the interaction between radiation and atoms can begin with a multipole expansion of the Hamiltonian. The lowest-order term is just from the point charges α located at $\vec{r}_\alpha$, and vanishes for neutral atoms. The next-order term is the electric dipole interaction $\vec{\mu}^{\mathcal{E}} \cdot \vec{\mathcal{E}}$ that arises from the non-isotropic distribution of charges induced by the electric field of the light. This term produces the electric dipole transitions (E1) that have been described in some detail in Chap. 3.

The next-order terms arise from the magnetic moments produced by moving charges within an atom and also from the quadrupole contribution to the charge distribution induced by the light field. The magnetic moments $\vec{\mu}^B$ interact with the magnetic field of the light $\vec{B}$ with about the same strength as that deriving from the interaction of the electric field gradient $\vec{\nabla} \times \vec{\mathcal{E}}$ with the induced quadrupole moment $\vec{Q}$. Thus the total Hamiltonian operator is

$$\mathcal{H}' = -\vec{\mu}^{\mathcal{E}} \cdot \vec{\mathcal{E}} - \vec{\mu}^B \cdot \vec{B} - \sum_{ij} Q_{ij} \nabla_i \mathcal{E}_j \tag{4.1}$$

where

$$\vec{\mu}^{\mathcal{E}} \equiv \sum_\alpha q_\alpha \vec{r}_\alpha, \quad \vec{\mu}^B \equiv \tfrac{1}{2} \sum_\alpha q_\alpha \vec{r}_\alpha \times \frac{d\vec{r}_\alpha}{dt}, \quad Q_{ij} \equiv \tfrac{1}{2} \sum_\alpha q_\alpha\, r_{i,\alpha}\, r_{j,\alpha}, \tag{4.2}$$

and the subscripts i and j refer to different Cartesian components of $\vec{r}_\alpha$.

The M1 interaction contains the electronic motion in the time derivative term of Eq. (4.2) and therefore constitutes the current required to produce a magnetic moment. The optical interaction with the quadrupole moment is described by the last term of Eq. (4.2). The next section of this chapter is devoted to the quantum mechanical description of these higher-order interactions between atoms and light.

4.2 Extending the electric dipole approximation

Section 3.2.1 exploits the small size of atoms ($\sim a_0$) compared with optical wavelengths $\lambda = 2\pi/k$ by expanding $e^{i\vec{k}\cdot\vec{r}}$ in a power series and keeping only its first term (just unity). This can be done because the magnitude of the radial part of the atomic wavefunction at distances r larger than a few a_0 is tiny. Thus the contribution to the radial integral at $r \gg a_0$ from the higher terms in the expansion is negligibly small. The resulting E1 transition matrix elements are used to describe optical transitions and their selection rules. However, if the angular part of the transition matrix element vanishes exactly (see Sec. 3.5), it is necessary to take a larger view and consider the transition matrix element that results from the next term in the expansion of $e^{i\vec{k}\cdot\vec{r}}$, namely $i\vec{k} \cdot \vec{r}$. It is not hard to estimate *ab initio* the rate of these transitions relative to those of E1 by $(\vec{k} \cdot \vec{r})^2 = (2\pi a_0/\lambda)^2 = (\hbar\omega/mc^2\alpha)^2 \approx \alpha^2/4$ using $\hbar\omega = mc^2\alpha^2/2$. Thus the rates for these transitions are generally five orders of magnitude smaller than those of E1.

One reason for lifting the electric dipole approximation is to describe transitions that are "dipole forbidden", i.e., $\mu^{\mathcal{E}}_{ij} = 0$ because one of the E1 selection rules is violated. This may occur, for example, in electron spin resonance and nuclear magnetic resonance transitions between states of the same ℓ (no parity change). Note that the total angular momentum j does not have to change to satisfy the electric dipole selection rules, only ℓ needs to change by ± 1. This is because the optical electric field does not interact with the intrinsic magnetic moment of the electron $\mu_e \approx \mu_B$, but only with its orbital motion. Said another way, the readily observed atomic transitions between states that are split by Zeeman and/or hyperfine effects cannot occur under the E1 selection rules.

To go beyond the electric dipole approximation, again choose the Coulomb gauge, let $\vec{\mathcal{A}}$ be given by Eq. (3.30), and recall that the transition probability depends on the matrix elements $\langle j|\vec{\mathcal{A}} \cdot \vec{p}|i\rangle$ as in Eq. (3.5). The next order of approximation in the exponential expansion of Eq. (3.5) is the term $i\vec{k} \cdot \vec{r}$ written as $\mathcal{A}_0 \langle j|(\hat{\varepsilon} \cdot \vec{p})(i\vec{k} \cdot \vec{r})|i\rangle$.

There is a vector identity that relates the dot product of two cross products (a scalar) to the sum of two scalars, each of which is composed of dot products of the components. For the $(\hat{\varepsilon} \cdot \vec{p})(\vec{k} \cdot \vec{r})$ term needed here, it yields

$$(\hat{\varepsilon} \cdot \vec{p})(\vec{k} \cdot \vec{r}) = (\vec{k} \times \hat{\varepsilon}) \cdot (\vec{r} \times \vec{p}) + (\hat{\varepsilon} \cdot \vec{r})(\vec{k} \cdot \vec{p}). \tag{4.3}$$

In the sections below, it is shown how each term of Eq. (4.3) results in transitions of quite different kinds. The $\mathcal{A}_0(\vec{k} \times \hat{\varepsilon}) \cdot (\vec{r} \times \vec{p})$ term produces M1 transitions and the $\mathcal{A}_0(\hat{\varepsilon} \cdot \vec{r})(\vec{k} \cdot \vec{p})$ term produces E2 transitions. Important examples are given as well.

4.2.1 Magnetic dipole transitions

The first term on the right side of Eq. (4.3) is the dot product of the magnetic field of the light $\vec{\mathcal{B}} = \vec{k} \times \vec{\mathcal{A}}_0$ with the orbital angular momentum $\vec{\ell} = \vec{r} \times \vec{p}$, so its contribution to the transition matrix element is $\langle j|\vec{\mathcal{B}} \cdot \vec{\ell}|i\rangle$. Since the magnetic dipole moment $\vec{\mu}$ is proportional to ℓ, this can be written as $\vec{\mu}^\mathcal{B} \cdot \vec{\mathcal{B}}$ just as in the semi-classical case of Eq. (4.1). Of course, the magnetic field of the light can also interact with the electron's magnetic dipole moment μ_e (see Sec. 8.1) to cause transitions. With such M1 transitions, two states can be connected by the magnetic dipole interactions between the electron's magnetic moment and the magnetic field of the radiation. The transitions induced this way are called magnetic dipole transitions.

The transition rate between two states can be calculated in a similar way as discussed for the E1 transitions, where the electric dipole moment is replaced by the magnetic dipole moment and the electric field by the magnetic field. Since $\mathcal{B}(\omega) = \mathcal{E}(\omega)/c$, $\mathcal{B}^2(\omega) = 2S(\omega)/\varepsilon_0 c^3$ and one obtains in analogy with Eq. (3.16) the transition rate for M1 transitions from state $|i\rangle$ to $|j\rangle$:

4.2 Extending the electric dipole approximation

$$R_{ij}^{M1} = \frac{\pi |\langle j|\hat{\varepsilon}\cdot\vec{\mu}^{\mathcal{B}}|i\rangle|^2 S(\omega_{ji})}{\varepsilon_0 \hbar^2 c^3}. \tag{4.4}$$

By taking the ratio between this equation and Eq. (3.16) one finds for the ratio between the rates for M1 and E1 transitions:

$$\frac{R_{ij}^{M1}}{R_{ij}^{E1}} = \frac{\alpha^2}{4}, \tag{4.5}$$

using $\mu^{\mathcal{E}} = ea_0$ and $\vec{\mu}^{\mathcal{B}} = e\hbar/2m$. The magnetic transitions are proportional to the velocity of the electron, which is α times the speed of light. This reduces the rate of the M1 transitions with respect to the E1 transition by α^2 or four orders of magnitude, if they are both allowed by the selection rules.

In contrast to the E1 selection rules discussed in Sec. 3.5, these M1 transition matrix elements vanish unless the states have the same value of ℓ, and so their selection rule becomes $\Delta\ell = 0$. This is most easily understood in terms of the parity σ of the operators and of the wavefunctions given by $\sigma = (-1)^\ell$. The operator $\vec{r}$ of E1 transitions has odd parity and so its matrix elements vanish unless the two states it connects have opposite parity. By contrast, $\vec{\ell} = \vec{r}\times\vec{p}$ has even parity so the connected states must have the same parity. Since the coupling is proportional to the expectation values of ℓ, M1 transitions require that $\Delta\ell = 0$.

Similarly to the case of E1 transitions as discussed in Sec. 3.5.2, the polarization of the light couples with the magnetic quantum number m of the atom. So for linearly polarized light one has $\Delta m = 0$ and for circular polarization $\Delta m = \pm 1$ depending on the handedness of the light. As before, the role of the intrinsic magnetic moment of the electron has been left out but is straightforward to include. Table 4.1 summarizes these selection rules.

For the E1 transitions described in Sec. 3.5, there is no selection rule on n because the matrix element $\vec{\mu}_{ji}^{\mathcal{E}}$ is proportional to the operator $\vec{r}$, so that its radial part never vanishes unless $i = j$. By contrast, M1 transitions do not allow a change of n because the magnetic dipole interaction of Eq. (4.1) depends only on $\vec{\ell}$ and not on $\vec{r}$. So, for example, the decay of the 2S state in hydrogen to the 1S ground state cannot be attributed to an M1 transition (see Sec. 4.3.1).

By contrast, transitions between hyperfine levels in the ground state of Cs, whose ~9.193 GHz frequency constitutes the accepted definition of the second worldwide, is indeed an M1 transition. Similarly, the transitions between the hyperfine levels in the ground state of hydrogen at $\nu \approx 1.420$ GHz that constitute the famous 21 cm line so important in astronomy and cosmology are also M1 transitions. Similar nuclear spin transitions also occur in the hydrogen atoms of living tissue, and generate the signals used in MRI for medical diagnosis. A significant number of biological studies use similar transitions of the nucleus of ^{13}C, an isotope

4.2.2 Electric quadrupole transitions

For the second term of Eq. (4.3) one can write

$$\mathcal{H}'_{E2} = \frac{e\mathcal{A}_0}{m} \sum_{m,n} \varepsilon_m r_m k_n p_n, \qquad (4.6)$$

where the sum m, n is over the Cartesian coordinates for the dot product. Then one can use the Heisenberg equation of motion from Sec. 3.2.1 to write $\hbar p_n = im[\mathcal{H}_0, r_n]$. The (m, n) matrix element for the interaction becomes $-ie\omega_{ij}\mathcal{A}_0 \varepsilon_m \langle j|r_m r_n|i\rangle k_n$ because ε_m and k_n come out of the matrix element. For monochromatic light $\vec{\mathcal{E}}(t) = -d\vec{\mathcal{A}}/dt = (i\omega\vec{\mathcal{A}}(t) + c.c.)/2$, so this reduces to $-e\mathcal{E}_0 \varepsilon_m \langle j|r_m r_n|i\rangle k_n \equiv \mathcal{E}_0 \varepsilon_m Q_{m,n} k_n$, where $Q_{m,n}$ is one element of the quadrupole moment as defined in Eq. (4.2). Again, in analogy with Eq. (3.16) one finds for the E2 transitions

$$R_{ij}^{E2} = \frac{\pi \sum_{m,n} k_n^2 |\langle j|Q_{m,n}|i\rangle|^2 S(\omega_{ji})}{\varepsilon_0 \hbar^2 c}. \qquad (4.7)$$

Assuming $\mu^{\mathcal{E}} = ea_0$ and $Q = ea_0^2/2$, the ratio between the rates for E2 and E1 transitions becomes

$$\frac{R_{ij}^{E2}}{R_{ij}^{E1}} = \frac{k^2 a_0^2}{4}, \qquad (4.8)$$

as expected, since E2 transitions derive from the next higher-order expansion of $\exp(i\vec{k} \cdot \vec{r})$ compared with E1 transitions. For visible light this ratio is $\sim 10^{-7}$. Setting $\hbar\omega = mc^2\alpha^2/2$ as in Sec. 4.2, this ratio reduces to $\alpha^2/4$ and shows that under these conditions the rates for M1 and E2 transitions are comparable.

All the elements Q_{mn} are functions of x, y, and z, and can therefore be written in terms of spherical harmonics just as in Sec. 3.5, but unlike the case for those E1 transitions each term involves two of them. Then the spherical harmonics all have the form $Y_{2,q}(\theta, \phi)$, so their expansion is not described by Eqs. (C.21). Instead there are different relations, and the outcome is different selection rules, yielding $\Delta\ell = 0, \pm 2$ (although Δj can be $\pm 2, \pm 1, 0$), and $\Delta m_\ell = 0, \pm 1, \pm 2$. Also, the sum of the two ℓ-values must be at least 2. Unlike the case of M1 transitions produced by the first term on the right-hand side of Eq. (4.3), the presence of $r \cdot r$ removes any restrictions on Δn. Neither M1 nor E2 transitions can connect two states of different parity, but for M1 there can be no change of ℓ whereas E2 transition can allow $\Delta\ell = \pm 2$. Table 4.1 summarizes these selection rules.

		General		LS-coupling			Single electron	
	$\pi_a \pi_b$	ΔJ	ΔM_J	ΔL	ΔS	Δn_i	$\Delta \ell_i$	Δm_i
E1	-1	$0, \pm 1$ $(J+J' \geq 1)$	$0, \pm 1$	$0, \pm 1$ $(L+L' \geq 1)$	0	—	± 1	$0, \pm 1$
E2	$+1$	$0, \pm 1, \pm 2$ $(J+J' \geq 2)$	$0, \pm 1, \pm 2$	$0, \pm 1, \pm 2$ $(L+L' \geq 2)$	0	—	$0, \pm 2$ $(\ell + \ell' \geq 1)$	$0, \pm 1, \pm 2$
M1	$+1$	$0, \pm 1$ $(J+J' \geq 1)$	$0, \pm 1$	0	0	0	0	$0, \pm 1$

Table 4.1 Selection rules for allowed electric dipole (E1), electric quadrupole (E2), and magnetic dipole (M1) transitions [28]. The selection rules apply for three different cases: (a) general rules that apply for all atoms (left column), (b) L–S-coupling rules that apply for atoms that can be described in the L–S-coupling scheme (center column), where L and S are both good quantum numbers, and (c) one-electron excitation, where only one of the electrons makes a transition (right column). The rules in the left-hand column can be generalized for nuclear spins by replacing J with F. Those in the center column refer to cases where more than one electron makes a transition. Note that total spin is always unchanged. The ones in the right-hand column are the simplest, and correspond for E1-transitions to the rules discussed in Sec. 3.5.2.

70 "Forbidden" transitions

These M1 and E2 transitions play extremely important roles in atomic spectroscopy. One of the most famous of these was the "existence" of the element nebulium in nebular spectra for more than 60 years. The hypothesis for its existence arose from otherwise unidentified spectral lines. Once it was realized that these arose from the M1 transition $^1D_2 \to {}^3P_{2,1}$ of doubly ionized oxygen at $\lambda = 501$ nm, and the E2 transition $^1D_2 \to {}^1S_0$ of neutral oxygen at $\lambda = 558$ nm, the myth was dismissed. The beautiful bright green color of the aurora also arises from such transitions. In addition, most nuclear γ-ray transitions are also E2.

4.3 Extending the perturbation approximation

4.3.1 The next higher-order process

To extend the perturbation approximation of Sec. 3.2.2 to its next highest order, start with the Schrödinger equation Eq. (3.1) but expand the higher orders of the sums, labeling them by an integer λ to keep track of the order. Thus

$$\frac{dc_j^{(\lambda+1)}(t)}{dt} = +\frac{1}{i\hbar} \sum_n c_n^{(\lambda)}(t) \mathcal{H}'_{jn}(t) e^{i\omega_{jn}t}, \qquad (4.9)$$

where $\lambda = 0$ yields Eq. (3.10) with $\mathcal{H}' \propto \vec{p} \cdot \vec{A}$.

In the electric dipole approximation, $\mathcal{H}' = -\vec{\mu}^\mathcal{E} \cdot \vec{\mathcal{E}}_0 \cos(\omega t)$ as in Sec. 3.2.1. First, substitute this into Eq. (4.9) for $\lambda = 0$, choose $c_i^{(0)}(0) = 1$ with all other coefficients $c_j^{(0)}(0) = 0$ and restrict them to $|c_j^{(0)}(t)|^2 \ll 1$. Then integrate with respect to time to find

$$c_j^{(1)}(t) = \frac{\vec{\mu}_{ji}^\mathcal{E} \cdot \vec{\mathcal{E}}_0}{2\hbar} \left(\frac{e^{i(\omega_{ji}+\omega)t} - 1}{\omega_{ji} + \omega} + \frac{e^{i(\omega_{ji}-\omega)t} - 1}{\omega_{ji} - \omega} \right), \quad j \neq i. \qquad (4.10)$$

This is essentially the same procedure that led to Eq. (3.12). The more general case occurs when $|\omega_{ji}|$ can be very different from ω so that neither denominator $\omega_{ji} \pm \omega$ in Eq. (4.10) is necessarily small and therefore it is necessary to keep both terms. Then calculate the coefficients $c_k^{(2)}$ for the $\lambda = 1$ case by using Eq. (4.9) again. The result is

$$\frac{dc_k^{(2)}(t)}{dt} = -\sum_j \left[\frac{e^{i(\omega_{ji}+\omega)t} - 1}{\omega_{ji} + \omega} + \frac{e^{i(\omega_{ji}-\omega)t} - 1}{\omega_{ji} - \omega} \right]$$

$$\times \frac{(\vec{\mu}_{kj}^\mathcal{E} \cdot \vec{\mathcal{E}}_0)(\vec{\mu}_{ji}^\mathcal{E} \cdot \vec{\mathcal{E}}_0)}{4i\hbar^2} (e^{i\omega t} + e^{-i\omega t}) e^{i\omega_{kj}t}. \qquad (4.11)$$

4.3 Extending the perturbation approximation

The straightforward time integration is performed, and here the RWA is carefully applied to the result for only the case where $2\omega \approx \omega_{kj} + \omega_{ji}$ connecting the pair of levels i and k to find

$$c_k^{(2)}(t) = \frac{|\vec{\mathcal{E}}_0|^2}{4\hbar^2} \sum_j \frac{(\hat{\varepsilon} \cdot \vec{\mu}_{ji}^{\mathcal{E}})(\hat{\varepsilon} \cdot \vec{\mu}_{kj}^{\mathcal{E}})}{(\omega_{ji} - \omega)} \frac{e^{i(\omega_{kj} + \omega_{ji} - 2\omega)t} - 1}{\omega_{kj} + \omega_{ji} - 2\omega}, \qquad (4.12)$$

where $\hat{\varepsilon}$ is the unit polarization vector. Thus the transition probability from state $|i\rangle$ to state $|k\rangle$ in second order is

$$|c_k^{(2)}(t)|^2 = \frac{|\vec{\mathcal{E}}_0|^4}{4\hbar^4} \left[\frac{\sin^2((\omega_{ki} - 2\omega)t/2)}{(\omega_{ki} - 2\omega)^2} \right] \left| \sum_j \frac{(\hat{\varepsilon} \cdot \vec{\mu}_{ji}^{\mathcal{E}})(\hat{\varepsilon} \cdot \vec{\mu}_{kj}^{\mathcal{E}})}{\omega_{ji} - \omega} \right|^2, \qquad (4.13)$$

where ω_{ki} is independent of j so the first bracketed term above comes out of the summation. Needless to say, energy conservation requires $\omega_{ki} \approx 2\omega$. By contrast, if one of the frequencies ω_{ji} is not very different from ω there is a resonant denominator inside the squared bracket of Eq. (4.13) that enhances the transition rate. However, this would also create a resonant condition in the first-order transition probability so that this second-order perturbation treatment is not warranted. In fact, all these coefficients $c_j^{\lambda+1}$ must be very much less than unity for these perturbation calculations to be valid.

The summation in Eq. (4.13) vanishes unless both $\hat{\varepsilon} \cdot \vec{\mu}_{ji}^{\mathcal{E}} \neq 0$ and $\hat{\varepsilon} \cdot \vec{\mu}_{kj}^{\mathcal{E}} \neq 0$ for at least one of the terms, but does not require $\hat{\varepsilon} \cdot \vec{\mu}_{ki}^{\mathcal{E}} \neq 0$. Since $\Delta\ell = \pm 1$ for each of these dipole transition matrix elements to be non-zero, $c_k^{(2)} \neq 0$ requires $\ell_j - \ell_i = \pm 1$ and $\ell_k - \ell_j = \pm 1$ which means $\ell_k - \ell_i = 0, \pm 2$. Thus one of the selection rules for this second-order transition is $\Delta\ell = 0, \pm 2$. Although this is similar to the selection rule for E2 transitions, the underlying reason is quite different.

It is important to emphasize here that these second-order transitions are *not* the same as the M1 and E2 transitions described in Secs. 4.2.1 and 4.2.2. This second-order process depends on two E1 transition matrix elements given by the dipole moments $\vec{\mu}_{ji}^{\mathcal{E}}$ and $\vec{\mu}_{kj}^{\mathcal{E}}$, and the product of two intensities (in Eq. (4.13) it is the square of intensity, but see below). By contrast, M1 and E2 are first-order processes that depend on the spatial variation of the electromagnetic field across atomic sizes via the $e^{i\vec{k}\cdot\vec{r}}$ term in $\vec{A}$.

This transition probability is quadratic in time for $(\omega_{ki} - 2\omega)t \ll 1$ just as in Eq. (3.13) in Sec. 3.2.3. It is necessary to consider the appropriate "spectral width" and integrate over it, essentially calculating the probability for each value of ω in some limited range and summing them. This constitutes an integration over the spectral intensity $\mathcal{S}(\omega)$ and follows the same procedure as before, but now uses the square of the spectral intensity. Caution is necessary here because this is not truly a square of $S(\omega)$, but actually a second-order correlation of the field. The integration over ω yields

$$|c_k^{(2)}(t)|^2 = \frac{\pi t S^2(\omega_{ki}/2)}{8\hbar^4 \varepsilon_0^2 c^2} \left[\sum_j \frac{(\hat{\varepsilon} \cdot \vec{\mu}_{ji}^{\mathcal{E}})(\hat{\varepsilon} \cdot \vec{\mu}_{kj}^{\mathcal{E}})}{\omega_{ji} - \omega} \right]^2, \tag{4.14}$$

where the spectral intensity is evaluated at $\omega = \omega_{ki}/2$. The transition rate is given by $R^{(2)} = d|c_k^{(2)}(t)|^2/dt$. Recall that $S(\omega)\Delta\omega$ is the magnitude of the Poynting vector but for $S^2(\omega)$ the frequency interval of unit ω does not change. Therefore $S^2(\omega)$ has the dimensions of (intensity)^2 per unit ω with the frequency range necessarily restricted to the excited state linewidth $\gamma \equiv 1/\tau$. Then

$$R^{(2)} = \frac{\pi I^2}{8\hbar^4 \varepsilon_0^2 c^2 \gamma} \left[\sum_j \frac{(\hat{\varepsilon} \cdot \vec{\mu}_{ji}^{\mathcal{E}})(\hat{\varepsilon} \cdot \vec{\mu}_{kj}^{\mathcal{E}})}{\omega_{ji} - \omega} \right]^2 \tag{4.15}$$

If one estimates $\mu^{\mathcal{E}} \sim ea_0$, $(\omega_{ji} - \omega) \sim \omega/2$, and a typical lifetime $\tau = 10^{-8}$ s, then $R^{(2)} = 10^{-14} I^2$ s^{-1}, with I in W/m^2. The above discussion follows that of Ref. [29].

Study of this subject began with the 1930 Ph.D. thesis of Maria Göppert-Mayer, described in Ref. [30], who is famous for the nuclear shell model. She addressed the probability of double quantum jumps and laid the groundwork for the use of second-order perturbation calculations to address such problems. Although her calculation did not deal with the specific problem of the hydrogen atom, her work is mentioned here because of its importance and historical priority. In a later paper her method was applied to the $2^2S \rightarrow 1^2S$ decay of metastable atomic hydrogen, whose lifetime was found to be about $1/7$ s (see Ref. [31]). The decay cannot be via an M1 transition because n changes from 2 to 1, and cannot be via an E2 transition because the sum of the ℓ-values is less than 2.

The wavy lines of Fig. 4.1(a) show this decay schematically, but it is very important to realize that Eq. (4.14) does not have terms that correspond to such transitions. To understand this better, choose the state $|i\rangle$ to be the 2S state, and the state $|k\rangle$ to be the 1S ground state. Then in order for the two induced dipole moments $\vec{\mu}_{ji}^{\mathcal{E}}$ and $\vec{\mu}_{kj}^{\mathcal{E}}$ to be non-zero, the states $|j\rangle$ must all be P states because both $|i\rangle$ and $|k\rangle$ are S states. The largest such term in the sum of Eq. (4.14) has $|j\rangle$ is the 3P state as shown by the dashed lines, but of course, the atom never populates that state. There are common misconceptions to be avoided about population of "virtual" states during such transitions. Equation (4.14) shows that this is simply not so, but that the states $|j\rangle$ are merely terms in the summation that comes out of a perturbation expansion. Atoms can never be found in any P states.

4.3 Extending the perturbation approximation

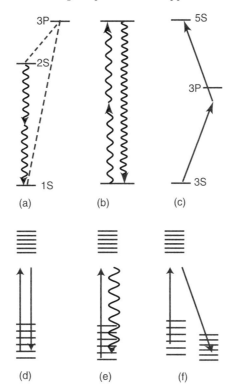

Figure 4.1 Possible transition schemes enabled by second-order processes. Part (a) shows the decay of the 2S state of H as based on the thesis of Göppert-Mayer in Ref. [30], and elaborated in Ref. [31]. The wavy lines are the same length here, suggesting the same frequency, but they do not have to be. The dominant term in the summation of Eq. (4.14) is the one with $|j\rangle = |3P\rangle$, indicated here by the dashed lines. Part (b) shows the second harmonic generating scheme in a non-linear optical medium. Atoms are excited to a high-lying state in second order, and emit frequency-doubled light in a first-order transition enabled by parity mixing from the electric field of the crystal host. Part (c) shows the excitation of the 5S state of Na as done in Refs. [32, 33] using light of wavelength 602 nm. The lower part of the figure shows processes in which the two electromagnetic fields are different, as described in Sec. 4.3.3. Part (d) shows Rayleigh scattering where the initial and final states may be the same, but the k-vectors or polarization of the light are different. Part (e) shows a Raman transition to a state lying above the initial state, thus producing Stokes fluorescence (lower frequency). Part (f) shows a stimulated Raman process. Energy is exchanged between two light beams, and the difference is imparted to the internal states of the atom (molecule).

Göppert-Mayer not only performed the predecessor of the calculation above, but also considered the possibility of two electromagnetic fields of different frequencies whose sum is the needed total energy (for hydrogen it is 2.46 PHz, 2.46×10^{15} Hz). The decay probability can be plotted against one or the other frequency thereby providing the emission spectrum of this decay. It spans the range from 0 to 2.46 PHz, and is symmetric about the broad maximum at the center frequency of 1.23 PHz.

4.3.2 Non-linear optics

Optical excitation of such transitions was first observed at the beginning of the laser era in 1960. It is clear that an E1 decay from a state whose energy is $2\hbar\omega$ could be in the ultraviolet for visible excitation, but it is also clear that such a decay from an excited state $|k\rangle$ is forbidden by the electric dipole selection rule since its ℓ-value must either be the same as that of the initial state, or differ from it by 2.

However, in a strong dc electric field, all the states involved in the transition are mixed by the Stark effect (see Chapter 12), and so each energy level has some component of states of opposite parity. Sufficiently strong fields are formed by the ions that compose crystal lattices, and UV generation in a quartz crystal with light from a ruby laser was first reported by Franken et al. [34]. A schematic diagram of the energy levels is shown in Fig. 4.1b. Equation (4.15) shows that such excitation is proportional to the square of the light intensity, and such processes are therefore labeled non-linear. The huge field of study called non-linear optics is based on Eq. (4.14).

After the advent of tunable lasers, excitation of free atoms by such transitions was dramatically demonstrated in the visible region of the spectrum in the mid-1970s. (Tunable lasers were required for energy conservation in the excitation process.) In a single issue of *Physical Review Letters*, three separate groups reported excitation of S or D states of Na from its ground S state by such "two-photon" transitions [32, 33, 35]. Figure 4.1c shows the excitation between two S states of Na as described in two of these papers; the third one excited a D state. To detect such excitations in an atom, for example by fluorescence from spontaneous cascade through lower-lying levels, it is reasonable to require $R \sim 10^4$. Then the needed intensity is $P \sim 1$ W focused into a spot of a few tens of μm diameter. In earlier times, radio frequency or microwave transitions of this kind were reported, but both matrix elements in the sum were M1.

4.3.3 Two different electromagnetic fields

As pointed out in Göppert-Mayer's work, there is no requirement for the optical field that connects the states i and j to be the same as the one that connects states j and s. The calculation above can be generalized to two different electromagnetic fields with a vector potential given by $\vec{\mathcal{A}} = \vec{\mathcal{A}}_1 + \vec{\mathcal{A}}_2$, where the time dependence of each vector potential is given by $\vec{\mathcal{A}}_n = \vec{\mathcal{A}}_{0,n} \cos(\omega_n t + \phi)$. Then, in the electric dipole approximation which again is kept for simplicity, $\mathcal{H}' = -\vec{\mu}^{\mathcal{E}} \cdot [\vec{\mathcal{E}}_{0,1} \cos(\omega_1 t) + \vec{\mathcal{E}}_{0,2} \cos(\omega_2 t)]$ for $\phi = -\pi/2$. If we substitute this into Eq. (4.9) for $\lambda = 0$ as above, choosing $c_i(0) = 1$ and all other coefficients $c_j(0) = 0$, subsequent time integration yields

$$c_j^{(1)}(t) = \frac{\vec{\mu}_{ji}^{\mathcal{E}} \cdot \vec{\mathcal{E}}_{0,1}}{2\hbar} \left[\frac{\left(e^{i(\omega_{ji}+\omega_1)t} - 1\right)}{(\omega_{ji} + \omega_1)} + \frac{\left(e^{i(\omega_{ji}-\omega_1)t} - 1\right)}{(\omega_{ji} - \omega_1)} \right]$$
$$+ \frac{\vec{\mu}_{ji}^{\mathcal{E}} \cdot \vec{\mathcal{E}}_{0,2}}{2\hbar} \left[\frac{\left(e^{i(\omega_{ji}+\omega_2)t} - 1\right)}{(\omega_{ji} + \omega_2)} + \frac{\left(e^{i(\omega_{ji}-\omega_2)t} - 1\right)}{(\omega_{ji} - \omega_2)} \right]. \quad (4.16)$$

Consider the case where $\omega_{ji} \pm \omega_1$ and $\omega_{ji} \pm \omega_2$ are not necessarily small and keep all the terms in Eq. (4.16) because they may be of comparable magnitude. Then the calculation for $\lambda = 2$ proceeds as above:

$$\frac{dc_k^{(2)}(t)}{dt} = \frac{i}{2\hbar} \sum_j c_j^{(1)}(t)$$
$$\times \vec{\mu}_{kj}^{\mathcal{E}} \cdot \left(\vec{\mathcal{E}}_{0,1}(e^{i\omega_1 t} + e^{-i\omega_1 t}) + \vec{\mathcal{E}}_{0,2}(e^{i\omega_2 t} + e^{-i\omega_2 t}) \right) e^{i\omega_{kj} t}. \quad (4.17)$$

There are $8 \times 4 = 32$ terms under the summation, and each of the subsequent time integrals will yield two terms, so there will be a total of 64 terms. These terms are found by performing the integration over time to find $c_k^{(2)}(t)$, and then squaring it to find the transition probability. As before, this will be quadratic in time, but an integration over the spectral intensity will again produce a linear dependence whose time derivative will be a desired constant transition rate. All these terms cannot be written out here, and so the procedure is to consider a few examples.

Examples of these cases are shown in the lower half of Fig. 4.1. One such case is shown in Fig. 4.1d, where the atom (molecule) is not excited at all, but remains in the original ground state via a transition at the same frequency. In this case, it is the $\vec{k}$-vectors of the electric fields $\vec{\mathcal{E}}_{0,i}$ that are different: the light is redirected, possibly with polarization change as well. This process is called Rayleigh scattering and is the mechanism that makes the sky blue. It is also responsible for many other light scattering processes by gases. Note that the frequency of the light does not satisfy the resonance condition with any of the excited states of the molecules in the atmosphere.

Another example is shown in Fig. 4.1e where the atom (molecule) is also returned to a low-lying state, but in this case by a spontaneous process. The energy of the final state $|k\rangle$ is different from that of the initial state $|i\rangle$, so the emitted light has a different frequency, and may also have a different direction and polarization. Such scattering from molecules may populate any number of rotational and vibrational states $|k\rangle$, resulting in a spectrum of the emitted light. Such processes are called Raman scattering, and can be used to map out vast numbers of molecular states that do not decay for various reasons and therefore cannot be found any other way.

The example of Fig. 4.1e has $|k\rangle$ lying above $|i\rangle$ so the emitted light is of longer wavelength. This is called Stokes radiation. It is also possible that molecules initially in higher-lying rotational and vibrational states can interact with the driving radiation and end up in lower states, thereby emitting shorter-wavelength radiation. Such anti-Stokes radiation can be produced if the Boltzmann distribution allows population of states above the ground state.

Figure 4.1f shows Raman processes that are driven by two different applied radiation fields. In such coherent processes, there is energy exchange between the light beams. These weak processes require rather strong light beams so their changes may be difficult to detect optically, but often the population of the final states can be detected instead. For example, very cold atoms can have their momenta changed by such processes, and their motion can be observed. Such stimulated Raman scattering processes can produce Stokes or anti-Stokes transitions.

Finally, the time reversal of Fig. 4.1b is a frequency division process. The output light is at half the frequency of the input, and the process is called downconversion. It can be parametric downconversion if the two frequencies are different and one of them is applied so that the other one is controlled by the two input frequencies. The process can be either stimulated or spontaneous, with appropriate acronyms. Because the emitted light at the two frequencies is correlated in time, this method is important for "single photon" experiments in quantum optics. By proper choice of crystal symmetries, the process can produce fields that are different in polarization or direction of $\vec{k}$ only, independent of their frequencies.

In all of these cases, and many others not discussed here, the transition rate is proportional to the product of two intensities rather than the square. For this reason, many people categorize these phenomena as still in the realm of non-linear optics.

4.3.4 Misconceptions about "intermediate states", resonances, and $\mathcal{A}^2$

It is very important to emphasize that the language of "intermediate states" or "virtual states" has no place in this discussion. Any of these non-linear processes occur when the driving radiation is not resonant with any of the states $|j\rangle$. The denominator inside the squared bracket of Eq. (4.14) (and its counterpart in the terms that would emerge from Eq. (4.17) suggests that a near-resonance would enhance the process. For example, in the case of exciting Rb from its ground $5^2S_{1/2}$ state to its $5^2D_{3/2}$ state at $\lambda \approx 778$ nm, there is a conveniently located $5^2P_{1/2}$ state that could be directly excited by $\lambda \approx 780$ nm light. Thus it is separated by only about 1 THz from exact resonance, and enhances this excitation probability to the point where it can be driven by low-power diode lasers.

However, when the driving light is tuned to within the natural width of resonance of a particular state $|j\rangle$, the problem needs to be described as a three-level system in a first-order perturbation calculation such as in Sec. 3.2.2. Resonant three-level systems are rich in sequential phenomena such as stimulated rapid adiabatic passage (STIRAP), Autler–Townes spectra, electromagnetically induced transparency (EIT), slow light, etc., but are not properly described in second-order perturbation calculations such as this one. These nearly resonant cases at the boundary need careful examination (see Chap. 23).

Another misconception that needs to be addressed here is the role of the $\mathcal{A}^2$ term in the Hamiltonian (Eq. (3.2)). It should be completely clear now that the non-linear processes described here cannot arise from that term. In fact, $\mathcal{A}^2$ has no atomic part so it cannot possibly change any atomic states and therefore cannot induce any transitions.

4.3.5 Still higher-order processes

It is quite clear that Eq. (4.9) with either one or two fields in the Hamiltonian can be expanded to still higher order in λ, and also that there could be three or more fields in the Hamiltonian. The number of terms in the final expansion increases still further, but again there are special cases that can be singled out. Often these involve multiple stimulated transitions and are therefore called coherent Raman effects.

Perhaps the most common of these is coherent anti-Stokes Raman spectroscopy (CARS) which is second order in the higher frequency of the two incident fields (ω_1) and first order in ω_2, resulting in an output at $2\omega_1 - \omega_2$. There is also higher order Stokes effect scattering (HORSES) whose output is at $3\omega_1 - 2\omega_2$. There is the inverse Raman effect (TIRE), the Raman induced Kerr effect (TRIKE), opto-acoustic Raman spectroscopy (OARS), and a seemingly endless list of acronyms, many of which are associated with transportation. One of the most widely used schemes is called four wave mixing (FWM) and it has many varieties that are exploited for generating additional frequencies of light, especially in the ultraviolet. Very many non-linear optics textbooks [36–39] and review articles are available to discuss these processes that are all described by higher-order calculations. It should now be clear why this chapter begins with the assertion that such higher-order expansions are far more than just a tedious exercise.

Appendix

4.A Higher-order approximations

In App. 3.C it is shown that the expansion of the electromagnetic field in its lowest order around the center of the atoms directly leads to the electric dipole (E1) interaction: the electric field interacting with the induced, oscillating dipole moment of the atom. In many cases this is the dominant interaction and taking this term into account is sufficient. However, in cases where the E1 transition is forbidden by selection rules, one has to take the next order into account. Here it is shown how the next order leads to two different terms: the electric quadrupole (E2) and the magnetic dipole (M1) interaction, as discussed in Sec. 4.2.

The next order can be obtained by expanding the vector and scalar potential as:

$$\mathcal{A}_i(\vec{r}_\alpha, t) = \mathcal{A}_i(0, t) + \sum_j r_j^\alpha \nabla_j \mathcal{A}_i(0, t), \qquad (4.18)$$

$$\Phi(\vec{r}_\alpha, t) \approx \Phi(0, t) + \vec{r}_\alpha \cdot \vec{\nabla}\Phi(0, t) + \frac{1}{2} \sum_{ij} r_i^\alpha r_j^\alpha \nabla_i \nabla_j \Phi(0, t).$$

where the center of the atom is placed in the origin. Note that the vector potential can be split into a symmetric (S) and anti-symmetric (T) part:

$$\mathcal{A}_i(\vec{r}_\alpha, t) = \mathcal{A}_i(0, t) + \frac{1}{2} \sum_j r_j^\alpha \left(S_{ij} + T_{ij} \right), \qquad (4.19)$$

given by

$$S_{ij} = \nabla_j \mathcal{A}_i(0, t) + \nabla_i \mathcal{A}_j(0, t), \quad T_{ij} = \nabla_j \mathcal{A}_i(0, t) - \nabla_i \mathcal{A}_j(0, t). \qquad (4.20)$$

The symmetric part S_{ij} leads to the E2 term, whereas the anti-symmetric part T_{ij} leads to the M1 term, as will be shown.

For the unitary transformation of the canonical momentum one has to include a term that compensates for the symmetrical part of the vector potential and the transformation term is given by $V(t) = \sum_{ij} Q_{ij} \nabla_i \mathcal{A}_j$, where $\vec{Q}$ is the electric quadrupole moment of the atom, as given by Eq. (4.2). After considerable algebra one arrives for the symmetric part at

$$\mathcal{H}'_{E2} = -\sum_{ij} Q_{ij} \nabla_i \mathcal{E}_j(0, t), \qquad (4.21)$$

which is the E2 interaction. Here the induced quadrupole moment of the atom interacts with the gradient of the electric field at the center of the atom. With this transformation the canonical momentum no longer reduces to the mechanical momentum, since the anti-asymmetric term remains.

This asymmetric term can be rewritten in terms of the magnetic dipole moment as defined in Eq. (4.2) and one obtains for the M1 interaction

$$\mathcal{H}'_{M1} = -\vec{\mu}^B \cdot \vec{B}. \tag{4.22}$$

The induced magnetic dipole moment interacts with the magnetic field component of the electromagnetic field. Thus the next order expansion of the electromagnetic field yields two interaction terms. Going to even higher orders yields two additional terms for each extra order, but in this book the description will be limited to the E2 and M1 terms.

Exercises

4.1 The electric dipole approximation deteriorates when $(\vec{k} \cdot \vec{r})$ becomes comparable to unity. What would have to be the frequency of light in this case? Compare the energy of such light with the energy R_∞ that holds the atom together.

4.2 Prove the vector identity of Eq. (4.3).

4.3 The electric quadrupole operator has terms in x^2, y^2, z^2, xy, xz, and yz (see Eq. (4.2)). Write these out in terms of the spherical harmonics $Y_{\ell,m}$ as defined in App. C.5 at the end of this book (recall the conversions between Cartesian and spherical coordinates). Show that they contain only terms of the form $Y_{2,q}$.

4.4 Work through the steps outlined after Eq. (4.6) and arrive at Eq. (4.7). Use Eq. (3.6) to replace the $\vec{p}$ in the second term of Eq. (4.6) with an appropriate matrix element of $\vec{r}$ and then rearrange the nine terms of the product $(\hat{\varepsilon} \cdot \vec{r})(\vec{k} \cdot \vec{p})$. Show that the transition operator can be written as

$$\begin{pmatrix} k_x & k_y & k_z \end{pmatrix} \begin{pmatrix} xx & xy & xz \\ yx & yy & yz \\ zx & zy & zz \end{pmatrix} \begin{pmatrix} \varepsilon_x \\ \varepsilon_y \\ \varepsilon_z \end{pmatrix} \equiv \vec{k}(\vec{r} \otimes \vec{r})\hat{\varepsilon}$$

4.5 The second-order time-dependent coefficient a_k has 32 terms in the general case of two incident fields at different frequencies. Write them all out and perform the integrals to find the resonant denominators. Then identify CARS, HORSES, FWM, and others. Draw diagrams to illustrate the transitions involved and the particular resonances that enhance the processes.

5
Spontaneous emission

5.1 Introduction

The preceding chapters have focused on atomic transitions induced by applied radiation. That is, there is an optical field illuminating the atoms, and their response is calculated. It is implicitly assumed that an excited atom left alone will spontaneously return to the ground state, and conserve energy by emitting light that satisfies $E_2 - E_1 = \hbar\omega_{21}$ in accordance with the Planck and Bohr pictures. But this spontaneous emission process is not addressed quantitatively in these earlier chapters.

Spontaneous emission is one of the most pervasive phenomena of atomic physics, and yet its origins remain among the least well-understood. The number of misconceptions is enormous, and they are to be found in textbooks, journal articles, and lecture notes. Because the natural decay of an excited atom adds energy to what is usually an empty mode of the electromagnetic field, the process is fundamentally quantum mechanical. Classical or even semi-classical descriptions cannot be assured to give correct answers. This is not to say that such discussion is to be avoided, but only that caution is needed in interpreting the results.

Toward the end of the nineteenth century, many experimental results made it clear that classical mechanics needed to be supplemented by new theories. The first venture into the new physics was Planck's hypothesis that the energy of classical radiating oscillators like those discussed in Chap. 1 would have to be quantized into integer multiples of $\hbar\omega$. This notion led to his famous formula for the spectrum of thermal radiation from a non-reflective object (black body), and this formula describes a radiation field in thermodynamic equilibrium with its environment. The result of Planck's hypothesis in the context of thermal equilibrium plays a vital role in the discussion of spontaneous emission.

5.2 Einstein A- and B-coefficients

Soon after Bohr proposed a model of atomic structure consisting of discrete energy levels, connected by radiation whose energy satisfied $\Delta E = \hbar\omega$ when atoms made

5.2 Einstein A- and B-coefficients

transition between them, Einstein published the first serious attempt to connect the absorption and emission processes. Absorption, he reasoned, could occur only in the presence of applied radiation, whereas spontaneous decay occurred without the application of any field. Accordingly, these two processes were separately labeled A for the rate of spontaneous emission and B for atomic properties that had to be combined with the characteristics of the applied field to calculate the absorption rate. It must be emphasized that these are fundamentally different phenomena.

He then argued that there were inconsistencies with this naive description of Bohr's idea. Well before the advent of quantum mechanics as it is known today, Einstein realized that restricting the transitions between discrete states to spontaneous emission and induced absorption of radiation conflicted with well-known laws of thermodynamics. Finding no solution within the framework of existing phenomena, in 1917 he made the bold stroke of postulating a totally new phenomenon that is now called stimulated emission.

If atoms could be excited by absorbing light, and decay only by spontaneously emitting it, then there could be conditions where the Boltzmann relation for the relative number of atoms in states of different energies would be violated. In thermal equilibrium this condition requires $n_e = n_g e^{-\hbar \omega_{eg}/k_b T}$, where $\hbar \omega_{eg}$ is the energy difference between states $|e\rangle$ and $|g\rangle$, and k_b is Boltzmann's constant. But thermal equilibrium also requires a radiation field whose spectral energy density $U(\omega)$ is given by

$$U(\omega) = \frac{\hbar \omega^3 / \pi^2 c^3}{e^{\hbar \omega / k_b T} - 1} \tag{5.1}$$

as discovered by Planck in 1901. Note that $U(\omega)d\omega$ has dimensions of J/m^3 (see App. 3.B.2). These two thermodynamic conditions are incompatible with the Bohr picture discussed above.

Einstein suggested that the conflict could be reconciled if there exists a third radiative process, namely stimulated emission. The radiation field could stimulate atomic transitions to lower states with the emission of light just as it could excite atoms to higher states by absorption of light. This idea is second nature to us now, but it was completely unexpected at that time.

5.2.1 Einstein's calculation

Einstein's idea can be illustrated with a two-level atom. The rate for absorption of light to raise atoms from the lower state $|g\rangle$ to the upper state $|e\rangle$ is $R_{eg} = B_{eg} n_g U(\omega)$. It depends on B_{eg} which is a property of the atoms alone; it requires a radiation field of spectral energy density $U(\omega)$; and since it is a rate for an ensemble of atoms, it is proportional to the population n_g of $|g\rangle$. The rate of

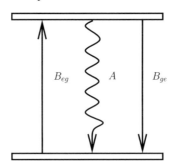

Figure 5.1 In Bohr's model of atomic transitions, atoms could be excited by absorption (B_{eg}) and decay by spontaneous emission (A). Einstein proposed the addition of a third process, namely return to the lower state by stimulated emission (B_{ge}).

emission from $|e\rangle$ to $|g\rangle$ is $R_{ge} = [A + B_{ge}U(\omega)]n_e$ where A is the spontaneous decay rate and is also a property of the atoms alone, and again is a rate for a system so it is proportional to n_e. The new process represented by the B_{ge} in the second term includes the contribution from stimulated emission, which also depends on the applied radiation field and the population of the upper state $|e\rangle$.

There are two relations required for thermal equilibrium: First, $U(\omega)$ is given by the Planck formula of Eq. (5.1), and second, $n_e = n_g e^{-\hbar\omega_{eg}/k_B T}$. A steady state between the atoms and the radiation field requires that $R_{eg} = R_{ge}$ and leads to

$$\frac{\hbar\omega_{eg}^3}{\pi^2 c^3} B_{ge} + A \left(e^{\hbar\omega_{eg}/k_b T} - 1 \right) = \frac{\hbar\omega_{eg}^3}{\pi^2 c^3} B_{eg} e^{\hbar\omega_{eg}/k_b T} \tag{5.2}$$

since n_g cancels. It should be emphasized that this relation describes thermal equilibrium, but the decay rate A and the two coefficients B are properties of the transition in the atoms alone, and do not depend upon the environment of thermal equilibrium. Since there can be thermal equilibrium at any temperature, the A and B coefficients can be found by solving Eq. (5.2) for any temperature. Equating the temperature-dependent and independent terms on the left- and right-hand sides leads to:

$$A = \frac{\hbar\omega_{eg}^3}{\pi^2 c^3} B_{ge} \quad \text{and} \quad B_{eg} = B_{ge}. \tag{5.3}$$

The first equation relates the spontaneous emission to stimulated emission, and the second relation shows that the two stimulated processes have the same rate.

5.2.2 Importance of this result

The relation $B_{eg} = B_{ge}$ deserves several comments. First, it is remarkable that this newly postulated process has the same magnitude as the absorption originally proposed by Bohr. It suggests that stimulated emission and absorption are closely related processes. This has already been made clear in the discussions of Chaps. 2 and 3.

Second, masers and lasers could not have been invented if their inventors had not known about stimulated emission. In fact, the two names are acronyms for "microwave/light amplification by stimulated emission of radiation", the process that enables their operation.

Third, the asymmetry between the total rates for excitation and decay, resulting from the presence of spontaneous emission in only one direction (down in atomic energy), results in the simple consequence that steady radiation alone can never invert the population of a two-level system in steady state. This has important consequences for the operation of masers and lasers.

Fourth, it is important to note that A depends on ω_{eg}^2 and not ω_{eg}^3 as is usually seen in calculations of the spontaneous emission rate (see below). This difference arises because A is calculated for a two-level atom whose properties need not appear in the calculation. The quantum mechanical result to be derived later depends on ω_{eg}^3 and also the dipole moment of the atom $e\langle \vec{r} \rangle$, and in the two-level limit these terms cancel. In the language of Sec. 3.4, the oscillator strength $(2m\omega_{eg}/\hbar)\langle \vec{r} \rangle^2 = 1$ for a two-level atom.

Fifth, this result can be used to calculate the the ratio of scattered power to incident intensity. The scattered power must equal the absorbed energy $\hbar\omega_{eg}$ times the absorption rate $BU(\omega_{eg})$ to conserve energy. Moreover, the incident intensity is c times the energy density $U(\omega_{eg})d\omega$. A reasonable estimate for $d\omega$ is the natural width A, perhaps multiplied by some constant β of order unity, as expected from the Fourier transform limit. Then

$$\sigma = \frac{\text{Scattered power}}{\text{Incident intensity}} = \frac{\hbar\omega_{eg}BU(\omega_{eg})}{cU(\omega_{eg})\beta A} = \frac{\hbar\omega_{eg}}{\beta c}\frac{B}{A} = \frac{3\lambda^2}{2\pi} \qquad (5.4)$$

for $\beta = \pi/6$ which is ~ 1. One has now calculated the cross-section σ for atoms scattering light in three different ways that all agree in the sense that they are a numerical factor of order unity times λ^2. In Sec. 1.3.2 one finds $\beta = 1$, near the end of Sec. 3.3.2 one finds $\beta = \pi/8$, and here one has $\beta = \pi/6$. It is indeed remarkable that these three results span only a factor of 2.5 in magnitude.

5.3 Discussion of this semi-classical description

The spontaneous emission rate A as introduced by Einstein is closely related to the classical damping rate γ of an electron, as discussed in Sec. 1.2. There are several insights to be gained from the results of Einstein's calculations. First, it is instructive to compare this result with the result of replacing $S(\omega)$ with $cU(\omega)$

in Eq. (3.16). One estimates $\langle \hat{\varepsilon} \cdot \vec{r} \rangle_{ji} = a_0/2$ and uses $\hbar\omega = mc^2\alpha^2/2$ to find $B = \pi^2 \alpha c a_0^2/\hbar$. Then using Eq. (5.3) yields $A = \alpha\hbar\omega^2/(2mc^2)$, a result surprisingly close to the classical value of $\gamma = 2\alpha\hbar\omega^2/3mc^2$ as shown in Eq. (1.1). It is quite surprising that with the invention of quantum mechanics ten years away, Einstein's assertion about stimulated emission was able to produce a result consistent with both the Lorentz classical picture and the nascent quantum mechanics. This intimate relation between spontaneous and stimulated processes also has several important consequences.

Second, there is a frequently quoted but incomplete argument that spontaneous emission is really stimulated by the zero-point energy $1/2\,\hbar\omega$ per mode of a quantized electromagnetic field (see App. 5.B). This argument begins by multiplying the mode density found from Eq. (5.10) by $1/2\,\hbar\omega$ to get the energy density $U(\omega)d\omega$ and using the Einstein relation between A and B to find the rate. Then the rate R of transitions driven by this energy density is

$$R = BU(\omega) = \left[\frac{\pi^2 c^3}{\hbar\omega^3} A\right] \frac{\hbar\omega^3}{2\pi^2 c^3} = \frac{A}{2} \neq A, \tag{5.5}$$

and this is not an error. To repeat some parts of the introduction to this chapter, spontaneous emission adds energy to what is usually an otherwise empty field mode so the process is fundamentally quantum mechanical. The semi-classical calculation is an inherently wrong approach because it accounts for neither the change of the field resulting from the emitted radiation nor its interaction with the emitting atom (radiative reaction). Simply inserting the zero-point energy into Einstein's semi-classical formula does not work because such descriptions cannot be assured to give correct answers, and in this case they do not.

The next section presents a more complete description of spontaneous emission as arising from the zero-point energy of the field by treating the interaction in a fully quantum mechanical way. In this model the emitted field is included as part of the "system" being studied, and the "atom–field system" is in a superposition state. In this way, the effect of the emitted field on the atom is taken into account. In the language of the density matrix (see Chap. 6), the coherences are no longer ignored as is done in the rate equation approach using the A and B coefficients.

5.4 The Wigner–Weisskopf model

Spontaneous emission cannot be properly described within the framework of a semi-classical description of the electromagnetic field as was done in the previous chapters. The famous Wigner–Weisskopf theory of the process is described in Ref. [40]. In this theory, it is shown how an atom in the excited state decays exponentially and produces spontaneous emission.

5.4 The Wigner–Weisskopf model

Consider an atom in the excited state at $t = 0$ and an empty radiation field. The system is in a pure state $|e; 0\rangle$, where the first parameter in the ket describes the state of the atom and the zero indicates the empty radiation field. The system makes a transition from the excited to the ground state by spontaneous emission, emitting light into the radiation field. Then the state is denoted by $|g; 1_S\rangle$ where $S = (\vec{k}, \hat{\varepsilon})$ is the mode of spontaneous emission and the direction of the emitted light is explicitly indicated by its wavevector $\vec{k}$ and its polarization by $\hat{\varepsilon}$. The state of the system can now be described by

$$\Psi(t) = c_{e0} e^{-i\omega_e t} |e; 0\rangle + \sum_S c_{g1_S} e^{-i(\omega_g + \omega)t} |g; 1_S\rangle, \tag{5.6}$$

where the sum is over all possible modes S. Note that the frequency ω in the exponent must be replaced by kc for the summation. Even though the summation runs over an infinite number of modes, this notation is sufficient for now.

To describe the evolution of the wavefunction in time, the Hamiltonian of the system has to be defined. This requires the quantization of the electromagnetic field, as done in App. 5.B. However, the only part of the Hamiltonian that couples the two states in Eq. (5.6) is the atom–field interaction: the atomic and field parts play no role. This coupling is analogous to its semi-classical counterpart, and the result for the time evolution of the two states is

$$\frac{dc_{e0}(t)}{dt} = -\sum_S ic_{g1_S}(t)\, \Omega_S\, e^{-i(\omega - \omega_{eg})t} \tag{5.7a}$$

and

$$\frac{dc_{g1_S}(t)}{dt} = -ic_{e0}(t)\, \Omega_S^*\, e^{i(\omega - \omega_{eg})t}. \tag{5.7b}$$

These equations are similar to Eqs. (2.4), where the coupling for each mode is given by $\hbar\Omega_S = -\vec{\mu} \cdot \vec{\mathcal{E}}_\omega$ and $|\Omega_S|$ is called the Rabi frequency. The electric dipole moment is $\vec{\mu} = e\langle e|\vec{r}|g\rangle$, and the electric field per mode is found from the classical expression for the energy density to be

$$\vec{\mathcal{E}}_\omega = \sqrt{\frac{\hbar\omega}{2\varepsilon_0 V}}\, \hat{\varepsilon}. \tag{5.8}$$

Here V is the volume used to quantize the field and it will eventually drop out of the calculation. The total energy of each mode of the electromagnetic field in the volume V is given by $\hbar\omega/2$, corresponding to the zero-point energy of the radiation field. By directly integrating Eq. (5.7)b and substituting the result into Eq. (5.7)a, the time evolution of $c_{e0}(t)$ is found to be

$$\frac{dc_{e0}(t)}{dt} = -\sum_S |\Omega_S|^2 \int_0^t dt'\, e^{-i(\omega - \omega_{eg})(t-t')} c_{e0}(t'). \tag{5.9}$$

Spontaneous emission

This represents an exponential decay of the excited state, and to evaluate the decay rate it is necessary to count the number of modes for the summation and then do the time integral.

To count the number of modes $S = (\vec{k}, \hat{\varepsilon})$, represent the field by the complete set of traveling waves in a cube of side L. Since the field is periodic with a periodicity L, the components of $\vec{k}$ are quantized as $k_i = 2\pi N_i/L$, with $i = x, y, z$. Then $dN_i = (L/2\pi)dk_i$ and therefore $dN = (L/2\pi)^3 d\vec{k}$. The frequency ω is given by $\omega = kc$, so (see also App. 21.B)

$$dN = 2 \times \frac{V\omega^2 d\omega}{8\pi^3 c^3} \int \sin\theta \, d\theta d\phi = \frac{V\omega^2 d\omega}{\pi^2 c^3}. \tag{5.10}$$

The factor of 2 in the center expression of Eq. (5.10) derives from the two independent polarizations $\hat{\varepsilon}$ of the fluorescent light. In a frequently quoted abstract [41], it was pointed out that Eq. (5.10) applies to free space but inside a cavity its mode structure resulting from the boundary conditions can have severe effects on mode counting. This notion was expanded in Ref. [42] and forms the basis of the widely studied field called cavity quantum electrodynamics (cavity QED or CQED).

The next step is to replace the summation in Eq. (5.9) by an integration over all possible modes, use Eq. (5.8) to evaluate $\Omega_S = -\vec{\mu} \cdot \vec{\mathcal{E}}_\omega/\hbar$, and insert the result of Eq. (5.10) to find

$$\frac{dc_{e0}(t)}{dt} = -\frac{1}{6\varepsilon_0 \pi^2 \hbar c^3} \int d\omega \, \omega^3 \mu_{eg}^2 \int_0^t dt' \, e^{-i(\omega - \omega_{eg})(t-t')} c_{e0}(t'), \tag{5.11}$$

where the volume V has dropped out, since $|\Omega_S|^2 \propto 1/V$. In this result, the orientation of the atomic dipole with respect to the emission direction has been taken into account, which yields a reduction factor of $1/3$ for a random emission direction.

The remaining time integral can be evaluated by assuming that the dipole moment μ_{eg} varies slowly over the frequency interval of interest, so it can be evaluated at $\omega = \omega_{eg}$. Furthermore, the time integral is peaked around $t = t'$, so that the coefficient $c_{e0}(t)$ can be evaluated at time t and taken out of the integral. The upper boundary of the integral can be shifted toward infinity, and the result becomes

$$\lim_{t \to \infty} \int_0^t dt' \, e^{-i(\omega - \omega_{eg})(t-t')} = \pi \delta(\omega - \omega_{eg}) - \mathcal{P}\left(\frac{i}{\omega - \omega_{eg}}\right), \tag{5.12}$$

where $\delta(x)$ is the delta function and $\mathcal{P}(x)$ is the principal value. The last term is purely imaginary and causes a shift of the transition frequency, which will not be discussed further. Substitution of the result of Eq. (5.12) into Eq. (5.11) yields the final result

$$\frac{dc_{e0}(t)}{dt} = -\frac{\gamma}{2} c_{e0}(t), \tag{5.13a}$$

where

$$\gamma = \frac{\omega^3 \mu^2}{3\pi\varepsilon_0 \hbar c^3}, \quad (5.13b)$$

which is the same as Eq. (1.1) for $\mu = ea_0$ and $\hbar\omega = mc^2\alpha^2/2$. Since the amplitude of the excited state decays at a rate $\gamma/2$, the population of the state decays with γ and the lifetime of the excited state becomes $\tau \equiv 1/\gamma$.

The decay of the excited state is irreversible. In principle, the modes of the spontaneously emitted light also couple to the ground state in Eqs. (5.7), but there is an infinite number of modes in free space. The amplitude for the reverse process has to be summed over these modes. Since the different modes add destructively, the probability for the reverse process becomes zero. The situation can be changed by putting the atom in a reflecting cavity with dimensions of the order of the wavelength λ. Then the number of modes can be changed considerably compared with free space. Several experiments have been carried out in which this effect has been studied in Rydberg atoms where $\lambda \sim 1$ cm [43].

Appendices

5.A The quantum mechanical harmonic oscillator

The Hamiltonian for a quantum mechanical harmonic oscillator with unit mass is given by

$$\mathcal{H} = \frac{1}{2}\left(\hat{p}^2 + \omega^2 \hat{q}^2\right), \quad (5.14)$$

where $\hat{p}$ is the momentum operator and $\hat{q}$ is the position operator. Here and in the next section the operators are explicitly identified with a hat. The resulting Schrödinger equation can be solved analytically and the eigenfunctions are proportional to the Hermite polynomials that are functions of position. If one knows these eigenfunctions, all properties of the system can be found. In many cases one needs to know only the expectation value of powers of the operators $\hat{p}$ and $\hat{q}$, and these can be obtained in a simpler way.

The method employs the commutation relation between $\hat{p}$ and $\hat{q}$:

$$[\hat{q}, \hat{p}] = i\hbar \quad (5.15)$$

which applies to any pair of canonically conjugate operators in quantum mechanics. It begins by defining the creation operator $\hat{a}^\dagger$ and the annihilation operator $\hat{a}$ as

$$\hat{a}^\dagger = \frac{\omega\hat{q} - i\hat{p}}{\sqrt{2\hbar\omega}} \quad \hat{a} = \frac{\omega\hat{q} + i\hat{p}}{\sqrt{2\hbar\omega}}. \quad (5.16)$$

So $2\hbar\omega\hat{a}^\dagger\hat{a} = \hat{p}^2 + \omega^2\hat{q}^2 + i\omega\hat{q}\hat{p} - i\omega\hat{p}\hat{q}$, and using Eq. (5.15) one finds $\hbar\omega\hat{a}^\dagger\hat{a} = \mathcal{H} - 1/2\hbar\omega$. Similarly $\hbar\omega\hat{a}\hat{a}^\dagger = \mathcal{H} + 1/2\hbar\omega$. Thus the commutation relation for $\hat{a}$ and $\hat{a}^\dagger$ is simply $\left[\hat{a}, \hat{a}^\dagger\right] = 1$, and more important, the Hamiltonian can now be written as

$$\mathcal{H} = \hbar\omega\left(\hat{a}^\dagger\hat{a} + 1/2\right). \tag{5.17}$$

Equation (5.17) can be used to find the eigenfunctions $|n\rangle$ with eigenvalues E_n. Thus $\mathcal{H}|n\rangle = \hbar\omega(\hat{a}^\dagger\hat{a} + 1/2)|n\rangle = E_n|n\rangle$. Now multiply this on the left by $\hat{a}^\dagger$

$$\hbar\omega\hat{a}^\dagger\left(\hat{a}^\dagger\hat{a} + 1/2\right)|n\rangle = E_n\hat{a}^\dagger|n\rangle \tag{5.18}$$

and use the commutation relation for the operators to find

$$\hbar\omega\left(\hat{a}^\dagger\hat{a} + 1/2\right)\hat{a}^\dagger|n\rangle = (E_n + \hbar\omega)\hat{a}^\dagger|n\rangle = \mathcal{H}\hat{a}^\dagger|n\rangle \tag{5.19}$$

which shows that $\hat{a}^\dagger|n\rangle$ is an eigenfunction of the Hamiltonian with an energy of $E_n + \hbar\omega$. Similarly, one can show that $\hat{a}|n\rangle$ is also an eigenfunction of the Hamiltonian with energy $E_n - \hbar\omega$. These operators are called creation and annihilation operators respectively since they increase or decrease the eigenvalue by $\hbar\omega$. Thus the eigenvalues are equally spaced by $\hbar\omega$.

The lowest state $|0\rangle$ must satisfy $\hat{a}|0\rangle = 0$ because it cannot be lowered, so $\mathcal{H}|0\rangle = \hbar\omega(\hat{a}^\dagger\hat{a} + 1/2)|0\rangle = 1/2\hbar\omega|0\rangle$ or $E_0 = 1/2\hbar\omega$ and then

$$E_n = (n + 1/2)\hbar\omega. \tag{5.20}$$

Thus the lowest state has non-zero energy $\hbar\omega/2$ (see Fig. 5.2).

For any eigenstate it must be that $\hat{a}|n\rangle = a_n|n-1\rangle$ and it is necessary to determine the coefficient a_n. The complex conjugate of this expression is $\langle n|\hat{a}^\dagger = \langle n-1|a_n^*$ and so their product is $\langle n|\hat{a}^\dagger\hat{a}|n\rangle = \langle n-1|n-1\rangle|a_n|^2 = |a_n|^2$. But $\langle n|\hat{a}^\dagger\hat{a}|n\rangle$ is simply n, so $a_n = \sqrt{n}$ and $\hat{a}|n\rangle = \sqrt{n}|n-1\rangle$. Simarly $\hat{a}^\dagger|n\rangle = \sqrt{n+1}|n+1\rangle$.

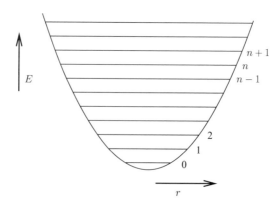

Figure 5.2 Energy levels of a harmonic oscillator.

5.B Field quantization

The earliest discussion of field quantization arose from Planck's description of black body radiation. He proposed that the "radiation oscillators" could be characterized by energies that were integer multiples of $\hbar\omega$, and other values were not allowed. Combining this idea with the sinusoidal nature of electromagnetic waves that satisfy Maxwell's equations gives a description of the radiation field as a collection of quantized simple harmonic oscillators. Each of them can be described simply by solving the Schrödinger equation for a harmonic potential, yielding the familiar zero-point energy of each oscillator, $E = 1/2\, h\nu = 1/2\, \hbar\omega$.

The process is based on a single oscillator with a particular frequency, but the problem of spontaneous emission addressed in this chapter requires a field whose zero-point energy is $1/2\hbar\omega$ per mode, and many modes need to be considered. Thus the usual approach is to consider quantization of the field in a cavity that supports many modes, or to use Fourier transforms to enumerate the accessible modes. Each mode is subscripted and evolves independently of all the others, but this way they can all be considered.

A brief discussion of field quantization is given here, but the reader is referred to standard texts such as Refs. [44–49]. It is helpful to consider a cavity that encloses the light field. The discussion is simplified by considering the cavity as a cube with length L as before, so the components k_i of the wavevectors inside the cavity are given by $k_{x,y,z} = 2\pi n_{x,y,z}/L$. This makes the modes denumerable, although there are an infinite number of modes possible. The vector field in the cavity is now formed by modes given by Eq. (3.30), where the components of $\vec{k}$ are given above. The corresponding equations for the electric and magnetic field are found from Eq. (3.31). For the energy of a particular mode k in the cavity one finds

$$E_k = \frac{\varepsilon_0}{2} \int_{\text{cavity}} \left(\vec{\mathcal{E}}_k^2 + c^2 \vec{\mathcal{B}}_k^2 \right) dV = 2\varepsilon_0 V \omega_k^2 \vec{\mathcal{A}}_k \cdot \vec{\mathcal{A}}_k^*, \tag{5.21}$$

where V is the volume of the cavity and $\vec{\mathcal{A}}_k$ and $\vec{\mathcal{A}}_k^*$ are the amplitudes of the vector field for the particular mode k. This can be rewritten by replacing the amplitudes with

$$\vec{\mathcal{A}}_k = \frac{1}{\sqrt{4\varepsilon_0 V \omega_k^2}} (\omega_k q_k + i p_k)\, \hat{\varepsilon}_k \qquad \vec{\mathcal{A}}_k^* = \frac{1}{\sqrt{4\varepsilon_0 V \omega_k^2}} (\omega_k q_k - i p_k)\, \hat{\varepsilon}_k, \tag{5.22}$$

where p_k and q_k are generalization of the momentum and position, respectively. The choices of the coefficients in front of p_k and q_k become clear in the next paragraph. The energy of the mode becomes

$$E_k = \frac{1}{2} \left(p_k^2 + \omega_k^2 q_k^2 \right), \tag{5.23}$$

which is very reminiscent of the energy of a harmonic oscillator. This is the starting point for the quantum mechanical description of the radiation field.

Using the analogy with the harmonic oscillator of App. 5.A, it is now clear that one can identify $\vec{\mathcal{A}}_k$ with the annihilation operator $\hat{a}_k$ and $\vec{\mathcal{A}}_k^*$ with the creation operator $\hat{a}_k^\dagger$ in the following way:

$$\hat{\vec{\mathcal{A}}}_k = \sqrt{\frac{\hbar}{2\varepsilon_0 V \omega_k}}\, \hat{a}_k \hat{\varepsilon}_k \qquad \hat{\vec{\mathcal{A}}}_k^* = \sqrt{\frac{\hbar}{2\varepsilon_0 V \omega_k}}\, \hat{a}_k^\dagger \hat{\varepsilon}_k, \qquad (5.24)$$

where the hat indicates that the two fields have become quantum mechanical operators to distinguish them from their classical counterparts. Similarly, one can find for the operator of the electric field:

$$\hat{\vec{\mathcal{E}}}_k = i\sqrt{\frac{\hbar \omega_k}{2\varepsilon_0 V}}\, \hat{\varepsilon}_k \left(\hat{a}_k e^{-i\omega_k t + i\vec{k}\cdot\vec{r}} - \hat{a}_k^\dagger e^{+i\omega_k t - i\vec{k}\cdot\vec{r}} \right) \qquad (5.25a)$$

and for the operator of the magnetic field:

$$\hat{\vec{\mathcal{B}}}_k = i\sqrt{\frac{\hbar}{2\varepsilon_0 V \omega_k}}\, \vec{k} \times \hat{\varepsilon}_k \left(\hat{a}_k e^{-i\omega_k t + i\vec{k}\cdot\vec{r}} - \hat{a}_k^\dagger e^{+i\omega_k t - i\vec{k}\cdot\vec{r}} \right). \qquad (5.25b)$$

But, most important, the Hamiltonian of the field in the cavity can now be written simply as

$$\mathcal{H}_R = \sum_k \hbar \omega_k \left(\hat{a}_k^\dagger \hat{a}_k + 1/2 \right), \qquad (5.26)$$

which is the generalization of Eq. (5.17) for multiple frequencies. By using the number operator $\hat{n}$ for all modes, one simply finds for the total energy in the cavity: $E = \sum_k (n_k + 1/2) \hbar \omega_k$. This equation makes the interpretation of n_k as the number of excitations in the mode k worthwhile. These excitations are called quanta of light, as Einstein referred to them, or more popularly, photons.

Once the electromagnetic field is quantized, the total Hamiltonian of the atom-light interaction can be written as

$$\mathcal{H} = \mathcal{H}_{\text{atom}} + \mathcal{H}_R + \mathcal{H}_{\text{int}}, \qquad (5.27)$$

where $\mathcal{H}_{\text{atom}}$ in the usual Hamiltonian describing the evolution of the atom, $\mathcal{H}_R$ the Hamiltonian for the field given by Eq. (5.26), and the last term $\mathcal{H}_{\text{int}}$ is the interaction term, which is in most cases the electric dipole interaction. Now there are many modes with frequency ω_k in the cavity, and in general all but one of these modes are far away from the frequency ω_{eg} that excites the atom. If one considers only the mode in the cavity that is the closest to the atomic transition frequency, in the RWA the total Hamiltonian becomes simply

$$\mathcal{H} = 1/2 \hbar \omega_{eg} \sigma_z + \hbar \omega_k \hat{a}^\dagger \hat{a} + \hbar \left(\Omega \hat{a} \sigma_+ + \Omega^* \hat{a}^\dagger \sigma_- \right), \qquad (5.28)$$

where the operators $\sigma_{\pm,z}$ are the Pauli projection operators (see App. 2.A). The first term is the atomic part, where the energy difference between excited and ground state is equal to $\hbar\omega_{eg}$. The second term is the field part, where there is only one mode k filled with $\langle \hat{a}^\dagger \hat{a} \rangle$ photons. The last term is the interaction between the atom and the light, where the first term destroys a photon in the cavity and raises the atom to the excited state and the second term creates a photon in the cavity and de-excites the atom to the ground state. The coupling strength of this mode is given by the usual Rabi frequency $|\Omega|$. This Hamiltonian is referred to as the Jaynes–Cummings model [9], which gives rise to quantum Rabi flopping and collapse and revival of the field mode.

5.C Alternative theories to QED

The ideas of field quantization and its implications have been a subject of active debate and discussion since their advent. There has been a great deal written about such calculations. In almost all cases, the main point of these discussions arises because the authors attempt to put a fundamentally quantum mechanical phenomenon into a classical framework. Attempting to "explain" spontaneous emission as a form of stimulated emission arising from either the zero-point field or the reaction to the radiated field or some combination of these is guaranteed to produce inconsistent results, simply because it is not an appropriate description of the phenomenon.

Much of the interest is centered on the zero-point energy (and perhaps its extraction for useful purpose, although most physicists are justifiably skeptical of this idea). For example, a well-known theory advanced by T. Boyer [50] and others that was initially called "random electrodynamics" but more recently called "stochastic electrodynamics" simply assumes the existence of the zero-point energy having energy $1/2\hbar\omega$ per mode. The theory shows that the predictions of traditional QED can be duplicated using this as a classical field (nothing quantum about it) along with classical mechanics and Maxwellian electrodynamics applied to electron behavior.

In such cases the density of the zero-point energy is always proportional to ω^3, just as is the Planck black body distribution, thereby giving it the special property of Lorentz invariance. That is, if viewed in a different (inertial) reference frame, there is as much radiation Doppler shifted into a given frequency interval as there is shifted out by uniform motion, leaving the distribution unchanged.

Another idea explored by Jaynes and co-workers [51, 52] and more recently by others [53], called neo-classical theory, makes no use of the zero-point energy. Instead it pays careful attention to the back action of the radiated field produced by the oscillating dipole of the radiating atom. Including this field in the semi-classical case might be described as the "radiation reaction" force. Such a description suggests that the radiated field acts back on the radiating atom, and that this interaction cannot be neglected. It depends upon the idea that an excited atom always has some component of ground state mixed into its wavefunction, and the result of the superposition is an oscillating dipole at the atomic frequency.

This radiation reaction can be treated as a completely classical effect, and predicts spontaneous emission at a rate consistent with the Einstein A-coefficient. It may be viewed another way: by including the zero-point energy as $1/2\hbar\omega$ per mode with the atom in the excited state, and $3/2\hbar\omega$ per mode after the atom has decayed by spontaneous emission, the average value becomes $\hbar\omega$ per mode and thereby eliminates the discrepancy of Eq. (5.5).

Exercises

5.1 Prove that, for the Einstein coefficients, $B_{jk} = B_{kj}$.

5.2 Statistical mechanics gives the ratio of excited atoms due to thermal excitation at a temperature T to those in the ground state to be
$$\frac{n_e}{n_g} = e^{-(E_e - E_g)/k_B T} = e^{-\hbar\omega_{eg}/k_B T}.$$
(a) Assuming the Einstein coefficient of induced emission to be equal to the Einstein coefficient of absorption, show that induced emission is equal to spontaneous emission at $T = \hbar\omega/(k_B \ln 2)$.
(b) Show that black body radiation at room temperature *cannot* be responsible for spontaneous decay of low-lying excited states of atoms (e.g. $n = 2$ of hydrogen). At what temperature would the stimulated rate from this radiation be comparable to the spontaneous rate? Calculate the numerical value of this temperature.

5.3 Show that the electromagnetic fields can be written in terms of the creation and annihilation operator as given in Eq. (5.25a), where the vector potential for a standing wave can be written as $\vec{\mathcal{A}} \equiv \hat{\varepsilon} \mathcal{A}_0 \sin(\vec{k} \cdot \vec{r}) \sin(\omega t)$.

5.4 The creation and annihilation operators are important for the discussion of the radiation field.
(a) Show that $a|n\rangle$ is indeed an eigenstate of the Hamiltonian $\mathcal{H} = \hbar\omega(\hat{a}\hat{a}^\dagger - 1/2)$ and calculate its eigenvalue.
(b) Show that $\mathcal{H} = \hbar\omega(\hat{a}^\dagger \hat{a} + 1/2)$ is equivalent to $\mathcal{H} = \hbar\omega(\hat{a}\hat{a}^\dagger - 1/2)$.
(c) Show that $\hat{a}^\dagger |n\rangle = \sqrt{n+1}\,|n+1\rangle$ whereas $\hat{a}|n\rangle = \sqrt{n}\,|n-1\rangle$.
(d) Show that the expression for γ given in Eq. (1.1) is equivalent to $\gamma = 2\alpha\hbar\omega^2/3mc^2$.
(e) Show that substituting the fields of Eqs. (5.25) into the expression for the total energy Eq. (5.21) indeed results in the Hamiltonian of Eq. (5.26).

6
The density matrix

6.1 Introduction

In Chap. 2, a description of the equations for the coherent evolution of an atom in a radiation field is presented. In addition to this traditional wavefunction picture, the internal state of the atom can be described by the density matrix. In Sec. 5.3 the effects of spontaneous emission on the internal state of the atom are not to be described in terms of a coherent evolution of the state because the system is not closed. Such situations are more readily described in terms of the density matrix. This chapter begins with an introduction to the density matrix and applies it in Sec. 6.3 below to the specific case of a two-level atom in a radiation field, as in Sec. 5.3. The resulting equations are solved and discussed in terms of the effects of spontaneous emission on the interaction between atoms and radiation fields. The reader is urged to consult the many standard references for applications of the density matrix to other topics.

6.2 Basic concepts

6.2.1 Pure states

In quantum mechanics all information about a system in a pure state is stored in the wavefunction $|\Psi\rangle$. However, in an experiment $|\Psi\rangle$ cannot be measured directly. Instead, one can only determine the expectation values of a set of quantum mechanical operators $\mathcal{A}$ given by

$$\langle \mathcal{A} \rangle = \langle \Psi | \mathcal{A} | \Psi \rangle, \tag{6.1}$$

when Ψ is normalized according to $\langle \Psi | \Psi \rangle = 1$. By proper arrangement of the experiment the wavefunction can be determined completely, except for one unnecessary parameter, the overall phase.

Alternatively, the state of the system can be described by the density operator ρ, which is given by $\rho = |\Psi\rangle\langle\Psi|$. The density operator ρ can be written in terms of the $n \times n$ density matrix, where n is the number of wavefunctions that completely spans the Hilbert space [54, 55]. In general, the wavefunction Ψ can be expanded in a basis set $\{\phi_n\}$ as in Eq. (2.1),

$$\Psi = \sum_{i=1}^{n} c_i \phi_i, \qquad (6.2)$$

so that the elements of the density matrix become

$$\rho_{ij} = \langle \phi_i | \rho | \phi_j \rangle = \langle \phi_i | \Psi \rangle \langle \Psi | \phi_j \rangle = c_i c_j^* \qquad (6.3)$$

and the normalization of the wavefunction yields $\text{Tr}(\rho) = \langle \Psi | \Psi \rangle = 1$. In the case of a two-level atom, $n = 2$ so that ρ is a 2×2 matrix.

Clearly the elements ρ_{ij} depend on the basis states $\{\phi_n\}$. The diagonal elements are the probabilities $|c_i|^2$ for the atom to be in state i, which are all between 0 and 1. The off-diagonal elements $c_i c_j^*$ are called the coherences, since they depend on the phase difference between c_i and c_j.

The expectation value of an operator given in Eq. (6.1) can be written as

$$\langle \mathcal{A} \rangle = \left\langle \sum_i c_i \phi_i \middle| \mathcal{A} \middle| \sum_j c_j \phi_j \right\rangle = \sum_{i,j} c_i^* c_j \langle \phi_i | \mathcal{A} | \phi_j \rangle \qquad (6.4)$$
$$= \sum_{i,j} \rho_{ji} \mathcal{A}_{ij} = \sum_j (\rho \mathcal{A})_{jj} = \text{Tr}(\rho \mathcal{A}).$$

Note that if the wavefunction Ψ is multiplied by an arbitrary phase factor $e^{i\alpha}$, there is no change of any observable of the system as shown by Eq. (6.4). Also ρ remains unchanged in this case, as required for an observable.

Since the density matrix contains n^2 complex elements, in principle it would have $2n^2$ real, independent parameters. Because ρ is Hermitian (see Eq. (6.3)), $\rho_{ij} = \rho_{ji}^*$ and there remain n^2 independent elements. By contrast, the wavefunction Ψ is completely specified by the expansion coefficients c_i, which contain only $2n-1$ independent parameters apart from its overall phase. This reduction in the number of parameters arises because the system under discussion here is in a pure state, which means that there is a fixed relation between the diagonal and off-diagonal elements. This relation is found from Eq. (6.3) to be $\rho_{ij} \rho_{ji} = \rho_{ii} \rho_{jj}$.

6.2.2 Mixed states

The alternative to such a pure state is a statistical mixture of several states $\{\Psi_n\}$ that can no longer be specified by just a single wavefunction. In that case the state is represented by a density operator of the form

$$\rho = \sum_i p_i |\Psi_i\rangle\langle\Psi_i|. \qquad (6.5)$$

This relation has the intuitive meaning that the system is in state i with a certain probability p_i. It can easily be checked that there is no longer a fixed relation between diagonal and non-diagonal elements, but instead $\rho_{ij}\rho_{ji} \leq \rho_{ii}\rho_{jj}$. The complete information on the system now requires n^2 independent elements of the density matrix.

The advantages of the density matrix formalism compared with the wavefunction approach can be summarized as follows: (1) It eliminates the arbitrary overall phase, (2) it establishes a more direct connection with observable quantities, and (3) it provides a powerful method for doing calculations. In addition, it can handle pure states as well as mixed states, the last one being of importance in the case of spontaneous emission.

The distinction between pure states and statistical mixtures is of fundamental importance in quantum mechanics. Suppose that for a certain quantum mechanical system there is a complete set of commuting operators. The question of whether a set of commuting operators is complete depends on the system under study. Then one measurement with each operator completely determines the state. Any subsequent measurement with one of the operators yields the same outcome as before, since all operators commute with each other. In this way the system has been prepared in a pure state, also referred to as a state of "maximum knowledge". If there is no measurement with one of the operators of this complete set, there is no information on the outcome of such a measurement. The system will then be in a statistical mixture of states $\{|\Psi_n\rangle\}$ with probabilities p_i to be in a pure state Ψ_i, where i labels the eigenstates of the unmeasured operator.

From the definition of the density matrix in Eq. (6.3), it is easy to show that for a pure, normalized state $\rho^2 = \rho$, whereas for a statistical mixture $\rho^2 \neq \rho$. In a pure state, one of the eigenvalues of the density matrix is unity and all the others are zero. In the case of a statistical mixture there are several eigenvalues between 0 and 1, which are the probabilities for the state to be in a particular eigenstate. These properties make it possible to determine from a given density matrix whether the system is in a pure state or not.

Spontaneous emission results in a transition of the system from an initial to a final state and can convert a pure state to a statistical mixture. This can happen because statistical mixtures are not only a consequence of incomplete preparation of the system, but also occur if there is only partial detection of the final state. Suppose a system consists of two parts A and B, such as an atom and a radiation field that are coupled, but only part A is observed. Then information about part B is lost, and a statistical average over part B is necessary. Using the density matrix to describe the system, one has to take the trace over part B, or

$$\rho_A = \mathrm{Tr}_B (\rho_{AB}). \tag{6.6}$$

If the system was initially in a pure state, the incomplete detection process causes the pure state to evolve into a statistical mixture.

As an example, consider a two-level atom in the excited state. After a short time the atom has a probability to remain in the excited state or it can make a transition to the ground state by spontaneous emission. The evolution of this system is given by (see Eq. (5.6))

$$|\Psi\rangle = c_e(t)|e;0\rangle + \sum_S c_{g,S}(t)|g;1_S\rangle, \qquad (6.7)$$

where the state of the atom is indicated by e or g and the emitted light by $S = (\vec{k}, \hat{\varepsilon})$ with its wavevector $\vec{k}$ and its polarization $\hat{\varepsilon}$. Note that the light can be emitted in all directions with various polarizations, so the sum runs over all possible values of S. If one observes only the state of the atom and not the emitted light, then the atom will be found in either the excited state $|e\rangle$ or the ground state $|g\rangle$; however, it will no longer be in a pure state. The new state can be described by its density matrix ρ_{atom}:

$$\rho_{\text{atom}} = \text{Tr}_{\text{field}} |\Psi\rangle\langle\Psi| = |c_e(t)|^2 |e\rangle\langle e| + \sum_S |c_{g,S}(t)|^2 |g\rangle\langle g|. \qquad (6.8)$$

The pure state $|\Psi\rangle$ has evolved to a statistical mixture of $|e\rangle$ and $|g\rangle$ since the emitted light has not been observed. Equation (6.8) shows that phase information has been lost from Eq. (6.7).

6.3 The optical Bloch equations

It is straightforward to use Eq. (2.4) to show that the time dependence of the density matrix depends on the Hamiltonian simply as (see App. 6.A)

$$i\hbar \frac{d\rho}{dt} = [\mathcal{H}, \rho]. \qquad (6.9)$$

This relation points out the special role of ρ in quantum mechanics. Note that the sign on the right-hand side is *opposite* to the usual Heisenberg equation of motion for quantum mechanical operators. The rest of this section continues the analysis of the Rabi two-level problem using the density matrix, which is written for a pure state as

$$\rho = \begin{pmatrix} \rho_{ee} & \rho_{eg} \\ \rho_{ge} & \rho_{gg} \end{pmatrix} = \begin{pmatrix} c_e c_e^* & c_e c_g^* \\ c_g c_e^* & c_g c_g^* \end{pmatrix}. \qquad (6.10)$$

The effects of the coupling to the light field and spontaneous emission can be added independently [56]. The evolution equation for the terms ρ_{ij} in the case

6.3 The optical Bloch equations

of interaction with a laser can be found by applying the evolution equation for the amplitudes, given by Eqs. (2.6). For instance, in the case of ρ_{gg} this is

$$\frac{d\rho_{gg}}{dt} = \frac{dc_g}{dt}c_g^* + c_g\frac{dc_g^*}{dt} = -i\frac{\Omega^*}{2}\tilde{\rho}_{eg} + i\frac{\Omega}{2}\tilde{\rho}_{ge}, \qquad (6.11)$$

where $\tilde{\rho}_{ge} \equiv \rho_{ge}e^{i\delta t}$. In the same manner, equations for the time derivative of the other elements of the density matrix can be obtained. Solving these equations gives the same solutions as Eqs. (2.8). The identification of ρ_{ij} in terms of $c_i c_j^*$ is valid for a pure state, but loses its meaning for a statistical mixture.

Spontaneous emission leads to a decay of the excited-state population with a decay rate γ, as shown in Chap. 5. This can now be described by an exponential decay of the coefficient $\rho_{eg}(t)$ with a constant rate $\gamma/2$,

$$\left(\frac{d\rho_{eg}}{dt}\right)_{\text{spon}} = -\frac{\gamma}{2}\rho_{eg}. \qquad (6.12)$$

The ground state is stable against spontaneous emission, but the population of the ground state still changes because of the spontaneous emission process, since the excited state decays to the ground state. The loss of population of the excited state leads to a gain of population in the ground state. This leads to the following equations for the two-level system, including spontaneous emission:

$$\begin{aligned}
\frac{d\rho_{gg}}{dt} &= +\gamma\rho_{ee} + \frac{i}{2}\left(\Omega\tilde{\rho}_{ge} - \Omega^*\tilde{\rho}_{eg}\right) \\
\frac{d\rho_{ee}}{dt} &= -\gamma\rho_{ee} + \frac{i}{2}\left(\Omega^*\tilde{\rho}_{eg} - \Omega\tilde{\rho}_{ge}\right) \\
\frac{d\tilde{\rho}_{ge}}{dt} &= -\left(\frac{\gamma}{2} + i\delta\right)\tilde{\rho}_{ge} + \frac{i}{2}\Omega^*\left(\rho_{gg} - \rho_{ee}\right) \\
\frac{d\tilde{\rho}_{eg}}{dt} &= -\left(\frac{\gamma}{2} - i\delta\right)\tilde{\rho}_{eg} + \frac{i}{2}\Omega\left(\rho_{ee} - \rho_{gg}\right).
\end{aligned} \qquad (6.13)$$

These equations are called the optical Bloch equations (OBE), in analogy to the Bloch equations for nuclear magnetic resonance. Note that $d\rho_{ee}/dt = -d\rho_{gg}/dt$, in accordance with the requirement of a closed two-level system where the total population $\rho_{gg} + \rho_{ee} = 1$ is conserved. Figure 6.1 shows the results from numerical integration of Eqs. (6.13) for different values of δ using the same parameters as used in Fig. 2.1, and setting $\gamma = |\Omega|$. As the plots show, the Rabi oscillations of Fig. 2.1 are quickly damped by the spontaneous emission process and after a short time the system is in steady state at values given by Eq. (6.17).

Furthermore, it is explicitly assumed that the decay of the coherences and the decay of the excited state are described by a single parameter γ. This will always be the case in the systems discussed within the framework of this book. However,

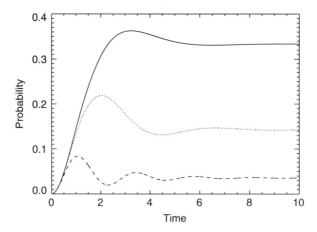

Figure 6.1 Plot of the probability ρ_{ee} for an atom to be in the excited state for $\gamma = |\Omega|$ and $\delta = 0$ (solid line), $\delta = |\Omega|$ (dotted line), and $\delta = 2.5|\Omega|$ (dashed line). Time is in units of $1/|\Omega|$.

in cases where interactions between atoms play a role, the decay of the coherences and the populations are described by different decay parameters, and in those cases parameters T_1 and T_2 are introduced to account for this difference. For details regarding this issue the reader is referred to several books on this topic [6, 46, 57].

6.4 Power broadening and saturation

The steady-state solutions of the OBE can be found by setting the time derivatives of Eq. (6.13) to zero and exploiting certain relationships among the $n^2 = 4$ real, independent parameters of ρ for a two-level system. The conservation of the population given by $\rho_{gg} + \rho_{ee} = 1$ eliminates one of these parameters, and two of the others are complex conjugates. Using the population difference $w \equiv \rho_{ee} - \rho_{gg}$ and the optical coherence $\rho_{eg} = \rho_{ge}^*$ in the OBE gives

$$\frac{d\rho_{eg}}{dt} = -\left(\frac{\gamma}{2} - i\delta\right)\rho_{eg} + \frac{iw\Omega}{2} \qquad (6.14a)$$

and

$$\frac{dw}{dt} = -\gamma w + i\left(\Omega^* \rho_{eg} - \Omega \rho_{eg}^*\right) - \gamma. \qquad (6.14b)$$

The steady-state case has $d\rho_{eg}/dt = 0$ and $dw/dt = 0$, and the resulting equations can be solved for w and ρ_{eg}:

$$w = \frac{-1}{1+s} \qquad (6.15a)$$

and

$$\rho_{eg} = \frac{-i\Omega}{2(\gamma/2 - i\delta)(1+s)}. \tag{6.15b}$$

Here the saturation parameter s is given by

$$s \equiv \frac{|\Omega|^2}{2|(\gamma/2-i\delta)|^2} = \frac{|\Omega|^2/2}{\delta^2 + \gamma^2/4} = \frac{s_0}{1+(2\delta/\gamma)^2}, \tag{6.16a}$$

where the last step defines the on-resonance saturation parameter

$$s_0 \equiv 2|\Omega|^2/\gamma^2 = I/I_s \tag{6.16b}$$

with the saturation intensity given by

$$I_s \equiv \pi hc/3\lambda^3 \tau. \tag{6.16c}$$

For the case of a low saturation parameter, $s \ll 1$, the population is mostly in the ground state ($w = -1$), whereas in the case of high s the population is equally distributed between the ground and excited state ($w = 0$). The population ρ_{ee} of the excited state is given by

$$\rho_{ee} = \frac{1}{2}(1+w) = \frac{s}{2(1+s)} = \frac{s_0/2}{1 + s_0 + (2\delta/\gamma)^2}, \tag{6.17}$$

and for $s \gg 1$, ρ_{ee} approaches $1/2$. Since the population in the excited state decays at a rate γ, and in steady state the excitation (or pump) rate and the decay rate are equal, the total scattering rate γ_p of light from the laser field is given by

$$\gamma_p = \gamma \rho_{ee} = \frac{s_0 \gamma/2}{1 + s_0 + (2\delta/\gamma)^2}. \tag{6.18}$$

At very high intensities, where $s_0 \gg 1$, γ_p saturates to $\gamma/2$. This equation can be rewritten as

$$\gamma_p = \left(\frac{s_0}{1+s_0}\right)\left(\frac{\gamma/2}{1+(2\delta/\gamma')^2}\right), \tag{6.19a}$$

where

$$\gamma' = \gamma\sqrt{1+s_0} \tag{6.19b}$$

is called the power-broadened linewidth of the transition. Because of saturation, the linewidth of the transition as observed in an experiment, where the absorption of light is detected while scanning its frequency, is broadened from its natural linewidth γ to its power-broadened value γ'.

Figure 6.2 shows a plot of γ_p as a function of the detuning δ for several values of the saturation parameter s_0. For large values of s_0 there is a significant power

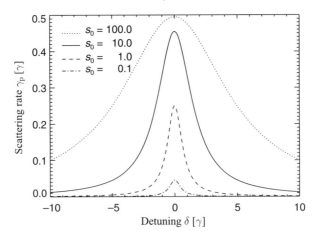

Figure 6.2 Excitation rate γ_p as a function of the detuning δ for several values of the saturation parameter s_0. Note that for $s_0 > 1$ the line profiles start to broaden substantially from power broadening. (Figure from Ref. [1].)

broadening of the spectral profile, which is a direct consequence of the fact that for large s_0, the absorption continues to increase with increasing intensity in the wings, whereas in the center, half of the atoms are already in the excited state. The absorption in the center of the profile is therefore saturated, whereas in the wings it is not.

Note that other line-broadening mechanisms, such as the Doppler effect, pressure broadening, and others, have been left out of the present discussion. However, they might also play a significant role under certain conditions, and their convolution with power broadening has to be considered carefully because of the different line shapes.

The scattering of light from a laser beam results in intensity loss when the beam travels through a sample of resonant atoms. The amount of scattered power per unit of volume is given by $\hbar \omega \gamma_p n$, where n is the density of the atoms. Thus $dI/dz = -\hbar \omega \gamma_p n$ for a laser beam of intensity I traveling in the z-direction. For low-intensity light tuned near the atomic resonance, the scattering rate is given by $\gamma_p \approx s_0 \gamma/2$, so the absorption rate is

$$\frac{dI}{dz} = -\sigma_{eg} n I, \tag{6.20a}$$

where the cross-section σ_{eg} for scattering light out of the beam on resonance is given by

$$\sigma_{eg} = \frac{\hbar \omega \gamma}{2 I_s} = \frac{3\lambda^2}{2\pi}. \tag{6.20b}$$

Exercises

Note that this cross-section is of the order of λ^2, which is much larger than the cross-section for atom–atom interactions, typically of the order of a_0^2.

Appendix

6.A The Liouville–von Neumann equation

The optical Bloch equations describe the time evolution of ρ. The derivation begins with the Schrödinger equation (Eq. (2.2)) and the expansion of its solutions as in Eq. (2.1) (here the oscillatory term is left out for clarity). The expression for two of these coefficients is written and one of them is conjugated to give

$$i\hbar \dot{c}_j = \sum_n c_n \mathcal{H}_{jn} \quad \text{and} \quad -i\hbar \dot{c}_k^* = \sum_n c_n^* \mathcal{H}_{nk} \tag{6.21}$$

Multiplying the left equation by c_k^* and the right one by c_j and subtracting yields

$$i\hbar \left(\dot{c}_j c_k^* + c_j \dot{c}_k^* \right) = \sum_n \left(\mathcal{H}_{jn} c_n c_k^* - \mathcal{H}_{nk} c_j c_n^* \right). \tag{6.22}$$

The left side is clearly $i\hbar \, d\rho_{jk}/dt$ so that

$$i\hbar \frac{d\rho}{dt} = [\mathcal{H}, \rho], \tag{6.23}$$

which is the Liouville–von Neumann equation for the time derivative of ρ.

Exercises

6.1 The elements of the density matrix are $\rho_{ij} = c_i c_j^*$, but it is also important to have a form for the density operator. Show that this is the operator $\hat{\rho} = |\Psi\rangle \langle \Psi|$.

6.2 Show that the density matrix satifies Eq. (6.23), which has the opposite sign from the Heisenberg relation for observables. Comment on what this means. (Hint: see App. 6.A.)

6.3 Use the evolution equation for the amplitudes given by Eq. (2.6) to find the time evolution of the populations $d\rho_{gg}/dt$ and $d\rho_{ee}/dt$ and of the coherences $d\tilde{\rho}_{ge}/dt$ and $d\tilde{\rho}_{eg}/dt$. Notice that the time evolution of the coherences depend on the populations, whereas the time evolution of the population depend on the coherences. This has to be contrasted with the Einstein relations, where the ground and excited are coupled directly by the field.

6.4 For a two-level atom the 2×2 density matrix has two real diagonal components ρ_{11} and ρ_{22}, and two complex off-diagonal components ρ_{ge} and ρ_{eg}. Since

$\rho_{ge} = \rho_{eg}^*$ there are actually four free parameters. Show that the constraints of normalization and $\rho^2 = \rho$ for a pure state reduce these to two free parameters. Discuss the meaning of this result in terms of the dressed atom and the Bloch sphere.

6.5 There is no real boundary between saturate and unsaturated transitions, but it is convenient to set it as $s_0 = 1$. For this intensity, what is the relation between the Rabi frequency $|\Omega|$ and the spontaneous decay rate γ in a two-level atom? What does this imply about the rate of Rabi flopping? What fraction of the atoms are on average in the excited state?

6.6 For an intensity $I = I_s$, what is the intensity in number of photons per second and per square centimeter? The optical cross-section for atomic absorption can be readily calculated to be $\sigma_{eg} = 3\lambda^2/2\pi$. If the natural unit of area for the problem is σ and the natural unit of time τ, then what is the intensity in number of photons per σ and per τ?

6.7 A careful calculation for a two-level atom gives the saturation intensity for $s_0 = 1$ to be $I_s = \pi hc\gamma/3\lambda^3$. What is the saturation intensity in convenient units (for example mW/cm^2) for a typical atomic transition, say Na for the yellow line at $\lambda = 589$ nm and $\tau = 16$ ns? If you shine a 0.5 mW helium–neon laser on the wall, making a spot with a diameter of a few mm, how does the intensity compare with I_s? How about sunlight on a bright day?

6.8 The rotating wave approximation (RWA) introduced in Sect. 2.3.2 is implemented by writing the field as the sum of a complex exponential and its conjugate, and then dropping the counter-rotating component after the rotating frame transformation because it is far off resonance. It results in an error of the resonant frequency that can be estimated quite well by simply considering the light shift of this neglected component using Eq. (2.10). Estimate the magnitude of this Bloch–Siegert shift for light tuned close to resonance with saturation parameter $s_0 \sim 1$ and compare it with the power-broadened linewidth γ' of Eq. (6.19b).

6.9 The states of the quantized light field $|n\rangle$ are introduced in App. 5.B in analogy to those of the harmonic oscillator in App. 5.A. For a light field in state Ψ the density matrix is simply $\rho_{ij} = \langle i|\Psi\rangle\langle\Psi|j\rangle$. Clearly for a light field having $\Psi = |n\rangle$, $\rho = \rho^2$ so it is a pure state. When light interacts with an atom the total wavefunction Ψ must include the atomic state so the density matrix for the light alone can be found only by taking the trace over the atomic states.
(a) Write the wavefunction for the system of a two-level atom plus light field. In this case the only possible states for the light field are $|n\rangle$ and $|n-1\rangle$.
(b) Write the density matrix for this system.

(c) Trace over the states of a two-level atom $|g\rangle$ and $|e\rangle$ to find the density matrix of the light field alone. Calculate ρ^2 and show that it is not equal to ρ, meaning that the light field is in a mixed state as a result of interacting with the atom.

6.10 The definitions of Eq. (2.13) clearly connect the components of the Bloch vector $\vec{R}$ with the elements of the density matrix given in Eq. (6.3). But the terms with γ in Eq. (6.13) result in $|\vec{R}| < 1$ in contrast to the statement in Sec. 2.5 where there is no damping. What is the meaning of this? Start with the w component of $\vec{R}$ and explain the meaning of $w = 0$ (hint: see Fig. 2.6a). When $w = 0$ in Chap. 2 it is clear that $|u|^2 + |v|^2 = 1$ so that $|\vec{R}| = 1$, but in Chap. 6 this is not so. Describe the state of the atom in such a case.

Part II

Internal structure

7
The hydrogen atom

7.1 Introduction

In 1885 Balmer discovered a simple arithmetic relation among the wavelengths of the spectral lines of hydrogen that led to the Rydberg formula. Balmer's formula was later to be found consistent with the Bohr model of the hydrogen atom, and for this reason hydrogen has served as the paradigm for the study of all atoms. The Bohr model serves to describe the energy of electrons in an atom, and also for the study of atoms in external fields, both optical and dc.

It can easily be shown that the behavior of electrons in the atom should be described quantum mechanically, and thus that the Bohr model using semi-classical arguments cannot be definitive. One of the triumphs of the establishment of quantum mechanics in the 1920s was that the Schrödinger equation for hydrogen could be solved exactly. Since the outcome of the theory corresponded completely with the experimental results known at that time, the quantum mechanical description of the hydrogen atom served as one of the first proofs of the validity of quantum mechanics.

The quantum mechanical model of the hydrogen atom is discussed in many textbooks about both quantum mechanics and atomic physics. It is discussed in this book because it serves as the basis for much of the remainder of the book, both in techniques and in notation. Furthermore, hydrogen is the only element (apart from its isotopes and single-electron ions) that can be solved exactly. More important, its solution will be discussed more in terms that are relevant for atomic physics, and not so much in terms of quantum mechanical aspects that are more mathematically oriented.

The system of units used in this book is the internationally accepted SI system. Atomic physics calculations are sometimes very cumbersome in this system, particularly for the internal structure, but for consistency SI units are used in this part as well. This usage is especially important for experimentalists so that formulas can be evaluated in terms of laboratory quantities. The atomic units can be found in App. B.1 at the end of this book.

7.2 The Hamiltonian of hydrogen

The Hamiltonian for the motion of the electron in hydrogen is given by

$$\mathcal{H} = -\frac{\hbar^2}{2\mu}\nabla^2 + V(r), \qquad (7.1)$$

where the electron mass m has been replaced by the reduced mass μ (see App. 7.A). The potential of the electron in the field of the nucleus depends only on the distance between the two particles and is given by the Coulomb interaction:

$$V(r) = -\frac{Ze^2}{4\pi\varepsilon_0 r}, \qquad (7.2)$$

with Ze the charge of the nucleus. For hydrogen, $Z = 1$.

The Schrödinger equation $\mathcal{H}\psi = E\psi$ that results from the Hamiltonian of Eq. (7.1) presents a calculus problem because $\vec{\nabla}$ comes from the momentum operator $p_x = -i\hbar\,\partial/\partial x$ (as well as the p_y and p_z terms), and as such is in Cartesian coordinates. But the Coulomb potential of Eq. (7.2) is radial, where $r = \sqrt{x^2 + y^2 + z^2}$, and is very difficult to deal with algebraically. Thus it is quite common to proceed by separating the Schrödinger equation into an angular part and a radial part, and solving the resulting equations. Appendix 7.B discusses this problem in the two most common coordinate systems in use for the hydrogen atom, the familiar spherical coordinates and also parabolic coordinates. Here a slightly different and more physical approach will be adopted.

The interaction potential of Eq. (7.2) depends only on the coordinate r, therefore constituting a central force problem, and so it presents a very intuitive physical reason to use spherical coordinates. For such a system (with no external forces), both the energy and angular momentum are conserved quantities, and classically one can parameterize the solutions in terms of these conserved quantities.

The Ehrenfest theorem of quantum mechanics shows that the expectation value of each of the operators that commute with the Hamiltonian do not change in time and thus are also conserved quantities (see App. 7.C). Moreover, it is shown there that commuting operators share the same set of eigenfunctions. Since the angular momentum operator commutes with the Hamiltonian for the hydrogen atom, one can first search for eigenfunctions of the angular momentum operator before looking for solutions of the full Hamiltonian. This simplifies the problem considerably, since the angular momentum $\vec{\ell}$ depends only on the angles θ and ϕ, whereas the Hamiltonian also depends on the radial distance r.

The Schrödinger equation for hydrogenic atoms in spherical coordinates is

$$\mathcal{H}\psi(r,\theta,\phi) = \left[\frac{-\hbar^2}{2\mu r^2}\frac{\partial}{\partial r}\left(r^2\frac{\partial}{\partial r}\right) - \frac{\vec{\ell}^2}{2\mu r^2} + V(r)\right]\psi(r,\theta,\phi) \qquad (7.3)$$

where the angular momentum operator $\vec{\ell}^2$ is defined in Eq. (7.30c). It would be convenient to find another operator that also commutes with the Hamiltonian, and in fact, each component of the angular momentum operator $\vec{\ell}$ commutes with $\mathcal{H}$. Thus the expectation value of each component $\ell_{x,y,z}$ is also a conserved quantity. However, these components do not commute with each other and therefore do not have a common set of eigenvalues. Nevertheless, one of them could be chosen, and usually it is ℓ_z. In spherical coordinates ℓ_z is given by

$$\ell_z = -i\hbar\frac{\partial}{\partial\phi}. \qquad (7.4)$$

There is no physical reason to choose the z-direction as a preferred direction, since there is no preferred direction in the hydrogenic problem. However, in spherical coordinates it leads to mathematically simple relations.

The Schrödinger equation is solved by first finding the eigenfunctions for ℓ_z and $\vec{\ell}^2$ and using them to obtain an ordinary differential equation that depends only on the radial coordinate r. Then this equation is solved by putting physically necessary conditions on the asymptotic solutions. This leads to a discretization of the acceptable solutions, which correspond to the solutions already found in Sec. 1.4.

7.3 Solving the angular part

As discussed in the previous section, $\vec{\ell}^2$ and ℓ_z commute and they share a common set of eigenfunctions. Since ℓ_z depends only on one coordinate (ϕ), its eigenfunctions $\Phi_m(\phi)$ can easily be found to be

$$\Phi_m(\phi) = \frac{1}{\sqrt{2\pi}}\exp(im\phi), \qquad (7.5)$$

with eigenvalues m. Here m is an integer since rotation of the system over 2π should lead to the same wavefunction.

The eigenfunctions of $\vec{\ell}^2$ must be products of the functions $\Phi_m(\phi)$ with some function of θ, and they are written as $Y_{\ell m}(\theta,\phi)$ where the ℓ will come from the θ-dependent part of $\vec{\ell}^2$. These eigenfunctions $Y_{\ell m}(\theta,\phi)$ of $\vec{\ell}^2$ must obey (see Eq. (7.30c))

$$-\hbar^2\left(\frac{\partial}{\sin\theta\,\partial\theta}\left(\sin\theta\frac{\partial}{\partial\theta}\right) + \frac{1}{\sin\theta}\frac{\partial^2}{\partial\phi^2}\right)Y_{\ell m}(\theta,\phi) = C_\ell\hbar^2 Y_{\ell m}(\theta,\phi), \qquad (7.6)$$

where $C_\ell\hbar^2$ is the eigenvalue of the operator $\vec{\ell}^2$.

ℓ	m	$\sqrt{4\pi}Y_{\ell\,\pm m}$
0	0	1
1	0	$\sqrt{3}\cos\theta$
1	± 1	$\mp\sqrt{3/2}\,\sin\theta\,e^{\pm i\phi}$
2	0	$\sqrt{5/4}\,(3\cos^2\theta - 1)$
2	± 1	$\mp\sqrt{15/2}\,\cos\theta\sin\theta\,e^{\pm i\phi}$
2	± 2	$\sqrt{15/8}\,\sin^2\theta\,e^{\pm 2i\phi}$
3	0	$\sqrt{7/4}\,(5\cos^3\theta - 3\cos\theta)$
3	± 1	$\mp\sqrt{21/16}\,(5\cos^2\theta - 1)\sin\theta\,e^{\pm i\phi}$
3	± 2	$\sqrt{105/8}\,\cos\theta\sin^2\theta\,e^{\pm 2i\phi}$
3	± 3	$\mp\sqrt{35/16}\,\sin^3\theta\,e^{\pm 3i\phi}$

Table 7.1 The lowest spherical harmonics $Y_{\ell m}(\theta, \phi)$ for $\ell \leq 3$ ($|m| \leq \ell$).

This differential equation was studied by Legendre more than two centuries ago, and its solutions are the associated Legendre polynomials $P_\ell^m(\cos\theta)$, whose eigenvalues are $C_\ell = \ell(\ell + 1)$. The common set of normalized eigenfunctions of ℓ_z and $\vec{\ell}^2$ are given by the spherical harmonic functions $Y_{\ell m}(\theta, \phi)$:

$$Y_{\ell m}(\theta, \phi) = (-1)^m \left[\frac{(2\ell + 1)(\ell - m)}{4\pi(\ell + m)!}\right]^{1/2} P_\ell^m(\cos\theta)e^{im\phi} \quad (m \geq 0) \tag{7.7a}$$

and

$$Y_{\ell m}(\theta, \phi) = (-1)^m Y_{\ell,-m}^*(\theta, \phi). \quad (m < 0) \tag{7.7b}$$

The lowest ten spherical harmonic functions are given in Tab. 7.1 and the lowest six functions are plotted in Fig. 7.1.

Summarizing these results, one has found simultaneous eigenfunctions $Y_{\ell m}(\theta, \phi)$ of ℓ_z and $\vec{\ell}^2$ with eigenvalues $m\hbar$ and $\ell(\ell + 1)\hbar^2$ respectively. These eigenfunctions will be used to help solve the radial Schrödinger equation for hydrogenic systems.

7.4 Solving the radial part

Since the angular momentum operators commute with the Hamiltonian, the solution of the Schrödinger equation can now be split in two parts: one depending only on the radial coordinate and one depending only on the angular coordinates,

$$\psi_{E\ell m} = R_{E\ell}(r)Y_{\ell m}(\theta, \phi), \tag{7.8}$$

where the solutions $Y_{\ell m}(\theta, \phi)$ of the angular part discussed above are used. The subscript E refers to the energy whose quantization is yet to be proven. Note that

7.4 Solving the radial part

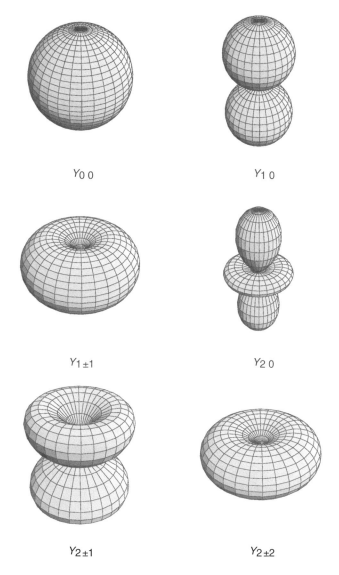

Figure 7.1 The lowest spherical harmonics $Y_{\ell m}(\theta, \phi)$ for $\ell \leq 2$ ($|m| \leq \ell$).

the Schrödinger equation depends on $\vec{\ell}^{\,2}$ but not on m, and by using the spherical harmonics as a solution of the angular part, the radial part of the Schrödinger equation does not depend on m.

With the angular part relegated to the eigenvalues of $Y_{\ell m}$, namely $\ell(\ell+1)\hbar^2$, the Schrödinger equation is changed from a partial to an ordinary differential equation that depends on only r. It is useful to write the solution as $u_{E\ell}(r) \equiv rR_{E\ell}(r)$, so that the resulting radial equation becomes

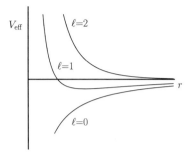

Figure 7.2 Effective potential for the radial wavefunction.

$$\frac{d^2 u_{E\ell}(r)}{dr^2} + \frac{2\mu}{\hbar^2}[E - V_{\text{eff}}(r)]u_{E\ell}(r) = 0, \tag{7.9a}$$

with

$$V_{\text{eff}}(r) \equiv -\frac{Ze^2}{4\pi\varepsilon_0 r} + \frac{\ell(\ell+1)\hbar^2}{2\mu r^2}, \tag{7.9b}$$

as plotted in Fig. 7.2. Now the angular part of the Schrödinger equation is completely embodied in the second term of V_{eff}, and it resembles a centrifugal barrier for $\ell \neq 0$. The probability of finding the electron in a spherical shell at radius r is proportional to $|rR_{E\ell}(r)|^2 = |u_{E\ell}(r)|^2$ (shell volume scales with r^2), and for this probability to be finite everywhere, $u_{E\ell}(r)$ must also always be finite.

7.4.1 Asymptotic properties

For large r the effective potential of Eq. (7.9b) can be neglected with respect to the kinetic energy term, and Eq. (7.9a) becomes

$$\frac{d^2 u_{E\ell}(r)}{dr^2} + k^2 u_{E\ell}(r) = 0 \qquad k = \sqrt{\frac{2\mu E}{\hbar^2}}. \tag{7.10}$$

For $E > 0$ the oscillatory solutions e^{ikr} describe the motion of the unbound electron in the field of the nucleus. For $E < 0$, the electron is bound to form an atom, and the resulting solutions are $u_{E\ell}(r) \propto \exp(\pm|k|r)$, which are exponentially growing or damped, depending on the sign in front of $|k|$. The solution with the (+) sign grows exponentially for large r, and since it cannot be normalized, it yields a physically unacceptable solution. The only acceptable solution is the one with the (−) sign.

The first step therefore forces the long-range behavior of $u_{E\ell}(r)$ to be a decreasing exponential, and substituting $u_{E\ell}(\rho) = e^{-\rho/2}f(\rho)$ with $\rho \equiv 2|k|r$ and $\lambda \equiv (Ze^2/4\pi\varepsilon_0\hbar)\sqrt{-\mu/2E}$ leads to

7.4 Solving the radial part

$$\left[\frac{d^2}{d\rho^2} - \frac{d}{d\rho} - \frac{\ell(\ell+1)}{\rho^2} + \frac{\lambda}{\rho}\right] f(\rho) = 0. \tag{7.11}$$

When $f(\rho) = \rho^\kappa g(\rho)$ (κ is integer) is substituted into Eq. (7.11), the resulting equation has solutions for only $\kappa(\kappa - 1) = \ell(\ell + 1)$, which means $\kappa = \ell + 1$ or $\kappa = -\ell$. The first case leads to physically acceptable solutions at short range for hydrogen that are called the regular solutions, whereas the second case leads to unacceptable solutions for hydrogen that are called irregular solutions. The resulting equation was studied by Laguerre in the nineteenth century, and its solutions can be found from a power series expansion for $g(\rho) = \sum_{j=0}^{\infty} c_j \rho^j$. For the powers of ρ one finds a recurrence relation for the coefficients c_j:

$$c_j = \frac{j + \ell - \lambda}{j(j + 2\ell + 1)} c_{j-1}. \tag{7.12}$$

This solution still diverges in the limit of $j \to \infty$ unless the series is truncated after a finite number of terms. Equation (7.12) shows that if any c_j is 0, all the coefficients following c_j are also zero so the series will terminate if λ is an integer. If the integer is denoted by n, the series terminates for $n > \ell + 1$, resulting in the Laguerre polynomials. They admit only energies given by

$$E_n = -\frac{R_\infty Z^2}{n^2} \frac{\mu}{m} = -\frac{mc^2 \alpha^2 Z^2}{2n^2} \frac{\mu}{m}, \tag{7.13}$$

where the Rydberg constant R_∞ is given by

$$R_\infty \equiv \frac{1}{2} m \left(\frac{e^2}{4\pi\varepsilon_0 \hbar}\right)^2 = \frac{1}{2} mc^2 \alpha^2 \tag{7.14}$$

and α is the fine-structure constant (see App. 8.A). Figure 7.3 shows the lowest four energies.

This is very reminiscent of the Bohr formula of Sec. 1.4. Note that E_n is independent of ℓ so states of the same n but different ℓ are degenerate. However, the kinetic energy can be split into a radial part T_r and an angular part T_ℓ as given by the first and second terms on the right-hand side of Eq. (7.3), respectively. Using the radial solutions for the hydrogenic wavefunctions (see Sec. 7.4.2) the expectation values of these terms become

$$\langle T_r \rangle = \left(n_r + \frac{\ell + 1}{2\ell + 1}\right) \frac{R_\infty}{n^3} \quad \text{and} \quad \langle T_\ell \rangle = \frac{2\ell(\ell + 1)}{(2\ell + 1)} \frac{R_\infty}{n^3}, \tag{7.15}$$

where $n_r = n - \ell - 1$ is the number of radial nodes in the wavefunction. Note that in the Bohr model of the atom the electron moves in a circular orbit and thus has no kinetic energy in the radial motion. Although the two terms of Eq. (7.15) both depend on n and ℓ, their sum depends only on n and not on ℓ. This is sometimes

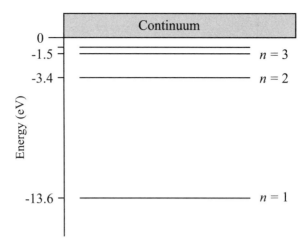

Figure 7.3 The energy diagram for the lowest states of hydrogen. Since the energy is proportional to $1/n^2$, the splitting of higher-lying states close to the ionization limit becomes small.

called an accidental degeneracy and holds only for a perfect $1/r$ potential (Kepler problem). In Exercise 7.13 this accidental degeneracy is explored in more detail.

From the Bohr postulate of Sec. 1.4.2 it follows that the difference between two discrete energies is

$$\Delta E_{12} = R_\infty Z^2 \frac{\mu}{m} \left(\frac{1}{n_1^2} - \frac{1}{n_2^2} \right), \tag{7.16}$$

This equation is very reminiscent of the Balmer formula of Eq. (1.8) if one adopts Planck's hypothesis $E = h\nu$ and uses $n_1 = 2$.

7.4.2 The radial solutions

Summarizing, the eigenfunctions of the radial equation are given by

$$R_{n\ell}(r) = -\left(\frac{2Z}{na_\mu} \right)^{3/2} N_{n\ell} \, e^{-\rho/2} \, \rho^\ell \, L_{n+\ell}^{2\ell+1}(\rho) \tag{7.17}$$

with

$$\rho \equiv \frac{2Zr}{na_\mu} \quad \text{and} \quad a_\mu = \frac{4\pi\varepsilon_0 \hbar^2}{\mu e^2} = \frac{m}{\mu} a_0 \tag{7.18}$$

Here $L_k^q(\rho)$ are the associated Laguerre polynomials and $N_{n\ell}$ is a normalization factor. For hydrogen, the first six eigenfunctions are given in Tab. 7.2 and shown in Fig. 7.4. On the right side of the figure the probability density $r^2 |R_{n\ell}(r)|^2$ is plotted and the functions are normalized such that the total probability is equal to 1.

7.4 Solving the radial part

n	ℓ	$R_{n\ell}(r) \times (na_0)^{3/2}$
1	0	$2\exp(-r/a_0)$
2	0	$2\left(1 - \frac{r}{2a_0}\right)\exp(-r/2a_0)$
2	1	$\frac{1}{3}\sqrt{3}\,(r/a_0)\exp(-r/2a_0)$
3	0	$2\left(1 - \frac{2r}{3a_0} + \frac{2r^2}{27a_0^2}\right)\exp(-r/3a_0)$
3	1	$\frac{4}{9}\sqrt{2}\left(1 - \frac{r}{6a_0}\right)(r/a_0)\exp(-r/3a_0)$
3	2	$\frac{4}{270}\sqrt{10}\,(r/a_0)^2\exp(-r/3a_0)$

Table 7.2 The lowest radial wavefunctions for hydrogen with $n \leq 3$ and $\ell \leq n - 1$.

A number of important properties can be derived from the solutions.

- For $\ell = 0$ states, the wavefunction is finite at $r = 0$ and thus the electron has a finite probability to be in the nucleus. For $\ell \neq 0$ the radial wavefunction at short range is proportional to r^ℓ and the probability of finding the electron in the nucleus becomes smaller for increasing ℓ. The square of the wavefunction in the origin is given by

$$|\Psi_{n\ell m}(0)|^2 = \frac{Z^3}{\pi n^3 a_0^3}\delta_{\ell 0}, \tag{7.19}$$

which is used in Sec. 8.4 to calculate the "Darwin term" for hydrogen.
- The Laguerre polynomials are of the order $n_r = n - \ell - 1$ and thus the number of nodes in the radial wavefunction becomes n_r as shown in Fig. 7.4. For $\ell = n - 1$ the radial wavefunction has no nodes and these "circular states" resemble the descriptions of Bohr. For these states the maximum probability of finding the electron at a certain distance as determined from $d|rR_{E\ell}(r)|^2/dr = 0$ is given by $r = n^2 a_0/Z$.
- The number of nodes in the angular wavefunctions is ℓ.
- The parity of the solutions is determined by reflecting the wavefunction through the origin: $\vec{r} \to -\vec{r}$. It can be shown that $\Psi_{n\ell m}(-\vec{r}) = (-1)^\ell \Psi_{n\ell m}(\vec{r})$ and thus states with ℓ even have even parity and states with ℓ odd have odd parity.
- The energy of the state depends only on the principal quantum number n and thus the degeneracy of the states is proportional to $2n^2$, where each state n has a $(n-1)$ degeneracy from ℓ and each ℓ state has a $(2\ell + 1)$ degeneracy from m. The factor 2 stems from the spin of the electron.

For historical reasons it is common in atomic physics to denote the orbital angular quantum number ℓ with special characters. These characters are derived from the appearance of spectroscopic lines in hydrogenic atoms and they are given by

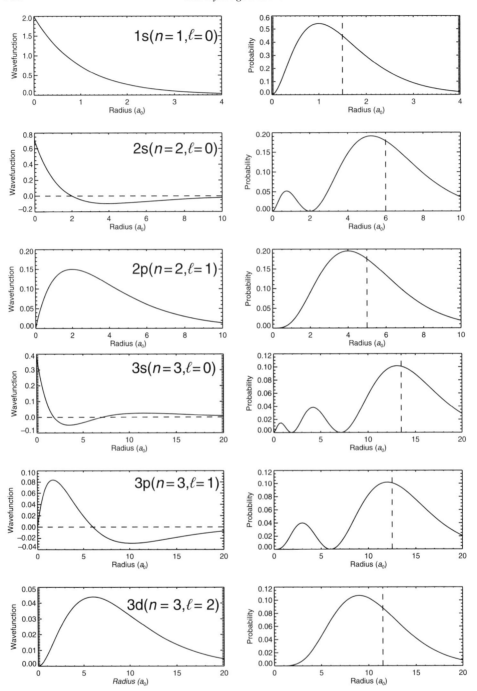

Figure 7.4 The lowest radial wavefunctions for hydrogen with $n \leq 3$ and $\ell \leq n - 1$. The plots on the left show the wavefunctions $R_{n\ell}(r)$, whereas the plots on the right show the probability distribution $r^2 |R_{n\ell}(r)|^2$. The dashed line indicates the average radius. Note the change of scale of the radial distance. The number of radial nodes of the wavefunction is $n_r = n - \ell - 1$.

7.5 The scale of atoms

ℓ	Notation	Description
0	s	sharp
1	p	principal
2	d	diffuse
3	f	fundamental

and continuing in alphabetical order for higher ℓ.

The hydrogenic wavefunctions will be used often in the description of all the other atoms in the periodic table. In most cases what is needed is the expectation value of an operator that is proportional to some power of the radius, and not the wavefunction itself. Table 7.6 shows the radial matrix elements for various powers of the radius. When $n \gg \ell$ these matrix elements reduce to simple forms that can be applied for the description of Rydberg states (see Chap. 13).

7.5 The scale of atoms

Solving the Schrödinger equation for the hydrogen atom in this chapter has a mathematical nature, but the solution also sets the scale for the atomic properties discussed in the following several chapters. Since this scale is very different from human scales, atomic physicists often perform their calculations in units that differ from the usual SI units that are used throughout this book. These "atomic units" are important because they provide a scale for the material here, so a summary of them can be found in App. B.1.

Since most processes in atomic physics deal with electronic interactions, it is natural to select the electronic mass m and charge e as the atomic units for these properties. Also, the natural scale for atomic angular momentum is Planck's constant divided by 2π, denoted $\hbar$. The discretization of electronic orbital size provides a natural length scale for atomic units, namely the Bohr radius $a_0 \equiv \hbar/m\alpha c \approx 5 \times 10^{-11}$ m for the electron in the lowest state of hydrogen. Note that a_0 is much smaller than the wavelength of visible light so individual atoms cannot be resolved using conventional microscopy. Such resolution requires more modern techniques such as scanning tunneling microscopy (STM) or atomic force microscopy (AFM). On the other hand, the Bohr radius is much larger than the classical radius of the electron $r_0 = \alpha^2 a_0$, which is found by setting the rest energy mc^2 of the electron equal to the potential of a charge e at r_0.

In addition, a_0 determines the potential energy from the Coulomb interaction, $e^2/(4\pi\varepsilon_0 a_0) = mc^2\alpha^2 \approx 27$ eV. The electron kinetic energy is proportional to the potential energy and thus the velocity v_0 of the electron in the lowest state of hydrogen is given by $v_0 = \alpha c$, and for higher orbitals the velocity becomes even smaller. Since it is two orders of magnitude smaller than the speed of light, relativistic effects do not have a dominant role in atomic physics.

The length a_0 and velocity v_0 set a scale for time $t_0 = a_0/v_0 < 10^{-15}$ s, and frequency $f_0 = v_0/(2\pi a_0)$. This timescale is so small that electronic processes are not easily resolvable in time using pulsed lasers, where the pulse duration is orders of magnitude larger than t_0. Thus atomic physicists have to use other methods to observe the electronic motion in real time.

Finally, a_0 sets the scale for the interactions between atoms since the interaction becomes strong when their electronic clouds start to overlap. The length of a chemical bond and the distance between atoms or molecules in a crystal is of the order of a_0. This size also sets the scale for atom–atom collision cross-sections, which are of the order of a_0^2. For the interaction between light and atoms it was shown in Chap. 6 that the cross-section is of the order of λ^2, determined by the wavelength λ of the light, and not the size of the electronic orbital a_0. Thus the cross-section for atom–light interactions is eight orders of magnitude larger than the atom–atom interactions. If the cross-section of the atom–light interaction were of the order of a_0^2, lasers would need a much higher intensity and all the virtues of atom–light interactions described in Part I would become impractical.

7.6 Optical transitions in hydrogen

7.6.1 Introduction

The optical transitions considered in Chap. 2 were restricted to the particularly simple case of a two-level atom, and these transitions can be described by a single Rabi frequency. Real atoms have more than two levels that can be coupled by the optical field, and furthermore, the relative strengths of their multiple transitions depend on the orientation of the atomic dipole moment with respect to the polarization of the light. The single Rabi frequency of Chap. 2 that describes the coupling is given by $\hbar\Omega = -\mu_{eg}\mathcal{E}_0$ (see Eq. (3.8)), where

$$\mu_{eg} \equiv -e\langle e|\hat{\varepsilon}\cdot\vec{r}|g\rangle \tag{7.20}$$

and $\hat{\varepsilon}$ represents the polarization of the light. The value of the dipole moment of Eq. (7.20) depends on the wavefunctions of the ground and excited states, and is generally complicated to calculate. It is often convenient to introduce the spherical unit vectors [58] given by Eq. (3.20) and to expand the polarization vector $\hat{\varepsilon}$ in terms of these vectors. In this notation the components of the dipole moment can be written as

7.6 Optical transitions in hydrogen

$$\hat{\varepsilon}_q \cdot \vec{r} = \sqrt{\frac{4\pi}{3}} r Y_{1q}(\theta, \phi) \qquad (7.21)$$

where the Y_{1q}s represent the simplest of the spherical harmonic functions, $q = 0, \pm 1$.

The matrix element of Eq. (7.20) can be broken up into two parts, one depending on all the various quantum numbers of the coupled states and the other completely independent of m, the projection of $\vec{\ell}$ on the quantization axis. This separation is embodied in the well-known Wigner–Eckart theorem discussed in App. C.4 at the end of the book. Here, the treatment will be somewhat different since this section treats the simplest case, namely, that fine and hyperfine structure are absent. The more general case will be treated in Sec. 10.5. Thus the hydrogenic wavefunctions for the ground and excited state can be used:

$$|g\rangle = |n\ell m\rangle = R_{n\ell}(r) Y_{\ell m}(\theta, \phi) \qquad (7.22a)$$

and

$$|e\rangle = |n'\ell'm'\rangle = R_{n'\ell'}(r) Y_{\ell'm'}(\theta, \phi). \qquad (7.22b)$$

Substitution of Eqs. (7.21) and (7.22) into Eq. (7.20) leads to

$$\mu_{eg} = -e \langle n'\ell'm' | \hat{\varepsilon} \cdot \vec{r} | n\ell m\rangle \qquad (7.23)$$

$$= -e \langle n'\ell' ||r|| n\ell\rangle \left\langle \ell'm' \left| \sqrt{\frac{4\pi}{3}} Y_{1q} \right| \ell m \right\rangle \equiv -e \mathcal{R}_{n'\ell',n\ell} \mathcal{A}^q_{\ell'm',\ell m}.$$

The following sections first treat the radial or physical part $\mathcal{R}_{n'\ell',n\ell}$, also known as the reduced or double-bar matrix element, and then the angular or geometric part $\mathcal{A}^q_{\ell'm',\ell m}$.

7.6.2 Radial part of the dipole matrix element

The radial part of the matrix element is given by

$$\mathcal{R}_{n'\ell',n\ell} = \langle R_{n'\ell'}(r) | r | R_{n\ell}(r)\rangle = \int_0^\infty r^2 dr \, R_{n'\ell'}(r) r R_{n\ell}(r), \qquad (7.24)$$

with $R_{n\ell}$ the radial wavefunction of the state. Here the $r^2 dr$ term in the integral originates from the radial part of $d\vec{r}$.

For the hydrogen atom the eigenfunctions of the bound states are known (see Eq. (7.17)) and the radial matrix elements can be calculated exactly [59]. For instance, for the first optical allowed transition in H, the 1s → 2p transition, the

(n,ℓ)	$(2,\ell+1)$	$(3,\ell+1)$	$(4,\ell+1)$	$(5,\ell+1)$	τ (ns)
1s	1.2902	0.5166	0.3045	0.2087	∞
2s	−5.1961	3.0648	1.2822	0.7739	∞^a
2p	−	4.7479	1.7097	0.9750	1.6
3s	0.9384	−12.7279	5.4693	2.2596	15.9
3p	−	−10.0623	7.5654	2.9683	5.4
3d	−	−	10.2303	3.3186	15.6
4s	0.3823	2.4435	−23.2379	8.5178	23.2
4p	−	1.3022	−20.7846	11.0389	12.4
4d	−	−	−15.8745	14.0652	36.5
4f	−	−	−	17.7206	73.0
5s	0.2280	0.9696	4.6002	−36.7423	36.1
5p	−	0.4827	3.0453	−34.3693	24.0
5d	−	−	1.6613	−30.0000	70.0
5f	−	−	−	−22.5000	140.0

Table 7.3 Radial matrix elements $\mathcal{R}_{n\ell,n'\ell+1}$ in units of a_0 for hydrogen for a transition $(n,\ell) \to (n',\ell+1)$. Note that $\mathcal{R}_{n\ell,n'\ell'}$ is symmetric with respect to interchange of n and ℓ, i.e. $\mathcal{R}_{n\ell,n'\ell'} = \mathcal{R}_{n'\ell',n\ell}$. In the last column the lifetime of the hydrogen states (n,ℓ) are indicated.
[a] See Sec. 7.6.4

radial wavefunctions involved are $R_{1s}(r)$ and $R_{2p}(r)$ as given in Tab. 7.2. Thus the integral becomes

$$\mathcal{R}_{2p,1s} = \int_0^\infty R_{2p}(r)\, r\, R_{1s}(r)\, r^2 dr = 2^7\sqrt{6}a_0/3^5 \approx 1.290\, a_0. \tag{7.25}$$

For other transitions in hydrogen, similar integrals can be evaluated. Substituting Eq. (7.17) into Eq. (7.24) and integrating over r with the help of standard integrals, the matrix element for any transition can be found. The results for $n \leq 5$ are given in Tab. 7.3.

7.6.3 Angular part of the dipole matrix element

The angular part $\mathcal{A}^q_{\ell'm',\ell m}$ of the dipole moment for atoms without fine and hyperfine structure ($s = 0$ and $I = 0$, see Chaps. 8 and 9) is defined by Eq. (7.23):

$$\mathcal{A}^q_{\ell'm',\ell m} = \sqrt{\frac{4\pi}{3}}\left\langle Y_{\ell'm'}\big|Y_{1q}\big|Y_{\ell m}\right\rangle = \sqrt{\frac{4\pi}{3}}\int \sin\theta d\theta d\phi\, Y_{\ell'm'}(\theta,\phi)\, Y_{1q}(\theta,\phi)\, Y_{\ell m}(\theta,\phi), \tag{7.26}$$

where $q = 0, \pm 1$ represents the polarization of the light and the integration is over the solid angle 4π. The integral can be expressed in terms of the Clebsch–Gordan

7.6 Optical transitions in hydrogen

ℓ'	$q = \pm 1$	$q = 0$
$\ell + 1$	$\left[\dfrac{(\ell \pm m')(\ell \pm m' + 1)}{2(2\ell + 1)(2\ell + 3)}\right]^{1/2}$	$\left[\dfrac{(\ell - m' + 1)(\ell + m' + 1)}{(2\ell + 1)(2\ell + 3)}\right]^{1/2}$
$\ell - 1$	$-\left[\dfrac{(\ell \mp m')(\ell \mp m' + 1)}{2(2\ell - 1)(2\ell + 1)}\right]^{1/2}$	$\left[\dfrac{(\ell - m')(\ell + m')}{(2\ell - 1)(2\ell + 1)}\right]^{1/2}$

Table 7.4 The angular part $\mathcal{A}$ for optical transitions $(\ell, m) \to (\ell', m')$ with the polarization of the light indicated by q, with $q = 0$ for linear and $q = \pm 1$ for right- and left-circular polarized light. Because of the selection rules, $\ell' = \ell \pm 1$ and $m' = m + q$.

coefficients (see Eq. (C.23)). Expressed in a $3j$-symbol (see Eq. (C.10) in the end of book appendix) the result is

$$\mathcal{A}^q_{\ell'm',\ell m} = (-1)^{\ell'-m'} \sqrt{\max(\ell, \ell')} \begin{pmatrix} \ell' & 1 & \ell \\ -m' & q & m \end{pmatrix}. \tag{7.27}$$

The symmetry of the $3j$-symbols dictates that they are only non-zero when the sum of the entries in the bottom row is zero, which means $m + q = m'$. Thus circularly polarized light couples only states that differ in m by ± 1, whereas linearly polarized light couples only states that have equal values of m. This result is thus identical to the result obtained in Sec. 3.5.2. Table 7.4 shows tabulated the values of $\mathcal{A}^q_{\ell'm',\ell m}$ for optical transitions.

7.6.4 Lifetime of the states

The emission of light determines the lifetime of the states in hydrogen. As shown in Eq. (5.13), the decay rate of a state is proportional to the square of the dipole moment μ^2. For transitions in hydrogen, μ has been calculated in the previous section, where in Sec. 7.6.2 the radial part of μ has been calculated and in Sec. 7.6.3 the angular part. In Sec. 5.4 there is decay from a state to only one other state, but in hydrogen, decay from state n, ℓ, and m is possible to many other states with different n', ℓ', and m'. This requires the summation of the decay rates γ_i over different polarizations q of the light and $\ell' = \ell \pm 1$ and $n' < n$ of the electron. Note that all decay branches have to be treated independently such that the total decay rate is given by the sum of the individual pathways, $\gamma_{tot} = \sum_i \gamma_i$, where i identifies a decay pathway. The lifetime of the state is then given by the inverse of the rate, namely $\tau = 1/\gamma_{tot}$.

The sum of the squares of the angular part $\mathcal{A}$ over q yields $\max(\ell, \ell')/(2\ell + 1)$ for any state (ℓ, m). This shows that the rate is independent of the orientation m of the state, which is reasonable since there is no preferred direction in space in the absence of an external field. The lifetime is then obtained by squaring the radial part of μ as given in Tab. 7.3 and summing over all the pathways. In Tab. 7.3 the lifetimes for hydrogen states are shown. From the table it is clear that the lifetime increases for increasing ℓ and also increases for increasing n. Clearly, the lifetime of the ground state (1s) is infinite, since there are no lower states to decay to. The lifetime of the first excited S state (2s) is also infinite in this treatment, since an S state can only decay to a P state and there is no P state below the 2s state. However, as shown in Sec. 4.3.1 there is a second-order process that couples the 2s state to the 1s state and this yields a lifetime of the 2s state of $1/7$ s.

Appendices

7.A Center-of-mass motion

The equations of motion of any two-particle system can be reduced to a single equation by separating the motion into two parts: one for the center-of-mass (CM) motion and the other for the motion of one particle relative to the other. For an atom consisting of a nucleus with an electron the position of the CM is $M\vec{R} \equiv m\vec{r}_e + M_N \vec{R}_N$, where m, $\vec{r}_e$ and M_N, $\vec{R}_N$ are the masses and laboratory coordinates of the electron and nucleus, respectively. The relative position $\vec{r}$ of the electron with respect to the nucleus is given by $\vec{r} = \vec{r}_e - \vec{R}_N$, and the Coulomb interaction between the electron and nucleus depends only on its absolute value r.

The total kinetic energy T of the atom in the laboratory frame is

$$T = \tfrac{1}{2} m v_e^2 + \tfrac{1}{2} M_N V_N^2 = \tfrac{1}{2} \mu v^2 + \tfrac{1}{2} M V^2, \tag{7.28}$$

where v_e, V_N, and V are the laboratory velocities of the electron, the nucleus, and the CM, respectively, v is the velocity of the electron relative to the nucleus, and the reduced mass μ is defined as

$$\mu \equiv \frac{m M_N}{m + M_N} \approx m\left(1 - \frac{m}{M_N}\right). \tag{7.29}$$

The second term on the right side of Eq. (7.28) is the kinetic energy of the CM in the laboratory frame, whereas the first term is the relative motion of a particle in the CM frame with reduced mass μ. In the absence of external forces the center-of-mass moves at a constant velocity V. The ratio m/μ for several elements is given in Tab. 7.5. The ratio becomes closer to one for heavier elements, and the shift in the energy levels caused by the finite mass of the nucleus becomes less important for heavier nuclei.

When atoms have more than one electron, the same transformation can be made and the relative motion of each electron in the CM frame can be described by

7.B Coordinate systems

Element	m/μ
^{1}H	1.000 545
^{2}D	1.000 272
^{3}D	1.000 182
^{4}He	1.000 137
^{133}Cs	1.000 004

Table 7.5 Ratio between electron mass and reduced mass for several elements.

the motion of a particle with reduced mass μ. However, the transformation for the kinetic energy leads to additional terms $\vec{\nabla}_i \cdot \vec{\nabla}_j$ between electron i and j, which have to be taken into account for a proper description for atoms with more than one electron. This leads to the so-called mass polarization shift. In the case of the alkali-metal atoms (see Chap. 10), which have one valence electron outside of a core formed by the other electrons, this shift can be accounted for by adding the mass of the core electrons to the mass of the nucleus in Eq. (7.29) to obtain the reduced mass.

7.B Coordinate systems

7.B.1 Spherical coordinates

The relation between Cartesian and spherical coordinates is given by

$$x = r\sin\theta\cos\phi, \quad y = r\sin\theta\sin\phi, \quad z = r\cos\theta \tag{7.30a}$$

and the operator ∇^2 is

$$\nabla^2 = \frac{1}{r^2}\frac{\partial}{\partial r}\left(r^2\frac{\partial}{\partial r}\right) - \frac{\vec{\ell}^2}{\hbar^2 r^2} \tag{7.30b}$$

with the angular momentum operator $\vec{\ell}^2$ defined as

$$\vec{\ell}^2 = -\hbar^2\left[\frac{1}{\sin\theta}\frac{\partial}{\partial\theta}\left(\sin\theta\frac{\partial}{\partial\theta}\right) + \frac{1}{\sin^2\theta}\frac{\partial^2}{\partial\phi^2}\right]. \tag{7.30c}$$

7.B.2 Parabolic coordinates

In parabolic coordinates these relations are

$$\xi = r + z, \quad \eta = r - z, \quad \tan\phi = y/x \tag{7.31a}$$

or

$$x = \sqrt{\xi\eta}\,\cos\phi,\ y = \sqrt{\xi\eta}\,\sin\phi,\ z = \frac{1}{2}(\xi - \eta) \tag{7.31b}$$

and so ∇^2 is

$$\nabla^2 = \frac{4}{\eta + \xi}\left[\frac{\partial}{\partial \xi}\left(\xi\frac{\partial}{\partial \xi}\right) + \frac{\partial}{\partial \eta}\left(\eta\frac{\partial}{\partial \eta}\right)\right] + \frac{1}{\eta\xi}\frac{\partial^2}{\partial \phi^2}. \tag{7.31c}$$

Parabolic coordinates will play an important role in the study of atoms in uniform dc electric fields because the eigenfunctions in such fields have the same symmetry.

7.C Commuting operators

If two quantum mechanical operators $\mathcal{A}$ and $\mathcal{B}$ commute, then they share the same set of eigenfunctions. The proof of this is most easily done for the case when the eigenvalues of the operators are non-degenerate. When they are not, it can also be done (see Refs. [60–63]). The eigenvalues a_n and eigenfunctions ψ_n of operator $\mathcal{A}$ are given as:

$$\mathcal{A}\psi_n = a_n\psi_n. \tag{7.32}$$

and similarly for $\mathcal{B}$. Now write the commutator $[\mathcal{A}, \mathcal{B}] = 0$ on the wavefunctions ψ_n:

$$[\mathcal{A}, \mathcal{B}]\psi_n = \mathcal{A}\mathcal{B}\psi_n - \mathcal{B}\mathcal{A}\psi_n = 0. \tag{7.33}$$

Since ψ_n is an eigenfunction of $\mathcal{A}$ with eigenvalue a_n, satisfying Eq. (7.33) requires

$$\mathcal{A}(\mathcal{B}\psi_n) = a_n(\mathcal{B}\psi_n). \tag{7.34}$$

Thus $\mathcal{B}\psi_n$ is an eigenfunction of $\mathcal{A}$ with eigenvalue a_n, and since there are no degenerate eigenvalues as stated above, it must be that $\mathcal{B}\psi_n \propto \psi_n$. The proportionality constant is just the eigenvalue b_n of the operator $\mathcal{B}$ on ψ_n. Thus $\mathcal{A}$ and $\mathcal{B}$ share the same set of eigenfunctions.

A special case arises if one of the operators is the Hamiltonian. The Ehrenfest theorem of quantum mechanics shows that the time derivative of the expectation value of an operator is proportional to the expectation value of the commutator of the operator with the Hamiltonian. Therefore operators that commute with the Hamiltonian are constants of motion since the commutator is zero. Thus the expectation value of the operator is constant.

k	$\langle\psi_{n\ell m}|r^k|\psi_{n\ell m}\rangle$
-6	$\dfrac{2^6 Z^6 \left[35n^4 + 5n^2(6\ell(\ell+1) - 5) + 3(\ell-1)\ell(\ell+1)(\ell+2)\right]}{a_\mu^6 n^7 \prod_{k=-3}^{5}(2\ell+k)}$
-5	$\dfrac{2^6 Z^5 \left[5n^2 - 3\ell(\ell+1) - 1\right]}{a_\mu^5 n^5 \prod_{k=-2}^{4}(2\ell+k)}$
-4	$\dfrac{2^4 Z^4 \left[3n^2 - \ell(\ell+1)\right]}{a_\mu^4 n^5 \prod_{k=-1}^{3}(2\ell+k)}$
-3	$\dfrac{2^3 Z^3}{a_\mu^3 n^3 \prod_{k=0}^{2}(2\ell+k)}$
-2	$\dfrac{2Z^2}{a_\mu^2 n^3 (2\ell+1)}$
-1	$\dfrac{Z}{a_\mu n^2}$
0	1
1	$\dfrac{a_\mu}{2Z}\left[3n^2 - \ell(\ell+1)\right]$
2	$\dfrac{a_\mu^2 n^2}{2Z^2}\left[5n^2 - 3\ell(\ell+1) + 1\right]$
3	$\dfrac{a_\mu^3 n^2}{8Z^3}\left[35n^4 - 5n^2(6\ell(\ell+1) - 5) + 3(\ell+2)(\ell+1)\ell(\ell-1)\right]$
4	$\dfrac{a_\mu^4 n^4}{8Z^4}\left[63n^4 - 35n^2(2\ell(\ell+1) - 3) + 5\ell(\ell+1)(3\ell(\ell+1) - 10) + 12\right]$

Table 7.6 Radial matrix elements $\langle\psi_{n\ell m}|r^k|\psi_{n\ell m}\rangle$ for various powers k, where $\prod_{k=i}^{j}(2\ell+k) \equiv (2\ell+i)(2\ell+i+1)\ldots(2\ell+j)$. The radius a_μ is defined in Eq. (7.18).

7.D Matrix elements of the radial wavefunctions

Important matrix elements for hydrogenic atoms can be found from

$$\langle\psi_{n\ell m}|r^k|\psi_{n\ell m}\rangle = \int_0^\infty dr\, r^{k+2}\,|R_{n\ell}(r)|^2 \tag{7.35}$$

where the 2 in the exponent of r derives from the integration element $d\Omega = r^2 \sin\theta\, dr\, d\theta\, d\phi$. The results for various values of k are given in Tab. 7.6 (see also Prob. 7.12). Note that for large n and $k < -1$ the matrix elements are proportional to $1/n^3$, whereas for $k > 1$ and $n \gg \ell$ they are approximately proportional to n^{2k}.

Exercises

7.1 From the radial part of any hydrogen wavefunction, one may obtain a probability density for r, the distance between the electron and the nucleus.

Calculate $\Delta r/r_{max}$, where r_{max} is the value of r that maximizes the probability distribution at P_{max}, and $\Delta r \equiv 1/P_{max}$ is a measure of the width of the probability distribution. Do the calculation for the cases $n = 1, 10, 100,$ and $1,000$. Use $\ell = n - 1$.

7.2 The wavefunction of an electron in the $|n\ell m\rangle = |210\rangle$ state of atomic hydrogen is given by

$$\psi_{210}(r, \theta, \phi) = \frac{1}{\sqrt{32\pi a_0^3}} \left(\frac{r}{a_0}\right) e^{-r/2a_0} \cos\theta.$$

For this state calculate:
(a) the most probable value of r
(b) the expectation value of r
(c) the expectation value of the potential energy
(d) the expectation value of the kinetic energy.

Hint: use $\int_0^\infty x^n e^{-cx} dx = n!/c^{n+1}$ where n is a positive integer and $c > 0$.

7.3 The components of the orbital angular momentum operator are given by Eq. (C.1) in the end-of-book Appendix C. Apply these operators to the spherical harmonics $Y_{\ell,m}(\theta, \phi)$ for $(\ell, m) = (3,2)$ to determine whether this state is a eigenstate for these operators.

7.4 A hydrogen atom is in a linear superposition as given by

$$\Psi(r) = R_{21}(r)\left(\sqrt{\tfrac{1}{3}}Y_{10}(\theta, \phi)\alpha + \sqrt{\tfrac{2}{3}}Y_{11}(\theta, \phi)\beta\right).$$

(a) If the angular momentum $\vec{\ell}^{\,2}$ is measured, what are the possible outcomes and what are the probabilities for each of these outcomes?
(b) Do the same for the projection ℓ_z of $\vec{\ell}$ on the z-axis.
(c) Do the same with the spin $\vec{s}^{\,2}$.
(d) Do the same with the projection of s_z of $\vec{s}$ on the z-axis.
The total angular momentum $\vec{j}$ is given by $\vec{j} = \vec{\ell} + \vec{s}$.
(e) If $\vec{j}^{\,2}$ is measured, what are the possible outcomes and what are the possibilities for each of the outcomes?
(f) Do the same for the projection j_z on the z-axis.
(g) What is the probability to detect the electron with spin up at the position (r, θ, ϕ)?

7.5 Determine by inspection the quantum numbers for an atom described by the wavefunction $\psi(r) \propto r^2 e^{-r/3a_0}(3\cos^2\theta - 1)$. Find the most probable value of r for the electron in this state and comment on the result.

7.6 Some of the wavefunctions for hydrogen in the $n = 3$ state are given by

$$\psi_{300}(r, \theta, \phi) = \frac{1}{3\sqrt{3\pi}} \left(1 - \frac{2r}{3} + \frac{2r^2}{27}\right) \exp(-r/3),$$

$$\psi_{310}(r, \theta, \phi) = \frac{2\sqrt{2}}{27\sqrt{\pi}} \left(1 - \frac{r}{6}\right) r \exp(-r/3) \cos\theta,$$

$$\psi_{320}(r, \theta, \phi) = \frac{1}{81\sqrt{6\pi}} r^2 \exp(-r/3) \left(3\cos^2\theta - 1\right).$$

(a) Determine the number of radial nodes for each of these wavefunctions. How many radial nodes are there in total?

(b) For which wavefunction is the probability to detect the electron at $r = 0$ the largest? Determine an estimate to detect the electron in the nucleus, given the radius of the nucleus of 10^{-15} m.

(c) Determines for one of the wavefunctions the expectation value $\langle r \rangle$ for the distance r. Determine for the same state the most probable distance r_{mp} to detect the electron. Explain possible differences between $\langle r \rangle$ and r_{mp}.

A hydrogen atom is at $t = 0$ in a arbitary linear superposition of the wavefunction given above.

(d) Is the atom in an eigenstate at $t = 0$ for one of the operators $\mathcal{H}$, ℓ^2, and/or ℓ_z? Discuss your answer for all three cases.

7.7 As examples for two-particle systems consider the following cases:

(i) ^{2}H; deuteron with one electron
(ii) ^{4}He$^+$; singly ionized helium
(iii) positronium; positron ($m = m_e$) and one electron
(iv) muonic hydrogen; proton and a negatively charged μ meson ($m \approx 207 m_e$).

Determine for each of these systems:
(a) The energy of the ground state.
(b) The effective radius r_{eff} for the ground state, as defined by $r_{eff} = \langle \psi_{100}|r|\psi_{100}\rangle$.
(c) The $(n = 2) \to (n = 1)$ transition frequency.
(d) The number of bound states.

7.8 In this exercise the current densities in the hydrogen atom are considered. For a wavefunction $\psi(\vec{r})$ the probability density $\rho(\vec{r})$ is given by

$$\rho(\vec{r}) = |\psi(\vec{r})|^2$$

and the probability current density $\vec{J}(\vec{r})$ is given by

$$\vec{J}(\vec{r}) = \frac{\hbar}{2im} \psi(\vec{r}) \vec{\nabla} \psi(\vec{r}) + c.c.$$

These expression are used to study the probability density and current around the hydrogen nucleus in a state

$$\psi_{n\ell m}(\vec{r}) = R_{n\ell}(r)Y_{\ell m}(\theta,\phi).$$

(a) If the wavefunction $\psi_{n\ell m}(\vec{r}) = f(\vec{r})\exp(i\phi(\vec{r}))$ can be split into an amplitude $f(\vec{r})$ and a phase $\phi(\vec{r})$, show that the density $\rho(\vec{r})$ depends only on the amplitude, whereas the current $\vec{J}(\vec{r})$ depends also on the phase.

(b) Determine $f(\vec{r})$ and $\phi(\vec{r})$ for the hydrogen atom in the state $\psi_{n\ell m}(\vec{r})$.
In cylindrical coordinates (ρ, ϕ, z) the components of $\vec{\nabla}U$ are given by

$$\left(\vec{\nabla}U\right)_\rho = \frac{\partial U}{\partial \rho} \quad \left(\vec{\nabla}U\right)_\phi = \frac{1}{\rho}\frac{\partial U}{\partial \phi} \quad \left(\vec{\nabla}U\right)_z = \frac{\partial U}{\partial z}.$$

(c) Determine $\vec{J}(\vec{r})$ in terms of $\rho(\vec{r})$. Discuss the physical implications of your result.

Classically the current $\vec{J}(\vec{r})$ yields a contribution $d\vec{L}$ to the angular momentum $\vec{L}$ given by

$$d\vec{L} = m\vec{r} \times \vec{J}(\vec{r})d\vec{r}$$

and the component in the z-direction is given by

$$L_z = m \int d\vec{r}\, r\, |J(\vec{r})| \sin\theta.$$

(d) Calculate L_z and explain your result.
(e) Indicate how large these components are in the x- and y-direction in this classical example. Why is there in this case a difference between the z-direction and the x- or y-direction?

7.9 Consider a tritium ^{3}H containing a nucleus and an electron. The trition nucleus, which consists of one proton and two neutrons, is unstable against beta decay to ^{3}He, which contains two protons and one neutron. This decay process occurs very rapidly with respect to characteristic atomic times and can be considered as instantaneous. As a result there is a sudden doubling of the Coulomb attraction between the electron and the nucleus. Assuming that the tritium atom is in the ground state when the decay takes place and neglecting recoil effects, find the probability that immediately after the decay the resulting He$^+$ ion can be found:

(a) In its ground state.
(b) In any state other than the ground state.
(c) In the 2s state.
(d) In a state with $\ell \neq 0$.

7.10 The Bohr model of the H atom in Chap. 1 gives the same energy levels as those found from the Schrödinger equation in this chapter. Since it is not completely obvious why this should be, discuss the relation between Bohr's calculation, angular momentum quantization, and orbital motion quantization.

7.11 Calculate the oscillator strength for the 2p→1s transition in a hydrogen atom and find the lifetime of the 2p level, checking the answer against Tab. 7.3.

7.12 The Feynman–Hellmann theorem [64, 65] will be used to derive elements $\langle 1/r^s \rangle$ for the hydrogenic wavefunctions, which will be used to evaluate the shift due to the fine structure interaction in Chap. 8.

Consider a Hamiltonian $\mathcal{H}(r,q)$, eigenvalues $E_n(q)$, and eigenfunctions $u_n(r,q)$ that depend on a parameter q. The eigenvalues $E_n(q)$ are given by

$$E_n(q) = \int dr\, u_n^*(r,q) \mathcal{H}(r,q) u_n(r,q).$$

(a) Prove that the following relation for the derivative of $E(q)$ is correct:

$$\frac{\partial E_n(q)}{\partial q} = \left\langle \frac{\partial \mathcal{H}(r,q)}{\partial q} \right\rangle.$$

Use the fact that the eigenfunctions $u_n(r,q)$ are normalized. This relation is called the Feynman–Hellmann theorem.

This theorem can be applied to the hydrogenic wavefunctions by choosing the radial Hamiltonian of Eq. (7.9) and the eigenvalues of Eq. (7.13).

(b) This theorem can be applied to find the expectation value $\langle 1/r \rangle$ by choosing $q = e^2/4\pi\varepsilon_0$. Check your result with Tab. 7.6.

(c) Find the expectation value $\langle 1/r^2 \rangle$ by choosing $q \equiv \ell$. Note that ℓ is related to n by the relation $n = n_r + \ell + 1$ with n_r the number of radial nodes. Again check your result with Tab. 7.6.

The force on the electron is given by $F(r) = -dV_{\text{eff}}(r)/dr$ with the effective potential $V_{\text{eff}}(r)$ given by Eq. (7.9b). In a stationary state $\langle F(r) \rangle = 0$.

(d) Use this relation to derive an expression for $\langle 1/r^3 \rangle$. Again check your result with Tab. 7.6.

7.13 It is possible to write the operators for the potential and kinetic energy of the H atom as

$$V = \frac{-e^2}{4\pi\varepsilon_0 r},\quad T_r = -\frac{\hbar^2}{2mr^2}\frac{\partial}{\partial r}\left(r^2 \frac{\partial}{\partial r}\right), \quad \text{and} \quad T_\ell = \frac{\ell(\ell+1)\hbar^2}{2mr^2}.$$

where radial and angular kinetic energies, T_r and T_ℓ respectively, have been separated.

(a) Use your favorite symbolic manipulation program to evaluate the expectation values of these three operators for the $n = 4$ states of H. There are four possible ℓ-values and three operators, so you have to do 12 calculations whose results are most easily presented in a table.

(b) What can you say about the ℓ-dependence of the results?

(c) Take the sum of $\langle T_r \rangle$ and $\langle T_\ell \rangle$ and comment on its ℓ-dependence.

(d) What can you say about the ratio $[\langle T_r \rangle + \langle T_\ell \rangle]/\langle V \rangle$ in each case?

8
Fine structure

8.1 Introduction

Under careful scrutiny the spectral features of many atoms were observed to be composed of a few closely spaced lines. The spectra showed shifts of the n-levels, and splittings within a single n-level, determined by the ℓ-value. The assumption was that the associated energy levels were doublets or multiplets of closely spaced levels, and the multiple lines arose from transitions among these multiple levels. This chapter addresses the origin of these small splittings and deviations from Eqs. (7.13) and (7.16) for hydrogen. These deviations arise from a few independent effects. Among these are the finite mass, non-spherical shape, and magnetic moment of the nucleus, but that topic is discussed in Chap. 9. Another derives from quantum electrodynamic effects, and these are briefly touched upon toward the end of this chapter.

The primary topics considered here derive from the effects of special relativity. One of these effects produces velocity-dependent corrections to the mass of the electron in its elliptical Kepler-like orbits, as given by Eq. (8.6). Another derives from motion in the electrostatic field of the nucleus, thereby producing a magnetic field. It was proposed that, in addition to charge and mass, electrons also have an intrinsic magnetic moment $\vec{\mu}_e$ that interacts with this magnetic field. The natural unit for this magnetic moment is the Bohr magneton

$$\mu_B \equiv \frac{e\hbar}{2m}. \tag{8.1}$$

Both of these effects give rise to shifts and splittings in the spectra of order of 10^{10} Hz, and are comparable to the measured splittings, at least in the lower states of hydrogen.

These are all embodied in the Lorentz-invariant Dirac equation whose eigenvalues in Eq. (8.22) depend on the Sommerfeld fine-structure constant α given by

$$\alpha \equiv \frac{e^2}{4\pi\varepsilon_0 \hbar c}, \qquad (8.2)$$

(see App. 8.A). Expanding the expression for its eigenvalues gives terms of different order in α, corresponding to the rest energy mc^2, the Bohr energies $\propto mc^2\alpha^2$ given by Eq. (7.13), and the $mc^2\alpha^4$ terms to be discussed here. This expansion of the energies does not distinguish among the physical origins of the energy shifts, but they can be inferred by expanding the Dirac Hamiltonian, a process different from expanding the eigenvalues. When the Hamiltonian is expanded, two of the $mc^2\alpha^4$ terms correspond to the mass and magnetic effects identified above, and there is a third term that has no classical counterpart. The magnitude of this term (known as the Darwin term) is proportional to the overlap of the atomic wavefunction with the nucleus, and thus is non-zero only for S states.

Even this fully relativistic description of Dirac failed under more precise comparison with the highly accurate measurements in hydrogen, most notably the hyperfine structure discussed in Chap. 9 and the Lamb shift, and these paved the way for the introduction of quantum electrodynamics into atomic structure.

8.2 The relativistic mass term

Although the velocity of the electron in hydrogen-like atoms, $v_n = Z\alpha c/n$, is considerably smaller than c, the relativistic corrections can still be compared with measurements. The relativistic formula for the kinetic energy associated with the electron's motion is $T = \sqrt{m^2c^4 + p^2c^2} - mc^2$, where m is the rest mass of the electron. Expanding this expression in terms of p^2 leads to

$$\frac{T}{mc^2} = \sqrt{1 + \left(\frac{p}{mc}\right)^2} - 1 = \frac{1}{2}\left(\frac{p}{mc}\right)^2 - \frac{1}{8}\left(\frac{p}{mc}\right)^4 \ldots \qquad (8.3a)$$

or

$$T \approx \frac{p^2}{2m} - \frac{p^4}{8m^3c^2}. \qquad (8.3b)$$

The first term on the right-hand side is incorporated in the Hamiltonian of Eq. (7.1), thus the second term is the first-order correction. The magnitude of this term can be estimated using $p_n = m\alpha c Z/n$ and this leads to a shift

$$E_{\text{rel}} \approx -\frac{\alpha^2 Z^4 R_\infty}{4n^4}, \qquad (8.4)$$

where R_∞ is the Rydberg constant (see Eq. (7.14)). This shows that the shift is a factor of α^2 smaller than the binding energy.

The small magnitude of the shift allows the use of first-order perturbation theory for its evaluation. Using the perturbation $\mathcal{H}' = -p^4/8m^3c^2 = -T^2/2mc^2$, expressing the kinetic energy operator as $T = \mathcal{H} - V$ with $V = -e^2/(4\pi\varepsilon_0 r)$ leads to

$$E_{\text{rel}} = \frac{-1}{2mc^2} \langle n\ell m | (\mathcal{H} - V)^2 | n\ell m \rangle \tag{8.5}$$

$$= \frac{-1}{2mc^2} \left[E_n^2 + 2E_n \left(\frac{e^2}{4\pi\varepsilon_0}\right) \left\langle \frac{1}{r} \right\rangle + \left(\frac{e^2}{4\pi\varepsilon_0}\right)^2 \left\langle \frac{1}{r^2} \right\rangle \right],$$

where $\mathcal{H} |n\ell m\rangle = E_n |n\ell m\rangle$ is used. The expectation values $\langle r^{-1} \rangle$ and $\langle r^{-2} \rangle$ can readily be found using Tab. 7.6 (see also Prob. 7.12) and rearranging all terms leads to

$$E_{\text{rel}} = -\frac{\alpha^2 Z^4 R_\infty}{n^3} \left(\frac{1}{\ell + 1/2} - \frac{3}{4n} \right). \tag{8.6}$$

For $n = 2$, $\ell = 0$ the bracketed term in Eq. (8.6) is $13/8$, and for $n = 2$, $\ell = 1$ it is $7/24$. Both of these shifts are negative.

Thus various values of ℓ give different energy shifts and from their differences the splittings can be calculated. For $\ell = 0$ and 1 the splitting is

$$\Delta E = E_{\ell=1} - E_{\ell=0} = \frac{4\alpha^2 Z^4 R_\infty}{3n^3}, \tag{8.7}$$

which decreases strongly for increasing n. The $1/n^3$ allows this to be described as a quantum defect (see Chap. 10).

8.3 The fine-structure "spin–orbit" term

8.3.1 The effect of the magnetic moment

Measurements by Zeeman of the effect of an applied magnetic field on the atomic spectra presented an enigma labeled the "anomalous Zeeman effect". The original view of the Zeeman effect by Lorentz was that the electron orbit constitutes a current loop and the concomitant magnetic moment interacts with the applied field, and results in an orientation-dependent energy shift. The Stern–Gerlach experiment suggested that there is a quantization of spatial orientation in addition to the known quantization of energy and angular momentum. However, both the number of energy levels and the magnitude of the splittings calculated this way were wrong.

The interactions between the electron magnetic moment $\vec{\mu}_e$ and various magnetic fields are often associated with a fictitious intrinsic angular momentum of magnitude $\hbar/2$ called spin, denoted by $\vec{s}$. Then the natural relation between angular momenta and magnetic moments requires a multiplicative factor called the Landé g-factor such that

$$\vec{\mu}_e = -\frac{g_e \mu_B \vec{s}}{\hbar}, \tag{8.8}$$

where $g_e \approx 2(1 + \alpha/2\pi + \ldots)$. There is no inherent reason for $g_e \approx 2$, it is simply an experimental result.

The name "spin" conjures up the notion of a classical angular momentum. Using the traditional definition $\vec{\ell} = \vec{r} \times \vec{p}$, it is impossible to have angular momentum $|\vec{s}| = \hbar/2$ without violating relativity for a uniformly charged, spinning sphere with the classical electron radius $r_0 = \alpha^2 a_0$. In fact, most experiments that are conventionally explained through the spin of the electron are really dependent on its magnetic moment only – they do not depend on the angular momentum associated with spin. For example, the torque arising from the angular momentum of absorbed or emitted light in an optical transition acts directly on only the orbital angular momentum ℓ, and on the "spin" only through its magnetic interaction with the orbital field (see center column of Tab. 4.1). However, the term "spin" will still be used where it is convenient and conventional. The pseudo-spin angular momentum of elementary particles, such as the spin $\vec{s}$ of electrons, can be described by a set of operators, namely the Pauli matrices of Eq. (2.16), and it is notable that both these and the general angular momentum operators of App. C obey the same commutation rules and have the same ladder operators.

The Einstein–deHaas experiment [66] is one of several that are purported to demonstrate the presence of angular momentum associated with electron spin. It was intended to measure Amperian currents and pre-dates both quantum mechanics and the Stern–Gerlach experiment, but it has been reinterpreted to address spin in several papers (a good summary can be found in Ref. [67]). It showed a torque on an iron rod resulting from a changing magnetic field, and the magnetization of iron results from only the spin S. The actual measured quantity is related to the gyromagnetic ratio, and later similar experiments are in agreement with both calculations and atomic spectroscopy.

8.3.2 The interaction energy

Consider an electron with magnetic moment $\vec{\mu}_e$ moving in the electric field of a nucleus $\vec{\mathcal{E}} = (Ze/4\pi\varepsilon_0 r^2)\hat{r}$. In the frame of the electron there is a magnetic field $\vec{\mathcal{B}}$ arising from its motion in the electric field of the nucleus, and there is an interaction between $\vec{\mu}_e$ and this field $E = -\vec{\mu}_e \cdot \vec{\mathcal{B}}$. For an electron moving uniformly in an electric field, the magnetic field is

8.3 The fine-structure "spin–orbit" term

$$\vec{\mathcal{B}} = -\frac{\vec{v} \times \vec{\mathcal{E}}}{c^2}. \tag{8.9}$$

The interaction between this "Lorentz field" and $\vec{\mu}_e$ gives rise to an energy shift (using $\vec{\ell} = \vec{r} \times \vec{p}$ and $g_e = 2$)

$$\mathcal{H}_{SO}^{(unc)} = -\vec{\mu}_e \cdot \vec{\mathcal{B}} = \frac{-g_e \mu_B \vec{s}}{\hbar} \cdot \frac{Ze}{4\pi\varepsilon_0 r^2} \frac{\vec{v} \times \hat{r}}{c^2} = 2\xi(r)\frac{\vec{\ell} \cdot \vec{s}}{\hbar^2} \tag{8.10a}$$

with

$$\xi(r) \equiv \frac{1}{2}\left(\frac{e\hbar}{mc}\right)^2 \frac{Z}{4\pi\varepsilon_0 r^3} = Z\alpha^2 R_\infty \left(\frac{a_0}{r}\right)^3 \tag{8.10b}$$

where $\vec{\ell}$ and $\vec{s}$ are measured in units of $\hbar$, "unc" means uncorrected, and the subscript "SO" refers to the origin of the interaction, "spin–orbit". Since r is of the order of a_0, the spin–orbit interaction is of the order α^2 smaller than the binding energy of the ground state of hydrogen.

An alternative view is to consider that, in the atomic reference frame, the electron's orbit constitutes a current $I = ev/2\pi r$ and results in a concomitant magnetic field $|\vec{\mathcal{B}}| = \mu_0 I/2r$. Calculation of this field gives the same result as Eq. (8.9) since $\mu_0 \varepsilon_0 c^2 = 1$, so the precession of the electronic magnetic moment $\vec{\mu}_e$ in this field gives the same energy as Eq. (8.10).

8.3.3 The Thomas correction to the fine-structure (spin–orbit) term

In the field after the Lorentz transform, the interaction of $\vec{\mu}_e$ with the relativistically induced field needs to be corrected for the non-inertial character of the orbital motion [5, 68]. This can best be done as a correction to the Lorentz transformation $\mathcal{T}(\vec{v})$ for velocity in the electron's orbit. In the presence of acceleration that can change $\vec{v}$ to $\vec{v} + d\vec{v}$, $\mathcal{T}(\vec{v} + d\vec{v})$ can be Taylor expanded as

$$\mathcal{T}(\vec{v} + d\vec{v}) = \mathcal{T}(\vec{v}) + d\vec{v} \cdot \nabla_v \mathcal{T}(\vec{v}) \tag{8.11}$$

where ∇_v is the gradient in velocity space. The 4×4 matrix for the transformation $\mathcal{T}(\vec{v} + d\vec{v})$ can be calculated directly from this expansion, and its spatial part comprises a rotation through angle

$$d\vec{\theta} = \frac{\gamma - 1}{v^2}(\vec{v} \times d\vec{v}), \quad \gamma \equiv \frac{1}{\sqrt{1 - (v/c)^2}} \tag{8.12}$$

Thus the correction to the precession frequency of $\vec{\mu}_e$ in the magnetic field of the electron's orbit is $(d\vec{\theta}/dt)$ and evaluation of it yields for $v \ll c$

$$\frac{d\vec{\theta}}{dt} = \frac{-1}{2m^2c^2}\left(\frac{Ze^2}{4\pi\varepsilon_0 r^3}\right)\vec{\ell}. \tag{8.13}$$

The interaction term becomes

$$\mathcal{H}_{SO}^{(corr)} = \vec{s} \cdot \frac{d\vec{\theta}}{dt} = -\xi(r)\frac{\vec{\ell} \cdot \vec{s}}{\hbar^2} = -\frac{1}{2}\mathcal{H}_{SO}^{(unc)} \qquad (8.14)$$

The corrected value is therefore

$$\mathcal{H}_{SO} = \mathcal{H}_{SO}^{(unc)} + \mathcal{H}_{SO}^{(corr)} = \xi(r)\frac{\vec{\ell} \cdot \vec{s}}{\hbar^2}. \qquad (8.15)$$

This gives rise to the misleading and erroneous name "Thomas factor". In fact, it is a subtractive correction of half the energy shift, *not* a factor of two. Precisely the same result can be derived in the "current" view by accounting for the precession of $\vec{\mu}_e$ relative to the orbital reference frame.

8.3.4 Evaluation of spin–orbit terms

Because of the presence of the spin angular momentum of the electron, the total angular momentum $\vec{j} = \vec{\ell} + \vec{s}$ of the electron becomes the sum of its orbital angular momentum $\vec{\ell}$ and its spin $\vec{s}$. In the absence of an external force acting on the atom, its total angular momentum $\vec{j}$ is a conserved quantity and thus it is used together with its projection m_j on the z-axis to characterize the atomic states. Still, the spin–orbit interaction involves the interaction between $\vec{\ell}$ and $\vec{s}$ through Eq. (8.15) and this requires the decoupling of the total angular momentum $\vec{j}$ into its components $\vec{\ell}$ and $\vec{s}$.

The spin–orbit interaction E_{SO} can be calculated by using the value of $\langle r^{-3} \rangle$ of Tab. 7.6 (see also Exercise 7.12), which yields

$$E_{SO} = \frac{A}{\hbar^2}\langle \vec{\ell} \cdot \vec{s} \rangle \qquad A = \langle \xi(r) \rangle \equiv \frac{\alpha^2 Z^4 R_\infty}{\ell(\ell + 1/2)(\ell + 1)n^3} \qquad (8.16)$$

The $\vec{\ell} \cdot \vec{s}$ factor is evaluated using the vector model that is discussed in App. C at the end of this book. The convention is $\vec{j} = \vec{\ell} + \vec{s}$ so that

$$\langle \vec{\ell} \cdot \vec{s} \rangle = \frac{1}{2}\langle \vec{j}^2 - \vec{\ell}^2 - \vec{s}^2 \rangle = \frac{\hbar^2}{4}(\pm(2\ell + 1) - 1) \qquad (8.17)$$

for $j = \ell \pm 1/2$. Substituting this into Eq. (8.16) leads to

$$E_{SO} = \frac{\alpha^2 Z^4 R_\infty}{4n^3}\frac{\pm(2\ell + 1) - 1}{\ell(\ell + 1/2)(\ell + 1)}, \qquad (8.18)$$

which shows that hydrogen has two energies corresponding to the ± sign, one shifted up and one shifted down. For $\ell = 1$ the two values of $\langle \vec{\ell} \cdot \vec{s} \rangle$ are $+\hbar^2/2$ and $-\hbar^2$. This leads to an energy splitting of the $j = \ell \pm 1/2$ states of $\Delta E_{SO}(\ell = 1) =$

$\alpha^2 Z^4 R_\infty/2n^3$. The $1/n^3$ dependence makes this splitting amenable to description as a contribution to the quantum defect (see Eq. (10.1)). Thus there is a correction to the Bohr energy of the same magnitude as the relativistic mass shift correction of Eq. (8.7), and it constitutes part of what is called "fine structure". This effect is also about 10^5 times smaller than the ground-state energy, and is roughly $h \times 10^{10}$ Hz or more precisely 10,969.12 MHz.

8.3.5 Spin–orbit interaction for other atoms

In this chapter, three comparable contributions to the fine structure of H are presented (see Secs. 8.2, 8.3, and 8.4). Since states of H with the same n but different ℓ are degenerate in the absence of this fine structure, these contributions are all quite important for understanding the atomic structure. In H the electron is attracted by the nuclear Coulomb potential $V_N(r)$, but for other atoms the valence electron also interacts with the core electrons via the core potential $V_c(r)$. Thus the short-range potential is increased by a factor of Z to $V_N(r) = -Ze^2/(4\pi\varepsilon_0 r)$, but outside the core it drops to $V_N(r) + V_c(r) = -e^2/(4\pi\varepsilon_0 r)$. The modified potential at intermediate distances lifts the degeneracy between the different ℓ states and causes states with low ℓ that penetrate the core region to have a lower energy than the high ℓ states that penetrate less (see Chap. 10). The relativistic and Darwin shifts discussed in this chapter for H are not as important for other atoms since those are dominated by the core potential.

However, the spin–orbit interaction is still important because the stronger $\vec{\mathcal{E}}$ field causes a larger splitting between states with $j = \ell \pm s$. This interaction for other atoms is similar to that of H apart from the electronic interaction with the modified potential $V(r) = V_N(r) + V_c(r)$ at intermediate distances. The electric field of the nucleus and core is $\vec{\mathcal{E}} = \vec{\nabla} V(r)/e$, and is radial for a spherically symmetric core. Using this electric field in the analysis of Sec. 8.3.2 changes only the radial part of the interaction and leads to

$$\xi(r) \equiv \frac{1}{2}\left(\frac{\hbar}{mc}\right)^2 \frac{1}{r}\frac{dV(r)}{dr}, \tag{8.19}$$

which modifies Eq. (8.10b) for atoms other than H. In principle, this term can be evaluated using a core potential $V_c(r)$ as obtained by methods discussed in Sec. 10.4 (see Fig. 10.2). However, in practice the value $A = \langle \xi(r) \rangle$ can be found from the experimentally observed splitting ΔE_{SO}, since this splitting is given by $\Delta E_{SO} = A(\ell + 1/2)$. Values for the first excited P state in the alkali-metal atoms are given in Tab. 9.1.

8.4 The Darwin term

The Darwin term has no classical analog and therefore no intuitive explanation. It arises when expanding the Dirac Hamiltonian to order v^2/c^2, and thus is a purely relativistic effect (although this expansion is sometimes called the non-relativistic limit). It appears as an integral involving the overlap of the wavefunction with the nucleus [69]

$$E_{\text{Dar}} = \frac{\pi\hbar^2 Z}{2m^2 c^2} \langle \delta(\vec{r}) \rangle = \frac{\alpha^2 Z^4 R_\infty}{n^3} \delta_{\ell 0}. \tag{8.20}$$

It is therefore non-zero only for S states. Curiously, the magnitude of this term is enough to shift any $S_{1/2}$ state upward into degeneracy with the $P_{1/2}$ state with the same n. There is no apparent reason why this is so and it occurs in all manifolds of hydrogen-like atoms for states that have the same j. The Darwin term will not be discussed further here.

Very precise spectral measurements in the 1930s suggested that this might not be exactly true, and the spectacular experiments of Lamb in the 1940s (see App. 8.B) that unequivocally measured their separation led to the development of quantum electrodynamics.

8.5 Summary of fine structure

The various values of the shifts of Eq. (8.6), Eq. (8.16), and Eq. (8.20) sum up to yield the total shift from the fine-structure interactions:

$$E_{\text{fs}} = -\frac{\alpha^2 Z^4 R_\infty}{n^3} \left(\frac{1}{j+1/2} - \frac{3}{4n} \right). \tag{8.21}$$

The terms are collected and summarized in Tab. 8.1 in the convenient units of $\alpha^2 Z^4 R_\infty / 24 n^3$.

The Bohr energies for all states having the same principal quantum number n are identical, but the degeneracy is lifted by both the spin–orbit and relativistic shifts. These are both "classical" in the sense that there are non-quantum mechanical pictures of the interactions that produce them. It seems a bit bizarre that the Darwin term should restore the degeneracy of the two terms with $j = 1/2$, as shown in Tab. 8.1. A plot of these energies is shown in Fig. 8.1.

8.6 The Dirac equation

In 1928 new ground was broken by Dirac's relativistic treatment of motion of an electron in an electromagnetic field. When applied to the Coulomb field of the hydrogen atom, this fully Lorentz-invariant formulation produced all of the phenomena discussed above. The derivation of his Hamiltonian and wavefunction is

8.6 The Dirac equation

Shift	$S_{1/2}$	$P_{1/2}$	$P_{3/2}$
Relativistic	−39	−7	−7
Spin–orbit	0	−8	+4
Darwin term	+24	0	0
Total	−15	−15	−3

Table 8.1 Table of contributions to the fine structure of atomic hydrogen for $n = 2$ in units of $\alpha^2 R_\infty/24n^3 = 0.9124$ GHz.

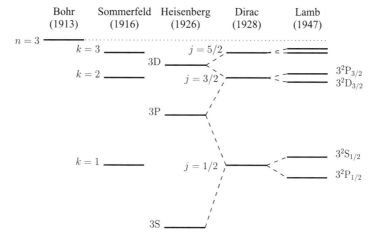

Figure 8.1 Evolution of the model of the fine structure of the $n = 3$ state in hydrogen. In the Bohr model all $n = 3$ states are degenerate (see Sec. 1.4.2). Sommerfeld included the relativistic correction into the Bohr model and obtained a shift as a function of $k = 1, \ldots, n$. After Schrödinger solved his equation for hydrogen reinstating the degeneracy of all $n = 3$ states, Heisenberg included the relativistic correction in the quantum mechanical model, where the states are shifted as a function of ℓ (see Eq. (8.6)). The solutions of the Dirac equation show that states with different ℓ but identical j are degenerate (see Eq. (8.22)). Finally, this degeneracy is lifted by the Lamb shift (see Sec. 8.7).

outside the scope of this book, but its energy eigenvalues have the rather simple form

$$E = mc^2 \left[1 + \left(\frac{\alpha Z}{n - k + \sqrt{k^2 - (\alpha Z)^2}} \right)^2 \right]^{-\frac{1}{2}} \tag{8.22}$$

where $k = j + \frac{1}{2}$. It is important to notice that Eq. (8.22) is independent of ℓ, thereby producing the j-degeneracy of Tab. 8.1. Since $\alpha \ll 1$, Eq. (8.22) can be expanded to find

$$\frac{E}{mc^2} = 1 - \frac{\alpha^2 Z^2}{2n^2} - \frac{(\alpha Z)^4}{2n^3}\left(\frac{1}{j+1/2} - \frac{3}{4n}\right) + \ldots \quad (8.23)$$

This formula produces the results of Tab. 8.1, but there are no separate terms to identify the intuitive effects described above. In first order this equation gives $E = mc^2$, the rest mass as might be expected in a fully relativistic calculation. The second-order term is exactly the Bohr energies of Eq. (7.13) (note the negative sign). The third-order term is the fine structure given in Eq. (8.21) and shown in Fig. 8.1.

Because these results incorporate the spin–orbit interaction, it is usually said that the Dirac equation predicts the existence of spin and the electron magnetic moment. Since the equation derives from a fully covariant treatment of electron motion, these properties are usually called relativistic.

8.7 The Lamb shift

The results of the high precision spectroscopy described in App. 8.B aroused interest in the possibility of a discrepancy between observations and the calculations from the Dirac equation. Among the many possibilities considered was the suggestion that the $2^2S_{1/2}$ and $2^2P_{1/2}$ levels were split by ~1 GHz, in contradiction to the Dirac eigenvalues of Eq. (8.23). This possibility inspired Lamb's famous confirming measurement using newly available radar instruments produced during World War II (see Ref. [70]) that was one of the main discussion topics at the legendary Shelter Island Conference of 1947. It led to Bethe's first description of the effect [71] showing that there was indeed such a splitting. The experiments used deuterium to reduce the Doppler broadening, and the measured value was 1,059.00(10) MHz [72] and the calculated value was 1,058.49(16) [73], absolutely phenomenal precision and agreement compared with the optical limitations arising from a linewidth 10^4 times larger.

In 1948 Welton [74] produced an elementary calculation based on the effect of the zero-point (vacuum) field on atomic energy levels. Welton's idea begins with the zero-point fields described in Chap. 5, namely, that in a vacuum the electromagnetic fields do not vanish. The effect of their residual values is to alter the orbit of an electron bound to a nucleus. The electron's motion (i.e. its wavefunction) is perturbed by the zero-point fields given in Eq. (5.8), and the resulting change causes a shift of the energy levels that depends on the spatial properties of the wavefunction. Thus the calculation proceeds by making a Taylor expansion of the Coulomb potential $V(\vec{r})$ around $\vec{r}$, given by

$$V(\vec{r} + \Delta\vec{r}) = V(\vec{r}) + 1/2(\Delta\vec{r}\cdot\vec{\nabla})^2 V(\vec{r}) + \ldots \quad (8.24)$$

8.7 The Lamb shift

where the even terms do not play a role because of the isotropy of the zero-point field. Welton notes that the expansion of Eq. (8.24) does not converge, but assumes that the curvature of the wavefunction is small. The second-order term evaluates to

$$\left\langle (\Delta \vec{r} \cdot \vec{\nabla})^2 \right\rangle_{vac} = 1/3 \left\langle (\Delta \vec{r})^2 \right\rangle_{vac} \vec{\nabla}^2, \tag{8.25}$$

where the factor $1/3$ derives again from the isotropy of the zero-point field. The energy shift ΔU can now be written as the product of two factors

$$\Delta U = \langle V(\vec{r} + \Delta \vec{r}) - V(\vec{r}) \rangle_{vac} = \frac{1}{6} \left\langle (\Delta \vec{r})^2 \right\rangle_{vac} \left\langle \vec{\nabla}^2 \left(\frac{-e^2}{4\pi\varepsilon_0 r} \right) \right\rangle, \tag{8.26}$$

where the first factor depends on the strength of the zero-point field and the second factor determines the effect of the field on the interaction potential.

The second factor of Eq. (8.26) can be evaluated by using $\vec{\nabla}^2(1/r) = -4\pi\delta(\vec{r})$ (see pg. 40 of Ref. [5]), which leads to

$$\left\langle \vec{\nabla}^2 \left(\frac{-e^2}{4\pi\varepsilon_0 r} \right) \right\rangle = \frac{e^2}{\varepsilon_0} |\psi_{n\ell}(0)|^2 = \frac{e^2}{\pi\varepsilon_0 n^3 a_0^3} \delta_{\ell 0}. \tag{8.27}$$

So in this model only the S states are shifted up and the shift decreases as $1/n^3$.

For the first factor of Eq. (8.26) the calculation of $(\Delta \vec{r})^2$ is done using the classical response of a charged particle to an applied electric field, where the zero-point field is given in Eq. (5.8). The square of the average of the displacement $(\Delta \vec{r})^2$ and is found using the classical equation of motion $d^2(\Delta \vec{r})/dt^2 = e\mathcal{E}_\omega(t)/m$, or $\Delta \vec{r} = e\mathcal{E}_\omega(t)/m\omega^2$, where $\mathcal{E}_\omega(t)$ is the sinusoidally oscillating zero-point field at frequency ω. Summing over all frequencies ω leads to

$$\left\langle (\Delta \vec{r})^2 \right\rangle_{vac} = \sum_\omega \left(\frac{e}{m\omega^2} \right)^2 \langle 0|\mathcal{E}_\omega^2|0\rangle = \sum_\omega \frac{\hbar e^2}{2m^2\varepsilon_0 V \omega^3}. \tag{8.28}$$

The sum can be rewritten as an integral replacing $\sum_\omega$ by $(1/2\pi^2) \int d\omega \, \omega^2$, and the result becomes

$$\left\langle (\Delta \vec{r})^2 \right\rangle_{vac} = \frac{\hbar e^2}{2\pi^2 \varepsilon_0 m^2 c^3} \int \frac{d\omega}{\omega}, \tag{8.29}$$

leading to a logarithmic divergence.

However, there are limits to the frequency range of the integral. For a dc (or very low frequency) field, the atom is simply polarized causing a Stark shift because the electron's orbital frequency is much larger than the oscillation rate of the field. This produces a lower limit on the zero-point frequencies that need to be considered, approximately given by the orbital frequency $\alpha c/a_0$. Moreover, there is a practical upper limit to the frequency of the effective zero-point field given by the Compton wavelength αa_0 where magnetic effects also become important. Thus the integral

142   Fine structure

is limited to the frequency range $\alpha c/a_0 < \omega < mc^2/\hbar$. The final result for the Lamb shift becomes

$$\Delta U = \frac{8\alpha^3 R_\infty}{3\pi n^3} \ln\left(\frac{1}{\alpha^2}\right) \delta_{\ell 0}, \tag{8.30}$$

showing that the shift scales with α^3.

For the 2S state of H, Eq. (8.30) yields $\Delta U/\hbar = 1336$ MHz, which is to be compared with the measured value of 1058 MHz. In Ref. [75] a better result is obtained by calculating more accurately the limits for the frequencies of the zero-point fields, and this changes the logarithmic term to $\ln(1/\alpha^2 K_0)$ where $K_0 \approx 16.64$ for the 2S state. Although a factor of ~17 seems like a huge change, it makes only a ~35% correction to the log term. Their final result is 1040 MHz, which is in much better agreement with the experimental value. There are other small corrections to this, and also a much smaller shift for the $2P_{1/2}$ state. Detailed information about these additional terms can be found in Ref. [75], and values for the Lamb shift in hydrogen are given in Ref. [76].

Although the early suggestions of a disagreement with the Dirac result were dismissed by some physicists as an experimental artifact, the experiments by Lamb were incontrovertible. The persistence in measuring things as accurately as could be done resulted in the development of a completely new theory of electromagnetism that is now called quantum electrodynamics (QED). Moreover, its development paved the way for other quantum-field theories which now dominate much of theoretical physics. In Fig. 8.2 the level shifts are shown for the $n = 2$ and 3 states of hydrogen.

Appendices

8.A The Sommerfeld fine-structure constant

The Sommerfeld fine-structure constant α given in Eq. (8.2) has been the subject of much lore and speculation since its introduction. Because the magnitude of its reciprocal, currently 137.035 999, is so close to an integer, there was speculation that future measurements would show it to be exactly an integer that would have fundamental implications. There have also been attempts to calculate it with various combinations of mathematical constants such as e and π, much like the ratio of proton to electron mass is very nearly $6\pi^5$. Over the decades of the twentieth century there have been numerous attempts to determine α to increasingly higher precision, and these efforts continue.

It is also convenient to define systems of measurement units such as mass, length, time, and charge in terms of naturally occurring quantities. For example, in atomic physics these might be the mass and charge of an electron, the speed of light, and

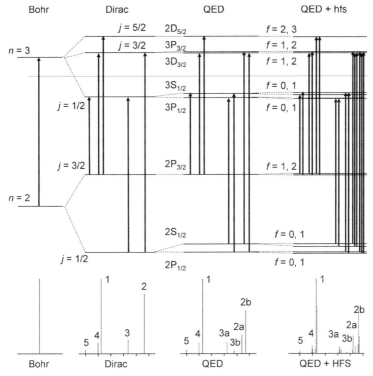

Figure 8.2 Level structure for the $n = 2$ and 3 states of hydrogen (not to scale). The degeneracy of the states as predicted by Bohr is lifted by the fine-structure interaction, as described by the Dirac equation. Subsequently the levels are shifted by the Lamb shift, which is important mostly for the S states. Finally, the levels are shifted by the hyperfine interaction, as discussed in Chap. 9. (Figure from Ref. [77].)

Planck's constant (see App. B.1). Because the α is dimensionless and $\neq 1$ exactly, there is no system of measurement in which all its constituent constants can become natural units and their magnitudes set to unity. For atomic units, the usual choice is $c = 1/\alpha$.

Because it is dimensionless it is often thought of as the ratio of two quantities of the same dimension, much like the ratio of two lengths. For example, the classical radius of the electron given by $e^2/(4\pi\varepsilon_0 mc^2)$ is just $\alpha^2 \times a_0$, where $a_0 \equiv \hbar/mc\alpha$ is the Bohr radius. It is often thought of as the ratio of the electromagnetic unit of action, $e^2/(4\pi\varepsilon_0 c)$, to the quantum unit of action, $\hbar$. This notion is supported by its ubiquitous role in quantum electrodynamics. In addition to the Bohr formula, it appears in the g-factor of the electron as $g_e = 2(1 + \alpha/2\pi + \ldots)$, similarly in the Lamb shift, fine- and hyperfine-structure calculations, and many other places. Thus it is considered as a coupling constant for electromagnetic interactions between particles.

Currently there are suggestions that it may vary with time on a cosmological scale (see Chap. 24). Thus there is considerable research by astronomers who are measuring the frequencies of light from very distant objects that is believed to have been emitted by their atoms in the distant past, perhaps not long after the "big bang". Such questions have yet to be resolved.

8.B Measurements of the fine structure

The doublet structure of many spectral lines had been known for a long time, and the earliest description of such splitting of the Balmer-α line of hydrogen at $\lambda \approx$ 656 nm was by Michelson and Morely in 1887 [78]. This transition connects the $n = 2$ and $n = 3$ states. The history of the fine structure of atomic hydrogen is a long story whose early part is well summarized in Ref. [79]. It illustrates how physicists' perseverance in seeking the highest attainable measurement precision and subsequent comparison with theory leads to new insights and even completely new theories.

The main obstacle to precision measurements on the Balmer-α line was the width resulting primarily from Doppler shifts Δv_D. At $T = 300$ K, the thermal velocity of hydrogen atoms is about 2800 m/s so that the full width of the spectral line is $2\Delta v_D = 2v/\lambda \sim 850$ MHz. Since the fine-structure separation between the $j = 3/2$ and $j = 1/2$ states of the $n = 2$ level of H is about 11 GHz, this is easily resolved, but precision measurements of the interval are limited by this width of about 8% of the separation.

There was a very small but subtle discrepancy arising not from the separation of the peaks, but from their asymmetrical shape. Improvements were made by cooling the vapor to liquid N_2 temperature and using deuterium instead because its higher mass resulted in a smaller Δv_D. The asymmetry was then unquestionable under these conditions as shown in Fig. 8.3, and led to the direct measurement of the Lamb shift (see Ref. [70]). The upward shift of the 2S level from degeneracy with the $2P_{1/2}$ is diagrammed in Fig. 8.4.

However, the discovery of the Lamb shift was only the beginning of the story. In an accompanying paper, Lamb and coworkers also described their measurement of the interval between the $2^2S_{1/2}$ and $2^2P_{3/2}$ levels, and added this to the previous measurement to determine the splitting between the $2^2P_{3/2}$ and $2^2P_{1/2}$ levels. Comparison of this result with theory was limited by the accuracy of the fundamental constants, for example as given in Eq. (8.16) and Tab. 8.1. Therefore the measurements were reinterpreted to establish a new value for the fine-structure constant α.

Comparison between the calculation of the hyperfine interval (hfs) in the ground state of hydrogen and high precision hfs measurements with the hydrogen maser

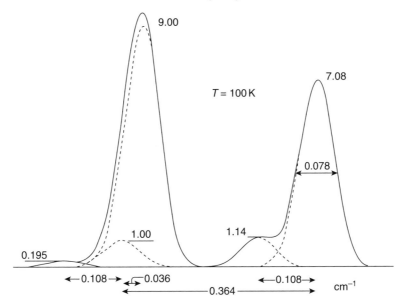

Figure 8.3 Spectrum of the Balmer-α line in deuterium at 100 K showing the Doppler-broadened lines. The scale in cm^{-1} can be correlated with that of Fig. 8.5 using 1 cm^{-1} = 30 GHz, and the horizontal axis spans about the same width as that of Fig. 8.5. (Figure from Ref. [78].)

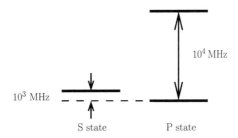

Figure 8.4 The Lamb shift measurements proved that the fine-structure calculations and the Dirac equation had left something out, giving wrong answers. This discrepancy gave birth to quantum electrodynamics and its subsequent application to the H atom resolved the discrepancy to extraordinary precision.

were also limited by the precision of α, and therefore these were also reinterpreted to extract a new value. This value and the one from Lamb's measurements disagreed by several times their uncertainties, and there was considerable controversy in the physics community about the discrepancy. In the late 1960s there was a direct measurement of the $2^2P_{3/2}-^2P_{1/2}$ interval by level crossing spectroscopy [27], and an extraction of the value of α from measurements with the Josephson effect. Both corroborated the value from hfs measurements, and the discrepancy with Lamb's

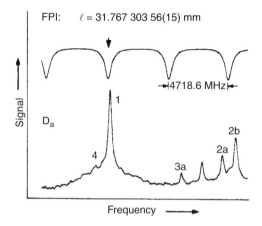

Figure 8.5 Spectrum of the Balmer-α line in deuterium using saturated absorption spectroscopy. The calibration marks are ~4.71 GHz apart, and the horizontal axis spans about the same width as that of Fig. 8.3. The two right-most peaks are separated by the clearly resolved Lamb shift between the $2^2S_{1/2}$ and $2^2P_{1/2}$ levels, and others are artifacts of the process known as crossover resonances. (Figure from Ref. [80].)

measurements was not resolved for another decade, when it turned out to be a very subtle effect from the non-Maxwellian atomic velocity distribution in his atomic beam.

The importance of such precision measurements provided the impetus for related ones in highly ionized but single electron systems (hydrogen-like ions up to 91 times ionized uranium), exotic atoms such as muonium, and non-hadronic systems such as positronium. All of these measured energies have different scaling rates with various QED processes, resulting in very important contributions to the field.

An important turnabout occurred with the development of tunable narrow-band lasers in the early 1970s. Various methods of spectroscopy using counter-propagating light beams resulted in signals much narrower than the Doppler width, and allowed the return of purely optical measurements that resulted in still higher precision (see for example Ref. [80]). An example is shown in Fig. 8.5 and is to be compared with Fig. 8.3.

In some sense, the necessity of two different light beams for such sub-Doppler spectroscopy as in Ref. [80] can be considered as a non-linear method, but even more so is the direct optical measurement of the 1S–2S interval in atomic hydrogen. The first important consideration arises because the electric dipole transition is forbidden by the $\Delta \ell = \pm 1$ selection rule so it must be a higher-order process as described in Sec. 4.3.1. In fact, this is precisely the transition studied by Maria

Göppert-Mayer in her Ph.D. thesis, but this time in excitation rather than decay. Since the transition rate for such processes depends on the square of the intensity (see Eq. (4.15)), this is also a non-linear phenomenon.

The second important consideration also derives from the second-order nature of the transition. Because excitation of the hydrogen 2P states from the ground state is dipole allowed, the 2P states also decay rapidly ($\tau = 1.6$ ns) making the natural width ~100 MHz. However, the lifetime of the 2S state against spontaneous decay is $1/7$ s, resulting in a natural width ~1 Hz (the Q of the transition is ~10^{15}) thereby enabling much higher-precision measurements.

A second-order process driving the 10 eV energy gap between 1S and 2S requires light of wavelength $\lambda = 2 \times 121.5$ nm, and this 243 nm light is produced by frequency doubling of $\lambda = 486$ nm light in a non-linear optical crystal. The combination of the two non-linear techniques, second-order interaction and frequency doubling to produce the light, allowed a direct measurement of this important 1S–2S interval in 1980 [81].

Since then, continued improvement has resulted in precision at the level of a few tens of Hz. The measurements currently allow direct determination of the 8.2 GHz Lamb shift in the ground state to a few parts per million, and their interpretation is limited by knowledge of the proton structure (charge distribution and its perturbing effect on the Coulomb potential). The current precision on this 1S–2S transition frequency is still much lower than the very small limit imposed by the natural width, and derives from velocity-dependent effects such as interrogation time limits and the spread of the second-order Doppler shifts of the atoms moving in an atomic beam. These can be reduced by cooling and trapping the atoms, but this may present other complications. The newest measurements are to be found in Ref. [82].

Exercises

8.1 Follow the suggestion at the end of Sec. 8.3.2 and calculate the magnetic field $\mathcal{B}_{\text{nuc}}$ seen by an electron in its own rest frame that derives from the apparent motion of the nucleus. Use $E = -\vec{\mu}_e \cdot \vec{\mathcal{B}}_{\text{nuc}}$ to find $\mathcal{H}_{\text{SO}}^{(\text{unc})}$ and compare with Eq. (8.10). Note its dependence on atomic number Z.

8.2 Consider the spin–orbit interaction for the $n = 2$ state of hydrogen. The interaction is given by Eq. (8.10).
(a) Identify all the states with $n = 2$. What is the total degeneracy of the $n = 2$ state?
(b) For one of these states the spin–orbit shift can be found without calculations. State which one and discuss why.

(c) Show that the calculation for the spin–orbit shift consists of two parts, namely the expectation value of $1/r^3$ and $\vec{\ell}\cdot\vec{s}$.

The functions $\Psi_{j\ell sm_j}$ are eigenfunctions of the operators $\vec{j}^2$, $\vec{\ell}^2$, $\vec{s}^2$, and j_z.

(d) Use the expression for $\vec{j}^2$ to determine $\vec{\ell}\cdot\vec{s}$.

(e) Use the properties of $\Psi_{j\ell sm_j}$ to calculate the expectation value of $\vec{\ell}\cdot\vec{s}$.

(f) Use Tab. 7.4 to determine the expectation value of $1/r^3$ for these states.

(g) Determine for each state with $n = 2$ the spin–orbit splitting and determine for each state j, ℓ, and s. Compare your results with Tab. 8.1. What is the degeneracy for each state?

8.3 Verify that the perturbation $\xi(r)\vec{\ell}\cdot\vec{s}$ does not connect the degenerate states with $m_\ell = +1, m_s = -1/2$ and $m_\ell = -1, m_s = +1/2$.

8.4 Expand the Dirac energy of Eq. (8.22) to three orders and identify the Coulomb and fine-structure terms.

8.5 The fine-structure splitting results in a pair of multiply degenerate levels, where the average shift weighted with the number of sub-states for each state is zero.

(a) Check this for the ^{2}P states of an alkali and for the ground-state hfs of atomic hydrogen.

(b) Show that it is also true for a level that has *three* fine-structure components, for example a ^{3}D state ($\ell = 2, S = 1$).

8.6 The calculation by Welton that results in Eq. (8.27) can also be done by considering the nucleus to be displaced instead of the electron.

(a) Use Gauss law to calculate the force on the nucleus if it is displaced by an amount r from the center, assuming the electron charge distribution is homogeneous.

(b) Use this result to calculate the total work done on the nucleus, if it is displaced from the center to a radius δr.

(c) Why does this lead to a non-zero result for S states only? Evaluate for S states its value.

(d) Compare the result with Eq. (8.27).

9
Effects of the nucleus

9.1 Introduction

Atomic spectroscopy provided a wealth of information about nuclear physics before the advent of huge accelerator facilities. The multiple, closely spaced atomic spectral lines could be ascribed to various nuclear properties, including the existence of isotopes, the magnitude of the nuclear magnetic moments, and even quadrupole and higher-order non-spherical aspects of nuclear structure. Some aspects of these effects are discussed below.

9.2 Motion, size, and shape of the nucleus

In Chap. 7 it is assumed that the nucleus of the atom is a point particle located at rest in the center of the coordinate system. Since the nucleus is much heavier than the electrons and the size of the nucleus is much smaller than the radius of the electron's orbit, this is a good approximation. In this section the effects of the nuclear motion, size, and shape are discussed.

9.2.1 Nuclear motion

The kinetic energy in the Hamiltonian of the Schrödinger equation is represented by the operator $p^2/2m$, whose constituent terms are the momentum operators $p_x = -i\hbar d/dx$, and similarly for p_y and p_z. But these operate only on the electron coordinates, and thus provide only its kinetic energy, and not that of the moving nucleus. As suggested in Chap. 7 the nuclear kinetic energy is small compared with that of the electron.

Consider that the motion of this two-body system may be divided into the overall center-of-mass motion and the motion of the constituent particles with respect to the center of mass. Then, in the center-of-mass rest frame, the total momentum of the constituents is zero so, for the two-body system of the hydrogen atom, $MV + mv = 0$, where upper (lower) case refers to the nucleus (electron). Then

$$T_N = \frac{P^2}{2M} = \frac{mv^2}{2}\left(\frac{m}{M}\right) \tag{9.1}$$

so the kinetic energy of the nucleus is smaller than that of the electron by the ratio of the masses, one part in 2,000 for hydrogen, and still smaller for heavier atoms. For this two-body system, the kinetic energy of the nucleus with respect to the center of mass can be included by making a simple correction by replacing the mass of the electron with the reduced mass $\mu = mM/(m+M)$ as in Eq. (7.29). For an atom, $M \gg m$ so $\mu \simeq m$ (see Tab. 7.5).

This transformation is exact for an one-electron system, but in case of more than one electron, corrections have to be made [69]. The transformation then leads to additional terms $\sum_{i>j} \vec{p}_i \cdot \vec{p}_j / M$, where the double sum is over all electrons and the term is referred to as mass polarization. This term is present only for more than one electron, and its observation in He is reported in Ref. [83].

Mass corrections made important contributions in the early history of nuclear physics. The notion of isotopes was suspected from fractional atomic weights by chemists long before the discovery of the neutron in 1932. But in 1931, Harold Urey concentrated deuterium in liquid hydrogen by evaporation, and observed the isotope shift spectroscopically. The signals appeared as weak lines separated from strong ones by exactly the calculated amount, a few parts in 10^4. This is noticeably larger than the experimental Doppler widths, but the signals were weak and so his Nobel prize for the discovery was well-deserved.

Before the advent of large accelerators, much information about isotopes was collected by atomic spectroscopy. At least, from the number and strength of components associated with a single electronic transition, the number and abundance of many isotopes could be deduced. With modern laser-atomic beam spectroscopy having resolution ~1 MHz, fractional mass shifts of the order $(m/A)^2 \sim 20$ MHz for mass number $A = 100$ are readily resolved and studied at the few percent level.

9.2.2 Nuclear size

Although the pure Coulomb potential $V \propto 1/r$ diverges near the origin, the hydrogen wavefunctions do not diverge. This is a general property of the $1/r$ potential, but is also true in a variety of similar potentials. In this section the deviation from a pure $1/r$ potential caused by the finite size of the nucleus is addressed. For spheres smaller than the nuclear radius, the enclosed charge is less than Z, and the altered potential makes small changes in the wavefunction. Thus it is necessary to include the effect of both the size and shape of the nucleus as corrections to the Coulomb potential at small distances.

For the simple case of a uniform spherical charge distribution of radius R_N and charge Z, any spherical surface of radius $r < R_N$ centered on it encloses a charge $Q_{enc} = Ze(r/R_N)^3$. Then Gauss's law gives $\mathcal{E} = Zer/(4\pi\varepsilon_0 R_N^3)$. The force is proportional to the distance from the origin, and so is a harmonic force. This yields the potential

$$V'(r) = \frac{Ze^2}{8\pi\varepsilon_0 R_N}\left[\left(\frac{r}{R_N}\right)^2 - 3\right] \qquad 0 \le r \le R_N, \tag{9.2}$$

where the -3 term arises because the potential at $r = R_N$ should be equal to the pure Coulomb potential $V(r) = -Ze^2/(4\pi\varepsilon_0 r)$. The potential difference $V'(r) - V(r)$ can be put into the Schrödinger equation for $r \le R_N$ and treated as a perturbation, since $R_N \ll a_0$. Since for $\ell \ne 0$ the wavefunction at $r = 0$ is zero, the term is important only for $\ell = 0$. Straightforward integration of the first-order perturbation for $0 \le r \le R_N$ leads to

$$E^{(1)} = \langle \psi_{n00}|V'(r) - V(r)|\psi_{n00}\rangle \tag{9.3}$$
$$= \frac{Ze^2}{4\pi\varepsilon_0}\frac{R_N^2}{10}|R_{n0}(0)|^2 = \frac{4R_\infty Z^4}{5n^3}\left(\frac{R_N}{a_0}\right)^2,$$

where $R_{n0}(0)$ is the radial wavefunction for $\ell = 0$ at the origin and in the last step its value is used. Note that for different isotopes the core radius R_N is different and the difference in the position of the energy levels between different isotopes can therefore be a measure of the core size.

In Fig. 9.1 the mean square radii for lead, mercury, and platinum isotopes are plotted vs. the number of neutrons. The results are compared to theory using the nuclear droplet model and show reasonable agreement. Because of the high precision of measurements such as those shown in Fig. 9.1, these experiments provide significant information about the size and shape of the nuclei of the different isotopes.

9.2.3 Nuclear shape

The effect of a non-spherical charge distribution of the nucleus, sometimes called a "field shift", can be calculated by expanding the charge distribution in multipole moments. These turn out to be the spherical harmonics, and their behavior is well described in several standard texts [18, 85]. Since the mass shifts described in Sec. 9.2.1 and the size dependence discussed in Sec. 9.2.2 above can be calculated quite accurately, subsequent comparison with spectroscopic measurements yields values for these field shifts and hence information about nuclear structure.

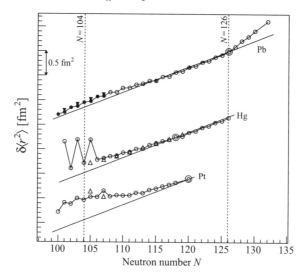

Figure 9.1 Mean square radii for lead, mercury, and platinum isotopes. The data for the different isotopes have been shifted arbitrarily with respect to each other for clarity. The measurement uncertainty is smaller than the symbol size because of the high resolution of the spectroscopic data. The solid lines are theory using the nuclear droplet model and show reasonable agreement. (Figure from Ref. [84].)

Although the effects of nuclear structure on atomic spectra are small and often subtle, almost all early nuclear physics was done with atomic spectroscopy.

This method of nuclear study may be old, but is still very much active. In 2004 a group at Argonne National Laboratory reported a study of the bizarre structure of ^{6}He by optical spectroscopy [86]. They deduced that it has the unusual structure of an alpha-particle and two orbiting neutrons 3.7 fm away.

9.3 Nuclear magnetism – hyperfine structure

Like the electron, the nucleus has a finite magnetic moment $\vec{\mu}_I \neq 0$. But unlike the electron, the nucleus is composed of many constituent nucleons and each nucleon has an intrinsic magnetic moment. The nuclear magneton is defined as $\mu_N = \mu_B(m/M)$, where M is the proton mass, and then the nuclear magnetic moment is $\vec{\mu}_I = g_I \mu_N \vec{I}/\hbar$ where $\vec{I}$ is a fictitious nuclear spin. The magnetic moment of the proton is $\mu_p = +2.79\mu_N$, so $g_p = +5.58$, and that of the neutron is $\mu_n = -1.91\mu_N$, so $g_n = -3.82$. The net total magnetic moment of heavy nuclei depends on how the nucleons are aligned with respect to one another. The energy shift of the atomic states arising from the interaction of this nuclear magnetic moment with the magnetic field caused by the electron is called the hyperfine shift, denoted E_{hfs}.

9.3 Nuclear magnetism – hyperfine structure

9.3.1 Atomic orbital angular momentum $\ell \neq 0$

Since an electron produces a magnetic field at the nucleus, there is an interaction between the nuclear magnetic moment $\vec{\mu}_I$ and this field that produces an energy shift given by $-\vec{\mu}_I \cdot \vec{\mathcal{B}}$. This shift is quite different from those discussed in Sec. 9.2 because it depends on the nuclear orientation with respect to the electron's magnetic moment, not simply on the particular nucleus in the atom. For an electron orbiting at some distance from the nucleus, there are two independent contributions to $\vec{\mathcal{B}}$ at the nucleus, one from the field $\vec{\mathcal{B}}_\ell$ produced at the origin by the orbiting electron whose motion constitutes a current, and the other $\vec{\mathcal{B}}_s$ from the intrinsic magnetic dipole moment of the electron $\vec{\mu}_e$.

The magnetic field at the nucleus that derives from the orbital part can be obtained from the Biot–Savart law. Since $i d\vec{s} = (-e\vec{v}/2\pi r)ds$,

$$\vec{\mathcal{B}}_\ell = \frac{\mu_0}{4\pi} i \oint \frac{d\vec{s} \times \vec{r}}{r^3} = \left(\frac{\mu_0}{4\pi}\right) \frac{e\vec{v} \times \vec{r}}{r^3} = -\left(\frac{\mu_0}{4\pi}\right) \frac{e\vec{\ell}}{mr^3} \tag{9.4}$$

where i is the current generated by the electron circling the nucleus. There are two minus signs here that cancel, one because the charge on the electron is negative, and the other because the radius vector $\vec{r}$ is defined as beginning at the origin, but here it is used as beginning at the electron since the current elements of the Biot–Savart law are defined this way.

The second contribution to the magnetic field arises from the dipole field produced by the intrinsic magnetic moment of the electron $\vec{\mu}_e$ and is given by

$$\vec{\mathcal{B}}_s = -\frac{\mu_0}{4\pi r^3} \left(\vec{\mu}_e - \frac{3\vec{\mu}_e \cdot \vec{r}}{r^2} \vec{r} \right) \tag{9.5}$$

These two magnetic components are not necessarily parallel to each other and their interaction produces complicated precession so that some components of their fields will average to zero.

The total field $\vec{\mathcal{B}}_j$ is given by

$$\vec{\mathcal{B}}_j = \vec{\mathcal{B}}_\ell + \vec{\mathcal{B}}_s = -\left(\frac{\mu_0}{4\pi}\right) \frac{2\mu_B}{\hbar r^3} \vec{G}, \quad \vec{G} \equiv \vec{\ell} - \vec{s} + 3\frac{\vec{s} \cdot \vec{r}}{r^2} \vec{r} \tag{9.6}$$

since $\vec{\mu}_e = -g_e \mu_B \vec{s}/\hbar$. The magnetic moment of the nucleus $\vec{\mu}_I = g_I \mu_N \vec{I}/\hbar$ interacts with the magnetic field $\vec{\mathcal{B}}_j$ leading to the magnetic dipole hyperfine interaction:

$$\mathcal{H}'_{\text{hfs}} = -\vec{\mu}_I \cdot \vec{\mathcal{B}}_j = E_N \frac{g_I \vec{G} \cdot \vec{I}}{\hbar^2 (r/a_0)^3}, \tag{9.7a}$$

with

$$E_N \equiv \left(\frac{\mu_0}{4\pi}\right) \frac{2\mu_B \mu_N}{a_0^3} = \alpha^2 R_\infty \frac{m}{M}. \tag{9.7b}$$

The shift is of the order of E_N/h, which evaluates to 95.409 MHz. Note that one part of the interaction depends on the radius r and the other part depends on the arrangement of the angular momenta $\vec{I}$ and $\vec{G}$. Because each of the angular momenta embodied in $\vec{G}$ precesses around the total angular momentum $\vec{j}$ (as shown in Fig. C.1 in the end-of-book appendix), the only conserved component of $\vec{G}$ is parallel to $\vec{j}$ and is given by

$$\vec{G}_j = \frac{\vec{G} \cdot \vec{j}}{j(j+1)\hbar^2} \vec{j} \qquad (9.8)$$

so that the $\vec{G} \cdot \vec{I}$ term is to be replaced by $\vec{G}_j \cdot \vec{I}$ in the interaction given by Eq. (9.7). The numerator in Eq. (9.8) can then be calculated directly as $\vec{G} \cdot \vec{j} = \vec{\ell}^2 - \vec{s}^2 + 3(\vec{s} \cdot \vec{r})^2/r^2 = \vec{\ell}^2$ since the second term cancels the third term [69].

As a result, the interaction Hamiltonian $\mathcal{H}'_{hfs}$ is proportional to $\vec{I} \cdot \vec{j}$. Since it represents the interaction between the nuclear magnetic moment represented by $\vec{I}$ and the total electron magnetic moment represented by $\vec{j}$, it can be described in terms of a total angular momentum $\vec{F}$ given by

$$\vec{F} \equiv \vec{I} + \vec{j}. \qquad (9.9)$$

This can be used to evaluate the $\vec{I} \cdot \vec{j}$ term by squaring Eq. (9.9) to find

$$\langle \vec{I} \cdot \vec{j} \rangle = \frac{1}{2} \langle \vec{F}^2 - \vec{I}^2 - \vec{j}^2 \rangle = \frac{\hbar^2}{2} \left(F(F+1) - I(I+1) - j(j+1) \right) \qquad (9.10)$$

This expression has to be evaluated in the basis in which $\vec{\ell}, \vec{s}, \vec{j}, \vec{I}, \vec{F}$, and M_F are good quantum numbers. Then the energy shift caused by the magnetic dipole interaction E_{hfs} can be evaluated using $\langle \mathcal{H}'_{hfs} \rangle$. The resulting $\langle 1/r^3 \rangle$ term can be found from Tab. 7.6 to yield

$$E_{hfs}^{(\ell \neq 0)} = \langle \mathcal{H}'_{hfs} \rangle = \frac{a}{\hbar^2} \langle \vec{I} \cdot \vec{j} \rangle \qquad (9.11a)$$

with

$$a \equiv \left(\frac{Z}{n}\right)^3 \frac{g_I E_N}{j(j+1)(\ell + 1/2)}, \qquad (9.11b)$$

and with $\langle \vec{I} \cdot \vec{j} \rangle$ given by Eq. (9.10).

9.3.2 Atomic orbital angular momentum $\ell = 0$

As shown in App. 9.A, the field inside a magnetic dipole is different from the external field given by Eq. (9.5). For $\ell \neq 0$ this internal field does not play a role because $\Psi(0)$ vanishes so the additional term given in Eq. (9.19) can be neglected. However, in the case of s states ($\ell = 0$) the electron magnetic moment also interacts

9.3 Nuclear magnetism – hyperfine structure

with the nuclear magnetic moment inside the nucleus because $\Psi(0) \neq 0$. Thus this term gives a contribution for s states only, in contrast to the previous section where the discussion following Eq. (9.8) suggests that only $\ell \neq 0$ states are shifted.

The argument of App. 9.A concludes that for s states only the last term of Eq. (9.19) needs to be evaluated, and this is easily done by substituting $\vec{\mu}_e = -g_e \mu_B \vec{s}$ and $\vec{\mu}_I = g_I \mu_N \vec{I}$ for $\vec{\mu}_1$ and $\vec{\mu}_2$. For $\ell = 0$ the interaction becomes

$$\mathcal{H}'_{\text{hfs}} = \frac{8\pi g_I a_0^3}{3\hbar^2} E_N \vec{s} \cdot \vec{I} \delta(\vec{r}). \tag{9.12}$$

The last two terms have to be evaluated in the basis where $\vec{\ell}, \vec{s}, \vec{j}, \vec{I}, \vec{F}$ and M_F are good quantum numbers. Since the two terms act on different coordinates, they can be evaluated separately. In the region where r is smaller than a few times R_N (see Sec. 9.2.2), the amplitude of the wavefunction can be approximated by $|\Psi(r)|^2 \approx |\Psi(0)|^2 = Z^3/\pi n^3 a_0^3$ because $|\Psi(r)|^2$ is constant over the range of a few times the nuclear size ($10^{-3} a_0$). Thus the expection value of the $\delta(\vec{r})$ term becomes $|\Psi(0)|^2$. The energy associated with this magnetic interaction then becomes

$$E_{\text{hfs}}^{(\ell=0)} = \langle \mathcal{H}'_{\text{hfs}} \rangle \equiv \frac{a}{\hbar^2} \langle \vec{I} \cdot \vec{s} \rangle = \left(\frac{Z}{n}\right)^3 \frac{8g_I}{3\hbar^2} E_N \langle \vec{I} \cdot \vec{s} \rangle \tag{9.13}$$

which is the hyperfine shift for $\ell = 0$ and is the same as Eq. (9.11) evaluated for $\ell = 0$ and $j = s = 1/2$. It is called the Fermi contact term, and is part of the hyperfine structure.

9.3.3 Hyperfine energies for hydrogen

The hyperfine shift for all values of ℓ can be rewritten as $E_{\text{hfs}} = a \langle \vec{I} \cdot \vec{j} \rangle /\hbar^2$ with a given by Eq. (9.11b). The energy shift E_{hfs} is called the hyperfine energy and the resultant level splittings are called hyperfine structure. The shifts are of the order $\alpha^2 m/M$ (see Eq. (9.7b)) smaller than the binding energy of the ground state. Furthermore, the hyperfine energies become quickly smaller for higher-lying states, since they are proportional to $1/n^3$.

The splitting of the states of hydrogen with $n = 1-3$ is shown in Fig. 9.2. Compared with the fine structure of Eq. (8.16) the hyperfine shift is of the order m/M smaller, as discussed in the beginning of this section. Note that a is independent of F and thus the energy difference between two adjacent F-states is given by

$$E_{\text{hfs}}(F) - E_{\text{hfs}}(F-1) = aF \tag{9.14}$$

which is known as the Landé interval rule.

For hydrogen the ground state with $F = 0$ is shifted down by $-3a/4$ and the state with $F = 1$ is shifted up by $+a/4$, where a for the ground state is given

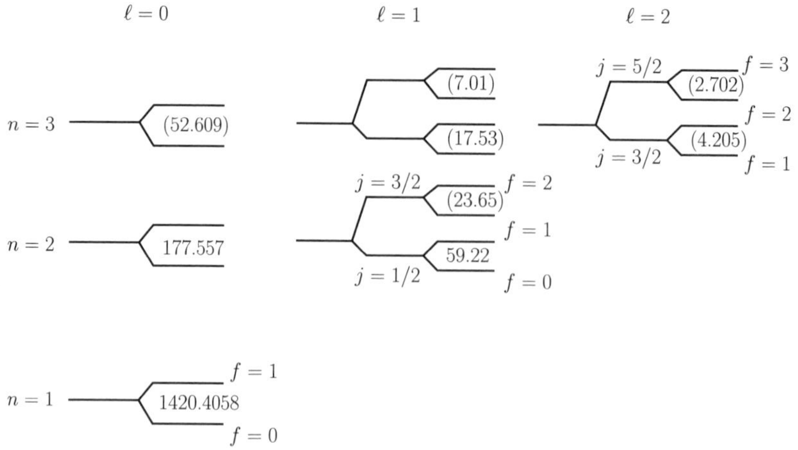

Figure 9.2

```
           ℓ = 0              ℓ = 1                    ℓ = 2
                                                              j = 5/2 ─── f = 3
                                          (7.01)              (2.702)
                                                                     ─── f = 2
n = 3  ─── (52.609)                                           (4.205)
                                          (17.53)   j = 3/2          ─── f = 1
                              j = 3/2           f = 2
                                         (23.65)
                                                f = 1
n = 2  ─── 177.557                       59.22
                              j = 1/2           f = 0

                              f = 1
n = 1  ─── 1420.4058
                              f = 0
```

Figure 9.2 The hydrogen energy levels are split by the hyperfine interaction as shown here (all values in MHz). There are twice as many states as in Fig. 8.1 because most of them are doublets, but the sublevel structure is also markedly changed. The spectrum becomes even more complicated for atoms with heavier nuclei. There are many more transitions, and new selection rules apply. Experimental hyperfine splittings are from Ref. [77], whereas values in parenthesis are calculated.

by $a = 8/3 \; g_I \alpha^2 R_\infty (m/M)$. The splitting between the states is a and the transition frequency between the two states becomes $\nu = a/h = 1422.8$ MHz. The experimental value is $\nu = 1{,}420.405\,751\,766\,7$ MHz [87] and the discrepancy of only 0.2% between this value and the calculation arises from quantum electrodynamical corrections.

This hyperfine splitting is the most accurately known quantity in physics. It is the frequency of the hydrogen maser, which is the best atomic clock ever built. The wavelength of this transition is 21 cm, which is the most pervasive and distinctive radiation in the universe. For the $2^2P_{1/2}$-state of hydrogen the splitting is a factor 24 smaller, where a factor 8 stems from the n^3-term in the denominator and a factor 3 from the $(\ell + 1/2)$ in the denominator. For the $2^2P_{3/2}$ state the hyperfine splitting is a factor 60 smaller than that of the ground state.

9.3.4 Hyperfine energies for other atoms

In the previous section the hyperfine structure of hydrogen caused by the magnetic dipole moment of the nucleus is discussed. In general, the electric and magnetic moments of the nucleus can be expanded in a similar way, as discussed for optical transitions in Sec. 4.1 for the induced electronic moments. This leads to higher-order contributions to the hyperfine shift. For hydrogen, the nucleus consists of only one proton and is assumed to be spherically symmetric, and thus only the magnetic dipole moment leads to a hyperfine shift. Fortunately, for other atoms only the electric quadrupole moment leads to a significant, additional shift. In this section the results for the hyperfine shift caused by the electric quadrupole moment are given without proof.

The interaction energy between the electric quadrupole moment of the nucleus and the potential V_e of the electron was first derived by Casimir, and the Hamiltonian is given by [88]

$$\mathcal{H}'_{\mathrm{hfs}} = \frac{3\vec{I}\cdot\vec{J}(2\vec{I}\cdot\vec{J}+1) - 2\vec{I}^2\vec{J}^2}{2I(2I-1)2j(2j-1)\hbar^4} Q_{zz} \frac{d^2V_e}{dz^2}, \quad (9.15)$$

with Q_{zz} the quadrupole moment in the z-direction. The shift caused by this interaction can easily be derived using the eigenfunctions $|FM_F\rangle$ and is given by

$$E_{\mathrm{hfs}} = \langle FM_F|\mathcal{H}'_{\mathrm{hfs}}|FM_F\rangle = b\frac{{}^3\!/_2 K(K+1) - 2I(I+1)j(j+1)}{2I(2I-1)2j(2j-1)}, \quad (9.16\mathrm{a})$$

with

$$b = \left\langle Q_{zz}\frac{d^2V_e}{dz^2}\right\rangle \quad (9.16\mathrm{b})$$

and

$$K = F(F+1) - I(I+1) - j(j+1). \quad (9.16\mathrm{c})$$

This shift has to be added to Eq. (9.11) to obtain the total hyperfine shift. Note that this term causes a departure from the Landé interval rule. Values of a and b for various alkali-metal atoms are shown in Tab. 9.1.

Appendices

9.A Interacting magnetic dipoles

The magnetic hyperfine interaction in atoms is caused by the interaction between the magnetic moments of the nucleus and the electron. A difficulty arises in the electrodynamical description of the interaction. Once this difficulty has been resolved, the resulting interaction can be treated in quantum mechanics as a small perturbation without difficulty. In this section the electrodynamics involved will be discussed following a very instructive treatment in Ref. [90].

The field created by an ideal magnetic dipole is given by

$$\vec{\mathcal{B}}(\vec{r}) = \frac{\mu_0}{4\pi r^3}\left(3(\vec{\mu}\cdot\hat{r})\hat{r} - \vec{\mu}\right). \quad (9.17)$$

Element	Abundance	I	F_g ($J_g = 1/2$)	a (MHz)	ΔE_{fs} (GHz)	F_e ($J_e = 1/2$)	a (MHz)	F_e ($J_e = 3/2$)	a (MHz)	b (MHz)
^{1}H	99.985	1/2	0,1	1420.405	10.968	0,1	59.18	1,2	23.67	–
^{6}Li	7.5	1	1/2, 3/2	152.137		1/2, 3/2	17.375	1/2, 3/2, 5/2	−1.155	−0.10
^{7}Li	92.5	3/2	1,2	401.752	10.091	1,2	45.914	0,1,2,3	−3.055	−0.221
^{23}Na	100	3/2	1,2	885.813	515.53	1,2	94.3	0,1,2,3	18.69	2.90
^{39}K	93.26	3/2	1,2	230.859	1730.4	1,2	28.85	0,1,2,3	6.06	2.83
^{40}K	0.0117	4	7/2, 9/2	−285.731		7/2, 9/2	–	5/2, 7/2, 9/2, 11/2	−7.59	−3.5
^{41}K	6.73	3/2	1,2	127.007		1,2	–	0,1,2,3	3.40	3.34
^{85}Rb	72.17	5/2	2,3	1011.910	7123.0	2,3	120.72	1,2,3,4	25.009	25.88
^{87}Rb	27.83	3/2	1,2	3417.341		1,2	406.2	0,1,2,3	84.845	12.52
^{133}Cs	100	7/2	3,4	2298.157	16611.8	3,4	291.90	2,3,4,5	50.34	−0.38

Table 9.1 Fine- and hyperfine-structure coefficients for the various alkali-metal atoms. The values for a and b are from Ref. [89].

9.A Interacting magnetic dipoles

Although this result is very familiar, it is not completely correct. As shown in Ref. [90] the average of the field given by Eq. (9.17) over a sphere of radius R is zero, whereas it is also shown in the same article that the average field of a magnetic dipole becomes $\vec{\mathcal{B}}_{av} = 2\mu_0\vec{\mu}/(2\pi R^3)$. The discrepancy arises since the magnetic field is obtained by taking the curl of the vector potential that diverges for $r = 0$.

Using a uniformly magnetized sphere of radius r_m with a magnetization $\vec{m}$, the magnetic dipole moment is given by $\vec{\mu} = 4/3\pi r_m^3 \vec{m}$ and the field inside the sphere is given by

$$\vec{\mathcal{B}}(\vec{r}) = \frac{\mu_0 \vec{\mu}}{2\pi r_m^3}, \quad r < r_m \tag{9.18}$$

By taking the limit of $r_m \to 0$ keeping $|\vec{\mu}|$ constant, the field goes to zero everywhere, except for $r = 0$. Thus for an ideal magnetic dipole this contribution to the field becomes

$$\vec{\mathcal{B}}(\vec{r}) = \frac{2}{3}\mu_0 \vec{\mu}\delta(\vec{r}) \tag{9.19}$$

and this term is usually neglected, since it only contributes at the position of the dipole itself. However, including this term in the field of Eq. (9.17) yields a contribution of the average field over a sphere of radius R and thus corrects for the discrepancy, as mentioned above.

The energy of a magnetic dipole in a magnetic field $\vec{\mathcal{B}}$ is given by $E = -\vec{\mu}\cdot\vec{\mathcal{B}}$. So two interacting magnetic dipoles $\vec{\mu}_1$ and $\vec{\mu}_2$ have an interaction energy of

$$E = -\frac{\mu_0}{4\pi r^3}\left(3\left(\vec{\mu}_1\cdot\hat{r}\right)\left(\vec{\mu}_2\cdot\hat{r}\right) - \vec{\mu}_1\cdot\vec{\mu}_2\right) - \frac{2}{3}\mu_0\left(\vec{\mu}_1\cdot\vec{\mu}_2\right)\delta(\vec{r}), \tag{9.20}$$

and the result shows that the interaction energy is symmetric in $\vec{\mu}_1$ and $\vec{\mu}_2$, as it should be. For the case of the interaction of the magnetic dipole moment of the electron $\vec{\mu}_e$ with the magnetic dipole moment of the nucleus $\vec{\mu}_I$ one has a choice. For the case of $\ell \neq 0$ in Sec. 9.3.1, first the magnetic field generated by the dipole moment of the electron is calculated, where both the term due to orbital angular momentum ℓ of the electron and its intrinsic magnetic moment are taken into account. The last term of Eq. (9.20) that is caused by the intrinsic magnetic moment of the nucleus can be neglected, since for $\ell \neq 0$ the electronic wavefunction for $r = 0$ is zero.

In Sec. 9.3.1 it is shown that the contribution of the intrinsic magnetic moment of the electron to the interaction energy becomes zero once averaged over the state and thus in Sec. 9.3.2 one can neglect this term. Since $\ell = 0$ in that section there is no contribution of the orbital angular momentum to the interaction energy and only the last term of Eq. (9.20) has to be evaluated, as is done in Sec. 9.3.2.

9.B Hyperfine structure for two spin-$1/2$ particles

The Hamiltonian and its expectation values for the interaction between the magnetic moments of an electron and a nucleus have been calculated in Sec. 9.3. The eigenfunctions are given by the states $|FM_F\rangle$. When the eigenfunctions in the uncoupled basis are needed, these states can be expanded in the uncoupled basis with the usual relations using Clebsch–Gordan coefficients (see Eq. (C.9)). Here a different approach will be used to calculate these eigenfunctions for the case of two interacting spin $1/2$ particles, as in the hydrogen atom ground state.

The Hamiltonian is a constant multiplying the operator $\vec{I} \cdot \vec{s}$ as given in Sec. 9.3, but the constant will be dropped here. Then the Hamiltonian operator can be written as

$$\vec{I} \cdot \vec{s} = I_z s_z + 1/2(I_+ s_- + I_- s_+) \tag{9.21}$$

where $I_\pm \equiv I_x \pm i I_y$ and similarly for $s_\pm$. Note that I_z and $I_\pm$, and correspondingly for $\vec{s}$, are operators with dimension $\hbar$. The action of these "ladder operators" $I_\pm$ on the states denoted by $|I, m_I\rangle$ is given by

$$I_\pm |I, m_I\rangle = \hbar \sqrt{(I \mp m_I)(I \pm m_I + 1)}\ |I, m_I \pm 1\rangle \tag{9.22}$$

and similarly for $s_\pm$.

The magnetic moments can be oriented parallel or antiparallel to the chosen axis, and labeled $m_I = \pm 1/2$ and $m_s = \pm 1/2$ for identification. There are four possible sublevels labeled by $|m_I, m_s\rangle = |+1/2, +1/2\rangle, |+1/2, -1/2\rangle, |-1/2, +1/2\rangle, |-1/2, -1/2\rangle$. Since these are eigenfunctions of I_z and s_z with eigenvalues $\pm \hbar/2$, the 4×4 Hamiltonian matrix can be found using Eq. (9.22) to be

$$\mathcal{H}'_{\text{hfs}} \propto \begin{pmatrix} 1 & 0 & 0 & 0 \\ 0 & -1 & 2 & 0 \\ 0 & 2 & -1 & 0 \\ 0 & 0 & 0 & 1 \end{pmatrix}. \tag{9.23}$$

Three of the four eigenvalues of this matrix are the same, and the fourth one is three times larger and of opposite sign, just as in Sec. 9.3 above. Two of the eigenvectors are $|a\rangle = |+1/2, +1/2\rangle$ and $|d\rangle = |-1/2, -1/2\rangle$, and the other two can be found from the central 2×2 submatrix to be $|b, c\rangle = \frac{1}{2}\sqrt{2}(|+1/2, -1/2\rangle \pm |-1/2, +1/2\rangle)$. For the various cases of I or s larger than $1/2$, the procedure is the same but the Hamiltonian matrix is larger.

By applying the central 2×2 submatrix to the vectors $|b\rangle$ or $|c\rangle$ it is easy to identify $|b\rangle$ with the eigenvalue $+1$ and $|c\rangle$ with -3. Calculating the value of F for all for states shows that the three with eigenvalue $+1$ belong to $F = 1$ (including $|b\rangle$) and the fourth belongs to $F = 0$ (namely $|c\rangle$), just what one might expect from the multiplicity associated with the possible values of M_F. These results are

State	Eigenstate	Energy ($\hbar^2 a/4$)	F	M_F
$\|a\rangle$	$\|+1/2, +1/2\rangle$	+1	1	+1
$\|b\rangle$	$\frac{1}{2}\sqrt{2}\left[\|+1/2, -1/2\rangle + \|-1/2, +1/2\rangle\right]$	+1	1	0
$\|c\rangle$	$\frac{1}{2}\sqrt{2}\left[\|+1/2, -1/2\rangle - \|-1/2, +1/2\rangle\right]$	−3	0	0
$\|d\rangle$	$\|-1/2, -1/2\rangle$	+1	1	−1

Table 9.2 The four hyperfine sublevels for the magnetic interaction between two spin-1/2 particles such as the electron and proton that constitute the hydrogen atom.

summarized in Tab. 9.2. Similar states exist for positronium to be discussed in Chap. 11. For larger angular momentum values, such as hydrogen in the 2P state with $j = 3/2$ or sodium in the ground state with $I = 3/2$ the resulting F-values are 1 and 2, the number of states is larger, but the method for finding them is the same as done here.

9.C The hydrogen maser

Although the ammonia beam molecular maser was the first one, the hydrogen maser described in Ref. [91] has demonstrated higher stability than any other candidate for an atomic clock. It is based on the ground-state hfs transition between the $F = 0$ and $F = 1$ states shown in Fig. 9.2 at a frequency near 1.42 GHz, corresponding to a wavelength of 21 cm. This is the transition that is so vitally important in radio astronomy.

The principle of operation has three major segments. First, H_2 has to be dissociated into H atoms, and this is done in a weak discharge of gas emerging from a source. The resulting atoms are then mechanically collimated by apertures to form a beam. Second, atoms in the highest-energy hfs sub-state have to be separated from the others to produce a population inversion, and this is done by exploiting the Zeeman energy shifts of Chap. 11 in an inhomogeneous field so that the shifts are position dependent resulting in a force. This is the principle of the Stern–Gerlach experiment, and the magnet is often called a Stern–Gerlach magnet. Third, the atoms enter a carefully constructed, high-Q cavity that is resonant at the hfs frequency. Radiation present in the cavity produces stimulated emission that more than compensates the losses until a relatively stable energy density is achieved. One of the loss mechanisms is the outcoupling of some energy by a judiciously placed aperture or antenna in the cavity, and this is the maser output.

Perhaps the most obvious problem to overcome in making such a maser results from collisions of the atoms with the cavity walls. Even at cryogenic temperature, the thermal velocity is ~500 m/s so the average time between wall collisions in a modest size cell is less than 0.5 ms. The Fourier transform limit of the linewidth would then be ~1 kHz, about 10^7 times larger than the measured frequency stability. The trick is to minimize the effect of the wall collisions, and this is done by lining the inside of the cavity with a quartz bulb whose inside walls are coated with a particular form of Teflon that has this special desired property, and the consequences are sub-Hz linewidths.

The frequency stability, as measured by a criterion called the Allan variance, is ~$1:10^{15}$ per hour corresponding to less than 1 ns/day. Nevertheless, the time standards of the world are based on a similar hfs transition in Cs atoms even though their stability is ten times worse, because the fundamental frequencies of different H masers vary in an uncontrolled way, probably resulting from minute differences in the nature of the cell wall coatings. In the atomic beam and fountain clocks using Cs, there are only inter atomic collisions whose rate can be controlled by changing the atomic density, and are therefore much more reproducible worldwide. There are many commercial manufacturers of atomic clocks of various kinds.

Exercises

9.1 Calculate the difference in wavelength for the hydrogen Balmer-α line at $\lambda = 656$ nm ($n = 3 \to n = 2$) that arises from the mass difference between hydrogen and deuterium. This was the resolution that Urey needed for his discovery of the deuteron.

9.2 Show by using the Biot–Savart law that the magnetic field from a classical dipole is given by Eq. (9.18).

9.3 Show that for an S state the scalar product of the electronic and nuclear spin angular momenta may be expressed in the form

$$\vec{I} \cdot \vec{J} = I_z J_z + (I_+ J_- + I_- J_+)/2$$

where $I_\pm = I_x \pm iI_y$ and $J_\pm = J_x \pm iJ_y$. The properties of the raising and lowering operators are given by Eq. (C.6). Then calculate all the matrix elements of the hyperfine interaction $\langle JIM_J M_I | \mathcal{H}_{\text{hfs}} | JIM'_J M'_I \rangle$ in the uncoupled representation for the case $I = 1/2$, $J = 1/2$. Finally, by solving the secular equation, determine the energies of the hyperfine sub levels in an applied magnetic field of arbitrary strength.

Exercises

9.4 In the hydrogen model the nucleus is considered as a point-particle with no spatial extension. In practice the nucleus has a finite extension of R_N, in which the motion of the electron at small r is modified. This effect will be estimated using two different approaches.

(a) Use the wavefunction $\psi_{100}(\vec{r})$ for the ground state of hydrogen to calculate the probability P that the electron is in the nucleus.

(b) Approximate this result using the inequality $R_N \ll a_0$.

As an alternative one can consider the wavefunction $\psi_{100}(\vec{r})$ as constant within the nucleus.

(c) Use this approximation to calculate P again and compare the result with the result of (b).

(d) Calculate P by using $R_N \approx 10^{-15}$ m.

(e) If we assume that within the nucleus the Coulomb potential $V(r) = 0$, estimate the shift of the energy of the ground state. Compare the result with the result of the unperturbed energy.

9.5 Consider a nucleus as a sphere of radius R_N with a uniform charge distribution. To estimate the shift of the atomic energy levels arising from this non-Coulombic part of the potential, start with the charge density $\rho_q = 3Ze/4\pi R_N^3$ and find the electric field inside the nucleus from Gauss' law using $Q_{enc} = \rho_r(4/3)\pi r^3 = Ze(r/R_N)^3 = \varepsilon_0 \int \vec{\mathcal{E}} \cdot d\vec{A} = 4\pi\varepsilon_0 \mathcal{E} r^2$. Do the integral $V = e \int \vec{\mathcal{E}} \cdot d\vec{r}$ and consider the resulting potential as a additive perturbation to the Coulomb potential, and use first-order perturbation theory to find the energy shift for the $n = 1$ and 2 states of hydrogen (see Tab. 7.6).

10
The alkali-metal atoms

The Schrödinger equation is exactly solvable for hydrogenic atoms and provides excellent results for the energies and wavefunctions. For all the other atoms, the Schrödinger equation can be written down and the Hamiltonian is known exactly, but the resulting equation cannot be solved analytically. The main problem for these systems is that the electron–electron interaction does not admit the central field solutions of the hydrogen atom. The solution is intractable, and one has to resort to approximations to describe these atoms.

A special case is the alkali-metal atoms, where all but one of the electrons are in a spherically symmetric core consisting of closed shells. This one remaining valence electron determines most of the atomic properties. Since the size of the core is much smaller than the orbital radius of the valence electron, the wavefunction is concentrated outside the core, where the field is purely Coulombic and the hydrogen solutions are appropriate. Only electrons that penetrate the core are strongly affected by it, and these electrons have low orbital angular momentum ℓ.

In the nineteenth century it was discovered that the energy of the states of the alkali-metal atoms can be accurately described by the Rydberg formula when the quantum number n is replaced by an effective quantum number n^* that differs from n by an ℓ-dependent constant denoted δ_ℓ. The shift was named the quantum defect by Schrödinger before the advent of quantum mechanics. It is remarkable that δ_ℓ depends only on ℓ and is nearly independent of n.

In this chapter many different techniques that are more or less standard in quantum mechanics will be used to describe the alkali-metal atoms. For each of these there is sufficient background provided to be able to understand the line of reasoning even without a formal introduction. For further details one should consult the standard textbooks on quantum mechanics. For example, a model will be introduced that can explain the physical background of the quantum defect and show how to calculate the various δ_ℓ values from first principles. The defects will subsequently be used to calculate the energy levels and the transition rates between several states.

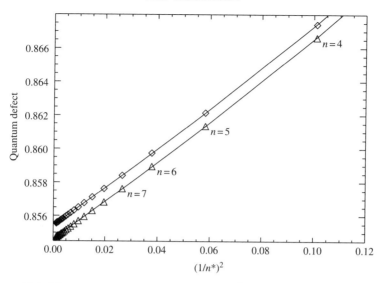

Figure 10.1 Measured quantum defects for the $^2P_{1/2}$ (diamonds) and $^2P_{3/2}$ (triangles) states of sodium as a function of $1/(n^*)^2$, where n^* is the effective quantum number. Note that the quantum defect is only slightly dependent on n (see App. B.2 for the meaning of the labels $^2P_{1/2,3/2}$). Data from Ref. [92].

10.1 Introduction

It was discovered in the early days of atomic spectroscopy that the energy levels of the alkali-metal atoms could be fitted accurately with the modified Rydberg formula:

$$E_n = -\frac{R_\infty}{(n-\delta_\ell)^2} \equiv -\frac{R_\infty}{n^{*2}}, \qquad (10.1)$$

where $n^* \equiv n - \delta_\ell$ contains the ℓ-dependent quantum defect. The spectra of the alkalis thus resemble that of hydrogen closely, apart from these shifts. It is large for the lowest ℓ states (s states), but becomes progressively smaller for higher ℓ states. For instance, for the d state of Na it is only 0.015 (see Tab. 10.4). For the two 3p states of sodium that are split by the fine structure (see Chap. 8), the measured quantum defects are plotted as a function of $1/(n^*)^2$ in Fig. 10.1. The figure shows that the quantum defects of both states are nearly independent of n, and in Sec. 10.2 below, the origin of the weak linear n-dependence will be described.

The non-zero quantum defect derives from the inner electrons so they need to be studied. Figure 10.2 shows the charge distribution of the electrons in the spherically symmetric core of sodium. The plots are based on Hartree–Fock calculations (see Sec. 15.6) where the wavefunctions of the electrons and the interaction potential are calculated in a self-consistent way. The first peak near $0.2a_0$ comes from the 1s

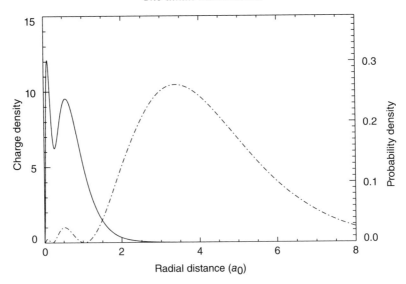

Figure 10.2 Distribution of the charge of the sodium core as a function of the distance from the nucleus, as determined from Hartree–Fock calculations. The solid line is for the core of the sodium atom with the valence electron present, which cannot be distinguished on this scale from the charge distribution for the Na^+ ion. The dashed-dotted line is the probability density of the valence electron in the 3s state. Note that more than 95% of the 3s wavefunction is located outside a radius of $r_c = 1.5\,a_0$, whereas about 98% of the charge distribution of the core is located within this range.

electrons and the second broader peak near $0.5a_0$ comes from the electrons in the $n = 2$ shell. Most of the charge of the core is thus located within a distance of $2a_0$ of the nucleus.

By contrast, the distribution of the probability density for the 3s electron is mostly located outside the core, where the nuclear charge is largely shielded by the core electrons and the potential is nearly Coulombic. Completely removing the valence electron leaves the charge distribution of the core nearly unaffected. Thus one can conclude that the properties of the outer electron are altered only in the small region inside the core. This suggests calculating the wavefunction in two different ranges: first at short range, where the complicated interactions with the core electrons have to be treated carefully, and second at long range, where the potential is nearly purely Coulombic. This is very much like the spatial separation of Sec. 12.3.4 for atoms in electric fields.

10.2 Quantum defect theory

The quantum defect method is based on separating the calculation of the radial wavefuncton into two distinct spatial regions (the angular part is not affected as

10.2 Quantum defect theory

long as the potential is central). At large range, where most of the wavefunction is concentrated (see Fig. 10.2), the potential is purely Coulombic and the solutions for the wavefunction have been described in Chap. 7. At short range, where the electron is less likely to be found, the interaction between the electron and the core is complicated and analytic solutions are not known. However, the effect of the interactions at short range can be described by a phase shift $\pi\delta$ of the wavefunction at a certain distance $r = r_m$. For $r > r_m$ the potential is taken to be Coulombic.

At short range, the potential energy $V(r)$ dominates the total energy E of the valence electron, especially for the higher-lying states (the Rydberg states). Thus the shift $\pi\delta$ of the wavefunction is nearly independent of n for states with the same angular momentum ℓ. However, the shift depends strongly upon the penetration of the valence electron in the core and this penetration becomes stronger for lower values of ℓ (note the factor ρ^ℓ in Eq. (7.17).

Using the WKB method to solve the radial Schrödinger equation shows how this arises. For bound states the radial equation can be rewritten as

$$\frac{d^2 u_{E\ell}(r)}{dr^2} + k(r)^2 u_{E\ell}(r) = 0; \qquad k(r) = \sqrt{\frac{2\mu(E - V(r))}{\hbar^2}}, \qquad (10.2)$$

where the potential $V(r)$ includes the centrifugal term $\ell(\ell+1)\hbar^2/2\mu r^2$ (see Eq. (7.9)). The WKB method exploits the slow variation of $k(r)$ with r in the region where r is far from the classically forbidden regions at both short and long range. When this slow radial dependence is neglected, the solution can be written as

$$u_{E\ell}(r) \propto \exp[ik(r)r]. \qquad (10.3)$$

By contrast, near the classical turning points the kinetic energy of the electron vanishes and the approximation breaks down. In the purely Coulombic case, these turning points are given by

$$r_{1,2} = \frac{n^2 a_0}{Z}\left(1 \mp \sqrt{1 - \frac{\ell(\ell+1)}{n^2}}\right), \qquad (10.4)$$

where r_1 (r_2) is the inner (outer) turning point.

The next step is to find the phase shift of the radial function at a suitable matching distance r_m because for $r > r_m$ the potential is dominated by the Coulomb part of $V(r)$ in Eq. (10.2). Integrating outward from $r = r_1$ leads to

$$u_{E\ell}(r) \propto \frac{1}{k(r)} \cos\left(\int_{r_1}^r k(r')dr' - \frac{\pi}{4}\right). \qquad (10.5)$$

Note that there is a phase shift of $\pi/4$ from integration of Eq. (10.2) over the classically forbidden region between $r = 0$ and $r = r_1$ [61]. At long range but not as far as r_2, Eq. (10.5) becomes

$$u_{E\ell}(r) \propto \sqrt{\frac{r}{2Za_0}} \cos\left(\sqrt{\frac{8Zr}{a_0}} - (\ell + 1/2)\pi - \frac{\pi}{4}\right). \quad (10.6)$$

The final step in the WKB calculation is to match the phase of Eq. (10.5) with a solution where the potential is purely Coulombic over the whole range. The difference between them at the matching distance $r = r_m$ is given by

$$\pi\delta = \int_0^{r_m} (k_0(r) - k_1(r))dr, \quad (10.7)$$

where $k_0(r)$ is the wavenumber in the case of a purely Coulombic potential, and $k_1(r)$ is the wavenumber for the non-Coulombic potential for $r < r_m$.

Since E is much smaller than $V(r)$ at distances below r_m, the square root for $k(r)$ can easily be expanded in powers of E and thus in powers of $1/(n^*)^2$ to give

$$\delta_\ell = \delta_\ell^{(0)} + \frac{\delta_\ell^{(1)}}{(n^*)^2} + \frac{\delta_\ell^{(2)}}{(n^*)^4} + \ldots \quad (10.8)$$

The coefficients for $\delta_\ell^{(i)}$ are given for the alkalis in App. 10.A up to order $i = 2$. They provide a method to determine the energies of the states to an accuracy of a few tens of MHz, making the quantum defect method very powerful tool for the description of alkali-metals. Apart from the energies, quantum defects can also be used to calculate the transition strength between different states (see Sec. 10.5).

10.3 Non-penetrating orbits

This section provides a description for the case where the outer electron does not penetrate the core so that it does not experience the core potential. However, the outer electron repels the inner electron cloud and the resulting core polarization creates a induced dipole moment that produces a force at the position of the outer electron. The induced dipole moment is $\propto 1/r^2$, and the field of a dipole $\propto 1/r^3$, so the induced electric field at the outer electron is $\propto 1/r^5$. Thus the induced polarization potential at the position of the valence electron is written as $V(r) = -\alpha_D e^2/8\pi\varepsilon_0 r^4$, where α_D is the dipole polarizability of the core [93]. Apart from the dipole polarizability, there are also contributions from higher orders, such as the quadrupole polarizability, so the total polarizability potential is given by

$$V_{\text{pol}}(r) = -\frac{e^2}{4\pi\varepsilon_0}\left(\frac{\alpha_D}{2r^4} + \frac{\alpha_Q}{2r^6} + \cdots\right). \quad (10.9)$$

Values for the polarizabilities of the alkali-metal cores are given in Tab. 10.1.

	α_D	α_Q	r_d	r_c	c_0	c_1
Li	0.1923	0.1134	0.47	0.47	+0.546266	−0.303243
Na	0.9459	6.9273	1.00	1.00	+0.38514	−0.10506
K	5.331	4.1026	1.20	1.20	+1.0793	+0.34470
Rb	8.976	113.18	1.40	3.75	−0.97872	+0.16158
Cs	19.06	581.60	1.50	4.00	−0.79710	+0.15052

Table 10.1 Dipole and quadrupole polarizabilities and other model potential parameters in atomic units for the alkali-metal core. The atomic unit for the dipole polarizability is $a_0^3 = 1.482 \times 10^{-31}$ m^3 and for the quadrupole polarizability $a_0^5 = 4.142 \times 10^{-52}$ m^5. The parameters for Li are from Ref. [94], whereas the parameters for the other elements are from Ref. [95].

Using the hydrogenic wavefunctions as a basis set in first-order perturbation theory enables calculation of the energy shift caused by the polarizability:

$$E_{\text{pol}} = \langle R_{n\ell}(r)|V_{\text{pol}}(r)|R_{n\ell}(r)\rangle. \tag{10.10}$$

This integral can readily be evaluated by using the results from Tab. 7.6. For $1 \ll \ell \ll n$ Eq. (10.10) becomes

$$E_{\text{pol}} = \frac{-e^2}{4\pi\varepsilon_0}\left(\frac{3\alpha_D}{4n^3\ell^5 a_0^4} + \frac{35\alpha_Q}{16n^3\ell^9 a_0^6} + \cdots\right). \tag{10.11}$$

Adding this shift to the hydrogenic energy $E = -R_\infty/n^2$ leads to

$$\begin{aligned}E_{n\ell m} &= -\frac{R_\infty}{n^2} - \frac{e^2}{4\pi\varepsilon_0}\left(\frac{3\alpha_D}{4n^3\ell^5 a_0^4} + \frac{35\alpha_Q}{16n^3\ell^9 a_0^6} + \cdots\right) \\ &\equiv -\frac{R_\infty}{n^2}\left(1 + \frac{2\delta_\ell}{n} + \cdots\right) \approx -\frac{R_\infty}{(n-\delta_\ell)^2},\end{aligned} \tag{10.12}$$

with the quantum defect δ_ℓ given by

$$\delta_\ell = \frac{3\alpha_D}{4\ell^5 a_0^3} + \frac{35\alpha_Q}{16\ell^9 a_0^5}. \tag{10.13}$$

Table 10.2 shows reasonable agreement between the calculated and experimental quantum defects for $\ell > 1$.

For any perturbation potential that decreases faster than $1/r$ at long range, the energy shift for $n \gg \ell$ given by the matrix elements in Tab. 7.6 is approximately proportional to $1/n^3$, so it can be expressed as a quantum defect independent of n [59]. This is a property of the hydrogen radial wavefunctions.

		δ_2	δ_3	δ_4	δ_5
Li	th	0.002 496	0.000 303	0.000 071	0.000 023
	exp	0.001 8	−0.009 7		
Na	th	0.012 762	0.001 572	0.000 369	0.000 120
	exp	0.014 9	0.001 6	0.000 4	0.000 1
K	th	0.071 237	0.008 455	0.001 969	0.000 642
	exp	0.277 5	0.009 8		
Rb	th	0.346 966	0.020 141	0.003 759	0.001 141
	exp	1.347 7	0.016 4		
Cs	th	1.465 791	0.061 732	0.009 407	0.002 613
	exp	2.470 6	0.033 4	0.007 0	

Table 10.2 Comparison of quantum defects δ_ℓ for large n and various values of ℓ between experiment (see Tab. 10.4) and theory (see Eq. (10.13)). The polarizabilities are taken from Tab. 10.1.

10.4 Model potentials

The non-Coulombic potential of the alkali-metal atom cores can only be approximated. Although it is not possible to measure the potential directly, the energy of the levels in the alkalis can be measured with high accuracy so that the potential can be determined. This requires a model potential that approaches the real potential at large range and mimics the potential at short range as much as possible. It is done by introducing some adjustable parameters for the short-range potential that are optimized in a least squares adjustment to reproduce the measured energies of the states. Then the wavefunction of the valence electron can be found, from which many other properties of the atom can be derived.

For the alkalis the following model potential has been introduced in Refs. [96, 97]:

$$V(r) = V_{\mathrm{HF}}(r) - \frac{\alpha_D e^2}{2r^4}\left(1 - e^{-(r/r_c)^6}\right) - \frac{\alpha_Q e^2}{2r^6}\left(1 - e^{-(r/r_c)^8}\right) + (c_0 + c_1 r)e^{-r/r_d}. \quad (10.14)$$

The first term on the right-hand side is the Hartree–Fock potential of the ionic core which approaches $-e^2/4\pi\varepsilon_0 r$ at large range and $-Ze^2/4\pi\varepsilon_0 r$ at short range. The second and third terms are the dipole and quadrupole polarizabilities which are cut off at the core radius r_d. The additional term is necessary to have sufficient flexibility to adjust the energy positions to the experimental values.

Of course, the wavefunction for the electron in this potential cannot be solved analytically, so the Schrödinger equation needs to be numerically integrated. There

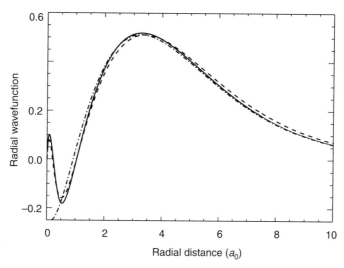

Figure 10.3 Radial wavefunction $u(r) = rR(r)$ for the 3s valence electron of Na. The wavefunction is calculated using three different methods: integrating the radial wave equation using the model potential of Eq. (10.14) (solid line); using the Hartree–Fock result of Ref. [98] (dashed line, see Sec. 15.6); and using the analytical result of the Bates–Damgaard model [99] (dashed-dotted line, see Sec. 10.5). The results agree rather well apart from differences at small distance.

are many methods available for numerical integration, but one of the simplest and most powerful ways is the Numerov method (see App. 10.B). In Tab. 10.1 values of the free adjustable parameters r_d, c_0 and c_1 are given for the alkalis together with the experimentally determined polarizabilities α_D and α_Q. In Fig. 10.3 the wavefunction is plotted for the 3s state of sodium using the model potential and is compared with the HF wavefunction. As the figure shows, the agreement between the two is reasonable.

10.5 Optical transitions in alkali-metal atoms

One of the most important capabilities of atomic theory is that it can predict the wavelength and strength of transitions between states in atoms. Such information is important for models in astrophysics, plasma physics and other fields of physics where atoms in excited states play an important role. The fingerprint of the emission spectrum of an atomic sample can serve for trace analysis for an atomic or ionic compound. However, this requires detailed knowledge not only of the wavelengths of the emission lines but also of the strength of the transitions involved.

Both the radial and the angular parts of alkali-atom wavefunctions are changed compared to hydrogen (see Sec. 7.6). The radial part changes because of the quantum defects of the states that can be described in the Bates–Damgaard model, as discussed in Sec. 10.5.1 below. The angular part also changes, since the fine and hyperfine interactions change the symmetry of the states. This is discussed in Sec. 10.5.2.

10.5.1 Radial part

In Ref. [99] the quantum defects are used to calculate the transition strengths for the alkali-metal atoms. In analogy with the radial wavefunctions of hydrogen, the radial wavefunctions can be defined as $R_{n^*\ell}(r) = u_{n^*\ell}(r)/r$, where the wavefunction $u_{n^*\ell}(r)$ satisfies

$$\frac{d^2 u_{n^*\ell}(r)}{dr^2} + \frac{2\mu}{\hbar^2}\left[\frac{e^2}{4\pi\varepsilon_0 r} - \frac{R_\infty}{n^{*2}} - \frac{\ell(\ell+1)\hbar^2}{2\mu r^2}\right] u_{n^*\ell}(r) = 0, \tag{10.15}$$

and the potential is taken to be exact outside the core. The solution can then be written as (note the Bohr radius a_0 should not be confused with the coefficients a_k):

$$u_{n^*\ell}(r) = \exp\left(\frac{-r}{a_0 n^*}\right)\left(\frac{2r}{a_0 n^*}\right)^{n^*} \sum_{k=0}^{\infty} \frac{a_k}{(r/a_0)^k}, \tag{10.16}$$

where the coefficients are given by the recurrence relation

$$a_k = a_{k-1} \frac{n^*}{2k}\left[\ell(\ell+1) - (n^*-k)(n^*-k+1)\right], \tag{10.17}$$

and the first coefficient $a_0 = 1$. Although Eq. (10.15) is the same as Eqs. (7.9), the solutions in Eq. (10.16) look quite different. In the case where n^* becomes an integer, these solutions necessarily become identical. In order to normalize the wavefunction properly, the result has to be divided by the normalization factor $n^* \sqrt{\Gamma(n^*+\ell+1)\Gamma(n^*-\ell)}$, where $\Gamma(x)$ is the gamma function.

The transition strengths need the radial matrix elements $\langle n\ell|r|n'\ell'\rangle$, but since the wavefunctions are given in powers of r multiplied by exponential factors, the integrals become standard. These matrix elements are given by

$$\langle \alpha' n'\ell'||r||\alpha n\ell\rangle = a_0 \sum_{k,k'} a_k(n^*,\ell) a_{k'}(n^{*'},\ell') \left(\frac{n^*+n^{*'}}{n^* n^{*'}}\right)^{-(n^*+n^{*'}+2-k-k')}$$

$$\times \Gamma(n^*+n^{*'}+2-k-k'). \tag{10.18}$$

In order to terminate the infinite sums, only terms with $k + k' \leq n^* + n^{*'} - 1$ are taken into account. The remaining terms are neglected.

10.5 Optical transitions in alkali-metal atoms

				Theory		Experiment
El.	Transition	n_s^*	n_p^*	μ (au)	γ (MHz)	γ (MHz)
H	1s→2p	1.000	2.000	0.745	99.52	99.47
Li	2s→2p	1.589	1.959	2.352	5.93	5.92
Na	3s→3p	1.627	2.117	2.445	9.43	10.01
K	4s→4p	1.770	2.235	2.842	5.78	6.09
Rb	5s→5p	1.805	2.293	2.917	5.78	5.56
Cs	6s→6p	1.869	2.362	3.093	4.99	5.18

Table 10.3 Dipole moment $|\mu|$ for the first optically allowed transition in the alkali-metal atoms using matrix elements $\langle ns|r|np \rangle$ that are calculated using the procedure of Bates and Damgaard [99]. The linewidth γ is calculated using Eq. (5.13), whereas the experimental value is derived from the lifetimes of the np states. A dipole moment in atomic units is 1 au = ea_0 = 8.475×10^{-30} C-m.

Table 10.3 shows good agreement between the calculated values and those derived from experiments for the first optically allowed transitions for the alkali-metal atoms.

10.5.2 Angular part

In the case of the fine and/or hyperfine interaction the situation is considerably different from Sec. 7.6.3. For the fine structure, the energy levels are split by the spin–orbit interaction and the atomic states are no longer specified by the quantum numbers $\vec{\ell}$ and m_ℓ. The states are now specified by $\vec{j}$, the vector sum of $\vec{\ell}$ and $\vec{s}$, and its projection m_j on the z-axis. However, the optical electric field still couples only to the orbital angular momentum $\vec{\ell} = \vec{r} \times \vec{p}$ of the states. In this situation the Wigner–Eckart theorem could also be applied to calculate the transition strength [58], but again this section will follow a different route that provides more insight into the problem. Although the formulas below may appear rather complicated, the principle is simple.

The atomic eigenstates are denoted by $|\alpha j m_j\rangle$ in the j-basis, α represents all the other quantum numbers, and m_j explicitly indicates for which angular momentum the magnetic quantum number m is the projection. In most cases, this is obvious from the notation, but in this section it is not. The dipole transition matrix element is therefore given by

$$\mu_{eg} = -e \left\langle \alpha' j' m_j' \middle| \hat{\varepsilon} \cdot \vec{r} \middle| \alpha j m_j \right\rangle. \tag{10.19}$$

Since the optical electric field couples only the $\vec{\ell}$ component of these $\vec{j}$ states, these eigenfunctions must first be first expanded in terms of the ℓ and s wavefunctions (see Eq. (C.9)):

$$|\alpha jm_j\rangle = \sum_{m_\ell m_s} \langle jm_j|\ell m_\ell; sm_s\rangle \, |\alpha\ell m_\ell\rangle |sm_s\rangle. \tag{10.20}$$

The fact that Eq. (7.27) for the integral of the product of three spherical harmonics and Eq. (10.20) both contain the Clebsch–Gordan coefficients is a result of the important connection between the spherical harmonics $Y_{\ell m}$ and atomic angular momenta.

Substitution of Eq. (10.20) into Eq. (10.19) twice leads to a double summation, which contains matrix elements in the (ℓ, s) basis of the form

$$\langle \alpha'\ell'm'_\ell|\langle s'm'_s|r|\alpha\ell m_\ell\rangle|sm_s\rangle = \langle \alpha'\ell'm'_\ell|r|\alpha\ell m_\ell\rangle \delta_{ss'}\delta_{m_s m'_s}. \tag{10.21}$$

The first term on the right-hand side is the matrix element that has been evaluated before (see Eq. (7.23)). The δ-functions reflect the notion that the light couples the orbital angular momenta of the states, and not the spin. The spin and its projection are not changed by the transition. Substitution of Eq. (10.21) into Eq. (10.19), expansion of the matrix elements in the ℓ-basis, and recoupling of all the Clebsch–Gordan coefficients in $3j$-symbols using standard angular momentum relations as discussed in Refs. [58, 100, 101], leads to

$$\mu_{eg} = -e(-1)^{\ell'+s-m'_j} \langle \alpha'n'\ell'||r||\alpha n\ell\rangle \tag{10.22}$$

$$\times \sqrt{(2j+1)(2j'+1)} \begin{Bmatrix} \ell' & j' & s \\ j & \ell & 1 \end{Bmatrix} \begin{pmatrix} j & 1 & j' \\ m_j & q & -m'_j \end{pmatrix}.$$

The array of quantum numbers in the curly braces is not a $3j$-symbol, but is called a $6j$-symbol. It summarizes the recoupling of six angular momenta. Values for the $6j$-symbols are also tabulated in Ref. [102]. For hydrogen and the alkalis the transition strengths of the D-lines are shown in Fig. 10.4. The strongest transition is that from the $m_j = j\,(-j)$ state using right-handed (left-handed) circularly polarized light. It is particularly convenient to separate the m-dependence of these matrix elements to find selection rules and relative transition strengths for different polarizations of light. The spherical basis set of Eq. (3.20) whose subscript is q corresponds to the usual choices of polarization, and the relative strength of different transitions therefore reduces to the square of the $3j$-symbol.

In the presence of hyperfine interactions the situation becomes even more complicated. However, the procedure is the same. First the eigenfunctions in the F-basis are expanded in the (j, I)-basis, where I is the nuclear spin, and a $6j$-symbol involving I, j, and F appears. Then the eigenfunctions of the j-basis are further reduced into the (ℓ, s)-basis. Since the procedure is similar to the procedure for the fine-structure interaction, only the result is shown:

10.A Quantum defects for the alkalis

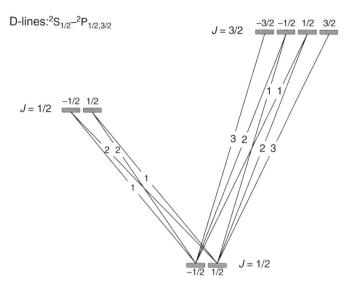

Figure 10.4 Transition strength for the D-lines of hydrogen and the alkali-metal atoms. The strength is normalized to the weakest allowed transition.

$$\mu_{eg} = -e(-1)^{1+\ell'+s+j+j'+I-m'_F} \langle \alpha'n'\ell'||r||\alpha n\ell \rangle \quad (10.23)$$

$$\times \sqrt{(2j+1)(2j'+1)(2F+1)(2F'+1)}$$

$$\times \begin{Bmatrix} \ell' & j' & s \\ j & \ell & 1 \end{Bmatrix} \begin{Bmatrix} j' & F' & I \\ F & j & 1 \end{Bmatrix} \begin{pmatrix} F & 1 & F' \\ m_F & q & -m'_F \end{pmatrix}.$$

The hyperfine interaction is important for the alkalis. For a system with nuclear spin $1/2$, such as H, the result is given in Fig. 10.5. Since s can be parallel or antiparallel to ℓ, $j' = 1/2$, $3/2$ and the fine-structure interaction is usually large compared with the hyperfine interaction. Results for transitions important for the alkalis are given in App. D.

Appendices

10.A Quantum defects for the alkalis

Table 10.4 shows the quantum defect parameters for the alkali-metal atoms. The parameters are derived from the energy levels for the states $n\ell_j$ with n in between $n_{\min}$ and $n_{\max}$ that are given in Ref. [92]. For each energy E_n the quantum defect is determined from

$$\delta_\ell(n) = n - \sqrt{\frac{R_X}{E_\infty - E_n}}, \quad (10.24)$$

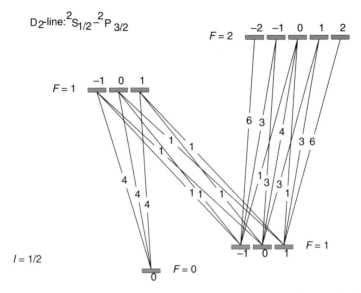

Figure 10.5 Transition strength for the first optical transition in an alkali-metal system with a nuclear spin of $1/2$, for example H. The strength is normalized to the weakest transition. For the D_1 lines see App. D. (Figure from Ref. [1].)

where E_∞ is the ionization potential and R_X the Rydberg constant for the particular element:

$$R_X = R_\infty \left(\frac{\mu}{m}\right), \tag{10.25}$$

with μ the reduced mass $\mu = mM/(m+M)$. The resulting values are fitted to the following expression:

$$\delta_\ell(n) = \delta_\ell^{(0)} + \frac{\delta_\ell^{(1)}}{\left(n - \delta_\ell^{(0)}\right)^2} + \frac{\delta_\ell^{(2)}}{\left(n - \delta_\ell^{(0)}\right)^4}. \tag{10.26}$$

The values of the table can be used to calculate the energies of any state using Eq. (10.1).

10.B Numerov method

The Numerov method is an effective numerical method to solve the Schrödinger equation and relies on the fact that the equation has only a second-order term but no first-order term. By discretizing the radial distance in steps of Δr, one can calculate the wavefunction at a position $r = r_{n+1}$ using the wavefunction at two gridpoints $r_n = r_{n+1} - \Delta r$ and $r_{n-1} = r_{n+1} - 2\Delta r$ with the following recipe:

El.	$n\ell_j$	$n_{\min}$	$n_{\max}$	$\delta_\ell^{(0)}$	$\delta_\ell^{(1)}$	$\delta_\ell^{(2)}$
Li	$ns_{1/2}$	2	9	0.3999	0.0227	0.0175
	$np_{1/2}$	2	7	0.0476	−0.0302	0.0138
	$np_{3/2}$	2	7	0.0476	−0.0302	0.0135
	$nd_{3/2}$	3	6	0.0018	−0.0030	−0.0055
	$nd_{5/2}$	3	6	0.0018	−0.0027	−0.0069
	$nf_{5/2,7/2}$	4	5	−0.0097	0.1508	0.0000
Na	$ns_{1/2}$	3	30	1.3479	0.0611	0.0188
	$np_{1/2}$	3	30	0.8554	0.1104	0.0824
	$np_{3/2}$	3	30	0.8546	0.1108	0.0825
	$nd_{3/2}$	3	30	0.0149	−0.0406	−0.0108
	$nd_{5/2}$	3	30	0.0149	−0.0402	−0.0132
	$nf_{5/2,7/2}$	4	30	0.0016	−0.0054	−0.0253
	$ng_{7/2,9/2}$	5	20	0.0004	−0.0017	−0.0408
	$nh_{9/2,11/2}$	6	10	0.0001	−0.0018	0.0057
K	$ns_{1/2}$	4	30	2.1804	0.1296	0.1095
	$np_{1/2}$	4	18	1.7139	0.2265	0.2874
	$np_{3/2}$	4	21	1.7108	0.2279	0.2883
	$nd_{3/2}$	3	29	0.2774	−1.0846	0.8062
	$nd_{5/2}$	3	30	0.2776	−1.0831	0.7998
	$nf_{5/2,7/2}$	4	11	0.0098	−0.0627	0.2438
Rb	$ns_{1/2}$	5	30	3.1316	0.1679	0.1890
	$np_{1/2}$	5	7	2.6534	0.3142	0.2521
	$np_{3/2}$	5	7	2.6400	0.3255	0.2301
	$nd_{3/2}$	4	7	1.3486	−0.6316	−1.3752
	$nd_{5/2}$	4	5	1.3469	−0.6221	−1.5439
	$nf_{5/2}$	4	8	0.0164	−0.0785	0.0320
	$nf_{7/2}$	4	8	0.0164	−0.0784	0.0279
Cs	$ns_{1/2}$	6	25	4.0493	0.2353	0.2818
	$np_{1/2}$	6	25	3.5917	0.3478	0.6422
	$np_{3/2}$	6	25	3.5589	0.3662	0.6280
	$nd_{3/2}$	5	25	2.4751	0.0570	−1.2841
	$nd_{5/2}$	5	25	2.4660	0.0593	−1.2440
	$nf_{5/2}$	4	25	0.0334	−0.1980	0.2644
	$nf_{7/2}$	4	25	0.0335	−0.1998	0.2685
	$ng_{7/2}$	5	25	0.0070	−0.0518	0.0667
	$ng_{9/2}$	5	11	0.0070	−0.0499	0.0256

Table 10.4 Experimental quantum defect parameters for the alkali-metal atoms.

$$(1-T_{n+1})u_{n+1} = (2+10T_n)u_n - (1-T_{n-1})u_{n-1}, \qquad T_n \equiv -\frac{1}{12}(\Delta r)^2 k(r_n)^2, \quad (10.27)$$

where $k(r)^2$ is given by Eq. (10.2). It can be shown that the error resulting from this procedure in one step is of the order of $(\Delta r)^6$ and thus small if the step size Δr is taken sufficiently small. The problem with this integration procedure is that it requires values of the wavefunction at two distinct points to start the integration.

In order to begin the integration, one starts in the classically forbidden regime at short range and integrates the wavefunction for a certain energy outwards. Because the starting energy is not necessarily the right energy, integrating the wavefunction all the way to large range is not a viable option, since in the classically forbidden range the wavefunction will inevitably start to diverge and grow exponentially to infinity.

A better way to perform the integration is to stop at a matching point in the classically allowed region and repeat the same procedure integrating inwards to the same matching point from the classically forbidden region outside. At the matching point the two solutions should match, which means that both the wavefunctions and their derivatives should be equal. Since one can always scale the wavefunction on both sides with a constant, the first requirement can always be fulfilled. However, to fulfill both requirements at the same time, it is always better to match the logarithmic derivative $(du_n/dr)/u_n$ from both sides. Care has to be taken that the formula used to calculate this derivative is also accurate to the same order as the integration technique. If the logarithmic derivative is equal on both sides, the two parts of the solution form the total solution over the whole range, when the values on both sides are scaled to be equal. The energy for which the integration has been performed is the energy of the state, and from the number of radial nodes of the wavefunction, the principal quantum number of the state can be derived.

Exercises

10.1 From the energy levels compiled by NIST[1], estimate the quantum defects δ_ℓ for the s, p, d levels for Li through Ne. How does the trend change for the first ion?

10.2 The Fermi model describes a simple way to see that the quantum defect formula of Eq. (10.1) is an appropriate description of alkali atom energies [103]. It models the charge of the core electrons as a spherically symmetric distribution with

---
[1] http://physics.nist.gov/asd

a maximum charge density near the nucleus and tapering off at larger distances. Suppose that outside the nuclear radius R_N the charge enclosed in a sphere of radius r is $Q_{enc}(r) = e(1 + (Z-1)R_N/r)$.

(a) Show that $Q_{enc}(r) = Ze$ at the edge of the nucleus and $= e$ at large r where the valence electron probability density is highest.
(b) Find an expression for the charge density of the core electrons in this model.
(c) Show that Gauss's law gives the electric field as

$$|\vec{\mathcal{E}}(r)| = (e/4\pi\varepsilon_0 r^2)(1 + (Z-1)R_N/r).$$

(d) Find the potential by integrating $e\vec{\mathcal{E}}(r)$ inward from ∞ where it is chosen to vanish, and show that it has two terms, proportional to $1/r$ and $1/r^2$.
(e) The $1/r$ term is precisely the first term of V_{eff} of Eq. (7.9b), and the $1/r^2$ term can be combined with the second term of V_{eff}. Then the recursion relation of Eq. (7.12) will cause the series to be truncated for λ not an integer, and writing it as $\lambda = n - \delta$ defines the quantum defect. Calculate δ for this model.
(f) Show that the resulting formula for the energies is indeed Eq. (10.1).

10.3 The Parsons–Weisskopf model [104] takes an opposite approach to that of Exercise 10.2, where the valence electron is barricaded by a potential wall at radius $r = r_c$. This hypothesis is supported with the argument that inside the core the de Broglie wavelength of the electron is short and therefore the amplitude of the wavefunction is small (see also Fig. 10.2). It is assumed that the potential outside this barricade is $-e^2/(4\pi\varepsilon_0 r)$ and becomes infinite for $r = r_c$. The solution of the Schrödinger equation, which is regular at infinity, is given by

$$u_{E\ell} \propto e^{-\rho/2}\rho^{\ell+1}U(\ell+1-\lambda, 2\ell+1, \rho) \qquad \lambda \equiv \frac{e^2}{4\pi\varepsilon_0\hbar}\sqrt{\frac{-m}{2E}},$$

where $U(k, m, x)$ is the confluent hypergeometric function of the second kind.
(a) Determine λ for the 3s state of Na using the experimental level energy of the NIST database.
(b) Use your favorite programming tool to locate the radius of the last zero crossing of the solution above and use this value in the remainder of the exercise as r_c.
(c) Determine the quantum defects for the ns states for Na with $n = 4-10$ both by using the solution above with the condition that for $r = r_c$ the wavefunction becomes zero, and by using the experimental level energies of the NIST database, and compare the outcomes.

(d) Calculate the wavefunctions for the 3s and 3p states assuming r_c is independent of ℓ, and calculate the oscillator strength of the 3s $\to$ 3p transition, as given by Eq. (3.18).

(e) Can you indicate why the oscillator strengths for the 3s $\to$ np transition are much smaller than for the 3s $\to$ 3p transition? Show by using Eq. (3.19) that this indicates that your result for the previous part should be close to unity.

11
Atoms in magnetic fields

11.1 Introduction

It is well known that light carries momentum and can exert a force through its radiation pressure, but it is also true that static electromagnetic fields have momentum. For example, an electron traveling through space in some galactic field follows a curved trajectory, and in order to satisfy momentum conservation, it must be that the electron exchanges momentum with the field. The momentum of the field is given by $+e\vec{A}$, where $\vec{A}$ is the vector potential that defines the field. Thus the Hamiltonian for an electron in a field must be modified to include this momentum, and it becomes

$$\mathcal{H} = \frac{1}{2m}\left(\vec{p} + e\vec{A}\right)^2 + V(r), \tag{11.1}$$

as in Eq. (3.2). Expanding the Hamiltonian leads to

$$\mathcal{H} = \frac{p^2}{2m} + V(r) + \frac{e}{2m}\left(\vec{A}\cdot\vec{p} + \vec{p}\cdot\vec{A}\right) + \frac{e^2}{2m}|\vec{A}|^2. \tag{11.2}$$

Using the Coulomb gauge, where $\vec{\nabla}\cdot\vec{A} = 0$ so that $\vec{A}$ and $\vec{p}$ commute, the Hamiltonian can be written as the familiar atomic Hamiltonian $\mathcal{H}_0$ plus terms to include the field

$$\mathcal{H} = \mathcal{H}_0 + \frac{e}{m}\vec{A}\cdot\vec{p} + \frac{e^2}{2m}|\vec{A}|^2. \tag{11.3}$$

For a uniform magnetic field the vector potential is given by $\vec{A} = 1/2(\vec{B}\times\vec{r})$, so the first-order term can be written as $2\vec{A}\cdot\vec{p} = (\vec{B}\times\vec{r})\cdot\vec{p} = \vec{B}\cdot(\vec{r}\times\vec{p}) = \vec{B}\cdot\vec{\ell}$. The atomic unit for magnetic fields is given by $\mathcal{B}_{au} = R_\infty/\mu_B = \hbar/ea_0^2$ or 2.35×10^5 T, where the Bohr magneton is defined in Eq. (8.1). This is large compared with magnetic fields in the laboratory, which are typically of the order of 1 T. The energy shift caused by magnetic fields is thus small compared with the unshifted energies of $\mathcal{H}_0$. Furthermore, the second-order term of Eq. (11.3) can generally be neglected with respect to the first-order term. The second-order term leads to the quadratic Zeeman effect and will be discussed in more detail in Sec. 13.3.1.

11.2 The Hamiltonian for the Zeeman effect

Up to now there have been four kinds of magnetic fields considered. There are the two connected with the intrinsic magnetic moments of the electron and the nucleus discussed in Chaps. 8 and 9 respectively, the one arising from the orbital motion of the electron discussed in Sec. 8.3, and an externally applied field as above. At the time of Zeeman's classic experiment in 1896, neither of the intrinsic magnetic moments were known, and in fact, there was essentially no knowledge about atomic structure. Lorentz had suggested that the motion of charged radiating particles would be perturbed by an applied field, and Zeeman wanted to measure the e/m of such particles, especially since he could get the sign of the charge from the expected circular polarization of the light.

But the experimental results posed some puzzles. The theory predicted two observable frequencies along the $\vec{\mathcal{B}}$-field, and three perpendicular to it. This was indeed observed in some of the lines of certain atoms such as Cd and Zn, but not in others of these same atoms. Furthermore, none of the spectra of the alkalis showed the predicted behavior. Thus there were some transitions that exhibited a "normal" Zeeman effect, but others that were labeled "anomalous". Even in the absence of any magnetic field, some of the spectral lines were already split into multiplets. These facts led to the proposal that the electron had an intrinsic magnetic moment $\vec{\mu}_e$, and the consequences were enormously important for the Zeeman effect.

As discussed in Chap. 8, an electron has a magnetic moment $\vec{\mu}_\ell$ deriving from the current associated with its orbit, as well as its intrinsic magnetic moment $\vec{\mu}_e$. The orbital magnetic moment is written as $\vec{\mu}_\ell \equiv -g_\ell \mu_B \vec{\ell}$, where $g_\ell = 1 - m/M \approx 1$. As can be found from the discussion following Eq. (11.3), the appropriate term for the Hamiltonian is $+\mu_B \vec{\mathcal{B}} \cdot \vec{\ell}/\hbar$, exactly the classical expression for the energy of a magnetic moment in a $\vec{\mathcal{B}}$-field.

The intrinsic magnetic moment of the electron $\vec{\mu}_e$ is written in terms of the fictitious angular momentum $\vec{s}$ called spin, and for the electron $|\vec{s}| = \hbar/2$. There is no reason to expect that the magnetic moment should be exactly μ_B, and so there is a Landé g-factor as before. By analogy to the orbital case, the magnetic energy added to the Hamiltonian from the intrinsic moment is $+g_e \mu_B \vec{s} \cdot \vec{\mathcal{B}}/\hbar$, and this result is borne out by more complete relativistic calculations.

In the presence of the electron magnetic moment, the spin–orbit interaction of Eq. (8.15) must be included. Then with the approximation of neglecting the quadratic Zeeman effect described in Sec. 13.3.1, and writing the spin–orbit interaction as in Eq. (8.15), the Hamiltonian becomes

11.3 Zeeman shifts in the presence of the spin–orbit interaction

$$\mathcal{H} = \mathcal{H}_0 + \mathcal{H}_{SO} + \mathcal{H}_Z \equiv \mathcal{H}_0 + \frac{\xi(r)\vec{\ell}\cdot\vec{s}}{\hbar^2} + \frac{g_\ell \mu_B \vec{\ell}\cdot\vec{\mathcal{B}}}{\hbar} + \frac{g_e \mu_B \vec{s}\cdot\vec{\mathcal{B}}}{\hbar}, \quad (11.4)$$

where $\xi(r)$ is given by Eq. (8.10b). This expression has to be evaluated to find the energies. In Eq. (11.4) the relativistic correction to the Hamiltonian has been neglected, which makes this analysis particularly well-suited for the alkali-metal atoms. For hydrogen, relativity can easily be included, since it shifts only the zero-field energies of the states.

11.3 Zeeman shifts in the presence of the spin–orbit interaction

Three regimes can be distinguished in the evaluation of Eq. (11.4). In the strong field regime the Zeeman shift caused by the magnetic field is much larger than the spin–orbit splitting so that the latter can be safely neglected. In that case the zero-field eigenfunctions of $\mathcal{H}_0$ can be used to evaluate Eq. (11.4) exactly. In the weak field regime the shift caused by the magnetic field is much smaller than the spin–orbit splitting. In that case m_ℓ and m_s are not good quantum numbers, and a vector model of the angular momenta is used to find the appropriate energy shifts caused by the magnetic field. In the third case, with intermediate fields, both terms have similar magnitudes and the eigenvalues have to be found by diagonalization of the full interaction matrix.

The choice of a coordinate system to evaluate Eq. (11.4) is arbitrary, and the results of *any* choice should be the same. The standard choice is to have the z-axis along the applied magnetic field $\vec{\mathcal{B}}$, so that $\vec{\mathcal{B}} = \mathcal{B}\hat{z}$. The same is true about the basis set in which the eigenfunctions are expanded, and with this choice of coordinate system, it seems natural to choose a basis set that has well-defined values of orbital angular momentum ℓ and its z-component m_ℓ, as well as the spin s and m_s. For the strong field case this is appropriate.

11.3.1 Strong fields

The case of fields strong enough to dominate the spin–orbit interaction is called the Paschen–Back regime, where the spin–orbit interaction can be neglected. Since both ℓ_z and s_z commute with $\mathcal{H}_0$, the eigenfunctions of $\mathcal{H}_0$ can be used to calculate the shift caused by the magnetic field exactly. The field-dependent Zeeman terms in Eq. (11.4) become

$$E_Z = \mu_B (m_\ell + 2m_s) \mathcal{B}, \quad (11.5)$$

where it is assumed that $g_\ell = 1$ and $g_e = 2$. Note that the states are still degenerate in $\vec{\ell}$ and $\vec{s}$, but that the degeneracy with respect to m_ℓ and m_s is broken since $\vec{\mathcal{B}}$ provides a preferred direction in space. The energy splitting for a one-electron

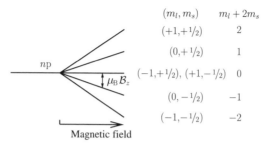

Figure 11.1 Zeeman shift for strong fields of a one-electron atom with the electron in a np-state. The separation between the states is $\mu_B B$.

atom with the electron in the np state is shown in Fig. 11.1. For optical transitions between Zeeman-split states, the selection rules for one-electron atoms are $\Delta m_\ell = 0, \pm 1$ and $\Delta m_s = 0$. Therefore the spectral lines are divided into three types with $\Delta m_\ell = -1, 0, +1$, called the "Lorentz triplet". The splitting between the lines is given by the Larmor frequency: $\omega_L \equiv \mu_B B / \hbar$.

11.3.2 Weak fields

In the case where the spin–orbit splitting is much larger than the Zeeman shifts, neither m_ℓ nor m_s are good quantum numbers and the Zeeman term in Eq. (11.4) cannot be evaluated easily. The eigenfunctions in the case of the spin–orbit interaction are eigenfunctions of n, ℓ, s, j, and m_j. We can write

$$\vec{\mu}_j = \vec{\mu}_\ell + \vec{\mu}_s = -\frac{\mu_B}{\hbar}\left(g_\ell \vec{\ell} + g_e \vec{s}\right) \approx -\frac{\mu_B}{\hbar}\left(\vec{j} + \vec{s}\right), \tag{11.6}$$

where in the last step $g_\ell = 1$ and $g_e \approx 2$ is used. Defining the pseudo-gyromagnetic ratio g_j as $\vec{\mu}_j \equiv -g_j \mu_B \vec{j}/\hbar$, Eq. (11.6) leads to $(g_j - 1)|\vec{j}|^2 = \vec{s} \cdot \vec{j}$ by taking the dot product with $\vec{j}$ on both sides and rearranging the terms. By squaring the relation $\vec{\ell} = \vec{j} - \vec{s}$, the pseudo-gyromagnetic ratio g_j for the coupled state becomes

$$g_j = 1 + \frac{j(j+1) + s(s+1) - \ell(\ell+1)}{2j(j+1)}. \tag{11.7}$$

The Zeeman shift caused by the magnetic field is then

$$E_Z = \mu_B g_j m_j B = \hbar g_j m_j \omega_L, \tag{11.8}$$

where ω_L is the Larmor frequency. The splitting for the weak field case is shown in Fig. 11.2. Note that the shift for weak fields is also linear as for strong fields, but the proportionality is different (see Eq. (11.5)).

11.3 Zeeman shifts in the presence of the spin–orbit interaction

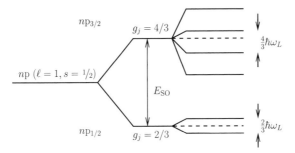

Figure 11.2 Zeeman splitting for weak fields of a one-electron atom with the electron in a np state including spin–orbit interaction. The separation between the states is $2/3\mu_B \mathcal{B}$ for the $n\text{p}_{1/2}$ state and $4/3\mu_B \mathcal{B}$ for the $n\text{p}_{3/2}$ state.

In the case of the hyperfine interaction, a similar analysis leads to the shift caused by the magnetic field as

$$E_Z = \mu_B g_F m_F \mathcal{B} = \hbar g_F m_F \omega_L, \tag{11.9}$$

with the gyromagnetic ratio given by

$$g_F = g_j \frac{F(F+1) + j(j+1) - I(I+1)}{2F(F+1)}, \tag{11.10}$$

where I is the nuclear spin and F the total angular momentum of the atom (see Eq. (9.9)). The nuclei of the alkali atoms discussed in Chap. 10 usually have large spins and hfs splittings (typically a few GHz). Their ground-state F-values are given by $I - j$ and $I + j$ with $j = 1/2$ since $\ell = 0$, so F can be considerably larger than the values $F = 0, 1$ of hydrogen. For $I = 3/2$ one finds $F = 1, 2$, and for $I = 5/2$ one finds $F = 2, 3$. The Zeeman splitting of the M_F sublevels of the lower-lying F state can lead to levels whose energies increase with increasing field, making them suitable for magnetic trapping (see Chap. 20). Unlike the case of hydrogen, trappable states in the lowest hfs sublevel are appropriate for the evaporative cooling necessary to make a Bose condensate (see Chap. 21). Moreover, for the $F = I - j$ case $g_F < 0$ so the states with $M_F < 0$ are the low-field seekers that are trappable.

11.3.3 Intermediate fields

For intermediate fields the spin–orbit and Zeeman interaction are comparable in size and the methods employed in the two subsections above do not apply. However, one can always choose one of the basis sets above and evaluate the matrix elements of the Hamiltonian in that basis. This leads to off-diagonal elements that necessitate the diagonalization of the interaction matrix. Here the basis functions

Table 11.1 Basis functions for the calculations of the Zeeman effect for intermediate field for the coupled (left) and uncoupled basis (right) for an np state. For the coupled basis the states are eigenfunctions of $\vec{j}$, $\vec{\ell}$, $\vec{s}$, and j_z. For the uncoupled basis the states are eigenfunctions of $\vec{\ell}$, ℓ_z, $\vec{s}$, and m_s. Note that $\ell = 1$ and $s = 1/2$ for an np state.

State	Coupled		Uncoupled	
	j	m_j	m_ℓ	m_s
1	3/2	−3/2	−1	−1/2
2	1/2	−1/2	−1	+1/2
3	3/2	−1/2	0	−1/2
4	1/2	+1/2	0	+1/2
5	3/2	+1/2	+1	−1/2
6	3/2	+3/2	+1	+1/2

$|j, m_j\rangle$ are chosen and $\mu_B(\ell_z + 2s_z)B$ needs to be evaluated in this basis. The states $|j, m_j\rangle$ can be expanded in the states $|\ell, m_\ell; s, m_s\rangle$ as

$$|j, m_j\rangle = \sum_{m_\ell, m_s} \langle \ell, m_\ell; s, m_s | j, m_j \rangle |\ell, m_\ell; s, m_s\rangle, \quad (11.11)$$

where the expansion coefficients $\langle \ell, m_\ell; s, m_s | j, m_j \rangle$ are known as Clebsch–Gordan coefficients. The coefficients can easily be evaluated using a symbolic manipulation package, and some of them are tabulated in Fig. C.2. The interaction matrix elements for the shift caused by the magnetic field are thus given by

$$(\mathcal{H}_Z)_{jj'} = \mu_B B \sum_{m_\ell, m_s} (m_\ell + 2m_s) \langle \ell, m_\ell; s, m_s | j, m_j \rangle \langle \ell, m_\ell; s, m_s | j', m'_j \rangle, \quad (11.12)$$

where the sum is over all possibilies for m_ℓ and m_s.

For the evaluation of the spin–orbit term $\xi(r)\vec{\ell} \cdot \vec{s}/\hbar^2$ in this basis one can find the angular part $\vec{\ell} \cdot \vec{s} = 1/2(\vec{j}^2 - \vec{\ell}^2 - \vec{s}^2)$ by squaring $\vec{j} = \vec{\ell} + \vec{s}$ (see end-of-book App. C) and since the basis functions are eigenfunctions of $\vec{j}^2$, $\vec{\ell}^2$, and $\vec{s}^2$ one finds

$$\frac{A}{\hbar^2} \langle \vec{\ell} \cdot \vec{s} \rangle = \frac{A}{2} \left(j(j+1) - \ell(\ell+1) - s(s+1) \right), \quad (11.13)$$

where A is given in Eq. (8.16).

For a one-electron atom with the electron in an np state, there are a total of six states, namely two $np_{1/2}$ states with $m_j = \pm 1/2$ and four $np_{3/2}$ states with $m_j = \pm 1/2, \pm 3/2$ (see Tab. 11.1). Note that the spin–orbit term is $-A$ for the $np_{1/2}$ state and

11.3 Zeeman shifts in the presence of the spin–orbit interaction

$A/2$ for the $np_{3/2}$ state. The Hamiltonian matrix can then be written as (using the coupled states of Tab. 11.1 in order)

$$\mathcal{H} = \begin{pmatrix} \frac{A}{2}-2\mu_B\mathcal{B} & 0 & 0 & 0 & 0 & 0 \\ 0 & -A-\frac{1}{3}\mu_B\mathcal{B} & -\frac{1}{3}\sqrt{2}\mu_B\mathcal{B} & 0 & 0 & 0 \\ 0 & -\frac{1}{3}\sqrt{2}\mu_B\mathcal{B} & \frac{A}{2}-\frac{2}{3}\mu_B\mathcal{B} & 0 & 0 & 0 \\ 0 & 0 & 0 & -A+\frac{1}{3}\mu_B\mathcal{B} & -\frac{1}{3}\sqrt{2}\mu_B\mathcal{B} & 0 \\ 0 & 0 & 0 & -\frac{1}{3}\sqrt{2}\mu_B\mathcal{B} & \frac{A}{2}+\frac{2}{3}\mu_B\mathcal{B} & 0 \\ 0 & 0 & 0 & 0 & 0 & \frac{A}{2}+2\mu_B\mathcal{B} \end{pmatrix}. \quad (11.14)$$

One can immediately see that the states with $m_j = \pm 3/2$ are not coupled to any other state and thus their energies can be written as $E_{a,d} = A/2 \pm 2\mu_B\mathcal{B}$ (subscripts as in Fig. 11.3). For those with $m_j = \pm 1/2$, the states with $j = 1/2$ and $3/2$ are coupled and this requires the diagonalization of a 2×2 submatrix:

$$\mathcal{H} = \begin{pmatrix} -A \pm \frac{1}{3}\mu_B\mathcal{B} & -\frac{1}{3}\sqrt{2}\mu_B\mathcal{B} \\ -\frac{1}{3}\sqrt{2}\mu_B\mathcal{B} & \frac{A}{2} \pm \frac{2}{3}\mu_B\mathcal{B} \end{pmatrix}. \quad (11.15)$$

The eigenvalues for the (+)-sign are given by

$$E_{b,e} = \frac{A}{2}\left(+\frac{\mu_B\mathcal{B}}{A} - \frac{1}{2} \pm \sqrt{\left(\frac{\mu_B\mathcal{B}}{A}+\frac{1}{2}\right)^2 + 2} \right), \quad (11.16a)$$

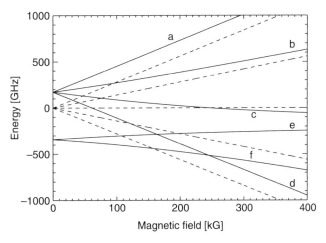

Figure 11.3 Energy shift for the 3p state in Na caused by the magnetic field. For small fields the linear shifts on the six states are given by Eq. (11.8). For strong fields the states with $(m_\ell, m_s) = (-1, +1/2)$ and $(+1, -1/2)$ become degenerate and also field-independent. The dashed lines indicate the shift of the $m_j = \pm 1/2$ states in absence of spin–orbit interaction and these states asymptotically connect to the same states with spin–orbit interaction. The labels a–f correspond to the labels in Fig. 3.5.

and for the (−)-sign by

$$E_{c,f} = \frac{A}{2}\left(-\frac{\mu_B B}{A} - \frac{1}{2} \pm \sqrt{\left(\frac{\mu_B B}{A} - \frac{1}{2}\right)^2 + 2}\right). \quad (11.16b)$$

Note that the ±-signs lead to four non-degenerate energies. The shift caused by the magnetic field lifts the two-fold degeneracy of the $np_{1/2}$ state and the four-fold degeneracy of the $np_{3/2}$ state. In Fig. 11.3 the results are shown as a function of the magnetic field B for the 3p state of Na. For small fields the shift is linear for all six states, but for larger fields two states become nearly degenerate and there is a splitting into five states, as found from Eq. (11.5). The average shift of all the levels E_{a-f} is zero. Thus, the "center of gravity" or "weighted energy" is unchanged from the Bohr energy. This is an example of the principle of spectroscopic stability.

It is instructive to do the same calculation by choosing eigenfunctions of the operators $\vec{\ell}$, ℓ_z, $\vec{s}$, and s_z (see Tab. 11.1). In this basis the Zeeman term $\mu_B(\ell_z+2s_z)B/\hbar$ is diagonal and the diagonal elements then become $\mu_B(m_\ell+2m_s)B$. For the spin–orbit interaction one can write $\vec{\ell} \cdot \vec{s} = \ell_z s_z + 1/2(\ell_+ s_- + \ell_- s_+)$, which can easily be derived using $\ell_\pm = \ell_x \pm i\ell_y$ and $s_\pm = s_x \pm is_y$ (see App. C.4). The term $\ell_z s_z$ is diagonal with diagonal elements $m_\ell m_s$. The term $\ell_+ s_-$ couples the states 2 to 3 and 4 to 5, whereas the term $\ell_- s_+$ couples 3 to 2 and 5 to 4. In all cases the coupling element is $\sqrt{2}$. The interaction Hamiltonian can then be written as (using the uncoupled states of Tab. 11.1 in order)

$$\mathcal{H} = \begin{pmatrix} +\frac{A}{2}-2\mu_B B & 0 & 0 & 0 & 0 & 0 \\ 0 & -\frac{A}{2} & \frac{1}{2}\sqrt{2}A & 0 & 0 & 0 \\ 0 & \frac{1}{2}\sqrt{2}A & -\mu_B B & 0 & 0 & 0 \\ 0 & 0 & 0 & \mu_B B & \frac{1}{2}\sqrt{2}A & 0 \\ 0 & 0 & 0 & \frac{1}{2}\sqrt{2}A & -\frac{A}{2} & 0 \\ 0 & 0 & 0 & 0 & 0 & A/2+2\mu_B B \end{pmatrix}. \quad (11.17)$$

Although the matrix looks very different from the matrix in Eq. (11.14), its eigenvalues are identical. This is one example of the principle discussed in the second paragraph of Sec. 11.3.

Appendices

11.A The ground state of atomic hydrogen

There are many cases of the Zeeman effect in atomic states that have $\ell = 0$, but by far the most important one is the ground state of hydrogen. Because $\ell = 0$ there is no spin–orbit interaction, but still there remains a coupling between the nuclear and electronic intrinsic magnetic moments, as discussed in Sec. 9.3.2.

The energy associated with this "Fermi contact term", given in Eq. (9.13), can be written as $a\vec{I} \cdot \vec{s}/\hbar^2$ so that it looks very much like the spin–orbit energy $A\vec{\ell} \cdot \vec{s}/\hbar^2$, but $a/A \sim 10^{-3}$ for typical atoms (see Eqs. (9.11) and (8.16)). In principle, one can include both kinds of effects, but the calculation is not very instructive and becomes unwieldy. With $\ell = 0$ (no spin–orbit interaction) the Hamiltonian is

$$\mathcal{H} = \frac{a\vec{I}\cdot\vec{s}}{\hbar^2} + \frac{(2\mu_B\vec{s} + g_I\mu_N\vec{I})\cdot\vec{\mathcal{B}}}{\hbar} = a\left(\frac{\vec{I}\cdot\vec{s}}{\hbar^2} + x\frac{(2s_z + 2\varepsilon I_z)}{\hbar}\right), \quad (11.18)$$

where $x \equiv \mu_B \mathcal{B}/a$ and $\varepsilon \equiv mg_I/2M \ll 1$.

One can choose any basis, so the uncoupled basis written as (m_I, m_s) suffices, and the $\vec{I}\cdot\vec{s}$ term is evaluated using $\vec{I}\cdot\vec{s} = I_z s_z + (I_+ s_- + I_- s_+)/2$. As above, this gives rise to a Hamiltonian matrix given by

$$\mathcal{H} = a \begin{pmatrix} 1/4 + x(1+\varepsilon) & 0 & 0 & 0 \\ 0 & -1/4 - x(1-\varepsilon) & 1/2 & 0 \\ 0 & 1/2 & -1/4 + x(1-\varepsilon) & 0 \\ 0 & 0 & 0 & 1/4 - x(1+\varepsilon) \end{pmatrix} \quad (11.19)$$

with eigenvalues

$$E_{1,2} = a\left(1/4 \pm x(1+\varepsilon)\right) \quad (11.20)$$
$$E_{3,4} = a\left(-1/4 \pm \sqrt{1/4 + x^2(1+\varepsilon)^2}\right).$$

Figure 11.4 shows these energy eigenvalues as a function of field strength. The alkali-metal atoms discussed in Chap. 10 all have $\ell = 0$ ground states, various nuclear spins I, and Zeeman energy-level structures similar to that of hydrogen. In these cases the dimension of the Hamiltonian matrix of Eq. (11.A) is

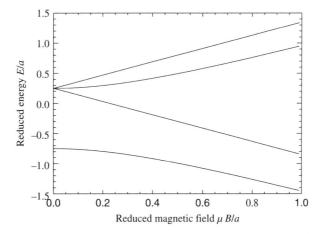

Figure 11.4 Energy-level diagram showing the Zeeman splitting of the ground state of hydrogen in a uniform field. This is the original Breit–Rabi diagram and describes the states of the hydrogen maser. The zero-field splitting is $h \times 1.420$ GHz corresponding to a wavelength of 21 cm, the most important spectral feature of radio astronomy.

11.B Positronium

As a final example, consider the ground state of positronium (Ps), a positron–electron pair. At first thought, this may seem impossible because these two antiparticles annihilate, but they do so slowly enough that their bound states can be studied in some detail. This purely leptonic system has been a fundamental testing ground for quantum electrodynamics because it has no baryonic contributions. The eigenstates for this system are given in Tab. 9.2. In this section the electron is labeled as particle 1 and the positron as 2.

The Hamiltonian is

$$\mathcal{H} = a_P \vec{s}_1 \cdot \vec{s}_2 - \vec{\mu} \cdot \vec{\mathcal{B}} = a_P \vec{s}_1 \cdot \vec{s}_2 - \mu_B \vec{\mathcal{B}} \cdot (g_1 \vec{s}_1 + g_2 \vec{s}_2) \quad (11.21)$$

where subscript P indicates positronium, and the $\vec{s}_1 \cdot \vec{s}_2$ term is evaluated with the vector model as in App. C,

$$\langle \vec{s}_1 \cdot \vec{s}_2 \rangle = \frac{1}{2} \langle \vec{S}^2 - \vec{s}_1{}^2 - \vec{s}_2{}^2 \rangle = \frac{1}{4} \hbar^2 (2S^2 - 3) \quad (11.22)$$

with $\vec{S} = \vec{s}_1 + \vec{s}_2$. The quantity $a_P = (2/3)\alpha^2 R_\infty$ is analogous to the a of the hyperfine splitting in Eq. (9.11) because it derives from the interaction of two intrinsic magnetic moments, but the positronium subscript is to indicate that its order of magnitude is 10^3 times larger because both magnetic moments are of the order of μ_B.

There is a special additional term in the Hamiltonian that applies only to the triplet state of Ps and to nothing else. It derives from the matter–antimatter constitution of the atom, it is of comparable magnitude to a_P [59], and it adds $(1/2)\alpha^2 R_\infty$ to the energy, making the total ground state splitting in zero field become $(7/4)a_P = (7/6)\alpha^2 R_\infty$.

The Zeeman energy term is $\mu_B \mathcal{B}(s_{2z} - s_{1z})$, and it is convenient to define $x_P \equiv 4\mu_B \mathcal{B}/7a_P$. The minus sign on the term $\mu_B \mathcal{B}(s_{2z} - s_{1z})$ above comes from the opposite charge of the electron and positron, and hence the opposite sign of magnetic moment. For small fields ($x_P \ll 1$) the energies are $7a_P/16$ and $-21a_P/16$. In the coupled basis (a–d) (see Tab. 9.2) one obtains

$$\mathcal{H} = 7/4 a_P \begin{pmatrix} 1/4 & 0 & 0 & 0 \\ 0 & 1/4 & -x_P & 0 \\ 0 & -x_P & -3/4 & 0 \\ 0 & 0 & 0 & 1/4 \end{pmatrix} \quad (11.23)$$

11.B Positronium

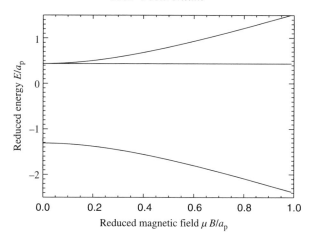

Figure 11.5 Energy-level diagram showing the Zeeman splitting of the ground state of positronium in a uniform field. The energy axis is in units of $a_P \approx 116$ GHz and the field is in units of $\mu_B \mathcal{B}/a_P$.

The eigenvalues are

$$E = {}^7/_4 a_P \left(-1/4 \pm \sqrt{1/4 + x_P^2}\right) \tag{11.24}$$

and Fig. 11.5 shows the energy-level splitting as a function of field strength.

The 203 GHz splitting of Ps at zero field is much larger than the 1.4 GHz splitting of H and presented significant experimental difficulties at the time of the measurement of Ref. [105]. It was too low for infrared technology and much too high for microwave methods. In spite of this, the experimenters cleverly exploited the different annihilation rates of the singlet and triplet states to enable the measurement.

To understand why these are so different, consider the annihilation $e^+ + e^- \to 2\gamma$. In the rest frame, $\sum_i P_i = 0$ so the two γ-rays must go in opposite directions and have equal energies to conserve energy and momentum ($p_\gamma = E/c$). Conservation of angular momentum also dictates certain limits. Since each γ has intrinsic spin 1, the total spin can be either 2 or 0. For annihilation out of the singlet state there is no problem, but the triplet state has $S = 1$ so a two-γ annihilation is forbidden. The decay must proceed as $e^+ + e^- \to 3\gamma$ with the emission directions split at angles that conserve energy, momentum, and angular momentum. This process is much weaker than $e^+ + e^- \to 2\gamma$ and so proceeds much more slowly, resulting in a longer lifetime for the triplet state.

Notice also that the matrix of Eq. (11.23) is *not* diagonal. This means that the singlet and one of the triplet states are not the eigenstates of the system for $x \neq 0$, but are mixed by the magnetic field. Each component of these states has a given

11.C The non-crossing theorem

Consider again the simplest case of an atom with a fine structure in a magnetic field as in Sec. 11.3. The fine-structure splitting is given by $A\vec{\ell}\cdot\vec{s}/\hbar^2$ and the simplest non-zero case has $\ell = 1$ and $s = 1/2$. This results in six states as in Sec. 11.3, so the Hamiltonian is a 6×6 matrix and the field dependence of the energy levels is shown in Fig. 11.3. Consider the region near the level "crossing" $\mathcal{B}_0$ and for simplicity assume that in this region the lines are straight and have the same magnitude of slope α, as shown in Fig. 11.6. Then choose E_0 to be the energy at their "crossing point" so that $E_1 = E_0 - \alpha(\mathcal{B} - \mathcal{B}_0)$ and $E_2 = E_0 + \alpha(\mathcal{B} - \mathcal{B}_0)$. The Hamiltonian matrix for this system is

$$\mathcal{H} = \begin{pmatrix} E_0 - \alpha(\mathcal{B} - \mathcal{B}_0) & C \\ C^* & E_0 + \alpha(\mathcal{B} - \mathcal{B}_0) \end{pmatrix} \quad (11.25)$$

where C arises from the spin-orbit term, as in the $\ell_+ s_-$ term of Eq. (11.17). The eigenvalues of the matrix in Eq. (11.25) are $E_\pm = E_0 \pm \sqrt{\alpha^2(\mathcal{B} - \mathcal{B}_0)^2 + |C|^2}$ so their splitting is $E_+ - E_- = 2\sqrt{\alpha^2(\mathcal{B} - \mathcal{B}_0)^2 + |C|^2}$.

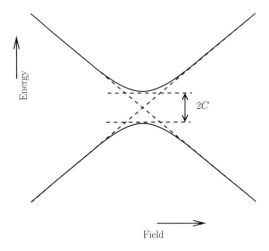

Figure 11.6 Plot of the energy levels near their "crossing" in Fig. 11.3. In that figure the horizontal axis is the magnetic field, but it could be any other parameter that shifts the levels, for example the detuning from resonance of light that couples two atomic states. The levels anti-cross here, and their smallest separation is $2C$, where C is the off-diagonal matrix element, or coupling between the levels. In general, C is never zero, so the levels never cross.

11.C The non-crossing theorem

The first thing to notice is that at the "crossing point" where $E_1 = E_2$, the splitting is non-zero. In the closeup look at this crossing of Fig. 11.6, it is shown that the levels do not actually cross. In this example the fine structure is the cause of a level anti-crossing ($\ell_+ s_-$ term as in Eq. (11.17)). There will be an anti-crossing whenever there is a coupling C of two states, and for any pair of states, there is always some coupling between them, unless the coupling is forbidden because of symmetry.

There can be a crossing if $C = 0$, or if E_1 and/or E_2 are complex. But E_1 and E_2 cannot be complex, because $\mathcal{H}$ is Hermitian. There are some cases where a complex energy is used to mimic decay such as $E = E_0 + i\gamma/2$, but there is generally no justification for doing this [106].

In 1929 it was shown in a more analytical way that there are rarely enough degrees of freedom available to a quantum system to make any of the levels degenerate [107, 108]. The proof starts with a $n \times n$ Hamiltonian matrix that has n^2 elements, and in general, the off-diagonal elements are complex. The total number of free parameters to determine such a matrix is n, as discussed below. The diagonal elements are real and thus each are specified by one parameter. The off-diagonal elements $\mathcal{H}_{ij}$ have the property $\mathcal{H}_{ij} = \mathcal{H}_{ji}^*$ and thus an off-diagonal element and its complement are fully specified by two parameters. The total number of free parameters of an $n \times n$ Hamiltonian matrix is thus n^2.

However, there is a unitary matrix U that diagonalizes the Hamiltonian to $\mathcal{H}_D = U^\dagger \mathcal{H} U$. It has n^2 elements but it also has n constraints because of unitarity, leaving $n^2 - n$ free parameters. Since the number of parameters in $\mathcal{H}$ given above is n^2, but it can be diagonalized by $n^2 - n$ choices, the number of truly free parameters is $n^2 - (n^2 - n) = n$, leaving n as the number of free parameters required to determine the Hamiltonian. This is not surprising because a diagonalized Hamiltonian has only n elements, namely the n real energy eigenvalues.

Now, in order to have a level crossing, at least two of the energies need to be degenerate so the diagonalized Hamiltonian matrix has the form

$$\mathcal{H}_D = \begin{pmatrix} \ddots & & & & \\ & E_{i-1} & & & 0 \\ & & E_i & & \\ & & & E_i & \\ & 0 & & & E_{i+1} \\ & & & & & \ddots \end{pmatrix} \tag{11.26}$$

where some of the states have the same energy. Here the degeneracy is only two-fold for E_i, but in general it could be m-fold. Then the number of free parameters required to specify this Hamiltonian matrix is $n - m + 1$.

The unitary matrix that diagonalizes $\mathcal{H}$ to the form $\mathcal{H}_D$ with an m-fold degeneracy has an $m \times m$ submatrix that is a multiple of the identity matrix, and is therefore trivial. So now the number of free parameters for U is $n^2 - n - (m^2 - m)$. Thus the number of free parameters required to specify $\mathcal{H}$ is reduced to

$$N = [n - m + 1] + \left[n^2 - n - (m^2 - m)\right] = n^2 - (m^2 - 1), \qquad (11.27)$$

where the first term between brackets is for $\mathcal{H}_D$ and the second term for U. For a simple crossing of two levels, $m = 2$ so the number of free parameters becomes $n^2 - 3$. This requires control of three independent parameters in order to get a degeneracy between two levels. In the example given above, the magnetic field constitutes only one of these. Thus there can be no degeneracy so the levels cannot cross. In the presence of both an electric and magnetic field there are the two field magnitude parameters, and the orientation between them is a third one, thus allowing a degeneracy and an exact level crossing.

11.D Passage through an anti-crossing: Landau–Zener transitions

If an atom is prepared in one of the states at the edge of Fig. 11.6, intuition suggests that sweeping the parameter of the horizontal axis slowly toward the other edge would cause its energy to follow the curve it started on, and this is indeed the case even though the consequence is clearly a change of the atomic state. However, it is not quite as intuitive that sweeping rapidly can cause the atom to jump to the other curve, thereby undergoing a sudden change of energy but maintaining its state. In fact, such non-adiabatic transitions do occur, and were studied independently by Landau and Zener in the 1930s for special cases. An excellent description is to be found in Ref. [109]. Such Landau–Zener transitions (LZ) are of great importance in many experiments.

The probability for a non-adiabatic LZ transition P_{LZ} can be calculated directly from the equations of motion of the c_js given by Eqs. (2.4). For a constant sweep rate of the energy separation between the two curves $\hbar d\Omega'/dt \equiv \hbar \dot{\Omega}'$, and energy levels of the form given by Eq. (2.10), the transition probability can be found exactly by projecting the wavefunction onto $|g\rangle$ or $|e\rangle$. The result is

$$P_{LZ} = e^{-\pi \Omega^2 / \dot{\Omega}'}. \qquad (11.28)$$

Clearly sufficiently small values of $\dot{\Omega}'$ can produce arbitrarily small values of P_{LZ}. Other special cases that can be solved analytically are the Rosen–Zener model and the hyperbolic model. Moreover, there are sophisticated methods for calculating non-adiabatic transition probabilities in many other cases [110, 111].

The eigenvalues of Eq. (2.9) are given in Eq. (2.10) which shows that they are never degenerate except when $\delta = 0$ and $\Omega = 0$ (see Fig. 2.13). Sweeping δ

through zero when $\Omega \neq 0$ cannot cause the energy curves to cross: their minimum separation is $\hbar\Omega$. This crossing prohibition is quite general, as shown in Ref. [109].

It is clear from Eqs. (2.11) and Fig. 2.13 that the eigenstates of the Hamiltonian in Eq. (2.9) are determined by the sign of δ, and that the higher-energy dressed state is $|e\rangle$ for $\delta < 0$ and $|g\rangle$ for $\delta > 0$ (this is most evident when $\delta \gg \Omega$ so that the mixing is weak). As discussed above, the state of the system can be changed from $|g\rangle$ to $|e\rangle$ by sweeping δ from some large value δ_0 to $-\delta_0$, and it is clear that this can be done even if Ω is constant. The path would then be a straight line across Fig. 2.13 and corresponds to a cut with constant Ω in Fig. 2.13. Along such a path, the dressed state energies are shown in Fig. 11.6.

Exercises

11.1 Consider the fine-structure splitting of a hydrogen atom in a magnetic field.
(a) Find the eigenfunctions of $\vec{\ell} \cdot \vec{s}$ for a ^{2}P and ^{2}D state in terms of $|\ell m_\ell\rangle |s m_s\rangle$.
(b) Find the eigenfunctions for a ^{2}P and a ^{2}D state in terms of the same basis set as in part (a) in the limit of very high magnetic field in the z-direction ignoring the $\mathcal{A}^2$ term.

11.2 In an applied electromagnetic field, the Hamiltonian for an atomic electron is given by Eq. (11.3). Show that for an atom of typical size $\sim 10^{-10}$ m in an ordinary laboratory magnetic field ($\mathcal{B} < 10^5$ Gauss), the term $(e^2/2m)|\vec{\mathcal{A}}|^2$ is negligibly small compared with the term $(e/m)\vec{\mathcal{A}} \cdot \vec{p}$ that gives the ordinary Zeeman effect.

11.3 An electron has an orbital magnetic moment $\vec{\mu}_\ell = -g_\ell \mu_B \vec{\ell}$ and intrinsic magnetic moment $\vec{\mu}_e = -g_e \mu_B \vec{s}$. For a single-electron atom in an external magnetic field $\vec{\mathcal{B}}$, there are interactions between these internal magnetic moments, called the spin–orbit coupling, as well as between each of these and $\vec{\mathcal{B}}$. How large does an applied $\vec{\mathcal{B}}$ have to be in order that the interactions between these magnetic moments and the magnetic field become comparable with the spin–orbit interaction?

11.4 Consider the Zeeman effect of positronium (Ps) in the ground state.
(a) Find the eigenfunctions and eigenvalues.
The magnetic field mixes the singlet and triplet states having $m = 0$, resulting in a shortened lifetime for the triplet component.
(b) If the positronium wavefunction is $\psi \propto \alpha\psi_s + \beta\psi_t$ with $\psi_{s,t}$ the singlet and triplet wavefunction, respectively, find β as a function of field.
(c) If the decay rate is $\gamma = \beta\gamma_s + \alpha\gamma_t$, find the lifetime as a function of field.
(d) At what field is the lifetime reduced by a factor of 2 using $\gamma_s \sim 10^3 \gamma_t$? For a numerical answer, use the hyperfine shift of the ground state of 203 GHz.

11.5 Show that if a magnetic field $\vec{\mathcal{B}}$ satisfies $\vec{\mathcal{A}} = \frac{1}{2}\vec{\mathcal{B}} \times \vec{r}$, then $\nabla \vec{\mathcal{A}} = 0$.

11.6 Construct an energy-level diagram to show how the Zeeman splitting of the hyperfine levels of an S-state having $J = 1/2$ and $I = 5/2$ changes as the external magnetic field is varied from zero to some large value.

11.7 Two energy levels brought together by an external field are weakly coupled by V (off-diagonal term).
(a) Show that there will be an anti-crossing of minimum separation $2V$. (Hint: you only need a 2×2 matrix.)
If one of the states decays its energy can be written as $E = \hbar(\omega + i\gamma)$.
(b) Show that it is now possible that there will be a crossing for some appropriate relation between V, ω, and γ.
(c) Plot the energy levels in the crossing region for several values of V/γ. (Hint: see Lamb's 1950s papers in *Physical Review*, e.g. refs. [70, 72, 106].)

11.8 The Hamiltonian for an atom in a magnetic field includes the Zeeman term $-\vec{\mu} \cdot \vec{\mathcal{B}}$ and the spin–orbit term $A\vec{\ell} \cdot \vec{s}$, and it is often convenient to write the $\vec{\ell} \cdot \vec{s}$ part in terms of the raising and lowering operators: $\vec{\ell} \cdot \vec{s} = \ell_z s_z + (\ell_+ s_- + \ell_- s_+)/2$.
(a) Show that this is true.
(b) For an atomic state $\ell = 1$ and $s = 1/2$ write out the decoupled basis, show that the Hamiltonian matrix is mostly empty, and then calculate the remaining matrix elements.
The large matrix can be simplified by rearrangement into 2×2 matrices.
(c) Do it and then calculate the eigenvalues for one of the matrices.

11.9 Find the field dependence of the eigenfunctions of the 2×2 matrices in a $\mathcal{B}$ field.

11.10 Write down the Hamiltonian matrix for Ps in a magnetic field. Diagonalize the matrix to find the field dependence of the eigenstates.

11.11 Landau–Zener transitions can be interpreted as resonantly driven by a Fourier component of the rate of energy change caused by the sweep of the external parameter.
(a) Show that Eq. (11.25) is equivalent to Eq. (2.9) when sweeping δ instead of B with Ω fixed, using a shift of the zero of energy, and making a change of variables.
(b) The dominant Fourier component of the energy sweep illustrated in Fig. 11.6 can be approximated using the time it takes to sweep from energy separation $2C \times \sqrt{2}$ through $2C$ at closest approach, and back to $2C \times \sqrt{2}$ as a half-period. Find the range of such a sweep, δ_0. Determine this frequency in terms of the δ and Ω values of (a) above.

(c) What is the condition on the sweep rate $d\delta/dt$ for this frequency to be comparable to the gap at closest approach of the dressed states? This is found to be Ω from Eq. (2.10).

(d) When does this mean for the exponent of Eq. (11.28)? Be careful not to evaluate $\dot{\Omega}'$ at $\delta = 0$ where it vanishes, but, for example, where $\delta = \delta_0$.

12
Atoms in electric fields

12.1 Introduction

In 1913 Stark and Lo Surdo independently made observations of the splitting of spectral lines caused by an applied uniform dc electric field [112]. Stark was the first to publish his results, in *Nature* [113], and since then such energy shifts have been called Stark shifts. The experimental results of Stark and Lo Surdo (and also of Zeeman) led Bohr to calculate the effects of external fields on atomic structure [114] within the framework of his model of the atom presented earlier in Chap. 1.

To describe these shifts it is convenient to choose the z-direction along the applied electric field so that the Hamiltonian can be written as the sum of $\mathcal{H}_0$ and the Stark term from the scalar potential $e\mathcal{E}z$ as

$$\mathcal{H} = \frac{p^2}{2\mu} + V(r) + e\mathcal{E}z \equiv \mathcal{H}_0 + e\mathcal{E}z. \tag{12.1}$$

The sign of the electric field term is positive because the charge of the electron is $-e$.

The separation of the Schrödinger equation into radial and angular parts in spherical coordinates of Chap. 7 required that the potential $V(r)$ depend on r only. In a uniform dc electric field $\mathcal{E}$ the potential includes $e\mathcal{E}z$, so that it is not such a central potential. Thus the Schrödinger equation in spherical coordinates cannot be separated in such a field, and solutions cannot easily be found.

The effect of an external electric field is therefore often treated as a perturbation because laboratory fields are usually much smaller than the intrinsic fields in a atom. For an electron at distance a_0 from the nucleus the field is $\mathcal{E}_{\text{au}} = 2R_\infty/ea_0 = e/4\pi\varepsilon_0 a_0^2 \sim 5 \times 10^{11}$ V/m and this is the atomic unit of field. Usual laboratory fields are of the order of 10^5 V/m, which is more than six orders of magnitude smaller.

In the previous chapter it was shown that the Zeeman term commutes with $\mathcal{H}_0$ in the absence of spin–orbit interaction so that its eigenfunctions can be used to find the energy shift caused by the magnetic field (see App. 7.C). Here the Stark term does not commute with the atomic Hamiltonian and in general it is not easy to find the energy shift caused by the electric field.

The discussion is simplified by considering the symmetry of the Stark operator $e\mathcal{E}z$ under the parity operator $\mathcal{P}$. For eigenfunctions of definite parity one has $\mathcal{P}^2\psi = \psi$ so the eigenvalues of $\mathcal{P}$ can be only ± 1 and the parity is given by $(-1)^\ell = \pm 1$. Thus $\mathcal{P}\psi_{n\ell m} = (-1)^\ell \psi_{n\ell m}$. The atomic Hamiltonian $\mathcal{H}_0$ commutes with $\mathcal{P}$ because ∇^2 is even and r is always positive so that the eigenfunctions of $\mathcal{H}_0$ are also eigenfunctions of $\mathcal{P}$.

Since the Stark term $e\mathcal{E}z$ changes sign with z, its diagonal elements all vanish: $\langle\psi_{n\ell m}|e\mathcal{E}z|\psi_{n\ell m}\rangle = 0$. These terms constitute the first-order energy shifts in a perturbation expansion so that there is no linear Stark effect for all atomic states of definite parity, and higher orders have to be taken into account to calculate the effect of the electric field on the energy of the states. For atoms other than hydrogen the eigenfunctions of $\mathcal{H}_0$ may be different from $\psi_{n\ell m}$, but still they have a well-defined parity and thus no first-order Stark shifts.

However, for hydrogen states with the same n are degenerate (neglecting fs, hfs, and QED effects) but have opposite parity. In that case one can make eigenfunctions of the Stark term $e\mathcal{E}z$ by taking linear combinations of these states, and the shifts of such states are indeed linear in the field because they are superpositions of mixed parity. In the next section this problem is solved in familiar spherical coordinates, and in Sec. 12.3 it is solved in the less conventional parabolic coordinates.

12.2 Electric field shifts in spherical coordinates

12.2.1 Stark effect in hydrogen, $n = 2$

The $n = 2$ state of hydrogen has four degenerate states with energy $E_2 = -R_\infty/4$ (neglecting fine structure and hfs): $2s_0$, $2p_{-1}$, $2p_0$, and $2p_{+1}$, where the subscripts $m = 0, \pm 1$ indicate the projection $m\hbar$ of ℓ_z. Because the m-values ± 1 are unique in this set, the selection rules above show that the $2p_{\pm 1}$ states cannot be coupled to other $n = 2$ states and thus are not shifted by the electric field. However, the $2s_0$ and $2p_0$ can be coupled since they have opposite parity and the same value of m.

The coupling element can easily be evaluated using $z = r\cos\theta$ as

$$\mathcal{H}'_{12} = e\mathcal{E}\int \psi^*_{210}\, z\, \psi_{200}\, d\vec{r} \tag{12.2}$$

$$= \frac{e\mathcal{E}Z^3}{16\pi a_0^3}\int_0^\infty dr\, r^3 \frac{Zr}{a_0}\left(1 - \frac{Zr}{2a_0}\right)e^{-Zr/a_0} \int_0^\pi d\theta\, \sin\theta\, \cos^2\theta \int_0^{2\pi} d\phi$$

$$= -\frac{3e\mathcal{E}a_0}{Z},$$

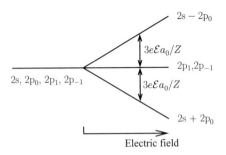

Figure 12.1 Stark shift for the $n = 2$ state of hydrogen.

where the eigenfunctions of Tab. 7.1 and Tab. 7.2 are used. The Hamiltonian matrix can then be written as

$$\mathcal{H} = \begin{pmatrix} E_2 & -3e\mathcal{E}a_0/Z \\ -3e\mathcal{E}a_0/Z & E_2 \end{pmatrix} \tag{12.3}$$

and the eigenvalues are simply given by $E_\pm = E_2 \pm 3e\mathcal{E}a_0/Z$. Note that the shifts for these two states are linear in the applied electric field. The eigenfunctions of the states are

$$\psi_\pm = \frac{1}{2}\sqrt{2}\left(\psi_{200} \mp \psi_{210}\right), \tag{12.4}$$

and it is easily checked that these are eigenfunctions of neither the parity operator $\mathcal{P}$ nor the angular momentum operator $\vec{\ell}^{\,2}$. The splitting for the $n = 2$ state of hydrogen is shown in Fig. 12.1. The $2p_{0,\pm 1}$ states can decay to the ground state by spontaneous emission, and their lifetime is $\tau_{2p} \approx 1.6$ ns. The $2s_0$ state is metastable because it cannot decay to the ground state by ordinary spontaneous emission (see Sec. 4.3.1), and its lifetime is $\tau_{2s} \approx 0.14$ s.

12.2.2 The non-linear Stark effect

In general, atomic states are non-degenerate and thus do not have a linear Stark effect. For the next higher-order calculation one can employ second-order perturbation theory, but in most cases this is rather complicated since a sum has to be taken over all other states, including the continuum. In case of non-degenerate second-order perturbation theory, the shift caused by an interaction $\mathcal{H}'$ is

$$E_S^{(2)} = \sum_{n'\ell'm' \neq n\ell m} \frac{|\langle \psi_{n\ell m}|\mathcal{H}'|\psi_{n'\ell'm'}\rangle|^2}{E_{n\ell m} - E_{n'\ell'm'}} \tag{12.5}$$

where $(n\ell m)$ represent all the quantum numbers of the system. Unlike the excited states, the ground state of hydrogen is non-degenerate and is the example chosen here for a second-order perturbation calculation. The Stark shift is

12.2 Electric field shifts in spherical coordinates

$$E_S^{(2)} = e^2 \mathcal{E}^2 \sum_{n\ell m \neq (100)} \frac{|\langle \psi_{100}|z|\psi_{n\ell m}\rangle|^2}{E_1 - E_{n\ell m}}, \qquad (12.6)$$

where $\mathcal{H}' = e\mathcal{E}z$ is used. Since $E_{n\ell m} > E_1$ for all $n\ell m$, the shift is negative.

It is possible to find a upper bound for the magnitude of this shift (actually lower bound because it is <0) using the following approximations. First replace the terms in the denominator of Eq. (12.6) with the term for the first excited state and take the denominator out of the sum. Since this value is smaller than all the other terms, the shift will be larger than the expression evaluated in this approximation. Next, add the $n\ell m = 100$ term back into the sum, since it does not contribute because $\langle\psi_{100}|z|\psi_{100}\rangle = 0$. The expression can now be evaluated as

$$E_S^{(2)} > \frac{-e^2\mathcal{E}^2}{E_2 - E_1} \sum_{n\ell m} |\langle\psi_{100}|z|\psi_{n\ell m}\rangle|^2 \qquad (12.7)$$

$$= \frac{-e^2\mathcal{E}^2}{E_2 - E_1} \sum_{n\ell m} \langle\psi_{100}|z|\psi_{n\ell m}\rangle\langle\psi_{n\ell m}|z|\psi_{100}\rangle$$

$$= \frac{-e^2\mathcal{E}^2}{E_2 - E_1} \langle\psi_{100}|z^2|\psi_{100}\rangle$$

where the last step uses the closure relation $\sum_{n\ell m}|\psi_{n\ell m}\rangle\langle\psi_{n\ell m}| = 1$. Use Tab. 7.6 to find

$$\langle\psi_{100}|z^2|\psi_{100}\rangle = \frac{1}{3}\langle\psi_{100}|r^2|\psi_{100}\rangle = \frac{a_0^2}{Z^2}. \qquad (12.8)$$

The final result is $E_S^{(2)} > -8/3(4\pi\varepsilon_0)\mathcal{E}^2 a_0^3/Z^4$ which has to be compared with the exact value $E_S^{(2)} = -9/4(4\pi\varepsilon_0)\mathcal{E}^2 a_0^3/Z^4$ for $n = 1$ (see Sec. 12.3). Such second-order shifts are much smaller than the binding energy of the ground state E_1 because they scale approximately as $E_1(\mathcal{E}/\mathcal{E}_{au})^2$ where $\mathcal{E}_{au}$ is the atomic unit of field (see Sec. 12.1). Thus for a field $\mathcal{E} = 10^8$ V/m it is $|E_S^{(2)}/E_1| \simeq 10^{-7}$.

The physical interpretation of the quadratic Stark shift can be understood from the results of perturbation theory. In first-order, the perturbed wavefunction is given by

$$\psi_{n\ell m}^{(1)} = \psi_{n\ell m} + \sum_{n'\ell'm' \neq n\ell m} \frac{\langle\psi_{n'\ell'm'}|\mathcal{H}'|\psi_{n\ell m}\rangle}{E_{n\ell m} - E_{n'\ell'm'}}\psi_{n'\ell'm'} \qquad (12.9)$$

and for $\mathcal{H}' = e\mathcal{E}z$ and $n\ell m$ representing the ground state this becomes

$$\psi_{100}^{(1)} = \psi_{100} + e\mathcal{E}\sum_{n'\ell'm' \neq 100}\frac{\langle\psi_{n'\ell'm'}|z|\psi_{100}\rangle}{E_{100} - E_{n'\ell'm'}}\psi_{n'\ell'm'}. \qquad (12.10)$$

The electric field induces an electric dipole moment μ in the z-direction for the ground state given by

$$\mu = -e \langle \psi_{100}^{(1)} | z | \psi_{100}^{(1)} \rangle = -(4\pi\varepsilon_0)\alpha\mathcal{E} \tag{12.11a}$$

with

$$\alpha = -\frac{2e^2}{4\pi\varepsilon_0} \sum_{n'\ell'm' \neq 100} \frac{|\langle \psi_{n'\ell'm'} | z | \psi_{100} \rangle|^2}{E_{100} - E_{n'\ell'm'}}, \tag{12.11b}$$

where α is the polarizability (not to be confused with the fine-structure constant) given in units of m^3. Thus the induced electric dipole moment is proportional to $\mathcal{E}$, and this dipole moment interacts with the applied electric field $\mathcal{E}$ to yield a shift that is proportional to $\mathcal{E}^2$. For the ground state the polarizability $\alpha = 9a_0^3/2$ or just 9/2 in atomic units.

The Lamb shift lifts the degeneracy of the 2s$_0$ and 2p$_0$ states of H with $j = 1/2$ and thus provides another case where the non-degenerate calculation applies. An electric field mixes some 2p$_0$ component into the 2s$_0$ state (see Eq. (12.9)) so that its lifetime is shortened considerably in an electric field thus quenching its metastability. For this case of the $n = 2$ states of hydrogen, a straightforward first-order perturbation calculation (non-degenerate when the Lamb shift is included as described in Sec. 8.7) shows that the 2s lifetime is shortened from its field-free value to $\tau'_{2s} = \tau_{2p}(\mathcal{E}_\tau/\mathcal{E})^2$ where $\mathcal{E}_\tau = 4.75 \times 10^4$ V/m [75]. The two lifetimes become comparable for $\mathcal{E} = \mathcal{E}_\tau$. An estimate similar to the one above of the second-order Stark shift in this case, using the term with $(n, \ell, m) = (2,0,0)$ as having the smallest denominator, gives a shift of $\sim 10^9$ Hz, which is comparable to the Lamb shift that separates the states and is the field value where the Stark mixing is nearly complete.

12.3 Electric field shifts in parabolic coordinates

12.3.1 Hydrogen in parabolic coordinates

Although more can be done with electric fields in spherical coordinates [115], the problem is much more tractable in parabolic coordinates. This is because the Schrödinger equation for hydrogen is separable, even in the presence of an electric field. The connection between Cartesian, spherical, and parabolic coordinates is (see App. 7.B.2)

$$\xi = r + z = r(1 + \cos\theta), \quad \eta = r - z = r(1 - \cos\theta), \quad \phi = \phi. \tag{12.12}$$

Since the Schrödinger equation is separable in parabolic coordinates, the solution $\Psi(\xi, \eta, \phi)$ can be written as

$$\Psi(\xi, \eta, \phi) = f(\xi) g(\eta) \Phi(\phi) \tag{12.13}$$

The ϕ-dependence of the Schrödinger equation is the same in either coordinate system (see Sec. 7.3) so the ϕ-dependence of the solution has the same form as before: $\Phi_m(\phi) = \exp(im\phi)$. In the case of the hydrogen atom with $V = -Ze^2/4\pi\varepsilon_0 r$, the Schrödinger equation can be separated into two independent equations for ξ and η, even in the presence of an electric field $\mathcal{E}$. This separation requires $V \propto 1/r$, not simply $V = V(r)$, and so applies only to hydrogenic systems. The result is

$$\frac{d}{d\xi}\left(\xi \frac{df}{d\xi}\right) + \frac{\mu}{\hbar^2}\left(\frac{E\xi}{2} + \nu_1 - \frac{\hbar^2 m^2}{4\mu\xi} - \frac{e\mathcal{E}\xi^2}{4}\right)f = 0,$$

$$\frac{d}{d\eta}\left(\eta \frac{dg}{d\eta}\right) + \frac{\mu}{\hbar^2}\left(\frac{E\eta}{2} + \nu_2 - \frac{\hbar^2 m^2}{4\mu\eta} + \frac{e\mathcal{E}\eta^2}{4}\right)g = 0$$

(12.14)

where m is the projection of $\vec{\ell}$ along $\vec{\mathcal{E}}$, and ν_1, ν_2 are two separation constants that satisfy

$$\nu_1 + \nu_2 = \frac{Ze^2}{4\pi\varepsilon_0}. \quad (12.15)$$

For $\mathcal{E} = 0$ the solution of these equations gives $E_n = -R_\infty Z^2/n^2$ as before, where the quantum number n is given by

$$n = n_1 + n_2 + |m| + 1. \quad (12.16)$$

The quantum numbers n_1 and n_2 are positive integers (or zero) and are referred to as the electric quantum numbers. The eigenfunctions are given by $\Psi_{n_1 n_2 m}(\xi, \eta, \phi) = f_{n_1}(\xi) g_{n_2}(\eta) \Phi_m(\phi)$, and Eq. (12.16) determines n.

Equations (12.14) are eigenvalue equations for $\mathcal{E} = 0$. However, in the limit of large $\mathcal{E}$ or high n, the second one (for η) is not an eigenvalue equation because there are unbound states degenerate with the "bound" states, making them quasi-bound. That is, electrons far enough downfield from the nucleus can have arbitrarily large negative energy, so for large enough field and n it describes free motion. Even though the Schrödinger equation separates, the resulting equations do not have analytical solutions, so perturbation methods are necessary. Of course, the equations can be integrated numerically rather simply because they have only one variable, and a review is given in Refs. [116, 117].

12.3.2 Linear Stark effect in parabolic coordinates

Perturbation theory can be used to calculate the linear Stark effect in parabolic coordinates. In spherical coordinates, the first-order term leads to zero shift for non-degenerate states since the Stark term evaluated with the eigenfunctions in spherical coordinates vanishes. In parabolic coordinates the Schrödinger can be separated only for a potential $\propto 1/r$, and this leads to degenerate states as discussed above. Then the terms with $\mathcal{E}\xi^2$ and $\mathcal{E}\eta^2$ can be used as perturbations leading to non-zero Stark shifts. However, the quantization condition of Eq. (12.15) remains in effect. Using the solution of the Schrödinger equation in parabolic coordinates with $\mathcal{E} = 0$ as the unperturbed wavefunctions, the Stark shift becomes [75]

$$E_S^{(1)} = +\frac{3ne\mathcal{E}a_0}{2Z}(n_1 - n_2). \tag{12.17}$$

For $n = 2$ one can have $m = 0$ and $m = \pm 1$. In the first case the electric quantum numbers can be found for Eq. (12.16) as $n_1 = 1$ and $n_2 = 0$, or $n_1 = 0$ and $n_2 = 1$. This leads to Stark energies of $E_\pm = E_2 \pm 3e\mathcal{E}a_0/Z$, which is the same as the Stark shift of the superposition states of Eq. (12.4) found for spherical coordinates. These superpositions are eigenstates of Eq. (12.14). For $m = \pm 1$ the states are not shifted, which is also the same as the case for the $2p_{\pm 1}$-states using spherical coordinates.

In general, the largest shifts are for states with $m = 0$ and either $n_2 = 0$ or $n_1 = 0$, where in the first case the shift is positive and in the second case the shift is negative. The largest splitting occurs at these two extremes $\Delta E_S^{(1)} = 3e\mathcal{E}a_0n(n-1)/Z$ and it scales with n^2. Comparison between the first-order Stark effect and the interval between two states in the Bohr atom leads to

$$\frac{E_S^{(1)}}{\Delta E_n} \simeq \frac{3e\mathcal{E}a_0n^2/Z}{2R_\infty Z^2/n^3} = \frac{3e\mathcal{E}a_0n^5}{2R_\infty Z^3} \tag{12.18}$$

Ordinary laboratory fields have only a small effect for low-lying levels, but for $n = 20$, even ordinary laboratory fields can cause mixing of the states (see Chap. 13). For typical laboratory fields of $\mathcal{E} \approx 1$ kV/cm, the ratio for $n = 2$ is 10^{-5}, but for $n = 20$ the ratio is ~ 1. Thus it does not require a very high principal quantum number to cause a marked effect in the nature of atomic structure.

12.3.3 Quadratic Stark effect in parabolic coordinates

Since the eigenfunctions for the unperturbed Hamiltonian in parabolic coordinates are known, one can perform perturbation theory to any order. For second-order perturbation theory, Eq. (12.5) can be used to find that the quadratic Stark shift is given by [75]

$$E_S^{(2)} = -\frac{(4\pi\varepsilon_0)\mathcal{E}^2 a_0^3 n^4}{16Z^4}\left[17n^2 - 3(n_1 - n_2)^2 - 9m^2 + 19\right] \tag{12.19}$$

The coefficient of the square-bracketed term is $R_\infty(\mathcal{E}/\mathcal{E}_{au})^2(n/2Z)^4$, where $\mathcal{E}_{au}$ is the atomic unit of electric field. For the ground state of hydrogen $n = 1$, $n_1 = n_2 = m = 0$, and $Z = 1$ so the shift becomes $E_S^{(2)} = -9/4(4\pi\varepsilon_0)\mathcal{E}^2 a_0^3$. This has been used in Sec. 12.2.2 to compare to the estimated second-order Stark shift in spherical coordinates. Here, the calculation is exact and even allows one to go to higher orders.

12.3.4 Atoms other than hydrogen

When there is no electron core so there is a purely $1/r$ potential, and the Schrödinger equation separates in parabolic coordinates for $\mathcal{E} \neq 0$. For the case of atoms with core electrons such as alkalis, the Schrödinger equation does not separate, and other methods are needed. One method that has been very successful was presented in Ref. [118]. In the vicinity of the core, the nuclear potential dominates that of the applied external field because the distance is short, and so phase-shifted, quantum defect wavefunctions in spherical coordinates are appropriate. Far from the core the atomic potential is essentially a pure $1/r$, so parabolic wavefunctions can be used and the Schrödinger equation can be separated and solved. The bridge between the inner and outer regions is made by using the WKB method to "splice" the two parts of the wavefunction in the intermediate region.

12.4 Summary

In summary, the description of the Stark effect is considerably more complicated than for the Zeeman effect. Special cases abound, and different methods are used for different ones. In Fig. 12.2 there is a schematic diagram of the different routes that can be followed. The primary division follows the organization of this chapter, namely the choice of coordinate systems. The various blocks on the left side under spherical coordinates follow the discussion in Sec. 12.2 above, and those on the right side under parabolic coordinates follow the discussion in Sec. 12.3.

For spherical coordinates, the hydrogen atom is the paradigm and the left side of Fig. 12.2 mostly follows the text. However, the case of alkali-metal atoms is special because there were so many early experiments on alkali Rydberg atoms in electric fields (see Chap. 13). Careful numerical studies that simply truncated the huge Hamiltonian matrices obtained results that agreed extremely well with measurements, and the rightmost box on the bottom row (near the center) of this spherical case refers to these calculations [115].

For parabolic coordinates, the separated equations for hydrogen have no analytic solutions so energies and eigenfunctions have to be obtained by approximation methods, either numerical or perturbation theory, as indicated. However, for alkalis (or other atoms) the Schrödinger equation cannot be separated in any coordinate system, so a completely different approach is needed. Even though the results of Ref. [115] agreed well with measurements, a more satisfactory method was found using different coordinate systems in different regions, and then joining these with

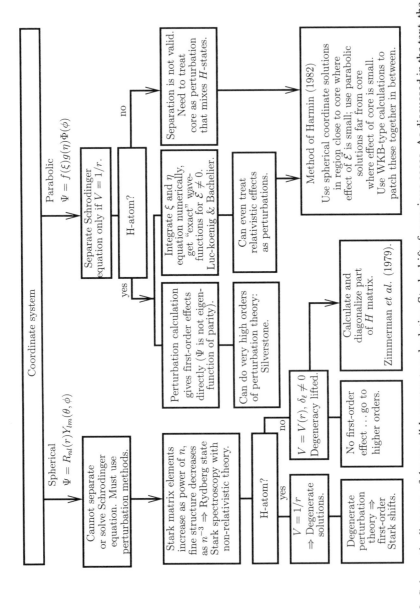

Figure 12.2 Schematic diagram of the possible routes to calculating Stark shifts for various cases. As discussed in the text, the primary organization is by coordinate system, and the subsequent boxes in the diagram identify various steps toward the calculations. The names in the diagram refer to the following publications: Luc-Koenig & Bachelier [116, 117], Silverstone [119], Zimmerman et al. [115], Harmin [118].

Exercises

WKB calculations as in Ref. [118]. This more sophisticated technique has not been discussed here.

Exercises

12.1 The $2^2S_{1/2}$ and $2^2P_{1/2}$ states of atomic hydrogen are said to exhibit a linear Stark effect because they are degenerate, but this is not true. Calculate the energy of all sublevels in an electric field, considering that the S and P manifolds are separated by the Lamb shift of 1057 MHz. Calculate the mixing coefficients of the states, and identify a dimensionless parameter which is small for the quadratic Stark effect and large for the linear Stark effect.

12.2 Show that in the linear Stark effect the $n = 3$ level of a hydrogen atom is split into five equally spaced components, and obtain the level separation in eV.

12.3 Suppose that a hydrogen atom is in a superposition of the 2s and 2p ($m = 0$) states at $t = 0$ when a constant, uniform electric field of 10^7 V/m is applied along the z-axis. The 2p component decays by Lyman-α emission at a rate $1/(2\pi\tau) \approx 100$ MHz but the mixing of the 2s state by the electric field keeps feeding population into the 2p state. Compare the rate of this population transfer with the decay rate.

13
Rydberg atoms

13.1 Introduction

Series of atomic spectral lines had been studied by emission from excited vapors even before Balmer described the regularity of the visible lines of hydrogen in 1884. However, direct excitation had been restricted to the few lowest-lying levels, particularly those accessible by electric dipole transitions from the ground state using classical light sources. The advent of tunable lasers in the 1970s changed such spectroscopic limitations enormously, allowing excitation to a very much larger number of states. Atoms in highly excited states, typically with principal quantum numbers $n \geq 10$, are called Rydberg atoms. They are oversize, fragile states of significant intrinsic interest. They are very rare in nature because of their size and fragility, but they are found in interstellar space (H I regions). One of the best sources for comprehensive descriptions of Rydberg atoms and references to the literature describing their exploration is Ref. [120]. It describes their characteristic properties and behavior both classically and quantum mechanically, and especially their strongly enhanced interactions with external fields.

The first thing to note is that Rydberg atoms are usually characterized by a single, highly excited electron while all the other electrons remain in their original states. Thus their properties are very nearly classical, and the atoms are much closer to the ideal hydrogen-like model of Chap. 7, along with appropriate quantum defects (see Chap. 10). The spectroscopy of such states is especially interesting because the active electron is much less perturbed by the other electrons in the atom than when it is in lower states. Moreover, there are some special circumstances in which the energy of a Rydberg state is degenerate with an atomic state having two electrons excited to lower levels, and such configurations yield further new information.

Second, Rydberg atoms do not usually decay back to the ground state by emission of characteristic radiation, so their study is mediated by other techniques that reveal new and interesting properties. Such properties arise because Rydberg atoms have a huge dipole moment and very low binding energy resulting from the large average distance between the excited electron and the nucleus-plus-core of the rest of the atom with net charge e.

Third, transitions between Rydberg states are quite different from those connecting to the ground state for at least three reasons. One is that the energy differences are so small that they are in the radio or microwave frequency range. The resulting large wavelengths produce huge cross-sections (see Sec. 3.3.2 and other calculations of σ) that make Rydberg atoms exquisitely sensitive to very low levels of radiation. These can even include black body radiation at room temperature, the zero-point energy of the quantized field, and tiny resonant fields in cavity quantum electrodynamics experiments [121]. A second reason is that spontaneous emission is extraordinarily slow because of the small value of ω^3 in Eqs. (5.3) and (5.13) relative to that of optical transitions. This long lifetime against spontaneous decay, coupled with the huge absorption cross-section resulting from the large wavelength, makes Rydberg atoms into a wonderful laboratory for quantum information studies and quantum non-demolition demonstrations. Finally, the approximations such as those leading to Eqs. (3.4) and (13.2), and several others, have to be re-examined.

Fourth, gaseous samples of cooled and trapped atoms can be excited to Rydberg states quite efficiently, and their huge sizes enable a high collision probability. Since the binding energy is so low, a collision can ionize one of the partners while dropping the other one to a slightly lower Rydberg state. Thus interactions between atoms in a cold Rydberg gas can produce a cold neutral plasma whose parameters are unlike those of any plasmas previously studied. Moreover, the resulting slow electrons can readily recombine with the ions, and slow oscillations between the plasma state and the cold gas of neutral atoms state have been observed and studied. Thus Rydberg atoms provide access to plasma studies in new domains.

13.2 The Bohr model and quantum defects again

Many properties of Rydberg atoms can be understood correctly in terms of the relatively simple notions used to describe hydrogen atoms (see Chap. 7). For Rydberg atoms, Eq. (10.1) gives a very good approximation to the energy levels. The energies are negative and approach one another and $E_n \approx 0$ at large values of n. The energy separation between adjacent levels is $\Delta E_n = 2R_\infty/(n^*)^3$, the speed of the electron is $\alpha c/n^*$, and the radius of the orbit is $r_n = a_0(n^*)^2$. The picture that emerges of a Rydberg atom is one of an ion core and an isolated electron very far away, floating lazily around in a slow orbit, much like a distant planet of the solar system. For $n \sim 100$ the orbit is as large as a living cell and the electron speed is only 2×10^4 m/s.

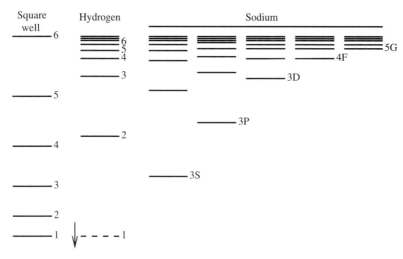

Figure 13.1 This shows the energy levels for various quantum mechanical systems. The left edge corresponds to a square well, beside it is the hydrogen atom, and the right part is the sodium atom. (Figure adapted from Ref. [122].)

Although the Bohr energy formula gives very good values for the hydrogen atom, it does not do as well as for other atoms because the valence electron's orbit may take it inside the closed-shell core where the potential descends very much more steeply than that of hydrogen (see Chap. 10). Such an electron therefore has much lower energy, as shown by the S and P levels in the central part of Fig. 13.1. The low angular momentum states (penetrating orbits) of alkali metals are therefore much more tightly bound than either the high angular momentum states (non-penetrating orbits) that see a nearly Coulombic potential or the corresponding n states of hydrogen. The differences between the high angular momentum states of alkalis and those of hydrogen may also arise because the ion core is composed of mobile electrons whose orbits can be perturbed by the valence electron (see Sec. 10.3).

These energy-level differences are characterized by the quantum defects as discussed and tabulated in Chap. 10. Equation (10.1) for the energies of the alkali energy levels agrees very well with spectroscopic data obtained from high-resolution studies of Rydberg atoms [123], and especially those with particularly simple cores such as helium and lithium [124]. These provide an excellent basis for comparison between atomic theory and experiment.

The electric field produced by the distant Rydberg electron at the ion core may not be uniform as is assumed in Sec. 10.3. Therefore the induced charge distribution need not correspond to a pure dipole moment, but may have quadrupole, octopole, and higher-order terms. For each of these there is an addition to the

Hamiltonian that can be treated similarly to the dipole term. It is a very curious property of the hydrogen radial wavefunctions that for all potentials $V(r)$ that vary as an inverse integer power of r of the form $1/r^k$ for $k > 1$, the leading term for the energy shift is proportional to $1/n^3$ [75, 125]. The result is that the quantum defect formula of Eq. (10.1) provides a very accurate prediction for the energy levels from interactions of different types (penetration as well as core polarization) because multipole terms are each proportional to some inverse power of r. Finally, neither the electron nor the core is at rest, resulting in various dynamic contributions to the polarizability (see discussion in Ref. [93]).

Equations (10.11) and (10.12) can be used to find approximate values for the quantum defects of non-penetrating states as given by Eq. (10.1), and the dependence is consistent with spectroscopic data as exhibited by the entries in Tab. 10.1 and Tab. 10.2. Therefore the tables of spectral data can be used to determine the dipole polarizabilities of the ion cores. The deviations from a perfect fit to equation Eq. (10.12) (or the unapproximated results of equation Eq. (10.11) can be used to determine the quadrupole polarizability or other higher-order effects (such as dynamic effects and retardation).

The discussion above has been restricted to Rydberg states of alkali-metal atoms. It is possible to study the Rydberg states of molecules or atoms with two or more valence electrons, such as alkaline earth atoms, and the added richness of the spectra provides new information. For example, after one electron is promoted to a Rydberg state, a second electron can be raised to one of the ion core's excited states. The atom now has too much internal energy to remain bound and autoionizes, emitting an "Auger electron" (see Sec. 14.5). This process may be enhanced by configuration interaction with another nearby state, and precise studies can be made. A theoretical structure called multichannel quantum defect theory (MQDT) has been developed to describe these processes [126] which can be measured carefully and cleanly by Rydberg spectroscopy.

13.3 Rydberg atoms in external fields

The description of Rydberg atoms in small external fields is very similar to that of ground-state atoms, but the field strengths that can be called small are very restricted. As highly excited electrons are weakly bound, moderate external fields can have large effects on them. For example, Sec. 13.3.1 below shows that the last term in Eq. (11.3) can usually be neglected, but also that it scales with n^4. Thus modest fields can induce considerable diamagnetism in Rydberg atoms as a result of their large size.

The strength of the Coulomb field that binds the Rydberg electron to the parent ion scales as $r^{-2} \propto n^{-4}$, so for $n = 30$ it is only 6 kV/cm, a value easily achieved

Rydberg atom property	Magnetic field	Electric field
Linear term	Independent of n	$\propto n^2$
Quadratic term	$E^{(1)} \propto n^4$	$E^{(2)} \propto n^6$
Lin. term ≈ quad. term	$n = 18$ at $\mathcal{B}_{lab}$	$n = 60$ at $\mathcal{E}_{lab}$
Field ionization $\mathcal{E}_{FI}$	none	$\mathcal{E}_{FI} = \mathcal{E}_0/(2n)^4$
Ratio $E^{(2)}$ to ΔE_n, ΔE_{FS}	$\propto n^7$	$\propto n^9$
$E^{(2)} \sim \Delta E_n$	$n = 44$ at $\mathcal{B}_{lab}$	$n = 27$ at $\mathcal{E}_{lab}$
$E^{(2)} \sim \Delta E_{FS}$	$n = 9$ at $\mathcal{B}_{lab}$	$n = 6$ at $\mathcal{E}_{lab}$
Selection rules for diagonal matrix elements	$\begin{cases} \Delta\ell = \pm 2, 0 \\ \Delta m = 0 \\ \Delta n \text{ none} \end{cases}$	$\begin{cases} \Delta\ell = \pm 1 \\ \Delta m = 0 \\ \Delta n \text{ none} \end{cases}$

Table 13.1 A summary of a few of the field-dependent properties of Rydberg atoms that are discussed in the text. The quantities used for numerical examples are $\mathcal{E}_{lab} = 1$ kV/cm ≈ $2 \times 10^{-7} \mathcal{E}_{au}$ and $\mathcal{B}_{lab} = 1$ T ≈ $4 \times 10^{-6} \mathcal{B}_{au}$. Lin., linear; quad., quadratic.

in the laboratory. Under such conditions, any approximation of the effects of an applied field as being small compared with the basic atomic structure is destined for failure. The energy-level scheme of Rydberg atoms in electric fields having even a few percent of this size looks very different from the familiar zero-field scheme, and in fact, the Stark shifts can easily exceed the separation between levels in adjacent n-manifolds. The discussion below describes some of these surprising features.

The scaling of various energies of Rydberg atoms with n determines many of their characteristics. However, it is important to note that, even though $1/n^3$ scaling applies to both the interval between n-manifolds of hydrogen given by $\Delta E_n = 2R_\infty/n^3$, and the fine-structure interval $\Delta E_{SO} = \alpha^2 R_\infty/(2n^3)$ as given in Eq. (8.23) for hydrogen, their magnitudes differ by a factor of α^2. Nevertheless, this strong n-dependence has important consequences for Rydberg atoms as summarized in Tab. 13.1.

13.3.1 Rydberg atoms in magnetic fields

The description of the Zeeman effect for Rydberg atoms begins with the second term of Eq. (11.3). It uses eigenfunctions that are the same as those of the field-free case with Hamiltonian $\mathcal{H}_0$ because the interaction term there becomes $g\mu_B \mathcal{B} L_z$. Since this commutes with both the ∇^2 term and $V = V(r)$ terms of the Hamiltonian $\mathcal{H}_0$, there are indeed simultaneous eigenfunctions of L_z and $\mathcal{H}_0$ and thus the Schrödinger equation can still be separated in spherical coordinates into solvable radial and angular parts.

13.3 Rydberg atoms in external fields

The quadratic term of Eq. (11.3) that does not commute with $\mathcal{H}_0$ is given by

$$\mathcal{H}' = \frac{e^2}{2m}|\vec{\mathcal{A}}|^2 = \frac{e^2\mathcal{B}^2 r^2 \sin^2\theta}{8m}, \tag{13.1}$$

where in the last step it is assumed that the magnetic field is in the z-direction, as in Chap. 11. This term is generally much smaller than the linear term, as can be seen by estimating their ratio. The linear term is of the order $\mu_B \mathcal{B}$, whereas for the quadratic term the expectation value of the r^2 factor can be found from Tab. 7.6 using $\ell \ll n$. Assuming the $\sin^2$ term is of order unity, the ratio then becomes

$$\frac{(e^2/2m)|\vec{\mathcal{A}}|^2}{(e/m)\vec{\mathcal{A}}\cdot\vec{p}} \sim \frac{e^2\mathcal{B}^2\langle r^2\rangle/m}{\mu_B\mathcal{B}} = \frac{2n^4}{Z^2}\frac{\mathcal{B}}{\mathcal{B}_{au}}, \tag{13.2}$$

where $\mathcal{B}_{au} \equiv R_\infty/\mu_B \approx 2.35 \times 10^5$ T is the atomic unit of magnetic field. Since typical laboratory magnetic fields do not exceed a few T, in general this ratio is $\ll 1$. However, for atoms in highly excited states with large n or large fields, the second-order shift becomes comparable to or larger than the first-order shift and thus it is evaluated below (see Ref. [127]).

Since the quadratic Zeeman effect term of Eq. (13.1) is generally small compared with the other terms of the Hamiltonian of Chap. 11, its energy shift can be calculated by treating it as a perturbation. Using $\mathcal{H}'$ of Eq. (13.1) with the hydrogenic wavefunctions, the first-order energy shift $\Delta E^{(1)}$ is

$$\begin{aligned}E^{(1)} &= \langle\psi_{n\ell m}|\mathcal{H}'|\psi_{n\ell m}\rangle\\ &= \frac{e^2\mathcal{B}^2}{8m}\langle R(r)|r^2|R(r)\rangle\langle Y_{\ell,m}(\theta,\phi)|\sin^2\theta|Y_{\ell,m}(\theta,\phi)\rangle\end{aligned} \tag{13.3}$$

Even though this term is proportional to $\mathcal{B}^2$, it is the result of a first-order perturbation calculation as indicated by the label $E^{(1)}$.

The radial integral above is found from Tab. 7.6. For the angular part, one can use the relation $\sin^2\theta \propto (Y_{2,0} + bY_{0,0})$, where b is a constant, so the angular integral becomes $\langle Y_{\ell m}|(Y_{2,0} + bY_{0,0})|Y_{\ell m}\rangle$. Since a product of two spherical harmonics can be re-written as the sum of two spherical harmonics as given in App. C.5, the angular part is also readily evaluated. The final result is

$$\begin{aligned}E^{(1)} &= \frac{n^2 a_0^2 e^2 \mathcal{B}^2}{16mZ^2}(5n^2 - 3\ell(\ell+1)+1)\left(\frac{\ell^2+\ell-1+m^2}{(2\ell-1)(2\ell+3)}\right)\\ &\approx \frac{n^4 R_\infty}{3Z^2}\left(\frac{\mathcal{B}}{\mathcal{B}_{au}}\right)^2\end{aligned} \tag{13.4}$$

for $n \gg \ell$.

Atomic diamagnetism scales with n^4 as in Eq. (13.4) for the case of the large n considered here. The ratios of $E^{(1)}$ to ΔE_n and to ΔE_{FS} both scale as n^7, so that $E^{(1)} \sim \Delta E_n$ for $n \sim 44$ and $E^{(1)} \sim \Delta E_{FS}$ for $n \sim 9$, both for $\mathcal{B} \sim 1$ T (see Tab. 13.1). Moreover, for $n \sim 18$ the diamagnetic energy is comparable to the Zeeman energy at the same 1 T field. Some spectroscopic measurements were reported in Ref. [128].

More accurate calculations of diamagnetic energy shifts require the off-diagonal matrix elements of the $\mathcal{B}^2$ part of the Hamiltonian. It is clear from the $\sin^2\theta$ dependence that the only non-vanishing angular integrals are those for which $\Delta \ell = \pm 2, 0$: $\mathcal{B}^2$ can only connect states of the same parity and must satisfy a "triangle condition" on the ℓ values. It is also clear that $\Delta m = 0$ and that there is no selection rule on n. For strong fields and/or large values of n, perturbative calculations are not satisfactory and other methods of calculations are needed.

Some very interesting results were reported on the diamagnetic spectra of barium and lithium. A laser pulse tuned to excite Ba atoms to a low-lying excited state was followed by a broad-spectrum light pulse whose absorption spectrum was measured in a field of $\mathcal{B} \approx 2$ T [129]. It revealed a series of transitions whose frequencies corresponded well with the expected results. A similar experiment with Li presented confusing results [130] that were later interpreted as the effect of the electric field seen by the atoms as a result of their motion in the magnetic field. The effect was much more pronounced in Li because its smaller mass resulted in a higher thermal speed.

13.3.2 Rydberg atoms in electric fields

A quantum mechanical description of Rydberg atoms in electric fields begins by separating the Schrödinger equation in parabolic coordinates as done in Sec. 12.3. In contrast to the weak magnetic field case where the zero field eigenfunctions are also eigenfunctions of $\vec{p} \cdot \vec{\mathcal{A}}$, the resulting radial differential equation for an electric field is not exactly solvable. Nevertheless, having the angular part separated makes the energy and eigenfunction calculations much easier. The lowest-order energy shifts are given in Eqs. (12.17) and (12.19), which scale as n^2 for the first order and n^6 for the second order [75]. With such strong dependence on n it is easy to see why Rydberg atoms are so very sensitive to electric fields. One can do higher-order perturbation theory [119], diagonalization with a truncated basis set [115], or other approximate calculations [75]. The definitive solution for many cases of practical importance was given in Ref. [118].

13.3.3 Energy estimates and quantum defects

In the case of applied magnetic fields, the relevant term in the Hamiltonian derived from the canonical momentum of the (atom plus field) system by replacing the

13.3 Rydberg atoms in external fields

vector potential $\vec{A}$ by $\vec{p} + e\vec{A}$ in Eq. (11.1). By contrast, for electric fields the Hamiltonian term comes from the scalar potential term Φ, and is simply $e\mathcal{E}z$ for $\vec{\mathcal{E}}$ in the z-direction as in Eq. (12.1). This term connects states having ℓ-values differing by ± 1, and for a particular n-manifold their zero-field energies are given in terms of a quantum defect δ_ℓ (see Chap. 10). The energy shift caused by the electric field can be found by evaluating $\mathcal{H}' = e\mathcal{E}z$ in this basis using the non-degenerate energies $E_{n,\ell}$. Consider the case where two of these energies $E_{n,\ell}$ and $E_{n,\ell'}$ are much closer together than all other energies. Since these are states of definite parity (eigenfunctions of $\mathcal{P}$), the diagonal elements of $\mathcal{H}'$ are zero. Denoting the off-diagonal element between these two nearest states as $C = \langle \psi_{n\ell m}|\mathcal{H}'|\psi_{n\ell'm}\rangle$, the Hamiltonian matrix becomes

$$\mathcal{H} = \begin{pmatrix} E_{n,\ell} & C \\ C & E_{n,\ell'} \end{pmatrix}. \tag{13.5}$$

Diagonalizing this Hamiltonian matrix yields eigenvalues $E'_{n,\ell}$ and $E'_{n,\ell'}$

$$E'_{n,\ell;n,\ell'} = 1/2\,(E_{n,\ell} + E_{n,\ell'}) \pm \sqrt{C^2 + 1/4\,(E_{n,\ell} - E_{n,\ell'})^2} \tag{13.6}$$

where the $+$ $(-)$ sign is for the state n, ℓ (n, ℓ'). For the two states considered these energies are exact, but in general other nearby states that are not taken into account make an additional contribution to the energies.

One can take an estimated value of $C \sim e\mathcal{E}a_0 n^2$ that is consistent with a classical result because $a_0 n^2$ is simply the average separation between the electron and the core. In the limit of small $\mathcal{E}$, Eq. (13.6) can be expanded to find the shift (again $+$ $(-)$ sign is for the state n, ℓ (n, ℓ'))

$$E^{(1)}_{n\ell;n,\ell'} \approx \pm \frac{e^2\mathcal{E}^2 a_0^2 n^7}{(\delta_{\ell'} - \delta_\ell)R_\infty}. \tag{13.7}$$

This is quite different from the second-order perturbation calculation of Eq. (12.19) because this is a first-order calculation on a basis of only two states, and the $\mathcal{E}^2$ dependence arises from expanding the radical of Eq. (13.6). It is to be emphasized that the shift $E^{(1)}$ scales as n^7. Note that if $E_{n,\ell} < E_{n,\ell'}$ the state ℓ shifts downwards and the state ℓ' shifts upwards, since $\delta_\ell > \delta_{\ell'}$. The states seem to repel each other, which is generally true in quantum mechanics, and this leads to the non-crossing of the states (see App. 11.C).

The Zeeman effect has no such spectacular enhancement. The magnetic shift of atomic energy levels is proportional to the Landé g-factor which depends only on the spin and the angular part of the wavefunction $Y_{\ell,m}(\theta, \phi)$, and varies by no more than a factor of three over almost all atomic states.

13.3.4 Numerical calculations

An extension of this kind of approximation is to calculate a larger submatrix of the infinite Hamiltonian and diagonalize it. This was done very successfully in the early work on the Stark effect in Rydberg states (see Ref. [131]). Because of the selection rules as discussed in the beginning of Sec. 12.2 one has to take into account only the coupling between states where $\Delta \ell = \pm 1$ and $\Delta m = 0$. Since the energy shifts are very large and scale with a high power of n (*i.e.*, see Eq. (13.7)), and the separation between the principal manifolds that scales with $1/n^3$ becomes small at high n, modest electric fields can strongly split these levels, can thoroughly mix the ℓ states so that ℓ is no longer a good quantum number, and can bring together states of different n. Of course, these states cannot cross, according to App. 11.C, and so the multitudinous number of sublevels quickly develop into a tangled set of loci. The principle is illustrated in Fig. 13.2. This energy-field region is called the "spaghetti region".

In some cases the quantum defects may even place states of higher n with small ℓ between manifolds of $n-1$ and $n-2$ so there is quite a complicated level structure. Adding this feature to the already strongly mixed ℓ states makes a truly complicated energy-level diagram worthy of the name "spaghetti region". There are many such diagrams in the literature of Rydberg spectroscopy (see Ref. [115]). There is no such enhancement of the Zeeman splitting at high n because the magnetic moment

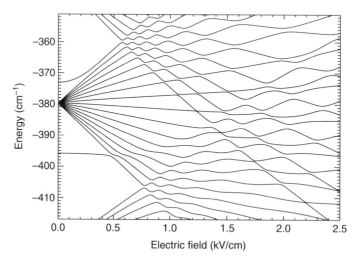

Figure 13.2 The spaghetti region begins at quite low fields when states of $\ell \sim n$ begin to split strongly to the point where states of different ℓ are brought together and strongly mixed. These do not cross, and the large multitude of states makes several sequential anti-crossings, resulting in a complicated diagram. Here the calculation for the $n = 16$ state of Na is shown for $m = 0$.

13.3 Rydberg atoms in external fields

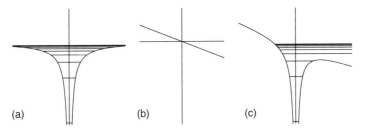

Figure 13.3 Part (a) shows the Bohr levels in a Coulomb potential and part (b) shows the potential of a uniform electric field. Part (c) shows that the sum of these has unbound states degenerate with each bound state. The peak on the right side is actually a saddle-point in three dimensions.

does not depend on the distance between electron and nucleus as does the electric dipole moment.

13.3.5 Field ionization

A simple, semi-classical description of the effects of a dc electric field $\mathcal{E}$ on Rydberg atoms shows that there are no perfectly bound states even for very weak fields. For a field in the z-direction there is always a finite distance at large enough positive z from the atom where the potential energy $e\mathcal{E}z$ is lower than the electron's binding energy (see Fig. 13.3). Moreover, the result of this large electron–nucleus separation of the Rydberg states makes them extremely sensitive to modest electric fields because of their high polarizability. Some of these large responses to electric fields are also summarized in Tab. 13.1.

Because of their relatively small binding energy, Rydberg atoms can be readily ionized by electric fields. The potential for a Rydberg electron in the field $\mathcal{E}$ can be written as $V = -e/4\pi\varepsilon_0 r - \mathcal{E}z$, and it has a local maximum value $V_{max} = -2\sqrt{e\mathcal{E}/4\pi\varepsilon_0}$ at $z = \sqrt{e/4\pi\varepsilon_0\mathcal{E}}$ (see Fig. 13.3). Clearly an electron with energy $E_n > eV_{max}$ can escape. The condition for field ionization becomes

$$\mathcal{E}_{FI} > \frac{e}{4\pi\varepsilon_0 a_0^2} \frac{1}{16n^4} \equiv \frac{\mathcal{E}_{au}}{16n^4} \tag{13.8}$$

where $\mathcal{E}_{au} \equiv e/(4\pi\varepsilon_0 a_0^2) = 2R_\infty/ea_0 \approx 5.1 \times 10^{11}$ V/m is the atomic unit of electric field. For $n = 15$, field ionization occurs at $\mathcal{E} \approx 6$ kV/cm, a readily attainable laboratory field.

The model used to describe the field ionization above is rather simplified. One reason is that some electrons occupy states with wavefunctions concentrated away from the saddle-point produced in the three-dimensional potential by the applied electric field. They may have enough energy to field ionize, but do not sample the potential at the escape route. Another arises because electrons with insufficient energy to ionize classically may escape by tunneling under the barrier. Further complications arise from the mixing of the wavefunctions by the field, resulting in excited states that are superpositions of zero-field eigenfunctions. Field ionization is a separate topic of serious investigation [132].

13.4 Experimental description

The apparatus used for Rydberg experiments can be simple for many atoms, especially the alkalis. A small stainless steel oven is loaded with a few grams of the metal and heated to a few hundred degrees. The oven is mounted in a modest vacuum system ($P \sim 10^{-6}$ torr) of characteristic linear dimension of 0.5 m resulting in a thermal atomic beam. Although some metals require a heat pipe oven, and some Rydberg spectroscopy has been done on rare gases that require a discharge for excitation but no oven, the system described above is the most common. The atomic beam is collimated by a series of apertures or by transverse laser cooling (see Chap. 19) and crossed by one or more carefully tuned laser beams that enter the vacuum system through appropriately placed windows (see Fig. 13.4a).

Population of the Rydberg states is generally done by laser excitation in two steps. Figure 15.2 shows that the ionization potential of most atoms is several eV, and is approximately 5 eV for all the alkalis. Since this corresponds to UV light, it is easier to populate the Rydberg states that are near the ionization limit with two steps as shown in Fig. 13.4b, using visible light from readily available tunable lasers. Selection rules then require $\Delta \ell = 0$ or 2.

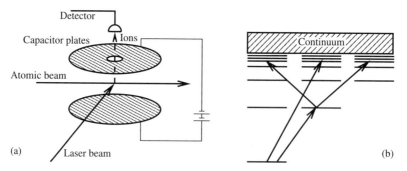

Figure 13.4 Part (a) is a schematic diagram of a typical setup for producing a beam of Rydberg atoms. The atomic beam is crossed by one or more laser beams between two metal plates that apply the electric field. One of them has a hole through which ions produced by field ionization can pass to the detector. Part (b) shows a typical two-step excitation scheme to move atoms to energies ~5 eV above their ground state, using visible light that excites the appropriate transitions.

At first it might seem that detection of Rydberg states could be very much the same as detection of other excited atomic states: one simply looks at the radiation emitted when the atom decays to its ground state. More careful consideration reveals that Rydberg atoms do not decay very well by radiation, and in fact may have lifetimes long enough for the atoms to hit the wall at the far end of the apparatus without ever emitting light.

Transitions down to adjacent Rydberg levels Δn away are very weak and slow because their frequencies are very low ($\Delta E_n = 2R_\infty \Delta n/n^3$) and spontaneous transition probabilities are proportional to the cube of the frequency ($1/n^9$). Transitions down to low-lying or ground states that have much higher frequencies are very weak because the radial wavefunctions of Rydberg levels have very many oscillations and the integral of such oscillatory functions multiplied by one with only very few nodes is a small fraction of a_0. The result is that detection of Rydberg atoms by spontaneously emitted radiation is ineffective.

The usual procedure is to mount flat plate electrodes on either side of the atomic beam and apply a sufficiently strong dc electric field to ionize the atoms (see Fig. 13.4a). Because the Rydberg electron is far away from the nucleus and core, it can easily be pulled away by the applied field and detected. The laser beams are tuned to the transitions of interest and directed into the apparatus where they cross the atomic beam between the field plates. The optical excitation is performed with the desired voltage on the field plates, and shortly afterward the voltage is stepped up sufficiently to ionize a selected n-state ($\mathcal{E}_{FI} > \mathcal{E}_0/16n^4$). The negatively charged plate has an array of small holes, and many of the ions accelerated toward it pass through the holes to a detector mounted on the opposite side. Such field ionization, as described earlier, is very efficient for Rydberg atom detection.

13.5 Some results of Rydberg spectroscopy

When atoms are subjected to extremely strong electric or magnetic fields, that is, those fields that produce forces comparable to the Coulomb binding force, their structure is markedly changed. When the external field is so large that it can no longer be considered a perturbation, the ordinary solutions to the Schrödinger equation are not even approximately right. Theoretical studies of atoms in such conditions have been stimulated by interest in plasma and astrophysical problems, but experimental measurements have not been done because laboratory fields cannot be made large enough [133].

Rydberg atoms have provided a testing ground for the theories of atoms in strong fields because laboratory fields can easily be made to satisfy the strong field condition. The signals are typically studied as a function of laser frequency, applied dc electric [115] or magnetic fields [123, 134, 135], applied rf or microwave fields

220   *Rydberg atoms*

that induce transitions between fine-structure levels or to other Rydberg levels [128, 130], background gas pressure, or combinations of these and other influences.

13.5.1 Stark spectroscopy

Soon after the advent of tunable lasers the spectroscopy of Rydberg atoms in weak and strong fields became a subject of great interest. Figure 13.5 shows signals resulting from step-wise excitation of Na as in Fig. 13.4b. Each vertical trace represents a laser frequency scan at a given electric field much smaller than the ionizing field [131]. A pulsed yellow laser excited atoms into the 3P level between the electric field plates and a pulsed violet laser subsequently populated the Rydberg state when its frequency was resonant with the transition. A few μs later the applied electric field was pulsed to a high enough value to ionize the Rydberg states, producing the signal. The measured Stark spectrum of the sodium atom is strikingly laid out in Fig. 13.5, and agrees well with the calculations, as shown in Fig. 13.2.

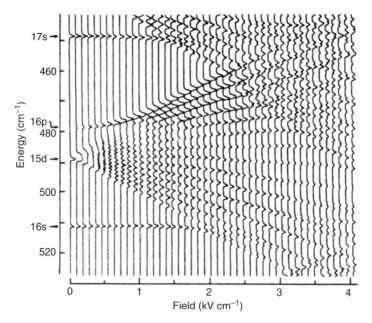

Figure 13.5 Each vertical trace in this figure is a plot of the ion current from a laser frequency scan at a given electric field much smaller than the ionizing field (horizontal axis). A few μs after the laser pulses that excited the atom through an intermediate state, the applied electric field was pulsed to a high enough value to ionize the Rydberg states, producing the signal. In this case the first laser pulse (yellow) excited Na from its ground 3S state to its first excited 3P state, and a second (blue) laser pulse, whose frequency was scanned, populated the Rydberg states. (Figure from Ref. [131])

13.5 Some results of Rydberg spectroscopy

One of the interesting features of Fig. 13.5 can be seen at low fields just above $E = -480$ cm^{-1}. Selection rules forbid excitation of the P states from the 3P state as shown on the trace for zero field, but even the next trace shows signs of a peak at ~100 V/cm, and there is a clear, strong peak at a few hundred V/cm. The transition is allowed because the electric field mixes the $n = 15$ states with higher angular momentum into the 16P state, and the resulting D-state component is readily excited from the 3P state. The curved locus of the eigenenergy plot soon straightens out as a result of the strong mixing that causes a linear Stark shift, as shown in Fig. 13.2.

Spectra similar to that of Fig. 13.5 can also be obtained at field values higher than that of Eq. (13.8) [132]. At first this may seem surprising because the ionization rate is so high that the states should be broadened into the photoionization continuum. However, it must be remembered that the total orbital angular momentum ℓ is no longer conserved in such a strong field so that every n state contains a mixture of higher and lower ℓ-values up to $\ell = n - 1$. Some of these have eccentric orbits that stay very close to the nucleus on the downfield side so that they do not sample the saddle-point area that would permit their escape. Thus their energy is above the limit corresponding to Eq. (13.8) but they are still quite well bound, so they result in narrow spectral features. By contrast, the rapidly ionizing orbital states are so broadened that the excitation to them is not noticeable. Observation of these resonant excitations does not require a delayed ionizing pulse, but instead the detector is gated to record only events that happen within a few μs after the laser pulse. This scheme is complementary to the data acquisition method of Fig. 13.5.

13.5.2 Precision measurements on high-ℓ states

Rydberg atoms are exquisitely sensitive to small effects, and for that reason are ideal for exploring them. One example is the use of precision microwave spectroscopy between neighboring Rydberg sublevels to measure retardation in the interaction between core electrons and the excited one [136]. Helium is chosen for these high-precision measurements because the wavefunction of the one-electron core can be known to extraordinary accuracy.

The excitation scheme is quite different. Ions from a discharge are accelerated through ~10 kV and then impinge on a thin foil where the violent interactions excite very many different states. Among these are those in the $n = 10$ manifold, and these are readily excited to higher n-values by a CO_2 laser at $\lambda \approx 10$ μm for easy detection. The resolution of the process easily distinguishes among the $n = 10$ sublevels so that microwave transitions among them are readily detected. For states with $\ell > 5$, the electron wavefunction is so well separated from the core by the centrifugal potential of Eq. (7.9b) that spectroscopy is sensitive to effects

222   *Rydberg atoms*

normally neglected, such as retardation. In Ref. [136] the reported precision was better than 2 kHz out of 500 MHz. Systematic effects such as Stark shifts from random stray electric fields were carefully minimized.

Another feature of such methods is the application to many other atoms and molecules simply by changing the material put into the discharge source. Thus studies have been conducted on simple atoms with $Z < 10$ [137, 138], H_2 and D_2 [139], and more complicated atoms as far down the periodic table as the actinides [140].

13.5.3 Electric field calibration

Precise measurements and mapping of inhomogeneities of dc magnetic fields are readily enabled with nuclear magnetic resonance (NMR). The gyromagnetic ratio and chemical shifts of many nuclei are known on the sub-ppm level and this enables high accuracy of the absolute field calibration. However, there was no counterpart for electric fields until the use of Rydberg atoms for this purpose. Before this, dc electric fields were measured by constructing electrodes as flat as possible, measuring their separation as well as possible, and applying a well-calibrated voltage across them. Stark shifts were measured and then served as secondary standards. However, the accuracy in the knowledge of the effective applied voltage is limited by contact potentials, imperfections on the electrode surfaces (e.g. patch effects), and possibly impurities deposited on the surfaces (e.g. charged dielectric contamination such as pump oil), while the accuracy of the spacing is limited by flatness and mechanical contacts as well as by deformations of the plates.

Rydberg atom spectroscopy has enabled direct atomic calibration of fields in two different ways. The width of optical excitation signals to Rydberg states in an applied electric field is often dominated by their field ionization rates. The ionization results from mixing of the "purely" hydrogenic Rydberg state with continuum states by the ion core. But the optical excitation could excite two nearby Rydberg states that are strongly mixed by the dc electric field as shown by the anti-crossings in Fig. 13.2. Then the ionization rate of one component of the mixture may nearly vanish because the ionization rates of each of the two original Rydberg states may interfere, resulting in a *very* narrow resonance. Such resonances are shown in Fig. 13.6 in the vicinity of an anti-crossing. These can be accurately calculated and sensitively measured.

A more direct and absolute field calibration method is obtained by accurate calculations of the Stark energy levels using a variety of methods that can be compared and fine tuned for accuracy as described in Ref. [138]. Careful measurements of Stark resonances in Li, whose core can be well-characterized, were combined with

13.5 Some results of Rydberg spectroscopy

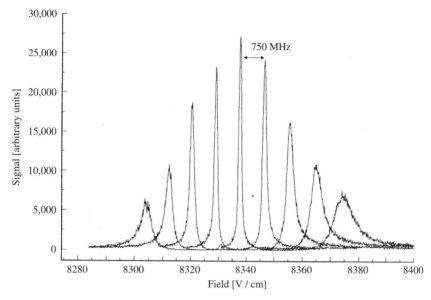

Figure 13.6 A series of narrow ionization resonances taken at various excitation laser frequencies that consequently appear at different electric field values in the vicinity of an anti-crossing. The central peak has an energy of $-2482.89\,825(4)$ cm^{-1} that can be accurately found and calibrated. The x axis has been calibrated in units of field. The resonances near the minimum width are very nearly symmetric. (Figure from Ref. [141].)

detailed calculations of the eigenvalues to produce absolute field calibration at the level of ± 2 ppm [141].

13.5.4 Circular Rydberg states

Rydberg atoms in circular states, namely $\ell = m_\ell = n - 1$, have very special properties. For example, they decay only through a millimeter wave transition to the nearest less-excited circular state so that, for example, for n = 50, the lifetimes can typically exceed 10 ms (note the ω^3 term in Eq. (5.13)). Furthermore, most decay paths are forbidden so they represent excellent models of a two-level system (see Chap. 2). Also, their Stark shifts are extremely small so the transition frequencies are quite insensitive to stray electric fields. Thus they are superb candidates for the careful testing of a variety of effects, especially those connected with cavity QED [121].

Circular states of high n cannot be populated by direct optical excitation from the ground state because each dipole transition can make a change of ℓ by only unity, so even a three-step process can make ℓ no larger than 3 starting from a ground s state. There are a number of ways to produce these states, and one of the most efficient of them uses an electric field sweep that essentially constitutes a multistep adiabatic rapid passage (see App. 2.D) through a series of ℓ-states in a microwave field oriented to produce Δm_ℓ steps of $+1$, so that m_ℓ tracks the increasing ℓ value. Such experiments can be done in a superconducting microwave cavity that both has a high Q and therefore low losses [121], and also minimizes the effects of thermal black body radiation (see Ref. [142]).

Such exploitation of the special properties of Rydberg atoms has yielded realization of experiments that had previously been believed to be impossible. For example, the Rabi frequency of a two-level system is proportional to the field amplitude, and in the limit of very low field amplitudes, field quantization should cause it to increase in discrete steps as the driving power is increased. Measurements show clear indication that the Rabi frequencies indeed increase in such discrete steps as the square root of sequential integers [143]. In another experiment there was a measurement of the vacuum Lamb shift, essentially the light shift caused by a field with energy of $1/2\hbar\omega$ per mode that provided direct evidence for the zero-point field (see Ref. [144]).

There have been measurements of entanglement of the states of two separate atoms, direct observation of the Wigner function, entanglement of two separate cavity modes, the quantum Zeno effect, and a host of other observations of states of atoms or of fields that have no classical analog. The application of such capability using Rydberg atoms to quantum information seems very promising indeed [121].

13.5.5 Coulomb blockade

The very strong dependence of the Stark shifts of Rydberg atoms makes them extremely sensitive to the electric fields produced by their excited neighbors. Thus a single Rydberg atom with its large dipole moment can shift the energies of the Rydberg states of other nearby atoms sufficiently that they cannot be excited by the same light field that excited the original atom. This is easily understood by recognizing that the dipole moment of a Rydberg atom scales with its orbital radius $n^2 a_0$, as does the electric field it produces in its neighborhood. Then the energy shift of the Rydberg states of nearby atoms, which scales with $\mathcal{E}^2 n^7$ (see Eq. (13.7)), results in an n^{11} scaling, obviously an enormous sensitivity. The consequence is called Coulomb blockade because the field shifts the transition out of resonance, and it is under careful study in many laboratories. A good review is in Ref. [145].

The primary interest in Coulomb blockade derives from the application to quantum information technology. One of the key components of proposed devices is a "quantum controlled *not* gate" (QCN) that consists of a qubit whose response to an incoming signal depends on the state of a second qubit. It operates by performing the *not* operation on the second qubit when the first qubit is in one state, but leaves it unchanged when the first qubit is in the opposite state. Its classical counterpart is called the *exclusive OR* gate, abbreviated XOR, and has the same truth table. Of course, classical bits are either zero or one, whereas a qubit can be a superposition of these states. The truth table for a QCN operating on a superposition state via a superposition control qubit are easily found.

Details of the utility of neighboring Rydberg atoms as both the control and the target qubit are subtle but have been worked out carefully. Perhaps one of the most important features derives from the relative insensitivity to atomic separation over a fairly large range, and nearly complete independence of orientation. Thus precise control over experimental variables is not required, and QCNs have recently been demonstrated on pairs of atoms held loosely in optical traps [146, 147].

Exercises

13.1 For Rydberg atoms, the transition moments between adjacent states are very large because z_{kj} is so large ($\sim n^2 a_0$ for $\Delta n \cong 1$). Calculate the lifetimes against stimulated decay or excitation for adjacent states of Rydberg atoms in a black body radiation bath at $T \cong 300$ K.

13.2 Consider circular states of hydrogen. The solution of the radial wavefunction are given by Eq. (7.17), where the Laguerre functions are given by $L_k^q(\rho) = \sum_{k=0}^{\infty} a_k \rho^k$, and $\rho = 2r/na_0$. The coefficients a_k are given by the recurrence relation Eq. (7.12) with $\lambda = n$.
(a) Prove that for $\ell = n - 1$ the radial wavefunction can be reduced to $R_{n\,n-1}(r) = N_n r^{n-1} e^{-r/na_0}$, with N_n a normalization constant.
(b) Determine the number of radial nodes for these states.
(c) Determine the expectation value of r and r^2.
(d) Prove that the spread in the radius defined by $\sigma_r = \sqrt{\langle (r - \langle r \rangle)^2 \rangle}$ is given by $\sigma_r = \langle r \rangle / \sqrt{2n + 1}$. Discuss the physical implications of the result.
(e) Show for large n how these states reflect the properties of the electrons in the Bohr model, in which the electrons move in circular orbits around the nucleus. Also show what properties do not correspond to the Bohr model.

13.3 Consider the Stark effect for the $n = 15$ state in hydrogen, where the unperturbed Hamiltonian and the perturbation are given by Eq. (12.1).
(a) What is the degeneracy of the $n = 15$ state?
Degenerate perturbation theory is needed to calculate the shift caused by an electric field. The shift $E_k^{(1)}$ in degenerate perturbation theory is given by

$$\det | \langle \psi_i | \mathcal{H}' | \psi_j \rangle - E_k^{(1)} \delta_{ij} | = 0,$$

with $\mathcal{H}' = e\mathcal{E}z$. Before diagonalizing this large matrix, the number of coupled states need to be reduced.

Consider an operator $\mathcal{O}$ that commutes with both the unperturbed Hamiltonian $\mathcal{H}_0$ and the perturbation $\mathcal{H}'$. The eigenstates ψ_i and ψ_j of $\mathcal{H}_0$ with different eigenvalues o_i and o_j of $\mathcal{O}$ are not coupled in the sum above. For each eigenvalue of $\mathcal{O}$ the determinant above can be diagonalized independently.

(b) Prove that from $[\mathcal{H}', \mathcal{O}] = 0$ follows $\langle \psi_i | \mathcal{H}' | \psi_j \rangle = 0$, if $\mathcal{O}\psi_i = o_i\psi_i$ and $\mathcal{O}\psi_j = o_j\psi_j$ with $o_i \neq o_j$.

The eigenstates of $\mathcal{H}_0$ are given by $\psi_{n\ell m}$, with $\ell^2 \psi_{n\ell m} = \ell(\ell+1)\hbar^2 \psi_{n\ell m}$ and $\ell_z \psi_{n\ell m} = m\hbar \psi_{n\ell m}$.

(c) Determine $[\ell^2, \mathcal{H}']$ and $[\ell_z, \mathcal{H}']$. Use the relation $\vec{\ell} = \vec{r} \times \vec{p}$ and the usual commutation rules $[r_i, p_j] = i\hbar \delta_{ij}$.

(d) Use the result of (b) and (c) to determine whether states $\psi_{n\ell m}$ and $\psi_{n\ell' m}$ are coupled by the electric field ($\ell \neq \ell'$). Do the same for $\psi_{n\ell m}$ and $\psi_{n\ell m'}$ ($m \neq m'$).

(e) Determine the total number of determinants $\mathcal{H}'_{ij}$ that need to be diagonalized for $n = 15$ and indicate for each determinant the number of elements.

13.4 For the purposes of quantum defect calculations, the matrix elements of $1/r^4$ are needed. Evaluate the diagonal elements for the 3s and 3p hydrogen wavefunctions. (Reference [75] is one good place to find the wavefunctions.)

14
The helium atom

14.1 Introduction

Helium is the natural connection between the hydrogen-like atoms (one electron outside a spherical core, such as the alkali-metal atoms of Chap. 10) and all the others in the periodic table. For these one-electron atoms the field-free description of the electron motion in terms of three quantum numbers n, ℓ, and m suffices. The other electrons are considered to form the "core" and its effect on the energy levels is reduced to the quantum defects δ_ℓ, as discussed in Chap. 10. The case of helium is very different since the two ground-state electrons are in the same electronic state and differ only in the orientation of their spins. Thus spin is of paramount importance for the description of all but the hydrogenic atoms. Since helium has only two spins to deal with, their roles can be written explicitly.

Because the two electrons are identical, quantum mechanics requires that the wavefunction undergoes nothing more than an unobservable phase shift when they are exchanged. This consequence is called the Pauli symmetrization principle, and is sometimes referred to as "the source of our individuality". Moreover, it has a profound influence on the wavefunction and thus on the allowed energy states. It will be shown that this symmetry requirement dictates that two electrons cannot be in states with identical quantum numbers, and this topic will be discussed in detail in this chapter.

14.2 Symmetry

The profound difference between the hydrogen and helium wavefunctions arises because the two electrons are indistinguishable, and this results in a fundamental change of the energy-level structure. The concept of "spin", introduced as a shortcut for the intrinsic magnetic moment of the electron in Chap. 7, now becomes a dominating feature of the description. Before discussing the helium energy levels, the effects of this symmetry with respect to exchange of identical particles is presented.

14.2.1 The exchange operator

In quantum mechanics, identical particles are inherently indistinguishable, so the exchange of the two electrons cannot change any observable and there should be no measurable effect on the overall wavefunction. However, there can be an overall phase change whose nature can easily be found by defining an operator P_{12} that exchanges the electrons and applying it to the wavefunction $\Psi(\vec{r}_1, \vec{r}_2)$. The exchange applies to all coordinates of the electron including its spin coordinate, but for simplicity only the spatial coordinate $\vec{r}$ will be indicated here. One application results in $P_{12}\Psi(\vec{r}_1, \vec{r}_2) = \Psi(\vec{r}_2, \vec{r}_1)$, but after a second application, e.g. $P_{12}(P_{12}\Psi(\vec{r}_1, \vec{r}_2)) = \Psi(\vec{r}_1, \vec{r}_2)$, the original solution must reappear. Thus the square of the phase factor is 1 so it must be ± 1. Then

$$P_{12}\Psi(\vec{r}_1, \vec{r}_2) = \pm\Psi(\vec{r}_1, \vec{r}_2). \tag{14.1}$$

When the eigenvalue is $+1$ the solution is called symmetric, and when it is -1 the solution is called anti-symmetric. Note that any wavefunction can be split into symmetric and anti-symmetric parts, but the importance of the equation above is that the total eigenfunction of the system must be either symmetric or anti-symmetric. This will turn out to be important for the energies of the eigenstates.

The Pauli principle requires that the total wavefunction must be anti-symmetric under the exchange of two fermions (e.g. electrons). Although this is required for exchange of any two electrons in complicated systems such as metals containing large numbers of them, here its implementation is rather simple, since there are only two electrons.

This principle applies to the total wavefunction, and thus the orientation of the electron's magnetic moment (spin) must be considered. In the case of a single electron as discussed in Chap. 7, the orientation of the spin has no meaning since there is nothing to measure it against. But with the two electrons of helium there is the relative orientation of their two spins, and this can be described as a separate part of the wavefunction denoted by χ. Thus $\Psi(\vec{r}_1, \vec{r}_2)$ becomes a sum over products $\psi(\vec{r}_1, \vec{r}_2)\chi(1, 2)$, where $\psi(\vec{r}_1, \vec{r}_2)$ is the spatial part of Ψ. The spin does not influence the energies of the eigenstates directly, as there are no spin-dependent terms in the Hamiltonian in Eq. (14.10) below (neglecting spin–orbit coupling terms and hyperfine interactions).

The electron spin has a value $s = 1/2$, and with respect to some chosen spatial quantization axis it can have an orientation $s_z = \pm 1/2$ whose corresponding eigenfunctions are denoted as

$$\chi_{+1/2} = \alpha = \begin{pmatrix} 1 \\ 0 \end{pmatrix} \quad \text{and} \quad \chi_{-1/2} = \beta = \begin{pmatrix} 0 \\ 1 \end{pmatrix}. \tag{14.2}$$

14.2 Symmetry

For the two electrons of helium, the four possible arrangements of the spins are:

$$\alpha_1\alpha_2, \quad \alpha_1\beta_2, \quad \alpha_2\beta_1, \quad \beta_1\beta_2 \tag{14.3}$$

The symmetry for the first one of these is quite simple because exchange of the two electrons can be written as $\alpha_1\alpha_2 \to \alpha_2\alpha_1$. Then $\chi = \alpha_1\alpha_2$ is unchanged by this process and therefore ψ must change sign. Similarly for $\beta_1\beta_2$. But exchange of the two electrons in $\alpha_1\beta_2$ results in $\alpha_2\beta_1$ which bears no relation to the original state, and so $\alpha_1\beta_2$ cannot be a valid wavefunction.

The only way to satisfy the symmetry requirements for such mixed spin orientations is to make linear combinations, and the two that arise for helium are $\chi = \frac{1}{2}\sqrt{2}(\alpha_1\beta_2 \pm \alpha_2\beta_1)$. For χ with a (+) sign, ψ must change sign on exchange because χ does not, and for the (−), ψ must not change sign because χ does.

In a similar way, the spatial part of the wavefunction $\psi(\vec{r}_1, \vec{r}_2)$ is also forbidden unless the two wavefunctions for electron "1" and "2" are identical, because it has no exchange symmetry. Allowed choices are restricted to forms such as $\frac{1}{2}\sqrt{2}(\psi(\vec{r}_1, \vec{r}_2) \pm \psi(\vec{r}_2, \vec{r}_1))$ that have a well-defined symmetry under exchange.

14.2.2 The addition of two spins

The operators for the spin can be written in terms of the Pauli matrices of Eq. (2.16) (see App. 2.A), because they span the space of two-dimensional Hermitian operators. The representation is $s_j = \hbar\sigma_j/2$ for $j=x, y$, and z. Writing the spin states in Eq. (14.3) shows that

$$\begin{array}{lll} s_x\alpha = \hbar\beta/2 & s_y\alpha = i\hbar\beta/2 & s_z\alpha = \hbar\alpha/2 \\ s_x\beta = \hbar\alpha/2 & s_y\beta = -i\hbar\alpha/2 & s_z\beta = -\hbar\beta/2 \end{array} \tag{14.4}$$

The addition of two spin states is a special case of the addition of angular momenta (see App. C) and yields the operators

$$\vec{S} = \vec{s}_1 + \vec{s}_2 \qquad S_z = s_{1z} + s_{2z}. \tag{14.5}$$

The eigenfunctions $\chi(1, 2)$ for the total spin $\vec{S}$ can be chosen as eigenfunctions of the operators $\vec{S}^2$ and S_z, where the first of these is written as

$$\vec{S}^2 = \vec{s}_1^{\,2} + \vec{s}_2^{\,2} + 2\vec{s}_1 \cdot \vec{s}_2, \tag{14.6a}$$

and the dot product $\vec{s}_1 \cdot \vec{s}_2$ leads to three additional terms:

$$\vec{s}_1 \cdot \vec{s}_2 = s_{1x}s_{2x} + s_{1y}s_{2y} + s_{1z}s_{2z}, \tag{14.6b}$$

leading to a total of five terms.

14.2.3 The eigenfunctions

The eigenfunctions of $\vec{S}^2$ and S_z are composed of combinations of the wavefunctions χ_i of the individual electrons because they span the space, and the four possiblities are shown in Eq. (14.3). Clearly each of them is an eigenfunction of S_z with eigenvalues $M_z = +1, 0, 0$, and -1, respectively. However, they are not all eigenfunctions of $\vec{S}^2$ because operating on one of the middle terms in (14.3) with $\vec{S}^2$ gives the other middle one. It is easy to show that the linear combinations given by

$$\chi(1,2) = \frac{1}{2}\sqrt{2}\left(\alpha_1\beta_2 \pm \beta_1\alpha_2\right) \tag{14.7}$$

are indeed eigenfunctions of $\vec{S}^2$ with eigenvalues $2\hbar^2$ for the (+) sign, and 0 for the (−) sign. In summary, there is one eigenfunction with eigenvalue $S = 0$

$$\chi(1,2) = \frac{1}{2}\sqrt{2}\left(\alpha_1\beta_2 - \beta_1\alpha_2\right) \equiv |0,0\rangle, \tag{14.8a}$$

which is called the singlet state, and three eigenfunctions with eigenvalue $S = 1$

$$\chi(1,2) = \begin{cases} \alpha_1\alpha_2 \equiv |1,1\rangle & M_z = +1 \\ \frac{1}{2}\sqrt{2}\left(\alpha_1\beta_2 + \beta_1\alpha_2\right) \equiv |1,0\rangle & M_z = 0 \\ \beta_1\beta_2 \equiv |1,-1\rangle & M_z = -1 \end{cases} \tag{14.8b}$$

which are called the triplet states. These four functions form the complete set of spin eigenfunctions for two electrons. It is straightforward to show that the eigenfunctions of $\vec{S}^2$ are also eigenfunctions of P_{12}. For a general spin function χ one has

$$P_{12}\vec{S}^2\chi = P_{12}(\vec{s}_1 + \vec{s}_2)^2\chi = (\vec{s}_2 + \vec{s}_1)^2 P_{12}\chi = \vec{S}^2 P_{12}\chi, \tag{14.9}$$

where P_{12} operates to exchange the labels 1 and 2 in the middle step. Equation (14.9) shows that the two operators commute and thus share the same eigenfunctions (see App. 7.C).

14.3 The Hamiltonian for helium

This section presents the Hamiltonian for an atomic system with two electrons (H^-, He, Li^+, ...) in general, but it is specific only for the neutral He atom. The Hamiltonian for a two-electron system can be written as

$$\mathcal{H} = \mathcal{H}_1 + \mathcal{H}_2 + \frac{e^2}{4\pi\varepsilon_0 r_{12}}, \tag{14.10a}$$

where the first two terms are the hydrogenic Hamiltonians for the individual electrons j

14.3 The Hamiltonian for helium

$$\mathcal{H}_j = -\frac{\hbar^2 \nabla_j^2}{2m} - \frac{Ze^2}{4\pi\varepsilon_0 r_j} \tag{14.10b}$$

and the last term in Eq. (14.10)a is the electron–electron repulsion. Here the charge of the nucleus Z is 2 for helium. The motion of the nucleus in the center-of-mass frame has been explicitly neglected since its effect is many orders of magnitude smaller than the effects of the repulsion term.

14.3.1 The independent particle model

The presence of the repulsion term results in a total departure from the solutions of the hydrogenic system. This term does not depend on the coordinates of one-electron separately, and more important, does not depend on the distance between each electron and the nucleus. The separation of coordinates as discussed in Chap. 7 is not possible, requiring solution of the Schrödinger equation in the six-dimensional space spanned by the coordinates $\vec{r}_1$ and $\vec{r}_2$:

$$\left(\mathcal{H}_1 + \mathcal{H}_2 + \frac{e^2}{4\pi\varepsilon_0 r_{12}}\right)\psi(\vec{r}_1, \vec{r}_2) = E\psi(\vec{r}_1, \vec{r}_2). \tag{14.11}$$

But Chap. 7 led to a complete set of solutions, so it is useful to begin here by treating the electron–electron interaction using hydrogenic wavefunctions as a basis set for the individual electrons.

If there were no repulsion term the problem could be separated into two independent equations for electrons 1 and 2, thereby producing the solutions of Chap. 7 for each individual electron, but with $Z = 2$. This lowest-order approximation, called the independent particle model, simply neglects the repulsion. It seems reasonable as a first step because the electrons repel each other and therefore reside further apart than their distance to the nucleus. Moreover their charges are both unity as opposed to $Z = 2$ for the nucleus so that their repulsion is much smaller than the nuclear attraction. Then the two independent equations are $\mathcal{H}_j \psi(\vec{r}_j) = E_j \psi(\vec{r}_j)$ and the eigenvalues are simply the eigenvalues for the hydrogen case with $Z = 2$. Introducing $(n_j \ell_j m_j)$ for the quantum numbers for each electron leads to eigenfunctions

$$\psi(\vec{r}_1, \vec{r}_2) = \psi_{n_1 \ell_1 m_1}(\vec{r}_1)\psi_{n_2 \ell_2 m_2}(\vec{r}_2) \tag{14.12a}$$

and eigenenergies

$$E_{n_1,n_2} = E_{n_1} + E_{n_2} = -Z^2 R_\infty \left(\frac{1}{n_1^2} + \frac{1}{n_2^2}\right). \tag{14.12b}$$

The result for the ground state, $n_1 = n_2 = 1$, is $E = -108.8$ eV, a rough approximation to the measured energy of -79.005 eV.

14.3.2 The symmetrized wavefunctions

Clearly the functions of Eq. (14.12a) are not eigenfunctions of the exchange operator P_{12}, but they can be made so by taking linear combinations of $\psi(\vec{r}_1, \vec{r}_2)$ and $\psi(\vec{r}_2, \vec{r}_1)$ given by

$$\psi_{\pm}(\vec{r}_1, \vec{r}_2) = \frac{1}{2}\sqrt{2}\left(\psi_{n_1 \ell_1 m_1}(\vec{r}_1)\psi_{n_2 \ell_2 m_2}(\vec{r}_2) \pm \psi_{n_1 \ell_1 m_1}(\vec{r}_1)\psi_{n_2 \ell_2 m_2}(\vec{r}_2)\right). \quad (14.13)$$

The (+) sign leads to a spatially symmetric wavefunction, whereas the (−) sign leads to a spatially anti-symmetric wavefunction. In order to satisfy the Pauli principle, the spatially symmetric wavefunctions need to be combined with the anti-symmetric spin wavefunction (singlet) and these states are historically referred to as para-helium. The spatially anti-symmetric wavefunctions need to be combined with the symmetric spin wavefunctions (triplet), and these states are called ortho-helium. It will become clear that the singlet and triplet states are not coupled in optical transitions and thus exhibit different excitation schemes.

Of particular interest is the case when the two electrons are in identical hydrogenic states (n, ℓ, m). Then the anti-symmetric spatial wavefunction vanishes and the symmetric wavefunction is

$$\psi_+(\vec{r}_1, \vec{r}_2) = \psi_{n\ell m}(\vec{r}_1)\psi_{n\ell m}(\vec{r}_2), \quad (14.14)$$

which must combine with the anti-symmetric singlet spin state. A special case is the ground state where $n_1 = n_2 = 1$ so the ground state is a singlet. Since the independent particle result of $E_{11} = -108.8$ eV is off by nearly 30 eV ~30% (see Fig. 14.1), the conclusion is that it is not a good approximation for the helium atom. Although the situation becomes relatively better for higher charges Z, other methods are required for a better description of this system.

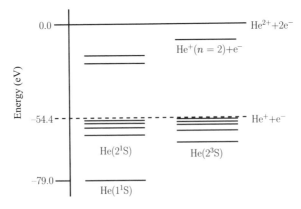

Figure 14.1 Energy levels of helium.

14.4 Variational methods

One frequently used approach to this kind of problem might be perturbation theory, but this is most useful when the perturbing term is small compared with the rest of the Hamiltonian. For the systems described in earlier chapters, the perturbations are very much smaller and so the technique works well. As shown above, the repulsive electron–electron interaction shifts the states by the same magnitude as their energies. Even though the first-order perturbation calculation gives better results (−74.8 eV) [60, 61, 148], it is not the right method for this problem.

14.4 Variational methods

A more appropriate route is the use of variational calculations. As shown in App. 14.A, any arbitrarily chosen trial wavefunction will have energy larger than the correct energy of the ground state. Choosing a trial wavefunction with one or several adjustable parameters and minimizing the energy by optimization of these parameters results in an energy closer to the experimental value.

One possible choice for the ground-state wavefunction is ψ_{100} for each electron, but with the value of Z as an adjustable parameter instead of 2. This reflects the notion that one electron can screen the nuclear charge for the other electron, depending on their positions. As shown in many textbooks (and also in App. 14.B), the result of this analysis is that the charge of the nucleus is screened by an amount $5/16$ that raises the energy of the ground state from −108.8 eV to −77.5 eV, which is rather close to the experimental value of −79.005 eV.

A drawback of variational calculations arises if the variational wavefunction does not resemble the "true" wavefunction, so its results do not lend themselves to a physical interpretation. In the case here, there is no simple, way to describe why the screening is $5/16$ instead of the intuitive $1/2$, as one might guess from the fact that one electron is closer to the nucleus only half the time. The next section describes the ground state of helium using another variational approach that lends itself better to a physical interpretation. Later in this chapter it is shown that such an approach also allows a description of excited states.

14.4.1 Variational method for the ground state

Even though the two electrons of helium are indistinguishable, suppose that in the ground state they are not in identical states. If they are well separated, the "outer" electron will feel the nucleus to be screened to a charge of $(Z − 1)e$ by the inner electron, whereas the "inner" electron is completely unscreened and feels the complete attraction of a charge of Ze. Note that the labels inner and outer depend on the actual configuration of the electrons, but the wavefunction still must be symmetrized with respect to the exchange of the two electrons.

As in the variational calculation outlined in the previous section, one uses the lowest hydrogenic wavefunctions for both electrons, but assumes that each electron will experience a screened charge $Z_{i,o}e$, where the index i, o refers to the inner and outer electron, respectively. Thus the variational wavefunction becomes

$$\psi(\vec{r}_1, \vec{r}_2) = \frac{1}{2}\sqrt{2}(1 + P_{12})\frac{Z_i^{3/2} Z_o^{3/2}}{\pi a_0^3} e^{-(Z_i r_1/a_0 + Z_o r_2/a_0)}$$

$$\equiv \frac{1}{2}\sqrt{2}(1 + P_{12})\phi(r_1, r_2). \qquad (14.15)$$

The angular part of the wavefunction is the same for both because both electrons are chosen to be in an s state. The spin part of the wavefunction has been suppressed in Eq. (14.15) since the Hamiltonian does not have spin-dependent terms. Furthermore, only the (+) sign before the exchange operator P_{12} appears because the spin part of the wavefunction is necessarily anti-symmetric (it is the ground state).

The kinetic and potential energy of the two electrons interacting with the nucleus can readily be calculated using the usual prescription for the operators, but it is important to realize that the variational wavefunction above needs proper normalization (see App. 14.B). However, the Hamiltonian also depends on the electron–electron interaction $e^2/4\pi\varepsilon_0 r_{12}$. The term $1/r_{12}$ can be evaluated using its expansion in Legendre polynomials, as shown in App. 14.B. For the electron–electron interaction it leads to

$$\left\langle \psi(\vec{r}_1, \vec{r}_2) \left| \frac{e^2}{4\pi\varepsilon_0 r_{12}} \right| \psi(\vec{r}_1, \vec{r}_2) \right\rangle = \mathcal{J} + \mathcal{K}, \qquad (14.16)$$

where the radial integrals $\mathcal{J}$ and $\mathcal{K}$ are given in terms of the radial wavefunction $\phi(r_1, r_2)$ of Eq. (14.15) by

$$\mathcal{J} = \left\langle \phi(r_1, r_2) \left| \frac{e^2}{4\pi\varepsilon_0 r_>} \right| \phi(r_1, r_2) \right\rangle \qquad (14.17a)$$

and

$$\mathcal{K} = \left\langle \phi(r_1, r_2) \left| \frac{e^2}{4\pi\varepsilon_0 r_>} \right| \phi(r_2, r_1) \right\rangle, \qquad (14.17b)$$

with $r_< = \min(r_1, r_2)$ and $r_> = \max(r_1, r_2)$. Here, the integral $\mathcal{J}$ is the regular Coulomb integral that is equivalent to the Coulomb energy of two classically interacting charge shells. The second integral $\mathcal{K}$ (note the exchange of r_1 and r_2 on the right side of the equation) is called the exchange integral and has no classical analog, since it depends on the exchange of the two electrons. The evaluation of these two integrals for the ground state using the variational wavefunction of Eq. (14.15) is given in App. 14.B.

14.4 Variational methods

The energy of the ground state can be written as

$$E(Z_i, Z_o) = \frac{E_k(Z_i, Z_o) + E_p(Z_i, Z_o) + J(Z_i, Z_o) + K(Z_i, Z_o)}{N(Z_i, Z_o)}, \quad (14.18)$$

where the functional dependence on the screened charges Z_i and Z_o of the kinetic energy E_k, the potential energy E_p, the Coulomb integral J, the exchange integral K, and the normalization N is given in App. 14.B. Figure 14.2 shows a contour plot of the total energy E as function of Z_i and Z_o. The energy has a minimum at $Z_i = 2.183$ and $Z_o = 1.189$ and the minimum energy becomes $E = -78.261$ eV. This can be compared to the experimental value for the ground-state energy, which is $E_{exp} = -79.005$ eV.

Several remarks can be made about this result. First, the variational result is close to the measured value, which shows that the trial wavefunction provides a proper description of the "real" wavefunction of the ground state. Second, both screened charges are close to the expectations of $Z_i \equiv 2$ and $Z_o \equiv 1$. That they are both a bit larger reflects the fact that the trial wavefunction is not precisely the "true" wavefunction. Third, the notion that a variational approach provides only an upper bound for the energy is corroborated by this result.

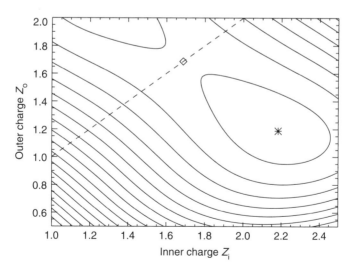

Figure 14.2 Contour plot of the energy of the ground state as a function of the variational parameters Z_i and Z_o. The minimum for the energy is indicated by a asterisk, the dashed line is for $Z_i = Z_o$, whereas the minimum for the case $Z_i = Z_o$ is indicated by a diamond. The contour lines are spaced in energy by $0.1R_\infty$ or ~ 1.36 eV.

14.4.2 Variational model for the singly excited states

Although there is always a possibility of exciting both the electrons out of the ground state, in this section only the singly excited states will be discussed. The spatial wavefunction of such a state can be written as

$$\psi(\vec{r}_1, \vec{r}_2) = \frac{1}{2}\sqrt{2}\left(\psi_{100}(\vec{r}_1)\psi_{n\ell m}(\vec{r}_2) \pm \psi_{n\ell m}(\vec{r}_1)\psi_{100}(\vec{r}_2)\right), \qquad (14.19)$$

where the sign +(−) refers to the symmetric (anti-symmetric) cases. Of course, they have to be combined with the anti-symmetric (symmetric) spin wavefunction, respectively. Since optical transitions couple only to the spatial part and not to the spin part, there is no coupling between singlet and triplet states, i.e. no coupling between para- and ortho-helium, so they appear separately in optical spectra. The notation for the excited states follows the notation for states of the alkali-metal atoms and can be written as $^{2S+1}L_J$, with L the orbital angular momentum, S the spin, and J the total angular momentum (see App. C.1).

The energies of the singly excited states (see Fig. 14.3) can be calculated either in the independent particle model or by perturbation theory, but both lead to rather poor correspondence with the experimental values. Moreover, the successful variational calculations for the ground state can be easily extended to singly excited

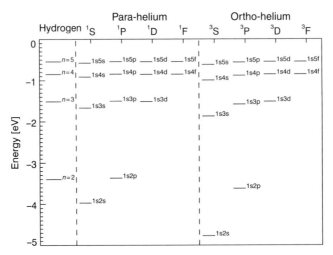

Figure 14.3 Singly excited states of helium. Since the inner electron nearly completely shields one of the two charges of the nucleus, the energy of the states are nearly equal to those of hydrogen, contrary to that of the ground state. However, the symmetrization resulting from the Pauli principle leads to a lower energy for the states of ortho-helium, where the spatial wavefunction is anti-symmetric with respect to the exchange of electrons, leading to a smaller repulsion between the electrons, compared with the states of para-helium, where the spatial wavefunction is symmetric. Note that the states with low ℓ penetrate the region of small distances more than states with high ℓ, which leads to a lower energy.

Conf.	Z_i	Z_o	Variational (eV)	Experimental (eV)
$(1s^2)^1S$	2.183 171	1.188 531	−78.251	−79.012
$(1s\,2s)^3S$	1.993 635	1.550 931	−58.958	−59.191
$(1s\,2s)^1S$	2.012 740	0.924 707	" −59.062"	−58.396
$(1s\,2s)^1S$	2.032 090	1.166 040	−58.266	−58.396
$(1s\,2p)^3P$	1.991 186	1.089 150	−57.980	−58.048
$(1s\,2p)^1P$	2.003 024	0.964 726	−57.754	−57.792
$(1s\,3d)^3D$	1.999 961	1.000 847	−55.935	−55.939
$(1s\,3d)^1D$	2.000 022	0.999 524	−55.935	−55.936
$(1s\,4f)^3F$	2.000 000	1.000 004	−55.274	−55.275
$(1s\,4f)^1F$	2.000 000	0.999 998	−55.274	−55.275

Table 14.1 Results of the variational calculations for the singly excited states of helium. Note that in many cases Z_i is close to 2, whereas Z_o is close to 1. This indicates that the outer electron is nearly fully screened by the inner electron, but the inner electron is not screened by the outer electron. The first line shows the result for the ground state. The third line shows the result for the first excited ^1S state, where the orthogonality with respect to the ground state is not taken into account. The next line shows the correct variational result, where this has been taken into account. As the table shows, the requirement for the orthogonality pushes the energy of the ^1S state above the energy of the ^3S state.

states. Each excited state has one more constraint than the previous one, since it must be constructed to be orthogonal to all the previous ones with lower energies. When seeking the lowest state for given symmetry, this is automatically achieved. If lower states of given symmetry exist, one can always subtract the overlap with the lower wavefunctions from the trial wavefunction. However, in that case one cannot rely on the property of variational calculations, namely that the energy calculated is a lower bound for the energy, since one does not know with certainty the lower-state wavefunctions.

In Tab. 14.1 the results of the variational calculations are shown for some singly excited states of helium together with the ground state. As the table shows, the inner electron feels a charge close to $Z_i = 2$, whereas the outer electron feels a charge close to $Z_o = 1$. The energies are very close to the experimental values, and the agreement improves considerably with increasing angular momentum. This is an intuitively satisfying result because electrons in these higher-ℓ states are primarily situated far from the core composed of the inner electrons.

14.5 Doubly excited states

When both electrons of the helium atom are in excited states, the total electronic energy is larger than the sum of the energies of the He^+ ion and a free electron

	E_{exc} (eV)	E_e (eV)
He $(2s^2)^1S$	57.870	33.283
He $(2s\,2p)^3P$	58.311	33.724
He $(2p^2)^3P$	59.674	35.087
He $(2p^2)^1D$	59.880	35.293
He $(2s\,2p)^1P$	60.126	35.539
He $(2p^2)^1S$	62.141	37.554

Table 14.2 Doubly excited states of helium with their excitation energy E_{exc}, and the energy E_e above the single ionization limit of helium, at which the auto-ionized electrons are detected.

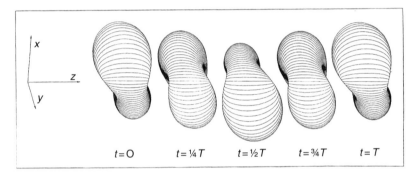

Figure 14.4 Oscillating charge clouds after the helium has been excited to auto-ionizing states resulting from a collision with a Li$^+$ ion. (Figure from Ref. [149].)

(see Fig. 14.1). Table 14.2 shows several doubly excited states of helium and their energies above the ionization level. Since the excitation energy is shared between the two electrons and neither electron has sufficient energy to ionize, the atom cannot ionize instantaneously. However, the strong interaction between the two electrons leads to a fast ionization and the process is called auto-ionization (AI). Since AI is very rapid ($<10^{-14}$ s), it dominates over radiative processes for the decay of the doubly excited states.

Doubly excited states can be produced optically or by electronic or ionic impact. In the last two cases, AI takes place in the vicinity of the incoming charged particle, and this leads to an energy shift of the emitted electron, since at the moment of ionization the incoming projectile and the target still interact by the Coulomb interaction. Such "post-collision interaction" allows the experimentalists to project the fast timescale of AI on the energy scale of the emitted electrons. Figure 14.4 shows such temporal oscillations of a He atom after excitation by a fast Li$^+$-ion. After excitation, the electron charge around the helium atom oscillates back and forth around the nucleus with a period T of 17 fs determined by the energy separation between the two excited states involved.

Appendices

14.A Variational calculations

When the Schrödinger equation cannot be solved directly, variational calculations can provide an upper bound for the energy of the ground state [60, 61]. One can choose an arbitrary wavefunction Ψ and calculate the energy of the system using this wavefunction. A simple analysis (see below) shows that the resulting energy is always larger than the energy of the ground state. To benefit from this property one can parameterize the wavefunction Ψ and then minimize the energy with respect to these parameters. Again the minimum found will still provide an upper bound for the ground-state energy. The more the wavefunction chosen reflects the real wavefunction of the ground-state, the better the energy found will match the correct energy.

To show that the energy calculated is always an upper bound for the ground-state energy, one can always decompose the trial wavefunction into a complete set of wavefunctions ϕ_n of the Hamiltonian $\mathcal{H}$, even if those wavefunctions are not known:

$$\Psi = \sum_n c_n \phi_n, \quad \text{with} \quad \mathcal{H}\phi_n = E_n \phi_n. \tag{14.20}$$

Here E_n are the eigenvalues of the eigenfunctions ϕ_n. Assuming the wavefunction is normalized:

$$\langle \Psi | \Psi \rangle = 1 = \sum_n |c_n|^2, \tag{14.21}$$

one finds for the expectation value of the Hamiltonian on Ψ:

$$\begin{aligned}\langle \mathcal{H} \rangle &= \langle \Psi | \mathcal{H} | \Psi \rangle = \left\langle \sum_m c_m \phi_m \middle| H \middle| \sum_n c_n \phi_n \right\rangle \\ &= \sum_{mn} c_m^* c_n E_n \langle \phi_m | \phi_n \rangle = \sum_{mn} c_m^* c_n E_n \delta_{mn} \\ &= \sum_n |c_n|^2 E_n \geq E_0 \sum_n |c_n|^2 = E_0 \end{aligned} \tag{14.22}$$

This completes the proof of the conjecture.

14.B Detail on the variational calculations of the ground state

The variational wavefunction of Eq. (14.15) resembles the "standard" choice for the variational wavefunction of the ground state for the case $Z_i = Z_o$. However, the wavefunction of Eq. (14.15) is not properly normalized and one finds, for instance, by using a symbolic manipulation package,

$$N = \langle \varphi | \varphi \rangle = 1 + \frac{64 Z_i^3 Z_o^3}{(Z_i + Z_o)^6} \tag{14.23}$$

where the result on the right applies for the "standard" trial wavefunction. For the kinetic energy one finds

$$E_k = \langle T \rangle = -\frac{\hbar^2}{2m} \left(\left\langle \frac{d^2}{dr^2} \right\rangle - 2 \left\langle \frac{d}{dr} \right\rangle \right) = R_\infty \left(Z_i^2 + Z_o^2 + \frac{128 Z_i^4 Z_o^4}{(Z_i + Z_o)^6} \right) \tag{14.24}$$

and for the potential energy

$$E_p = \langle V \rangle = \frac{-Ze^2}{4\pi\varepsilon_0} \left(\left\langle \frac{1}{r_1} \right\rangle + \left\langle \frac{1}{r_2} \right\rangle \right) = -2 R_\infty N Z (Z_i + Z_o) \tag{14.25}$$

For the electron–electron interaction the $1/r_{12}$ term in Eq. (14.10) can be expanded in Legendre polynomials [5]

$$\frac{1}{r_{12}} = \sum_{k=0}^{\infty} \frac{(r_<)^k}{(r_>)^{k+1}} P_k(\cos\theta)$$

$$= \sum_{k=0}^{\infty} \sum_{q=-k}^{k} \frac{4\pi}{2k+1} \frac{(r_<)^k}{(r_>)^{k+1}} Y_{kq}^*(\theta_1, \phi_1) Y_{kq}(\theta_2, \phi_2), \tag{14.26}$$

with $r_< = \min(r_1, r_2)$ and $r_> = \max(r_1, r_2)$. The angle θ in the middle part of the equation is the angle between the two directions (θ_1, ϕ_1) and (θ_2, ϕ_2). The last part of the equation is useful, since it decouples the angular coordinates of the two electrons. The introduction of $r_<$ and $r_>$ seems cumbersome, but in practice the radial integral over r_1 can always be split into two ranges, namely for $0 < r_1 < r_2$, where $r_< \equiv r_1$ and $r_> \equiv r_2$, and $r_2 < r_1 < \infty$, where $r_< \equiv r_2$ and $r_> \equiv r_1$.

One single term is replaced here by an infinite sum (or even a double sum), but because of the angular symmetry of the wavefunction, the double sum reduces to a single term. For the angular part of the interaction $1/r_{12}$ in the ground state, integration over (θ, ϕ) leads to

$$\int Y_{kq}^*(\theta, \phi) \sin\theta \, d\theta \, d\phi = \sqrt{4\pi} \, \delta_{k0} \, \delta_{q0}, \tag{14.27}$$

which follows directly from the orthogonality of the spherical harmonics. Thus, for the ground state the electron–electron interaction is simply $e^2/4\pi\varepsilon_0 r_{12} = e^2/4\pi\varepsilon_0 r_>$.

The Coulomb J and exchange K integrals become

$$J = \left\langle \phi(1,2) \left| \frac{e^2}{4\pi\varepsilon_0 r_>} \right| \phi(1,2) \right\rangle = \frac{2 R_\infty Z_i Z_o (Z_i^2 + 3 Z_i Z_o + Z_o^2)}{(Z_i + Z_o)^3} \tag{14.28}$$

and

$$K = \left\langle \phi(1,2) \left| \frac{e^2}{4\pi\varepsilon_0 r_>} \right| \phi(2,1) \right\rangle = \frac{40 R_\infty Z_i^3 Z_o^3}{(Z_i + Z_o)^5} \quad (14.29)$$

The variational energy is given by

$$E = \frac{E_k + E_p + J + K}{N}. \quad (14.30)$$

Figure 14.2 shows a contour plot of the energy as function of the charges Z_i and Z_o. The minimum can be found for $Z_i = 2.183$ and $Z_o = 1.189$, which is $E_{min} = -78.261$ eV.

Restricting the analysis by choosing $Z_i = Z_o \equiv Z'$ leads to

$$N = 2, \ \langle T \rangle = 4 R_\infty Z'^2, \ \langle V \rangle = -8 R_\infty Z' Z, \ J = \frac{5 R_\infty Z'}{4}, \ K = \frac{5 R_\infty Z'}{4}. \quad (14.31)$$

and thus

$$\frac{E'}{2 R_\infty} = Z'^2 - 2 Z' Z_0 + {}^5\!/_8 Z'. \quad (14.32)$$

The minimum can easily be obtained by differentiating this equation with respect to Z'. In that case one finds $E'_{min} = -R_\infty (Z - {}^5\!/_{16})^2/2 = -77.5$ eV. Both values have to be compared with the experimental value of $E_{exp} = -79.005$ eV, and it is clear that the discrepancy for the first, two-parameter variational result is a factor of 2 smaller than for the second, one-parameter result.

Exercises

14.1 Use the Pauli spin matrices of Eq. (2.16) on the states $\alpha = \begin{pmatrix} 1 \\ 0 \end{pmatrix}$ and $\beta = \begin{pmatrix} 0 \\ 1 \end{pmatrix}$ to find $\vec{S}^2$, S_z, and the eigenvalues of the exchange symmetry operator P for the two-particle states $\alpha_1\alpha_2, \beta_1\beta_2, \frac{1}{2}\sqrt{2}(\alpha_1\beta_2 + \beta_1\alpha_2)$, and $\frac{1}{2}\sqrt{2}(\alpha_1\beta_2 - \beta_1\alpha_2)$. Show that $S^2 = 1 + P$.

14.2 The independent electron model of the helium atom results in excited state energy levels given by $E_n = -R_\infty(4 + 1/n^2)$ where n is the principal quantum number of the excited electron. Calculate the wavelengths of several transitions and compare your results with the data of Tab. 14.1.

14.3 Consider the symmetry of the wavefunctions for the helium atom. The Hamiltonian is given by Eq. (14.9) and the eigenfunctions $\Psi(\vec{r}_1, \vec{r}_2) = \phi(\vec{r}_1, \vec{r}_2)\chi(1,2)$ can be written as productfunctions of the spatial part $\phi(\vec{r}_1, \vec{r}_2)$ and the spin part $\chi(1,2)$. Note that in this exercise the explicit form of the functions are not relevant, but the symmetry properties are.

(a) Why can the eigenfunctions of $\mathcal{H}$ always be split into two parts?

First the spatial part is considered and the operator P_{12} is defined as the permutation operator of electron 1 with electron 2.

(b) Determine the eigenvalue of P_{12} and the corresponding eigenfunctions $\phi(\vec{r}_1, \vec{r}_2)$.

(c) Are all eigenvalues of P_{12} permitted by the Pauli principle?

(d) Prove that the eigenfunctions of P_{12} are degenerate in energy, if we neglect the electron–electron interaction $e^2/4\pi\varepsilon_0 r_{12}$.

(e) Show that this term lifts the degeneracy and give an expression for the splitting between the two states. Can you provide a classical interpretation of your answer?

14.4 The metastable state of the helium atom has $J = 1$, which means there are three sublevels, $M_J = -1, 0, 1$. The lowest state excitable from this metastable state has three possible values of J, namely $J = 0, 1, 2$, and a total of 9 M_J sublevels. Draw a diagram of these states, and indicate which of the 27 possible transitions are forbidden by selection rules. Be sure to state which selection rule forbids each one of these.

14.5 The wavefunction of metastable helium in the 2^3S state is given by

$$\Psi(2^3S) = \frac{1}{2}\sqrt{2}\,[u_{1s}(r_1)u_{2s}(r_2) - u_{2s}(r_1)u_{1s}(r_2)] \begin{cases} \alpha(1)\alpha(2) \\ \frac{1}{2}\sqrt{2}\,[\alpha(1)\beta(2) + \beta(1)\alpha(2)] \\ \beta(1)\beta(2) \end{cases}$$

(a) What are the values of the total angular momentum L and the projection M_L for these three states?

(b) Show that these three wavefunctions are eigenfunctions of the operator $S_z = S_{1z} + S_{2z}$ and determine the eigenvalues.

(c) Prove this also for $\vec{S}^2 = (\vec{S}_1 + \vec{S}_2)^2$ and determine the eigenvalues.

(d) Determine the permutation symmetry P_{12} of both the spatial and spin part of the $\Psi(2^3S)$, where the permutation operator is given by $P_{12}\Psi(1, 2) = \Psi(2, 1)$.

There is another state constructed out of the same one-electron orbitals shown above, but with a different symmetry.

(e) What is the symmetry of this state?

(f) Why is the 2^3S state metastable, but the state under (e) not?

14.6 Calculate the action of the operator $\vec{S}^2 = 3/2 + 2\vec{S}_1 \cdot \vec{S}_2$ on the four spin functions of Eq. (14.7) and obtain the results quoted in the equation.

14.7 A helium atom is excited from the ground state to the auto-ionizing state 2s4p by absorption of ultraviolet light. Assume the 2s electron moves in the unscreened Coulomb field of the nucleus and the 4p electron in the fully screened Coulomb field $\propto -1/r$.

(a) Obtain the energy of this auto-ionizing level and the corresponding wavelength of the UV radiation.

(b) Find the velocity of the electron emitted in the auto-ionizing process in which the auto-ionizing level 2s4p decays into a free electron and a He$^+$ ion in the ground state 1s.

14.8 Although the eigenenergies for the hydrogen atom can be obtained analytically, variational calculations can also be performed for hydrogen. The Hamiltonian for hydrogen is given by

$$H = T + V = \frac{-\hbar^2}{2\mu r^2}\frac{\partial}{\partial r}\left(r^2 \frac{\partial}{\partial r}\right) - \frac{e^2}{4\pi\varepsilon_0 r}.$$

For the ground state, the energy in variational theory is given by

$$E_0 \leq \langle \phi | H | \phi \rangle,$$

with ϕ a trial function. For ϕ one can use

$$\phi(r) = C\left(1 - \frac{r}{r_0}\right) \quad r \leq r_0$$
$$\phi(r) = 0 \quad r > r_0$$

(a) Show that the normalization constant C is given by $C = \sqrt{30}/r_0^{3/2}$.
(b) Show that the expectation value of T for $\phi(r)$ is given by $\langle T \rangle = 5\hbar^2/\mu r_0^2$.
(c) Show that the expectation value of V for $\phi(r)$ is given by $\langle V \rangle = -5e^2/2(4\pi\varepsilon_0)r_0$.
(d) Find an upper bound for E_0 by variation of r_0. What is the value of r_0 for which $\langle H \rangle$ is minimal? Express your result in terms of the Bohr radius a_0 (see Eq. (7.18)).
(e) Compare your result of (d) with the exact result for the ground state of hydrogen (see Eq. (7.13)) and discuss the discrepancies.

14.9 Do the first-order perturbation calculation suggested near the end of Sec. 14.3.2 and check that the binding energy of the ground state is indeed -74.8 eV.

15

The periodic system of the elements

The description of atoms so far in this book has been restricted to the simplest ones, namely hydrogen, the alkali-metal atoms, and helium. In the first two cases there is only one active electron that is important, whereas in the last case there are two. However, nature provides many more atoms that all have more intricate electronic configurations, and this chapter deals with some of these elements.

For hydrogen, the Schrödinger equation can be solved exactly, and the agreement between the calculated and observed electronic levels is remarkable. For the alkali-metal atoms the results are not quite as good, but reasonable agreement can still be obtained if the interaction of all electrons apart from the valence electron is described by a model potential and appropriate quantum defects. For the helium atom, interaction between the electrons plays a key role, and a perturbative description does not provide such good agreement between theory and experiment. Although the variational approach used for He leads to reasonable agreement, such an approach does not work well for the description of many-electron systems.

The Hamiltonian for an atom with a number σ of electrons is given by:

$$\mathcal{H} = \sum_{i=1}^{\sigma}\left(-\frac{\hbar^2}{2m}\nabla_i^2 - \frac{Ze^2}{4\pi\varepsilon_0 r_i}\right) + \sum_{i=1,\,j>i}^{\sigma}\frac{e^2}{4\pi\varepsilon_0 r_{ij}}, \qquad (15.1)$$

where the first summation on the left-hand side is for the kinetic and potential energy of each of the σ electrons, and the second summation is for the mutual electron–electron interactions. Here it is implicitly assumed that the nucleus at the origin with charge Z is infinitely heavy compared with the electron. The description here is valid for ions, where $\sigma \neq Z$.

Since the Hamiltonian of Eq. (15.1) does not depend explicitly on χ or the total spin S and its projection S_z, the wavefunction of the electrons can be written as $\Psi(1\cdots\sigma) = \psi(\vec{r}_1,\ldots,\vec{r}_\sigma)\chi(1,\ldots,\sigma)$, which has been separated into a spatial part ψ and a spin part χ. The Pauli symmetrization principle discussed in Chap. 14

requires the total wavefunction to be anti-symmetric with respect to the exchange of any two electrons, or $\hat{P}_{ij}\Psi(1\cdots\sigma) = -\Psi(1\cdots\sigma)$, where $\hat{P}_{ij}$ is the operator that exchanges electron i with electron j. The spatial part $\psi(\vec{r}_1,\ldots,\vec{r}_\sigma)$ must satisfy

$$\left[\sum_{i=1}^{\sigma}\left(-\frac{\hbar^2}{2m}\nabla_i^2 - \frac{Ze^2}{4\pi\varepsilon_0 r_i}\right) + \sum_{i=1,\,j>i}^{\sigma}\frac{e^2}{4\pi\varepsilon_0 r_{ij}}\right]\psi(\vec{r}_1,\ldots,\vec{r}_\sigma) = E\psi(\vec{r}_1,\ldots,\vec{r}_\sigma). \quad (15.2)$$

However, the spatially dependent wavefunction $\psi(\vec{r}_1,\ldots,\vec{r}_\sigma)$ is a σ-particle wavefunction in 3σ dimensions, so it is not possible to solve this equation in a simple way.

For most of the elements in the Periodic Table the interaction between many electrons plays an important role. Although it is easy to write down the Hamiltonian for such a system, it cannot be solved exactly and it is necessary to use some approximations that will be introduced in this chapter. However, the "art" of describing such systems has progressed during the past century to a level that can no longer be adequately described in a few pages.

The ordering in the Periodic Table has long been established from their chemistry and from Moseley's law, that relates X-ray spectra to atomic number Z. One begins by choosing Z as the number of electrons, denoted herein by σ (for an atom, σ and Z are equal; for an ion they are not, hence the different symbols). To find the atomic ground states, begin filling the "shells" starting with $n = 1$ with $2(2\ell + 1)$ electrons in each shell.

For $Z > 18$ it is not obvious how to order this filling of the shells, but the "Aufbau" principle (see Sec. 15.3) provides an empirical recipe for the construction of an electronic wavefunction built from the wavefunctions of single electrons. Even so, it is not always clear which electron configuration lies lowest and thereby constitutes the ground state, so Hund's rules, also empirical, are used here. Then the coupling of the angular momenta of these individual electrons to form the total internal angular momentum of the atom is described, and finally the wavefunctions are anti-symmetrized with respect to electron exchange. The chapter ends with a summary of the arrangement of the Periodic Table.

15.1 The independent particle model

The problem can be simplified considerably by treating the electrons as independent particles and splitting the Hamiltonian into two parts $\mathcal{H} = \mathcal{H}_c + \mathcal{H}'$, with the central-field Hamiltonian

$$\mathcal{H}_c = \sum_{i=1}^{\sigma}\left(-\frac{\hbar^2}{2m}\nabla_i^2 + V(r_i)\right) \quad (15.3)$$

and the perturbation

$$\mathcal{H}' = \sum_{i=1,\,j>i}^{\sigma} \frac{e^2}{4\pi\varepsilon_0 r_{ij}} - \sum_{i=1}^{\sigma}\left(\frac{Ze^2}{4\pi\varepsilon_0 r_i} + V(r_i)\right)$$

$$\equiv \sum_{i=1,\,j>i}^{\sigma} \frac{e^2}{4\pi\varepsilon_0 r_{ij}} - \sum_{i=1}^{\sigma} S(r_i), \tag{15.4}$$

where $V(r_i)$ is the central potential that results from the screening of the nuclear charge by the other electrons. In Eq. (15.4) the difference between the interaction $-Ze^2/4\pi\varepsilon_0 r_i$ and $V(r_i)$ is treated as a perturbation. Note that the potential $V(r_i)$ depends only on r_i and thus allows the treatment of the term as a central-field Hamiltonian, whose solutions are discussed in Sec. 7.2. Such a treatment is useful only if $\mathcal{H}' \ll \mathcal{H}_c$. In the limit of small and large distances, one finds

$$\lim_{r \to 0} V(r) = -\frac{Ze^2}{4\pi\varepsilon_0 r} \qquad \lim_{r \to \infty} V(r) = -\frac{(Z-\sigma+1)e^2}{4\pi\varepsilon_0 r} \tag{15.5}$$

where in the last case assuming the total system is neutral ($\sigma = Z$) one finds $V(r) = -e^2/4\pi\varepsilon_0 r$.

In the central-field approximation one seeks eigenfunctions of $\mathcal{H}_c$, and the Schrödinger equation is separable into σ differential equations, one for each electron. The total one-electron wavefunction is given by $\psi_c = u_1(\vec{r}_1)u_2(\vec{r}_2)\cdots u_\sigma(\vec{r}_\sigma)$ where each u_i is the solution of the one-electron equation

$$\left[-\frac{\hbar^2}{2m}\nabla_i^2 + V(r_i)\right]u_{n\ell m}(\vec{r}_i) = E_{n\ell}u_{n\ell m}(\vec{r}_i). \tag{15.6}$$

The energy $E_{n\ell}$ now depends on n and ℓ, since the potential $V(r)$ is no longer precisely $1/r$. As shown below, this leads to specific rules for the construction of the electronic structure of atoms.

Since $V(r)$ is a central-field potential, the one-electron orbitals $u_{n\ell m}(\vec{r})$ are given by $u_{n\ell m}(\vec{r}) = R_{n\ell}(r)Y_{\ell m}(\theta,\varphi)$ where the $R_{n\ell}(r)$ are solutions of the differential equation

$$-\frac{\hbar^2}{2m}\left(\frac{\partial^2}{\partial r^2} + \frac{2}{r}\frac{\partial}{\partial r} - \frac{\ell(\ell+1)}{r^2} - \frac{2m}{\hbar^2}V(r)\right)R_{n\ell}(r) = E_{n\ell}R_{n\ell}(r). \tag{15.7}$$

The radial eigenfunctions $R_{n\ell}(r)$ are not the same Laguerre polynomials that were found for hydrogen since the potential $V(r)$ deviates from a pure Coulomb potential. The total energy of the atom in this description becomes $E_c = \sum_{i=1}^{\sigma} E_{n_i \ell_i}$. The perturbation $\mathcal{H}'$ can be used with these eigenfunctions as a basis to obtain the first-order shifts of the eigenvalues caused by the deviation from the central-field approximation.

15.2 The Pauli symmetrization principle

The Pauli symmetrization principle requires that the total wavefunction for any system of fermions be anti-symmetric with respect to the exchange of any two of them, as discussed in Sec. 14.2. A consequence, as shown later, is that no two fermions may occupy the same quantum state simultaneously. For the electrons in a single atom or molecule, this requires that no two electrons can have the same four quantum numbers, that is, if n, ℓ, and m_ℓ are the same, m_s must be different such that the electrons have opposite spins. The total one-electron wavefunction including spin is written as $\Psi_{n\ell m_\ell m_s} = u_{n\ell m_\ell}(\vec{r})\chi_{m_s} = R_{n\ell}(r)Y_{\ell m_\ell}(\theta,\phi)\chi_{m_s}$, where the total wavefunction is separated into a radial part, an angular part, and a spin part, respectively. Note that m_s can have the values $\pm 1/2$, also written as $\uparrow/\downarrow$ or spin up/down. Since these one-electron wavefunctions are eigenfunctions of a central-field Hamiltonian, the angular part is given by the usual spherical harmonics.

For a large number of electrons, constructing anti-symmetric wavefunctions by combining one-electron wavefunctions becomes tedious. Consider, for instance, the case of lithium, which has three electrons. Assuming the first two electrons are in the 1s-state and the third electron is in the 2s-state with spin up, the wavefunction can be written as

$$\Psi = 1s(1)1s(2)2s(3)\left(\alpha(1)\beta(2) - \beta(1)\alpha(2)\right)\alpha(3). \tag{15.8}$$

Although this wavefunction has the right exchange symmetry for electrons 1 and 2, it is not properly symmetrized with respect to the exchange of electrons 1 and 3 and of electrons 2 and 3. Introducing the right symmetry for the exchange of these electrons adds four more terms to the wavefunction and makes the expression quite cumbersome. It can be simplified by writing the wavefunction as the determinant of a matrix. This "Slater determinant" is necessarily anti-symmetric for the exchange of two electrons, and is given by

$$\Psi_c(1\ldots\sigma) = \frac{1}{\sqrt{\sigma!}}\begin{vmatrix} \Psi_{n\ell m m_s}(1) & \Psi_{n\ell m m_s}(2) & \ldots & \Psi_{n\ell m m_s}(\sigma) \\ \Psi_{n'\ell'm'm'_s}(1) & \Psi_{n'\ell'm'm'_s}(2) & \ldots & \Psi_{n'\ell'm'm'_s}(\sigma) \\ \vdots & \vdots & \vdots & \vdots \\ \Psi_{n''\ell''m''m''_s}(1) & \ldots & \ldots & \Psi_{n''\ell''m''m''_s}(\sigma) \end{vmatrix}, \tag{15.9}$$

Note that $\hat{P}_{ij}\Psi_c(1\ldots\sigma) = -\Psi_c$, since the effect of the exchange operator $\hat{P}_{ij}$ is the same as exchanging two columns i and j in the matrix. Moreover, if two rows or columns are the same, the determinant vanishes making the Slater determinant consistent with the principle of "exclusion". This is the origin of the term "Pauli exclusion principle". For lithium the matrix is a 3×3 matrix producing six terms for the fully symmetrized wavefunction, the same terms as would be produced if Eq. (15.8) were properly symmetrized.

In general one seeks eigenfunctions not only of $\mathcal{H}_c$, but also of the total orbital angular momentum $\vec{L}^2$, its projection L_z, the total spin angular momentum $\vec{S}^2$, and its projection S_z. The Slater determinants are not always eigenfunctions of these operators, and in general one needs linear combinations of them.

15.3 The "Aufbau" principle

Understanding the electronic structure of atoms begins by adding one electron at a time to a nucleus whose charge is incremented to maintain electrical neutrality. For hydrogen the energy of the electron depends only on n, but for all the other atoms this is not so because the potential is not simply $V(r) = -Ze^2/4\pi\varepsilon_0 r$ as in hydrogen. Instead, other electrons can partially shield the nuclear charge and the energy depends on the shape of their orbits through the angular momentum ℓ as well as on n, as described in Sec. 10.3 for the alkali metals. In the central-field case the energy of one-electron orbitals $E = E_{n\ell}$, and the following rules have been found empirically:

$$E_{n\ell} < E_{n'\ell} \text{ if } n < n' \tag{15.10}$$
$$E_{n\ell} < E_{n'\ell'} \text{ if } n + \ell < n' + \ell'$$
$$E_{n\ell} < E_{n'\ell'} \text{ if } n + \ell = n' + \ell' \text{ and } n < n'$$

These rules are shown graphically in Fig. 15.1. Each one-electron orbital is doubly degenerate because of the electronic spin, and also $2\ell + 1$ degenerate because of the projections of ℓ on the z-axis (determined by the spin axis in the absence of an external field). This leads to a total degeneracy of $4\ell + 2$. Thus there are two electrons in an s-orbital, six electrons in a p-orbital, ten electrons in a d-orbital and 14 electrons in an f-orbital. The electronic structure of the ground state of barium

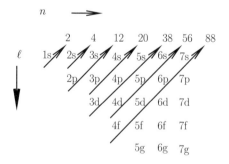

Figure 15.1 Diagram of the Aufbau principle. Beginning at the upper left, the 1s shell is filled with two electrons (H and He). Then proceed to the right for the 2s shell (second arrow), and then the third arrow begins with the 2p shell, followed by the 3s shell. Next are 3p and 4s, and the fifth arrow begins with 3d, then 4p, and then 5s. Other cases follow.

Z		Element	Elec. conf.	Term	M	IP (eV)	$r\,(a_0)$
1	H	hydrogen	1s	$^2S_{1/2}$	1.008	13.598	1.00
2	He	helium	$1s^2$	$^1S_0$	4.003	24.587	1.76
3	Li	lithium	[He]2s	$^2S_{1/2}$	6.939	5.392	2.87
4	Be	beryllium	$2s^2$	$^1S_0$	9.012	9.322	2.12
5	B	boron	$2s^2\,2p$	$^2P_{1/2}$	10.811	8.299	1.51
6	C	carbon	$2s^2\,2p^2$	$^3P_0$	12.011	11.260	1.46
7	N	nitrogen	$2s^2\,2p^3$	$^4S_{3/2}$	14.007	14.534	1.40
8	O	oxygen	$2s^2\,2p^4$	$^3P_2$	15.999	13.618	1.40
9	F	fluorine	$2s^2\,2p^5$	$^2P_{3/2}$	18.998	17.422	1.36
10	Ne	neon	$2s^2\,2p^6$	$^1S_0$	20.183	21.564	2.12
11	Na	sodium	[Ne]3s	$^2S_{1/2}$	22.990	5.139	3.52
12	Mg	magnesium	$3s^2$	$^1S_0$	24.312	7.646	3.02
13	Al	aluminum	$3s^2\,3p$	$^2P_{1/2}$	26.982	5.986	2.70

Table 15.1 Data for the ground state of the first 13 elements of the periodic system. M is the atomic mass, IP is the ionization potential, and r is the atomic radius.

with $Z = \sigma = 56$, for example, can be written as $1s^2\,2s^2\,2p^6\,3s^2\,3p^6\,4s^2\,3d^{10}\,4p^6\,5s^2\,4d^{10}\,5p^6\,6s^2$, where the superscript gives the number of electrons. Note that the first 54 electrons form completely filled shells having the same electronic structure as xenon, and often the electron structure of barium is written as $[Xe]6s^2$.

Table 15.1 shows the configuration of the electrons for the ground state of the first 13 atoms. For the noble-gas atoms (except helium) the scheme leads to the filling of a complete shell of electrons in p-orbitals. Since the shell is completely filled and all angular momenta are completely paired, it is difficult to remove an electron from the shell and thus the ionization potential of the noble-gas atoms is large. Figure 15.2 shows that the ionization potential as a function of the atomic number Z peaks for the noble gases. For the alkali-metal atoms, the valence electron is located outside a completely filled shell and since it is easy to remove this electron, the corresponding ionization potential for these atoms is thus small. If the ionization potential is small, the electron is weakly bound and so the atomic radius is large. This can be seen in Fig. 15.3, where the atomic radii are peaked for the alkali-metal atoms.

15.4 Coupling of many-electron atoms

Even though the discussion so far has been restricted to the eigenfunctions of the central-field Hamiltonian $\mathcal{H}_c$ of Eq. (15.3), it has still provided enough information

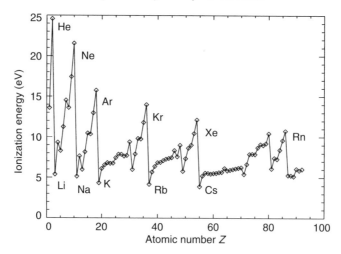

Figure 15.2 Ionization potentials for the different elements. Notice the large ionization potentials for the noble-gas atoms which have a closed, completely filled outer shell, and the small ionization potentials for the alkali-metal atoms which have only one valence electron.

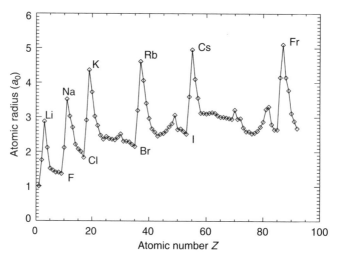

Figure 15.3 Atomic radius for the different elements. Notice the large radii for the alkali-metal atoms which have only one valence electron, and the small radii for the halogen elements which have one electron missing in their outer shell.

to determine the dominant features of atomic structure. However, corrections have to be made to the energies since this $\mathcal{H}_c$ does not fully describe the system. These corrections have two origins. The first one arises from the full, exact Coulomb interaction between the electrons, and the central-field potential $V(r_i)$ as given in Eq. (15.4):

15.4 Coupling of many-electron atoms

Symbol	Definition	Meaning
$\mathcal{H}$	Eq. (15.1)	Hamiltonian for many-electron atom
$\mathcal{H}_c$	Eq. (15.3)	Independent particle Hamiltonian
$\mathcal{H}'$	Eq. (15.4)	Electron–electron interaction Hamiltonian
$\mathcal{H}_{SO}$	Eq. (15.12a)	Spin–orbit Hamiltonian
$\mathcal{H}_{RS}$	$\mathcal{H}_c + \mathcal{H}'$	Hamiltonian for Russel–Saunders coupling
$\mathcal{H}_{jj}$	$\mathcal{H}_c + \mathcal{H}_{SO}$	Hamilton for j–j coupling

Table 15.2 It is convenient to summarize the definition and purpose of the many Hamiltonian symbols that are used in this chapter.

$$\mathcal{H}' = \sum_{i=1,\, j>i}^{\sigma} \frac{e^2}{4\pi\varepsilon_0 r_{ij}} - \sum_{i=1}^{\sigma} S(r_i). \tag{15.11}$$

The second correction does not arise from such electrostatic interactions, but instead from magnetic interactions described as spin–orbit in Chap. 8, namely

$$\mathcal{H}_{SO} = \sum_{i}^{\sigma} \xi(r_i)\vec{\ell}_i \cdot \vec{s}_i / \hbar^2 \tag{15.12a}$$

with

$$\xi(r_i) = \frac{\hbar^2}{2m^2c^2} \frac{1}{r_i} \frac{dV(r_i)}{dr_i}, \tag{15.12b}$$

where $\vec{\ell}$ and $\vec{s}$ are the orbital angular momentum and spin of each electron. Note that the potential in this case is not $-Ze^2/4\pi\varepsilon_0 r$, and therefore $\xi(r)$ is not strictly proportional to $1/r^3$. The vector sum of the orbital angular momenta of all electrons is $\sum_i^{\sigma} \vec{\ell}_i = \vec{L}$ and of their spins is $\sum_i^{\sigma} \vec{s}_i = \vec{S}$. In the case of a completely filled shell, $\mathcal{H}_{SO}$ does not contribute to the total energy because $|\vec{L}|^2 = 0 = |\vec{S}|^2$. There are two "constants of motion", namely the total angular momentum $\vec{J} \equiv \vec{L} + \vec{S} = \sum_i^{\sigma} \vec{J}_i$ and the total parity of the state.

When $\langle \psi | \mathcal{H}' | \psi \rangle \gg \langle \psi | \mathcal{H}_{SO} | \psi \rangle$ the electrostatic interaction dominates and it is appropriate to couple the orbital angular momenta $\vec{\ell}_i$ and spins $\vec{s}_i$ of the individual electrons separately to obtain a total orbital angular momentum $\vec{L}$ and spin $\vec{S}$, before coupling them together to form the total angular momentum $\vec{J} = \vec{L} + \vec{S}$. The resulting Hamiltonian has its largest matrix elements on the diagonal. This scheme is called Russel–Saunders coupling or L-S coupling.

On the other hand, for $\langle \psi | \mathcal{H}_{SO} | \psi \rangle \gg \langle \psi | \mathcal{H}' | \psi \rangle$, the magnetic interaction dominates so the Hamiltonian matrix is most nearly diagonal if $\vec{\ell}_i$ and $\vec{s}_i$ are first coupled to obtain the total angular momentum $\vec{j}_i$ of each individual electron, before they are coupled together to form $\vec{J} = \sum_i \vec{j}_i$. This is called j–j coupling. In the intermediate case $\langle\psi|\mathcal{H}_{SO}|\psi\rangle \approx \langle\psi|\mathcal{H}'|\psi\rangle$, one has intermediate coupling where the diagonal and off-diagonal Hamiltonian matrix elements are comparable, the solutions are superpositions, and this case is not discussed here. However, the values $\vec{J}^2$ and its projection J_z are the same in all coupling schemes.

15.4.1 Russel–Saunders coupling

For Russel–Saunders coupling it is best to seek eigenfunctions of $\mathcal{H}_{RS} \equiv \mathcal{H}_c + \mathcal{H}'$ and apply $\mathcal{H}_{SO}$ as a perturbation. Since $\mathcal{H}'$ commutes with $\vec{L}$, $\vec{S}$, and $\vec{J}$, the solutions are degenerate with L, S, and J, which leads to a degeneracy of $(2L + 1)(2S + 1)$. This means that superpositions of different J-states are also energy eigenfunctions, so J is not determined. Then the total state can be written with the term symbol $^{2S+1}L^\sigma$, with the spin multiplicity $2S + 1$ and the parity σ. The reader is referred to App. B.2 for the definitions associated with this term symbol. Note that multi-electron atoms use the same symbols as for one-electron orbitals to describe the orbital angular momentum L: S ($L = 0$), P ($L = 1$), D ($L = 2$), F ($L = 3$), G ($L = 4$), etc. Electrons outside fully filled shells are called valence electrons and only these contribute to the total angular momentum.

Using the eigenfunctions of $\mathcal{H}_{RS}$ one can exploit perturbation theory to find the energy shifts caused by the term $\mathcal{H}_{SO}$. Since $\mathcal{H}_{SO}$ commutes with $\vec{J}^2$, but not with $\vec{L}$ and $\vec{S}$, each value of L and S is split up into a "multiplet" of different values for J: $J = |L - S|, \ldots, L + S$. The term symbol then becomes $^{2S+1}L^\sigma_J$, with J the total angular momentum.

When there are two electrons in non-identical orbitals, for example ns and $n'p$, the possible electronic configurations are

$$ns\, n'p\quad L = 1 \begin{cases} S = 0 \quad ^1P \\ \\ S = 1 \quad ^3P \end{cases} \qquad \begin{cases} J = 1 \quad ^1P_1 \\ \begin{cases} J = 0 \quad ^3P_0 \\ J = 1 \quad ^3P_1 \\ J = 2 \quad ^3P_2 \end{cases} \end{cases} \qquad (15.13)$$

$$\mathcal{H}_c \qquad\qquad \mathcal{H}_c + \mathcal{H}' \qquad\qquad \mathcal{H}_c + \mathcal{H}' + \mathcal{H}_{SO}$$

In the independent particle model the energy is split into a multiplet of six states by the deviation from the central-field $\mathcal{H}'$, and these six are split further by the spin–orbit interaction $\mathcal{H}_{SO}$ into a multiplet of ten states as shown above.

Calculation of the spin–orbit interaction in multi-electron atoms begins with

$$\begin{aligned} E_{SO} = \langle\psi|\mathcal{H}_{SO}|\psi\rangle &= \langle\alpha LS\, M_L M_S|\mathcal{H}_{SO}|\alpha LS\, M_L M_S\rangle \\ &= A\langle\alpha LS\, M_L M_S|\vec{L}\cdot\vec{S}|\alpha LS\, M_L M_S\rangle \\ &= A\langle\alpha LS\, JM_J|\tfrac{1}{2}(\vec{J}^2 - \vec{L}^2 - \vec{S}^2)|\alpha LS\, JM_J\rangle \end{aligned} \qquad (15.14)$$

15.4 Coupling of many-electron atoms

where A is independent of L and S, but depends on the $\xi(r_i)$. The last line of Eq. (15.14) is written in terms of the eigenfunctions of $\vec{J}^2 = \vec{L}^2 + \vec{S}^2 + 2\vec{L}\cdot\vec{S}$. For any angular momentum operator $\vec{K}^2 = K(K+1)\hbar^2$ results in

$$\Delta E_{SO}(J+1, J) \equiv E_{SO}(J+1, L, S) - E_{SO}(J, L, S) = AJ \quad (15.15)$$

for the interval between two adjacent J-states and is known as Landé's interval rule. When the subshell is less than half full the result is a "normal multiplet" where $A > 0$ and the lowest J has lowest energy. Otherwise there is an inverted multiplet with $A < 0$ and the highest J has the lowest energy.

15.4.2 j–j coupling

The case of j–j coupling is usually important for high-Z atoms because the derivative $dV(r_i)/dr_i$ in Eq. (15.12) is so much larger than in low-Z atoms. Construction of the eigenfunctions begins with those of $\mathcal{H}_c + \mathcal{H}_{SO}$ that are also eigenfunctions of j_i^2 and are written as $|\alpha n_i \ell_i j_i\rangle$ for each electron. For one s electron and one p electron there are no restrictions on the spins so that the allowable configurations are

$$ns\,n'p \begin{cases} j_1 = 1/2 \quad j_2 = 1/2 & \begin{cases} J = 1 & (1/2, 1/2)_1 \\ J = 0 & (1/2, 1/2)_0 \end{cases} \\ j_1 = 1/2 \quad j_2 = 3/2 & \begin{cases} J = 1 & (1/2, 3/2)_1 \\ J = 2 & (1/2, 3/2)_2 \end{cases} \end{cases} \quad (15.16)$$

$$\mathcal{H}_c \qquad \mathcal{H}_c + \mathcal{H}_{SO} \qquad\qquad\qquad \mathcal{H}_c + \mathcal{H}_{SO} + \mathcal{H}'$$

This is denoted as $n\ell n'\ell'(j_1, j_2)_J$. In old literature the states are sometimes denoted by their Paschen notation, but this notation is no longer in use. For two equivalent electrons the Pauli principle restricts the choices so that for two identical p electrons one finds

$$np\,np \begin{cases} j_1 = 1/2 \quad j_2 = 1/2 & J = 0 \\ j_1 = 1/2 \quad j_2 = 3/2 & J = 1, 2 \\ j_1 = 3/2 \quad j_2 = 3/2 & J = 0, 2 \end{cases} \quad (15.17)$$

$$\mathcal{H}_c \qquad\qquad \mathcal{H}_c + \mathcal{H}_{SO}$$

i.e. a total of five states that have a distinct symmetry. Discussion of how this comes about is in the next section.

15.4.3 Possible combinations for two electrons

If the valence shell contains two electrons there are many different coupling schemes. When the two electrons are in non-identical states, the Pauli principle

Conf.	Non-identical electrons	Conf.	Identical electrons
ss'	^{1,3}S	s^2	^{1}S
sp	^{1,3}P		
sd	^{1,3}D		
pp'	^{1,3}S, ^{1,3}P, ^{1,3}D	p^2	^{1}S, ^{3}P, ^{1}D
pd	^{1,3}P, ^{1,3}D, ^{1,3}F		
dd'	^{1,3}S, ^{1,3}P, ^{1,3}D, ^{1,3}F, ^{1,3}G	d^2	^{1}S, ^{3}P, ^{1}D, ^{3}F, ^{1}G

Table 15.3 Configuration for the coupling of two electrons. For non-identical electrons, there are no restrictions, and this leads to a large number of states, as shown on the left. For identical electrons as shown on the right, the Pauli principle requires the two electrons to have different quantum numbers and this restricts the number of configurations considerably.

can be easily fulfilled using proper combinations of the states, and the angular momenta of the two electrons can be coupled in all possible arrangements. That is, the spins can combine to make $S = 0$ or $S = 1$ without regard to the ℓ-values. This leads to the allowed terms for the configurations as shown on the left in Tab. 15.3. However, if the two electrons are in identical states, the number of allowed configurations is restricted. Allowed configurations are given on the right of Tab. 15.3. As can be seen, the number of possible configurations is strongly reduced for identical electrons.

To discuss this in more detail for two identical p electrons, one first has to create a table of all possible combinations, excluding those combinations in which the electrons are in identical states, as shown in the top of Fig. 15.4. Since one knows all possible values of M_L and M_S, one has to split this table up into different configurations. The bottom of Fig. 15.4 shows that one can identify first the state ^{1}D on the vertical axis with $M_L = 0$. The remaining states can be split up into a ^{3}P and ^{1}S state, leading to the result shown on the right of Tab. 15.3 above for the p^2 configuration.

15.5 Hund's rules

The previous section shows how the different atomic energy states are constructed for various electron configurations. The energy ordering of these terms can be found from the empirical Hund's rules. There are some heuristic justifications for these rules, and they are correct for the vast majority of cases:

1. The term with the highest multiplicity ($2S + 1$) for a particular value of L has the lowest energy, and the energy goes up with decreasing S.
2. For a particular value of S, the term with the highest L has the lowest energy.

15.5 Hund's rules

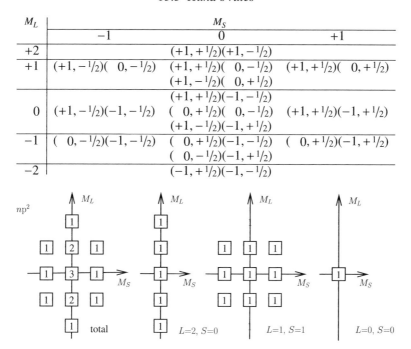

Figure 15.4 *L–S* coupling for two identical p-electrons. The chart at the top shows the number of ways (m_ℓ, m_s) of the two electrons can be coupled to the total projection of the angular momenta (M_L, M_S). The bottom graph shows that the 15 different ways of coupling can be split into three states, namely the states ^{1}D, ^{3}P, and ^{1}S.

For example, Fig. 15.4 shows that for two non-identical p electrons there are six possible terms, and Hund's rules give the following descending order of energy:

$$np\,n'p \begin{cases} L=0, S=0 & ^1S \\ L=1, S=0 & ^1P \\ L=2, S=0 & ^1D \\ L=0, S=1 & ^3S \\ L=1, S=1 & ^3P \\ L=2, S=1 & ^3D \end{cases} \qquad (15.18)$$

$$H_c \qquad\qquad H_c + H_1$$

A good example is the spectrum of helium. There are four $1s2\ell$ ($\ell = 0, 1$) states, two singlets and two triplets, and the ^{3}S lies lowest, followed by the ^{1}S, then ^{3}P, and finally ^{1}P. The S states lie below the P states because the $2s$ orbits penetrate into that of the $1s$ electron. Such penetration depresses the S-state energies resulting in their larger quantum defects (see Chap. 10).

For two identical p electrons the energy of He would exceed the ionization threshold, but not for higher Z atoms such as carbon. Figure 15.4 shows that in this case, there are only three possible terms, and Hund's rules give the energy ordering as:

$$np^2 \begin{cases} L = 0, S = 0 & {}^1S \\ L = 2, S = 0 & {}^1D \\ L = 1, S = 1 & {}^3P \end{cases} \quad (15.19)$$

$$H_c \qquad\qquad H_c + H_1$$

A good example is to identify the ground state of carbon. Its four lowest-energy electrons occupy s orbitals with $n = 1$ and 2, each pair with opposite spins as required by the Pauli principle. The lowest state for the other two electrons (each with $\ell = 1$) would have to be a triplet ($S = 1$) and as Fig. 15.4 shows, the only allowed triplet state for two p electrons has $L = 1$, making it ^3P, and this is indeed the ground state. The $2p^2\,^1D$ state is about 1 eV higher and the $2p^2\,^1S$ state is about 1.5 eV above that. These three $2p^2$ states are all composed of p electrons so their penetration into the core is all the same, thus producing equal quantum defects. This latter $2p^2\,^1S$ state in Ca is the same as in C but with a larger core, and it plays an important role in Sec. 24.4.

Hund's rules can be used to find the term with the lowest energy for a certain configuration of electrons. This term will then be the ground state for that particular atom.

15.6 Hartree–Fock model

In Chap. 7 considerable efforts are undertaken to retrieve the eigenvalues and eigenfunctions of a single electron in the Coulomb potential of the nucleus. This problem can be treated exactly, since the interaction potential is well-known and the Schrödinger equation can be solved analytically. For the alkali-metal atoms in Chap. 10 the situation becomes more complex, but by treating the effect of the core electrons only as an effective potential for the valence electron, accurate solutions for the wavefunction of the valence electron can be found. This does not work for helium, although the Hamiltonian for this two-electron system can be written down exactly as discussed in Chap. 14. The electron–electron repulsion cannot be treated as a perturbation and the solution of the problem requires variational methods.

For other atoms another approach is required, since more than one electron is active and using perturbative schemes for many electrons becomes rather cumbersome. One of the methods used is the Hartree–Fock model that was first presented by Hartree in 1928 [150]. The model assumes that each electron moves in a central

15.6 Hartree–Fock model

field, which is given by the potential of the nucleus in combination with the potential created by the charge densities of the other electrons. The Schrödinger equation is solved for each electron independently and the resulting wavefunctions are made self-consistent with the potential from which they are calculated. The total wavefunction of the atom can be written as a product of single-electron wavefunctions, as is the case for helium in Chap. 14:

$$\psi(\vec{r}_1,\ldots,\vec{r}_\sigma) = \phi_1(\vec{r}_1)\ldots\phi_\sigma(\vec{r}_\sigma), \tag{15.20}$$

with σ the number of electrons. The Schrödinger equation now becomes separable into σ independent one-electron equations

$$\mathcal{H}_k\phi_k(\vec{r}_k) = E_k\phi_k(\vec{r}_k) \quad \text{with} \quad \mathcal{H}_k = -\frac{\hbar^2\nabla_k^2}{2m} + V(\vec{r}_k), \tag{15.21}$$

where the potential for electron k combines the attraction by the nucleus with charge Z and the repulsion by the other electrons:

$$V(\vec{r}_k) = -\frac{Ze^2}{4\pi\varepsilon_0 r_k} + \sum_{j\neq k}^{\sigma} \int d\vec{r}_j\ |\phi_j(\vec{r}_j)|^2\ \frac{e^2}{4\pi\varepsilon_0|\vec{r}_j - \vec{r}_k|}. \tag{15.22}$$

If all electrons are in closed shells one can derive [151] that the potential $V(\vec{r}_k)$ is spherically symmetric, but in other cases the deviations from a spherically symmetric potential are so small that they can be neglected. This simplifies the solution of Eq. (15.21) considerably. Note that the spin–orbit interaction is completely neglected in the present discussion.

Solving Eqs. (15.21) and (15.22) self-consistently, one starts with an approximate central potential $V(\vec{r}_k)$ and finds the eigenfunctions from solving Eq. (15.21) for each electron k. Next an updated potential is obtained from solving Eq. (15.22), which can be used to reiterate the wavefunction $\phi(\vec{r}_k)$. If the potential does not change appreciably from one iteration to the next, a self-consistent solution has been found. The total energy of the atom is then found as the sum of the one-electron energies: $E = \sum_k E_k$.

The wavefunction of Eq. (15.20) is not anti-symmetric with respect to the interchange of two electrons, as required by the Pauli principle. A symmetrized version of Eq. (15.20) is found by using the Slater determinant of Eq. (15.9), which complicates the model considerably. This was first implemented in the model by Fock in 1930 [152] and a description of the complete Hartree–Fock model can be found in Ref. [60]. The model becomes so complicated because the exchange terms K as discussed in Sec. 14.4.1 have to be applied for any two electrons. For an in-depth discussion of the method the reader is referred to Ref. [151]. The symmetrization of the electrons as required by the Pauli principle can also be invoked in the Hartree model by requiring that two electrons are not in the same state and thus do not have

	s					p	
n_s	α_s	$c_k(1s)$	$c_k(2s)$	$c_k(3s)$	n_p	α_p	$c_k(2p)$
1	11.012 30	0.961 79	−0.234 74	0.035 27	2	5.549 77	0.464 17
3	12.660 10	0.040 52	−0.006 06	0.001 21	4	8.668 46	0.036 22
3	8.361 56	0.019 19	0.111 54	−0.018 89	4	5.434 60	0.292 82
3	5.738 05	−0.002 98	0.431 79	−0.068 08	4	3.555 03	0.316 35
3	3.612 87	0.001 91	0.517 Cl	−0.092 32	4	2.316 71	0.075 63
3	2.250 96	−0.000 49	0.041 47	0.000 76			
3	1.115 97	0.000 16	−0.003 24	0.407 64			
3	0.710 28	−0.000 07	0.001 26	0.644 67			

Table 15.4 Coefficients of the Slater orbitals for the Hartree–Fock solutions of the wavefunction for Na, as used in Fig. 10.2.

the same quantum numbers. The equations in the Hartree model are much easier to solve and the wavefunctions are often used as a first approximation in the solution of the Hartree–Fock model.

Since the solution of the Hartree–Fock model is very difficult and requires many iterations, their solutions are given in the literature as fits of the solutions to Slater orbitals that are given by

$$\chi_{n\ell m}(r, \theta, \phi) = N r^{n-1} e^{-\alpha r} Y_{\ell m}(\theta, \phi), \qquad (15.23)$$

where α is an adjustable parameter and N a normalization constant. The solutions of the Hartree–Fock equations are then given as

$$u_{n\ell}(r, \theta, \phi) = \sum_k c_k(n\ell) \chi_{n\ell m}(r, \theta, \phi). \qquad (15.24)$$

Note that the long-range behavior of the Slater orbitals is the same as the hydrogenic wavefunctions, but they do not have radial nodes. The values of α and the coefficients c_k are tabulated in Ref. [98]. Values used for the calculation as shown in Fig. 10.2 for Na are given in Tab. 15.4.

15.7 The Periodic Table

The cycles of the Periodic Table of the elements are built up from the addition of electrons to their various electron configurations as described in Sec. 15.3. In the upper rows of the table the scheme is straightforward since the Pauli principle simply dictates which quantum numbers are allowed for each succeeding orbital of the ground states of the atoms. These various arrangements are called "shells" and are labeled according to their quantum numbers. However, for the $n = 3$ shell

15.7 The Periodic Table

the filling scheme changes as shown in Fig. 15.1. Once the 3p shell has been filled with six electrons, producing argon with $Z = 18$, the next two electrons go into the 4s shell, making potassium ($Z = 19$) and calcium ($Z = 20$), instead of going into the 3d shell as the Bohr model might suggest. After that, the 3d shell is filled with the next ten electrons ($Z = 21–30$) before any electrons are placed into the 4p shell. Placing those six 4p-electrons next still leaves the $n = 4$ shell without its d or f electrons, but nevertheless forms the rare gas krypton ($Z = 36$). However, before the rest of the $n = 4$ electrons come, the two 5s electrons are placed to make rubidium ($Z = 37$) and strontium ($Z = 38$). The rest of the table gets somewhat more complicated, but the scheme illustrated in Fig. 15.1 gives the right result.

It is important to realize that the filling scheme of Sec. 15.3 for the atomic ground states is *not* a consequence of the Pauli principle. Of course, it must be followed for the ground and all the excited states of all atoms, but ascertaining the electron configuration that constitutes the lowest-lying state is a different matter. The energies are determined by the interactions among the electrons, both electrostatic and magnetic (spin). The energies of most of the states of most of the atoms are known from spectroscopic measurements. These are summarized in many places in books and on the web, and often presented in plots known as Grotrian diagrams. A good collection is found in Ref. [153].

Although the energies determine the sequence of filling the shells and hence the ordering of the rows, the columns of the Periodic Table tell a different story. These are arranged so that elements with similar electronic structures fall in the same column, hence the columns of rare gases, halogens, and so on. Atoms in the first column are called alkali metals and are headed by lithium, sodium, potassium, etc. Their excited-state energy-level arrangement is quite simple because the atoms have a single electron outside an otherwise full p shell, so they all have spectra similar to hydrogen. Of course, the n-degeneracies are lifted by the quantum defects as discussed in Chap. 10, but their spectra, fine and hyperfine structure, and interactions with external fields are all quite similar. Their heavier nuclei have greater charge so their fine structure splitting is much larger, and they have nuclear spin so their hyperfine structure splitting is more complicated, but their Zeeman and Stark effects resemble those of hydrogen.

The second column is topped by beryllium, and because its outermost shell is filled with two s electrons, it is quite similar to helium. Nevertheless, the absence of any p electrons makes it rather special, so the discussion here begins with the next atom in the column, magnesium. Other alkaline earth atoms with similar spectra are calcium, strontium, and barium, and all of these are frequent subjects of study in atomic physics. As in the case of helium, their ground state is the singlet with the two s electrons having opposite spin ($S = 0$), and there are higher excited states that are both singlet and triplet.

Atom	Ground state ($^1S_0$)	Excited singlet ($^1P_1$)			Excited triplet ($^3P_{2,1,0}$)		
			λ (nm)	τ (ns)		λ (nm)	τ (µs)
Mg	[Ne]$3s^2$	3s 3p	285	1.97	3s 3p	457	3900
Ca	[Ar]$4s^2$	4s 4p	423	4.60	4s 4p	657	385
Sr	[Kr]$5s^2$	5s 5p	461	4.98	5s 5p	659	21
Ba	[Xe]$6s^2$	6s 6p	554	8.36	6s 6p	791	0.01
Yb	[Xe]$4f^{14} 6s^2$	6s 6p	399	5.1	6s 6p	507–578	0.85

Table 15.5 Lowest singlet and triplet transitions of the alkaline earth elements. As discussed in the text, the lanthanide atoms have the structure of barium for their first 56 electrons and then their 4f shell starts filling (see Fig. 15.1). Thus the two outermost electrons are 6s, and their excitations resemble the other alkaline earths. The example of Yb is included here because it is among the simplest of these. Be warned that the accuracy of lifetimes is often only one significant figure.

Selection rules (Sec. 3.5) forbid the dipole transition decay of the triplet states of these alkaline earth atoms to the ground singlet state, so the lowest of these triplets is long-lived. However, unlike the case of helium these transitions do not terminate in the $n = 1$ level so their frequencies are much smaller, often in the visible. Moreover, the multi-electron interactions enhance the transition probabilities so these are easily driven by ordinary lasers, albeit with linewidths in the kHz domain rather than the MHz domain as is the case for the dipole allowed, strong transitions. Several of these are listed in Tab. 15.5. Note the rapid transition with increasing Z to shorter lifetimes of the lowest triplet state in the right-most column.

There are multitudinous transitions among the upper triplet states that terminate in the lowest-lying one, and of course, among the singlets as well. Moreover, the singlet–triplet coupling is stronger so that there are more intercombination transitions in the spectra. Finally, when they are singly ionized their states resemble those of first-column atoms and are therefore important in ion trapping studies.

The third column, topped by aluminum, typically has a rare gas core with two additional s electrons and one p electron. The usual excitations are from the p-orbital to higher states, and these can include both s and d states. Thus the spectra can be very complicated, and even more so for the higher-Z atoms. The singly ionized states of these atoms play an important role because their structure is essentially that of the alkaline earths with a narrow singlet triplet intercombination transition.

The rare-earth elements (lanthanides) form an interesting special case. As Fig. 15.1 shows, the 4f shell does not start to fill until the 6s shell is full, thereby making a barium core. These 4f electrons can have many complicated configurations, depending on how many electrons there are. For example, the ground state of erbium with twelve 4f electrons has $L = 5$ (H-state). Nevertheless, one of the two outermost 6s electrons can be excited in much the same way as for the alkaline earths, and therefore produce states that are quite similar. The lowest states of ytterbium are also included in Tab. 15.5. Note that yttrium shares a name similar to the other lanthanides but is not in this series. It has $Z = 39$, one more electron than strontium, and so is in the third column along with the other transition metals.

The end column consists of noble (or rare) gas atoms. These have their p and d shells filled and are therefore quite unreactive and also hard to excite optically. The first excitation typically requires elevating an electron from the filled p shell, and often its energy corresponds to the far UV. The resulting hole in the p shell leaves five p-electrons and the configurations can become quite complicated. Often a different scheme, is used called the Paschen notation for describing and labeling the states.

Appendices

15.A Paramagnetism

Paramagnetism is unlike most other topics in this book because it depends on thermal interactions among large numbers of atoms. It was first described by Langevin [154, 155] as the alignment of the unpaired electron spins by an external magnetic field. The magnetic interaction $-\vec{\mu} \cdot \vec{B}$ causes states with magnetic moments parallel to the magnetic field to have a lower energy than those aligned anti-parallel. Most of the atoms in a sample can settle in this lower state if thermal fluctuations are sufficiently small that the Boltzmann distribution allows them to reside there. Of course, some thermal interactions are required for atoms to reach this distribution leading to the magnetization of the medium. The typical interaction energy of an atom in a magnetic field is of the order of $\mu_B \mathcal{B}$, and if the ratio $\mu_B \mathcal{B}/k_B T$ becomes considerably smaller than 1, the medium becomes strongly magnetized.

The strength of the magnetization depends both on temperature and also on the number of spins in each atom that can be aligned. For instance, in noble-gas atoms all the electrons are in closed shells and thus cannot be aligned by an external field, so paramagnetism does not arise. For other atoms the valence electrons determine the magnitude of the magnetization. However, their angular momenta can be coupled in different ways and thus paramagnetism depends on the coupling of the electrons in the ground state. Here the rules as formulated by Hund are crucial (see Sec. 15.5).

The magnetic moment of an atom is determined by the sum of the magnetic moments arising from orbital motion and spin. It is given by $\vec{\mu} = -\mu_B(\vec{L} + 2\vec{S})/\hbar = -g_J\mu_B\vec{J}/\hbar$, where g_J is given by Eq. (11.7). The interaction energy caused by the Zeeman effect in a given state M_J is $\Delta E = g_J\mu_B M_J \mathcal{B}$ and the number N_M of atoms in state M_J is proportional to the Boltzmann factor $\exp(-\Delta E/k_B T)$. The magnetization is thus given by

$$M = \sum_{M_J=-J}^{J} -g_J\mu_B M_J N_M \qquad (15.25)$$

$$= N \frac{\sum_{M_J=-J}^{J} -g_J\mu_B M_J \exp\left(\frac{-g_J\mu_B M_J \mathcal{B}}{k_B T}\right)}{\sum_{M_J=-J}^{J} \exp\left(\frac{-g_J\mu_B M_J \mathcal{B}}{k_B T}\right)},$$

where the sum in the denominator provides the proper normalization of the sum in the numerator and N is the total number of atoms. In the case of high temperatures all exponential functions can be approximated by their first-order term, and for the susceptibility χ_P of paramagnetism this leads to

$$\chi_P = \frac{\mu_0 M}{\mathcal{B}} = \frac{N J(J+1) g_J^2 \mu_B^2 \mu_0}{3k_B T} = \frac{C}{T}, \qquad (15.26)$$

where C is called the Curie constant and is of the order of 0.1 K. The susceptibility is thus proportional to the effective magnetic moment $\mu_{\text{eff}} = g_J \sqrt{J(J+1)}\mu_B$, and this demands the evaluation of the configuration of the ground state of the atom under study.

For example, consider the case of the divalent iron ion Fe^{2+}. For this ion one has to couple six 3d electrons, and the resulting configurations are $^1S(2)$, $^1D(2)$, ^1F, $^1G(2)$, ^1I, $^3P(4)$, ^3D, $^3F(2)$, ^3G, ^3H, ^5D, where the number in parenthesis is the number of configurations with that symmetry. The energies of these states are shown in Fig. 15.5. As can be seen from the figure, the lowest states has a symmetry $^5D_4$. In order to determine the configuration with the lowest energy for any atom or ion, one requires Hund's rules.

Langevin's paramagnetism can be studied most conveniently for materials containing the ions of the rare-earth metals, also referred to as the lanthanide group. All these ions have an electron configuration [Xe] $4f^n$ $5s^2$ $5p^6$, where n is between 0 and 13. The outer 5s and 5p shells shield the 4f electrons from the surrounding ions in the crystal and thus the coupling of the electrons in each ion is nearly independent of its surrounding ions. The rules for determining the ground state are given in Sec. 15.5. In addition to the two Hund's rules mentioned in Sec. 15.5, there is a third Hund's rule, namely that for a shell that is less than half filled, the total angular momentum J is given by $|L - S|$, whereas for a shell that is more than half

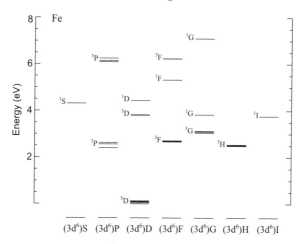

Figure 15.5 Energy levels for Fe^{2+} with the electronic configuration [Ar]3d^6. The splitting due to different J is in many cases too small to be visible. The lowest state is the $^5D_4$ term, as predicted by Hund's rules.

filled, J is equal to $L+S$. Using these rules in a consistent way leads to the known configuration of the ground states for the rare-earth metals.

Figure 15.6 shows the angular momenta of the ground state of the rare-earth metals. Using first rule Hund's, the maximum for the total spin S is just n times the spin of each electron, or $S = n/2$ with n the number of 4f electrons. For a less than half filled shell, this is correct. For a shell that is more than half filled with electrons having the same spin, this would cause the Pauli symmetrization principle to be violated since at least two electrons would have the same quantum numbers. So for a more than half filled shell, the spin becomes $S = (14-n)/2$. Using the second rule, the maximum for the orbital angular momentum is $L = \sum_{M=3}^{4-n} M$ for a less than half filled shell, and $L = \sum_{M=3}^{11-n} M$ for a more than half filled shell. In that case the Pauli principle dictates that L has a maximum value. Finally, the third rule can be used to determine J from the values of S and L. The resulting configuration can be seen on the second line of symbols above Fig. 15.6.

Using the values of S, L and J one can calculate g_J and thus the effective magnetic moment $\mu_{\text{eff}} = g_J \sqrt{J(J+1)}\mu_B$. The results are shown in Fig. 15.7 and are compared with the experimental values that are obtained by dividing the Curie constant C for the ions by $N\mu_0/3k_B$. As the figure shows, the comparison is rather good, supporting Langevin's idea that paramagnetism in these ions can be fully attributed to the alignment of the angular momenta of the valence electrons by the magnetic field. There is some discrepancy for Eu, but this can be attributed to the existence of different configurations with energies close to the lowest state, and thus to the occupation of these states at room temperature.

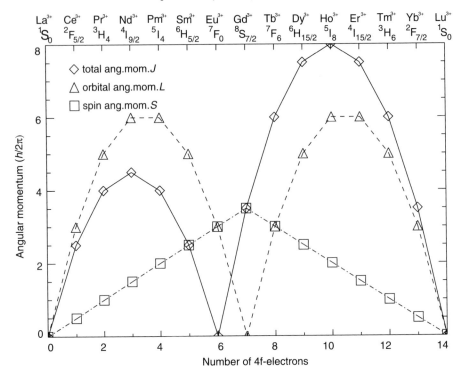

Figure 15.6 Total angular momentum J, orbital angular momentum L and spin angular momentum S for the trivalent ions of the rare-earth metals, as deduced from the Hund's rules. In all cases the configuration is [Xe] $4f^n\ 3s^2\ 3p^6$, where n is the number of 4f electrons. The first line on top of the figure shows the corresponding ion; the second line shows the term as derived from Hund's rules.

Similar results can be obtained for ions in the iron group, where the configuration is [Ar] $3d^n$. Here the 3d electrons are not shielded by electrons in higher shells and thus the presence of ions at nearest neighbor sites does have an effect on the alignment of the angular momenta. This effect is particularly strong for the orbital angular momenta: these angular momenta are often quenched by the crystal field. However, as shown in Fig. 15.8, there is good agreement for the effective magnetic moment using only the spin angular momenta for the magnetic moment. The rather good agreement shows that the orbital angular momenta play no role for paramagnetism for many of these ions in the iron group.

15.B The color of gold

Gold is the most noble of all metals. Because of its resistance to corrosion, it has been widely prized throughout the centuries. Furthermore, it is mechanically ductile and malleable, and because of its glittering color it is used for objects of art, jewelry, and coinage. More recently, it has become important in electronics because of its high electrical and thermal conductivity. In the field of nanotechnology,

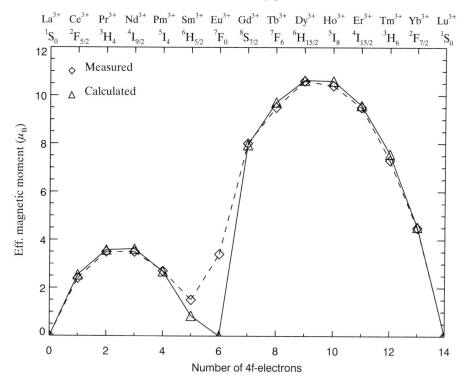

Figure 15.7 Calculated and measured effective magnetic moment μ_{eff} for the rare-earth metal ions.

nanoparticles of gold have become important because of their special optical properties since the absorption of its colloids is in the visible range. It is this special absorption property of gold that is the subject of this appendix.

Gold (Au) belongs to the group IB (column 11) or coinage elements in the periodic table, together with copper (Cu) and silver (Ag). All these elements have completed filled $(n-1)$ d shells containing 10 d electrons, and then one valence electron in the ns orbital, where $n = 4$ for Cu, five for Ag and six for Au. Therefore one might expect that all physical and chemical properties of these elements are similar. Here the focus is on the optical properties of these elements to show that they all behave quite differently. Since copper's 3d orbital has no radial node, this electron can penetrate the core of other electrons significantly, so its optical properties are strongly influenced. Thus this section concentrates on the differences between Ag and Au, and will not discuss the case of Cu further.

The coinage metals all have a single s valence electron so their ground state is $^2S_{1/2}$. Their excitation schemes are quite similar to that of the alkali-metal elements and proceed by promoting this valence electron from the ns orbital to the np orbital leading to the $^2P_{1/2,3/2}$ configurations. Excitation follows the usual selection rules for electric dipole transitions, as given in Tab. 4.1. The wavelengths of these transitions

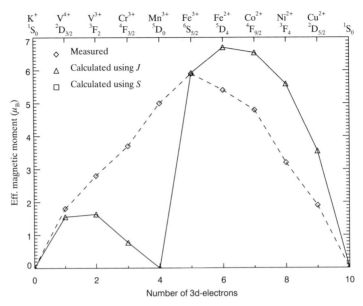

Figure 15.8 Calculated and measured effective magnetic moment μ_{eff} for the ions of the iron group. The calculated results using only the spin angular momentum S are in better agreement with the measured values as the results using the total angular momentum J. This shows that for these ions the orbital angular momentum L is quenched by the crystal field created by the surrounding ions.

are 338.29 and 328.06 nm for Ag, and 267.59 and 242.79 nm for Au, where the first (second) wavelength in each case is to the $^2P_{1/2}(^2P_{3/2})$ state.

This behavior is completely different for the solid state, where the optical properties of a metal are determined by the interactions with neighboring atoms. In a solid, the valence electrons of each atom are delocalized, forming bonds with other atoms that determine the mechanical properties of the metal. Thus the electronic levels of the atoms are shifted and broadened to form electronic bands. The excitation of electrons in a metal arise from transitions from the conduction band to higher-lying, non-occupied bands. Since the atomic levels are not isolated, the perturbations from nearby atoms cause mixing of electronic orbitals, and the selection rules for free atoms do not apply. It turns out that the optical properties of the coinage elements are mainly determined by the excitation of the electron in the $(n-1)$d band to the ns band. This implies a transition, $[(n-1)d^{10}ns]\,^2S_{1/2} \rightarrow [(n-1)d^9 ns^2]\,^2D_{3/2}$, that is parity-forbidden in the atomic case.

Despite the completely different optical excitations between the atomic and solid-state cases, the differences between the wavelengths for absorption in Ag and Au are still caused by atomic effects, mainly by relativistic effects of the valence electron. In Fig. 15.9 the reflectances of Cu, Ag, and Au are shown in the visible

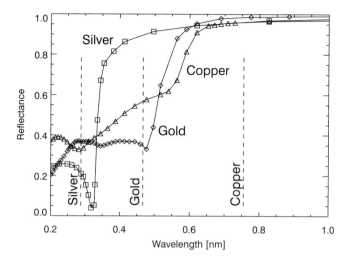

Figure 15.9 Reflectance of copper, silver, and gold in the visible region. The vertical dashed lines indicate the wavelength of the parity-forbidden $^2S_{1/2}$–$^2D_{3/2}$ transition in the different atoms. Data for the reflectance from Ref. [156].

range. Gold has a strong absorption below $\lambda = 500$ nm resulting in its characteristic yellow color. For silver the absorption starts at wavelengths below 350 nm, so nearly all light with wavelengths in the visible region is reflected, giving the metal a white color. The large shift of the absorption window between the two elements is remarkable given the similar electronic configuration of the atoms. However, Fig. 15.9 also shows the atomic transition wavelengths for the parity-forbidden $^2S_{1/2} \to {}^2D_{3/2}$ transitions, and although the upper edge of the absorption window is slightly shifted to larger wavelengths for both atoms, the large shift between the S $\to$ D transitions in the two atoms is obviously present in the atomic case. Note that the first allowed atomic excitation from the ground $^2S_{1/2}$ state is to the excited $^2P_{1/2,3/2}$ states, but these wavelengths play no role in the solid-state case.

Relativistic effects are proportional to Z^3 (see Sec. 9.3), where Z is the nuclear charge. Since $Z = 47$ for Ag and $Z = 79$ for Au, it might be that this causes the large difference between the transition wavelengths of the two elements. However, there can be quite significant relativistic effects but they are not always easy to identify. In the case of atoms with large Z there are three contributions to take into account.

First the velocity of electrons in orbitals leads to a relativistic mass correction:

$$m_{\text{rel}} = \frac{m}{\sqrt{1-(v/c)^2}}. \tag{15.27}$$

For hydrogen in the $n = 1$ state, the electron velocity $v = \alpha c$ ($\alpha \approx 1/137$) so the correction is small, although it plays an important role for the hyperfine shift (see Sec. 9.3.3). However, the electronic velocity is proportional to Z^2, and the relativistic correction becomes more important with increasing Z. Second, the relativistic correction to the mass leads to a shift of the orbital radius of the electron given by

$$\langle r \rangle_{\text{rel}} = \langle r \rangle \sqrt{1 - \left(\frac{\alpha Z}{n^*}\right)^2}, \qquad (15.28)$$

since the orbital radius $\langle r \rangle$ is proportional to $1/m$ (see Eq. (7.18)). Here n^* is the effective quantum number for the electronic orbital (see Eq. (10.1)). For s and p orbitals this leads to a reduction of radius of atomic orbits and thus a contraction of the core electron cloud. With the extent of the core electrons reduced by this relativistic effect, the charge of the nucleus is better screened for the outer d and f electrons and these orbitals expand. Finally, nuclear spin–orbit interaction is proportional to Z^3 (see Eq. (9.11)) and thus increases for increasing Z, but this effect will not be considered further here.

It is impossible to distinguish among these effects in atoms experimentally, so it is necessary to rely on theoretical calculations. One of the important aspects of these calculations is that the relativistic effects can be switched on and off by taking the speed of light $c = 1/\alpha$ in atomic units or $c = \infty$, respectively. The results for the 6s- and 5d-orbitals for $Z = 55$–100 are shown in Fig. 15.10. One clearly observes a contraction caused by relativistic effects for the 6s-orbital, whereas the 5d-orbitals do not show such a strong behavior. The dashed line in the figure shows the result for the 6s-orbital of Eq. (15.28) with $n^* = 1.6$. Although the formula agrees with the numerical results for lower Z-values, for $Z > 70$ the effects are much more pronounced than the formula predicts.

In particular, the relativistic ratio has a minimum for Au ($Z = 79$) and the effect does not become as strong as this again until $Z = 100$. The calculations are based on multiconfiguration Dirac–Fock (MCDF) calculations in the spirit of Sec. 15.6, and the results cannot easily be associated with particular effects. However, as is clear from the figure, the relativistic effects for Au are strongly increased by subtle interference between different terms. And it clearly marks the reduced energy between the 5d- and 6s-orbital leading to a shift of the transition from the UV to the visible range. Such a shift is nearly absent for Ag, and this effect is the main cause for the difference in color of Au and Ag.

This section has focused on one aspect of two elements, namely the optical properties of metallic Ag and Au, and how relativistic effects have a pronounced influence on them. However, relativistic effects are present in many other high-Z

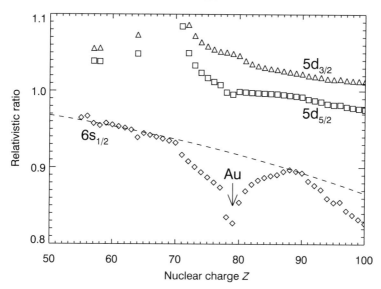

Figure 15.10 Ratio of the radius of the 6s- and 5d-orbital between the relativistic case $c = 1/\alpha$ and non-relativistic case $c = \infty$. The dashed line shows the effect of a simple scaling of the effect given by Eq. (15.28) with $n^* = 1.6$. Data from Ref. [157].

elements and have an influence on many physical and chemical properties of elements. The subject is part of the field of relativistic quantum chemistry, where complex, advanced numerical codes are used to explain the properties of atoms. By zooming in on only one aspect, it shows that such an advanced field of science can be used to explain everyday phenomena.

Exercises

15.1 Find the allowed terms of the configuration np^2 in L–S coupling. Find the largest value of the total orbital angular momentum for the configuration $n\ell^x$ of x equivalent electrons and plot it as a function of x, where $1 < x < 2(2\ell + 1)$.

15.2 Write the labels of the L-S coupled states, i.e., $^{2S+1}L_j$ that arise from the three configurations $nsnp$, $npnd$, and $(np)^2ns$.

15.3 Consider the Aufbau principle for sodium ($Z = 11$).
(a) Determine the electron configuration ($1s^2\ 2s^2\ \ldots$) of the ground state of Na. What is the degeneracy of this state? For this state determine L, S, and J, and show the corresponding spectroscopic notation.
The first excited states of Na are the $^2P_{1/2}$ and $^2P_{3/2}$ states.

(b) Determine the electron configuration of these states. What is the interaction that is mainly responsable for the splitting between these states? Why does this interaction not play a role for the ground state?

(c) What are the dipole selection rules for radiative excitations of these states from the ground state?

(d) Use these selection rules to determine all possible transitions among the lowest three states of Na.

15.4 The construction of the periodic system is described by the "Aufbau" principle. The first two rows of the periodic system are given by

Z	Element	Configuration	Term
1	H	...	...
2	He	...	...
3	Li	...	...
4	Be	...	...
5	B	...	...
6	C	...	...
7	N	...	...
8	O	...	...
9	F	...	...
10	Ne	...	...

(a) Use the "Aufbau" principle to determine the electron configuration of the ground states of the first 10 elements.

For the coupling of the angular momenta of the electron L–S coupling is used.

(b) Determine the term $^{2S+1}L_J$ for the ground state of the first four elements of the periodic system. Explain your results.

For the other elements the coupling of the angular momenta can lead to several possible ground-state terms.

(c) Determine all possible terms for the fifth element boron (B) and the sixth element carbon (C).

(d) Use Hund's rule to determine the term of the ground state of B and C.

15.5 To illustrate Hund's rules consider the ground state of zirconium. The electron configuration of zirconium (Zr, element 40) is given by

$$[Kr](5s^2 4d^2)^3 F_2.$$

(a) Give a short desciption of the Aufbau principle.

(b) Use the Aufbau principle to determine the total electron configuration of all 40 electrons of Zr.

(c) What are the spin angular momentum S, the orbital angular momentum L, and the total angular momentum J for the ground state of Zr?

(d) Determine all possible values of S by adding all individual electron spins s_i. According to Hund's first rule the ground state with the highest value of S has the lowest energy.

(e) Does the ground state of Zr meet the requirement of Hund's first rule? Explain your answer.

(f) Determine all possible values of L by adding all the individual angular momenta l_i.

(g) Determine all possible values of the total angular momentum J by coupling of L and S.

According to the second Hund's rule the lowest value of J has the lowest energy, if the valence shell is less than half filled. Otherwise, the highest value of J has the lowest energy.

(h) Does the ground state of Zr meet the requirement of the second Hund's rule? Explain your answer.

15.6 Consider the case of two identical d electrons. Naive considerations suggest that titanium's ground-state configuration of $1s^2\ 2s^2\ 2p^6\ 3s^2\ 3p^2\ 4s^2\ 3d^2$ provide singlet and triplet arrangements of the S, P, D, F, and G state. Construct a table like the top chart of Fig. 15.4. Select the largest value of M_L and identify it with the ^1G state. Remove these entries from your table and select from the remaining entries the largest M_L. Identify the state. Repeat the process until all states are identified. Compare your result with Tab. 15.3.

16
Molecules

16.1 Introduction

Up to here this book has considered in detail only neutral atoms, namely a single nucleus surrounded by its electrons. This chapter will consider molecules, systems with at least two nuclei bound together by their electrons. This change increases the number of possibilities considerably, since all atoms can be combined to form molecules in principle, and the number of atoms in a molecule is unlimited. The study and engineering of molecules forms the heart of chemistry, but this chapter does not intend to give a proper account of this field of science.

Moreover, the discussion here will be restricted to molecules having only two nuclei. The reason is that such diatomic molecules can be described with the techniques that have already been discussed earlier in this book, and therefore they are the atomic physicists' favorite kind of molecule. Furthermore, the atmosphere of the earth consists of more than 99% diatomic molecules, namely N_2 and O_2, and the most abundant molecule in the universe is H_2. Nevertheless, it is important to remember that polyatomic molecules give us life and reasons to live, even though they will not be discussed here.

The paired nuclei of diatomic molecules provide important additional degrees of freedom. In an atom, the center-of-mass and the nucleus nearly coincide, and in the center-of-mass frame the nucleus is nearly at rest. However, in a molecule the kinetic energy of the nuclei is not negligible since the center of mass is generally well separated from each of them, and the energy of their motion in the center-of-mass frame adds significantly to the total energy of the system.

Since molecules consist of at least three particles, the equations of motion cannot be solved exactly, neither in classical mechanics nor in quantum mechanics. In order to find solutions, one has to resort to approximations. One of the most important of these approximations is enabled by the large difference in mass between the nuclei and the electrons. This results in the nuclei moving much more slowly

16.1 Introduction

than the electrons in the center-of-mass frame. It therefore seems natural to treat the nuclei as fixed at a given separation R, calculate the energy of the system, and then repeat the calculations at a different value of R. The result provides a potential $V(R)$ created by the electrons in which the nuclei move. This approach is called the Born–Oppenheimer approximation and is of paramount importance for the description of molecules.

Consider two atoms at large internuclear distance. Since both atoms are neutral, there is no first-order, long-range force between them. However, any deviation away from a perfectly spherical charge distribution can cause each atom to induce a dipole moment in the other one, and the dipole-induced interaction, called the van der Waals interaction, can be attractive depending on the orientation of the dipoles with respect to each other. The interaction is weak, but is present even between widely separated atoms. The atoms can be drawn together until they are so close that the repulsion between their nuclei and electrons dominates. At very short distances the interaction between the atoms in a molecule is always repulsive.

At some intermediate internuclear distance, typically of the order of a few Bohr radii, there is a minimum in the potential where the two atoms can be bound to each other. The binding between the nuclei is caused by the interactions with the electrons, which are at least 2,000 times lighter than the nuclei. The forces between the electrons and the nuclei, and the mutual forces of the nuclei, are both Coulombic in nature and of the same order of magnitude, and this has important consequences.

The added degrees of freedom from the nuclear motion of molecules become evident from their observed spectra. Spectra of atoms contain sharp lines and atomic states are separated by energies of the order of a few eV, so radiation from transitions between them falls in the visible regime. Molecular spectra, as shown in Fig. 16.1, contain broad features, and closer inspection reveals that they are from groups of very closely spaced, sharp spectral lines. These bands of sharp lines are spaced much more closely and occur more frequently than atomic spectral lines. This chapter will lead to a better understanding of these spectral features.

The chapter begins with a heuristic description of the contributions of nuclear motion to molecular energy. This motion leads to vibrational and rotational energy and is crucial for understanding molecular structure. Next the quantum mechanical treatment of the nuclear motion is discussed in terms of the Born–Oppenheimer approximation that allows for a separation of the electronic and nuclear motion. Atoms can be bound in a molecule by different mechanisms, and three of them will be discussed in some detail. These are the van der Waals interaction mentioned above, the ionic bond, and the covalent bond. The chapter concludes with a description of optical transitions in molecules.

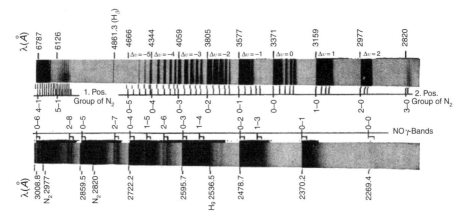

Figure 16.1 Molecular spectrum of N_2. Because of the large number of ro-vibrational levels for the molecule, its spectrum is rather dense. Note the narrow line for atomic hydrogen at 486.13 nm in the top spectrum. (Figure from Ref. [158].)

16.2 A heuristic description

As discussed in the previous section, the mass of the electron is much smaller than the mass of the nuclei. The mass ratio M/m is in the range of 2,000–200,000, and since their momenta in the center-of-mass frame are typically comparable, the nuclear speeds are much smaller than the electron speeds. Because the electrons are much faster, they can orbit the nuclei several hundred times before the nuclei move appreciably. The effect of the electrons on the motion of the nuclei can thus be treated in an averaged way by fixing the internuclear distance R and calculating the electronic energy as mentioned above. Finding the resulting molecular potential $V(R)$ by this Born–Oppenheimer approximation is a formidable task in the field of quantum chemistry, but for the molecules under consideration here it is assumed that this potential can be found or approximated.

Figure 16.2 shows the generic shape of a diatomic molecular potential having only one coordinate R in which the nuclei move. At long range the interaction is attractive because of the van der Waals interaction, and at short range it is repulsive because of the overlap of the electron clouds. The potential has a minimum at $R = R_{eq}$, called the equilibrium separation of the nuclei. Near this minimum the potential can be approximated harmonically with an effective spring constant k_{eq}, as indicated by the dashed curve in Fig. 16.2. The relative motion of the nuclei can then be treated as harmonic oscillation with a frequency $\omega_N = \sqrt{k_{eq}/\mu_N}$, where $\mu_N \equiv M_A M_B/(M_A + M_B)$ is the reduced mass of the nuclei (not to be confused with the nuclear magneton of earlier chapters). The energies of the lowest states are thus $E_{vib} = \hbar\omega_N(v + 1/2)$ where v is the vibrational quantum number. The electrons can be considered bound to the nuclei with an effective "spring constant" for the

16.2 A heuristic description

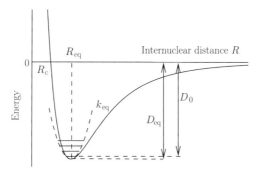

Figure 16.2 Generic shape of a molecular potential $V(R)$ in which the nuclei move. The potential has a minimum D_{eq} at the equilibrium distance R_{eq}, where the potential can be approximated by a harmonic potential $V(R) = -D_{eq} + 1/2 k_{eq}(R - R_{eq})^2$ where k_e is the effective spring constant. The dissociation energy D_0 of the ground state is reduced by $1/2\hbar\omega_N$ with respect to D_{eq} because of the zero-point motion of the vibration. In addition, the classical inner turning point R_c at threshold is indicated.

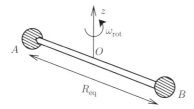

Figure 16.3 Rotation of a diatomic molecule around the axis going through the center-of-mass of the molecule, where the two nuclei are fixed at a distance R_{eq}.

orbiting electrons that is about the same as k_{eq} for the nuclei. This is not surprising since the forces are comparable, and it leads to a ratio of the kinetic energies for the lowest states of $\omega_{el}/\omega_N \sim \sqrt{\mu_N/m}$, which is of the order of 50–500. This ratio justifies *a posteriori* the approximation that starts this section. The total energies are related to the kinetic energy through the virial theorem.

In addition to the vibrational motion of the nuclei, diatomic molecules can also rotate about their centers of mass as shown in Fig. 16.3. For angular momentum $\vec{N}$ quantized in units of $\hbar$, the rotational kinetic energy is $E_{rot} = \hbar^2 N(N+1)/2I$, where the moment of inertia is $I = \mu_N R_{eq}^2$ with R_{eq} the distance between the nuclei. Since the kinetic energy of atomic electrons bound to a nucleus at distance R_{eq} can be written as $E_e \sim \hbar^2/mR_{eq}^2$, the ratio of these energies is then $E_e/E_{rot} \sim M/m$. This shows that E_{rot} is a factor 2,000–200,000 smaller than E_e and a factor 50–500 smaller than the vibrational energies.

Since a molecule can be in different vibrational and rotational states characterized by v and N respectively, as well as different electronic states, the number of possible transitions is much greater than for single atoms, although there are selection rules for some of these transitions, as discussed in Sec. 16.6.

Electronic transitions in molecules have optical frequencies of $\sim 10^{15}$ Hz as in atoms. For vibrational transitions, the frequencies are in the 10^{13} Hz range or a few thousand cm^{-1}, where cm^{-1} is a unit often used in molecular spectroscopy corresponding to ~ 30 GHz. Finally, rotational transitions are in the range of 1–100 cm^{-1} or 10^{10}–10^{12} Hz. The result is a large number of closely spaced transition frequencies that comprise the bands of Fig. 16.1.

In addition to the widely cited Ref. [2] where Bohr first presented his model of atomic structure, he published two other, lesser known papers in 1913 [159, 160]. In the third paper of this trilogy he proposed a model of molecules that adhered to his original ideas of atoms, namely the existence of stationary states and the requirement that transition involve light with frequency given by the Planck formula. He considered the nuclei to be fixed at an internuclear distance R and the two electrons moving in circular orbits around the internuclear axis at certain positions z_1 and z_2. Note that the fixation of the nuclei on the stationary internuclear axis can be considered as a precursor to the Born–Oppenheimer approximation discussed in Sec. 16.3.1.

For the circular orbits he used the quantization rule $mvr = n\hbar$ that relates the radius r and velocity v of the electron orbit. Since the nuclei are fixed, the total kinetic energy contains only that of the two electrons. The total potential energy contains six terms, namely the repulsions between the two nuclei and that between the two electrons, and four terms representing their respective attractions. The total energy is minimized as a function of R using the electron radius r as an optimization parameter. Bohr considered four different situations for the electron configurations with ϕ the azimuthal angle between them. The energies are shown in Fig. 16.4 as a function of R_{eq}. Although the model predicts a binding energy that is roughly half the actual binding energy in H$_2$, it does predict nuclear spacing rather accurately.

Recently, Bohr's model of the molecule has been revisited in Ref. [162]. It has also been shown that the model can be improved by inclusion of several perturbations such as the centrifugal barrier between each electron and its nearest nucleus [163, 164]. These extensions show that the results of such semi-classical models are in better agreement with the experimental values than the Heitler–London treatment of Sec. 17.3. However, it is clear that such models are based on ad-hoc assumptions and that a full quantum mechanical treatment of diatomic molecules is required.

16.3 Quantum description of nuclear motion

The quantum description of the nuclear motion of a diatomic molecule is based on the time-independent Schrödinger equation. Its eigenfunctions depend on the coordinates $\vec{R}_{A,B}$ of the nuclei, and $\vec{r}_i$ of the electrons with $i = 1, 2, \ldots, \sigma$ where σ is

16.3 Quantum description of nuclear motion

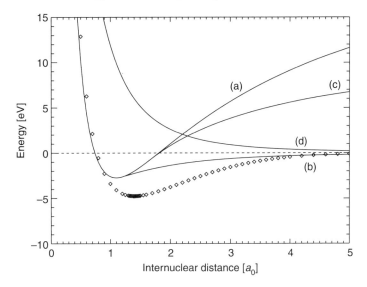

Figure 16.4 Potential energy of the H_2 molecule as a function of the internuclear distance in units of a_0 for four configurations of the electrons: (a) $z_1 = z_2 = 0$ and $\phi = \pi$, (b) $z_1 = -z_2$ and $\phi = \pi$, (c) $z_1 = z_2 \neq 0$, and $\phi = \pi$, and (d) $z_1 = -z_2$ and $\phi = 0$, where $z_{1,2}$ is the distance of the electrons 1 and 2 to the center of the molecule and ϕ the azimuthal angle difference between them. The symbols show the most recent calculations of the potential energy [161].

the number of electrons. For $\sigma = Z_A + Z_B$ the molecule is neutral. The translational kinetic energy of the molecule can then be treated separately by considering the motion of the nuclei in the center-of-mass frame. This allows replacing the six coordinates $\vec{R}_{A,B}$ of the nuclei by just the internuclear separation $\vec{R} \equiv \vec{R}_A - \vec{R}_B$. The Schrödinger equation then becomes

$$\mathcal{H}_{\text{mol}}\Psi(\vec{R},\vec{r}_1\ldots\vec{r}_\sigma) = E\Psi(\vec{R},\vec{r}_1\ldots\vec{r}_\sigma), \tag{16.1}$$

with the Hamiltonian given by

$$\mathcal{H}_{\text{mol}} \equiv T_N + T_e + V_{\text{int}} = -\frac{\hbar^2}{2\mu_N}\nabla_R^2 - \frac{\hbar^2}{2m}\sum_{i=1}^{\sigma}\nabla_i^2 + V_{\text{int}}. \tag{16.2}$$

Here T_N is the kinetic energy of the two nuclei in the center-of-mass frame, T_e is the kinetic energy of the electrons, and the interaction potential is

$$V_{\text{int}} = \frac{e^2}{4\pi\varepsilon_0}\left(\frac{Z_A Z_B}{|\vec{R}_A - \vec{R}_B|} + \sum_{i<j=1}^{\sigma}\frac{1}{|\vec{r}_i - \vec{r}_j|} - \sum_{i=1}^{\sigma}\frac{Z_A}{|\vec{R}_A - \vec{r}_i|} - \sum_{i=1}^{\sigma}\frac{Z_B}{|\vec{R}_B - \vec{r}_i|}\right). \tag{16.3}$$

This equation cannot be solved exactly, since it involves $3 \times (\sigma + 1)$ independent coordinates.

16.3.1 Born–Oppenheimer approximation

The Born–Oppenheimer approximation introduced in Sec. 16.2 uses the Schrödinger equation only for the electrons with a fixed $\vec{R}$, given by

$$[T_e + V_{\text{int}}]\phi_q(\vec{R};\vec{r}_1\ldots\vec{r}_\sigma) = E_q(R)\phi_q(\vec{R};\vec{r}_1\ldots\vec{r}_\sigma), \tag{16.4}$$

where the electronic eigenfunctions $\phi_q(\vec{R};\vec{r}_1\ldots\vec{r}_\sigma)$ are normalized over the electronic coordinates $\vec{r}_1\ldots\vec{r}_\sigma$. Here the semicolon after $\vec{R}$ signifies that the wavefunctions ϕ_q are calculated with $\vec{R}$ held constant so that they are only parametrically dependent on it. Although finding these wavefunctions can be a formidable task, it is assumed that this can be accomplished and that they form a complete set at each value of $\vec{R}$. For a diatomic molecule, the eigenvalues $E_q(R)$ depend on only the distance $R \equiv |\vec{R}|$.

The trial solution for the Schrödinger equation given in Eq. (16.1) can be written as

$$\Psi(\vec{R},\vec{r}_1\ldots\vec{r}_\sigma) = \sum_q F^{(q)}(\vec{R})\phi_q(\vec{R};\vec{r}_1\ldots\vec{r}_\sigma), \tag{16.5}$$

where $F^{(q)}(\vec{R})$ is the wavefunction for the motion of the nuclei and the superscript (q) signifies that the eigenvalue E_q needs to be invoked to find solutions.

The electronic part of Eq. (16.5) can be eliminated by substituting $\Psi(\vec{R},\vec{r}_1\ldots\vec{r}_\sigma)$ of Eq. (16.5) into Eq. (16.1), multiplying both sides by ϕ_s^* on the left, integrating over the electronic coordinates $(\vec{r}_1\ldots\vec{r}_\sigma)$, and using Eq. (16.4) for the electronic states, to find

$$\sum_q \int d\vec{r}_1\ldots d\vec{r}_N\, \phi_s^*\, T_N\, F^{(q)}(\vec{R})\, \phi_q = [E - E_s(R)]\, F^{(s)}(\vec{R}). \tag{16.6}$$

For the kinetic energy operator of the nuclei T_N one has to calculate

$$\nabla_R^2(F^{(q)}\phi_q) = \nabla_R^2(F^{(q)})\,\phi_q + 2\,\nabla_R(F^{(q)})\,\nabla_R(\phi_q) + F^{(q)}\,\nabla_R^2(\phi_q). \tag{16.7}$$

The heart of the Born–Oppenheimer approximation consists of omitting the last two terms of Eq. (16.7), because they are negligibly small compared with the first one, since ϕ_q depends only weakly on R. Then the Schrödinger equation for the nuclei reduces to

$$\left[-\frac{\hbar^2}{2\mu_N}\nabla_R^2 + E_s(R)\right]F^{(s)}(\vec{R}) = EF^{(s)}(\vec{R}). \tag{16.8}$$

Solution of this equation leads to rovibrational states in the potential $E_s(R)$.

16.3.2 Nuclear eigenfunctions

Apart from the potential $E_s(R)$, the remaining nuclear part of the Schrödinger equation, Eq. (16.8), can be written as a sum of two terms. The first one depends on the internuclear distance R only and the second one on the rotation of the molecule in the laboratory frame:

$$-\frac{\hbar^2}{2\mu_N}\nabla_R^2 = -\frac{\hbar^2}{2\mu_N}\frac{\partial}{\partial R}\left(R^2\frac{\partial}{\partial R}\right) - \frac{\vec{N}^2}{2\mu_N R^2}. \tag{16.9}$$

The term depending on the rotational angular momentum $\vec{N}$ constitutes a rotational barrier for the nuclei, similar to the rotational barrier for the electron in the hydrogen atom.

When the sum of the angular momenta of the electrons is zero, the total angular momentum of the molecule usually denoted by $\vec{J}$ reduces to just $\vec{N}$. In the absence of external torques, $\vec{N}$ is conserved so the solutions for the nuclear part are eigenfunctions of $\vec{N}^2$ and N_Z with eigenvalues $N(N+1)\hbar^2$ and $M_N\hbar$, respectively, and Z is the coordinate in the laboratory frame.

The radial part of the solutions depends on the detailed shape of the potential $E_s(R)$. However, the equation has the form of Eq. (7.1), with the reduced mass of the electron replaced by the reduced mass of the nuclei μ_N, and the Coulomb potential $V(r)$ replaced by the molecular potential $E_s(R)$. The solutions can therefore be written as

$$F^{(s)}(\vec{R}) = \frac{1}{R}\mathcal{F}^{(s)}_{v,N}(R)Y_{NM_N}(\Theta,\Phi), \tag{16.10}$$

so that the radial eigenfunction $\mathcal{F}^{(s)}_{v,N}(R)$ satisfies the equation

$$\left(-\frac{\hbar^2}{2\mu_N}\frac{d^2}{dR^2} + \frac{\hbar^2 N(N+1)}{2\mu_N R^2} + E_s(R)\right)\mathcal{F}^{(s)}_{v,N}(R) = E_{s,v,N}\mathcal{F}^{(s)}_{v,N}(R). \tag{16.11}$$

If $E_s(R)$ is known, this equation can be solved, at least numerically.

By contrast, when the angular momenta of the electrons $\vec{L}$ is non-zero, the total angular momentum $\vec{J}$ is given by $\vec{J} = \vec{N} + \vec{L}$, and therefore $\vec{N}$ is not a conserved quantity, as discussed in Sec. 16.5. For simplicity, the spin angular momenta will not be considered here, although they could be included as well, but even with this simplification some care is needed. The rotational barrier is now proportional to $\vec{N}^2 = (\vec{J}-\vec{L})^2$, and the nuclear part of the Schrödinger equation, Eq. (16.8), becomes dependent on the electronic coordinates through $\vec{L}$. This compromises the Born–Oppenheimer approximation that enabled a separation of the electronic and nuclear coordinates. In particular, $\vec{L}$ can produce coupling between the different states (s) and it becomes very complicated to find solutions for Eq. (16.5). By requiring that the individual terms $F^{(s)}(\vec{R})\phi_s(\vec{R}; \vec{r}_1 \ldots \vec{r}_\sigma)$ of Eq. (16.5) remain uncoupled, the separation between coordinates can be restored.

The expectation value of the rotational term $\vec{N}^2$ of the Hamiltonian needs to be evaluated using

$$\vec{N}^2 = (\vec{J} - \vec{L})^2 = \vec{J}^2 + \vec{L}^2 - 2(J_x L_x + J_y L_y + J_z L_z). \tag{16.12}$$

As discussed in Sec. 16.5, the component L_z along the internuclear axis in the molecular frame is a conserved quantity denoted as $\Lambda \hbar$, and since the rotational angular momentum $\vec{N}$ is perpendicular to the internuclear axis, the eigenvalue of J_z and L_z are the same and given by $\hbar \Lambda$. Then $-2J_z L_z$ from Eq. (16.12) combines with the L_z^2 term from $\vec{L}^2$ to make $-\Lambda^2 \hbar^2$. The other components L_x and L_y have zero expectation values. Since the operator $\vec{J}$ depends only on the nuclear angles (Θ, Φ), it does not work on the electronic wavefunctions $\phi_s(\vec{R}; \vec{r}_1 \ldots \vec{r}_\sigma)$. Then the expectation value of the rotational term of Eq. (16.12) on the electronic wavefunctions devolves to

$$\langle \phi_s | \vec{N}^2 | \phi_s \rangle = \vec{J}^2 + \langle L_x^2 + L_y^2 \rangle - \Lambda^2 \hbar^2. \tag{16.13}$$

Note that $\vec{J}^2$ only works on the angles Θ and Φ and not on the electronic wavefunctions ϕ_s, so its expectation value is just $\vec{J}^2$, since the wavefunctions ϕ_s are normalized. The middle term of Eq. (16.13) depends on the state (s) and can be included in the potential energy $E_s(R)$.

For $\vec{L} \neq 0$ the solutions to the nuclear part can now be expressed in terms of eigenfunctions of $\vec{J}^2$, J_Z, and L_z, and are given by

$$F^{(s)}(\vec{R}) = \frac{1}{R} \mathcal{F}_{v,J}^{(s)}(R) \mathcal{Y}_{JM_J}^{(\Lambda)}(\Theta, \Phi). \tag{16.14}$$

The radial eigenfunction $\mathcal{F}_{v,J}^{(s)}(R)$ satifies the equation

$$\left(-\frac{\hbar^2}{2\mu_N} \frac{d^2}{dR^2} + \frac{\hbar^2 \left[J(J+1) - \Lambda^2 \right]}{2\mu_N R^2} + E'_s(R) \right) \mathcal{F}_{v,J}^{(s)}(R) = E_{s,v,J} \mathcal{F}_{v,J}^{(s)}(R), \tag{16.15}$$

where the effective potential energy is given by

$$E'_s(R) = E_s(R) + \frac{\langle L_x^2 + L_y^2 \rangle}{2\mu_N R^2} \tag{16.16}$$

and the last term is added to $E_s(R)$ to simplify the notation. Since the reduced mass μ_N of the nuclei is much larger than the electron mass m, the last term is smaller than $E_s(R)$ by the ratio m/μ_N and therefore negligible. In the remainder of the chapter, only $E_s(R)$ will be used and the last term will no longer be discussed.

On the other hand, the angular wavefunctions $\mathscr{Y}_{JM_J}^{(\Lambda)}(\Theta,\Phi)$ of Eq. (16.16) are complicated functions, since they have to be simultaneous eigenfunctions of $\vec{J}^2$, J_z, and L_z. Solutions can be found in several textbooks on molecular physics (see for instance Ref. [165]). Moreover, in the simple case of $\Lambda = 0$ they reduce to the familiar spherical harmonics of Eq. (16.10).

16.3.3 Rovibrational energies

If $E_s(R)$ is known, Eq. (16.15) can be solved numerically with the same techniques as described in Sec. 10.4. However, the vibrational and rotational motions of the molecule are not strongly coupled and in a lowest–order approximation can be treated separately. Much can be learned from approximating the potential $E_s(R)$ near the equilibrium distance by a harmonic potential, as shown in Fig. 16.2. Writing

$$E_s(R) \approx E_s(R_{eq}) + \tfrac{1}{2}k_{eq}(R - R_{eq})^2, \quad \text{with} \quad k_{eq} = \left.\frac{d^2 E_s(R)}{dR^2}\right|_{R=R_{eq}}, \qquad (16.17)$$

the vibrational energy E_{vib} can be found from the harmonic oscillator model of App. 5.A and yields

$$E_{vib} = \hbar\omega_N(v + \tfrac{1}{2}), \quad \text{with} \quad \omega_N = \sqrt{\frac{k_{eq}}{\mu_N}} \qquad (16.18)$$

and v is an integer. The rotational energy E_{rot} can be found by treating the molecule as a rigid rotor and evaluating the moment of inertia at the equilibrium distance R_{eq} (see Fig. 16.3). This yields

$$E_{rot} = B\left(J(J+1) - \Lambda^2\right), \quad \text{with} \quad B \equiv \frac{\hbar^2}{2\mu_N R_{eq}^2} = \frac{\hbar^2}{2I}, \qquad (16.19)$$

where B is called the rotational constant and $J \geq \Lambda$. The total energy becomes

$$E_{s,v,J} = E_s(R_{eq}) + E_{vib} + E_{rot}. \qquad (16.20)$$

This relation shows that the dissociation energy D_0 of the molecule in its ground state is reduced from $E_s(R_{eq}) = D_{eq}$ by an amount $\tfrac{1}{2}\hbar\omega_N$, because of the zero-point energy of the vibrational motion.

As discussed in App. 16.A, the Morse potential of Eq. (16.42) is a better representation of the molecular potential $E_s(R)$ than the harmonic approximation. The additional advantage of a Morse potential is that nearly exact analytical solutions for the vibrational levels can be found [166]:

$$E_{vib} = \hbar\omega_N\left[(v+\tfrac{1}{2}) - \beta(v+\tfrac{1}{2})^2\right], \qquad (16.21)$$

with β a small parameter. The anharmonicity constant is given by $\beta = \hbar\omega_N/4D_{eq}$. Using the procedure outlined in App. 16.A, the parameters of the Morse potential can be found with only a few potential energy values, and Eq. (16.21) can subsequently be used to find the energies of the vibrational levels.

The analysis of this chapter so far assumes that the rotation and/or vibration does not influence the rotational frequency. However, because of the anharmonicity of the potential, higher rotational or vibrational states lead to an increase of the equilibrium distance between the two nuclei and thus to a lowering of the rotational energy. As shown in Ref. [69], the resulting energy of the rovibrational states can be corrected using

$$E_{s,v,J} = -D_{eq} + \hbar\omega_N\left[(v+1/2) - \beta(v+1/2)^2\right] + \frac{\hbar^2}{2\mu_N R_{eq}^2}\left(J(J+1) - \Lambda^2\right)$$
$$-a(v+1/2)J(J+1) - bJ^2(J+1)^2, \qquad (16.22)$$

where the terms on the right-hand side of the first line represent the dissociation energy, the vibrational energy including an anharmonic correction, and the rotational energy for the rigid rotator. The terms on the second line are corrections for the rotational energy, where the term proportional to a is the rotation–vibration coupling and the term proportional to b is a rotational correction. The terms a and b are given by [69]

$$a = \frac{3\hbar^3\omega_N}{4\mu_N\alpha R_{eq}^3 D_{eq}}\left(1 - \frac{1}{\alpha R_{eq}}\right) \quad \text{and} \quad b = \frac{\hbar^4}{4\mu_N^2\alpha^2 R_{eq}^6 D_{eq}}. \qquad (16.23)$$

Equation (16.22) provides an accurate analytical description of the energies of rovibrational states.

16.4 Bonding in molecules

Atoms are bonded to form molecules by their electrons. There are different types of bonding, and the electronic configuration of the atoms determines which type dominates, although more than one may be relevant. Core electrons are typically strongly bound to their nucleus and thus do not play a significant role in molecular bonding. Instead, the nature of the bonding depends on the amount of filling of the last shell of the valence electrons of each atom. The three types of bonding that will be discussed below are:

1. van der Waals molecules, where the attraction arises only from the induced polarization (e.g. dipole moments). These are the weakest types of bonds, and often the potential is so shallow that there are only a few bound rovibrational states.

2. Covalently bound molecules, where the attraction is mediated by electrons shared by both atoms. These bonds have intermediate strength and the potential depth is often a few eV with multitudinous bound states.
3. Ionic bonds, where there is an actual charge exchange and the resulting oppositely charged ions are bound by a Coulomb attraction. These form the strongest kinds of bonds and the deepest potential wells.

16.4.1 The van der Waals interaction

Consider the interaction between two noble-gas atoms to form a molecule such as He_2. As in any other molecule, the van der Waals interaction is attractive at long range. However, when the two nuclei approach each other, the four electrons cannot all remain in the $n = 1$ state because of the Pauli principle. The promotion of two electrons to the $n = 2$ state requires a great deal of energy and leads to a repulsive interaction between the two atoms. So in this case the induced dipole interaction is the only mechanism leading to binding.

The interaction between such closed-shell atoms is dominated by the van der Waals interaction at all distances, and it is clear that covalent or ionic bonding does not play a role. For instance, for He the first excitation is at 19.6 eV and the ionization potential (IP) is 24.6 eV, whereas the electron affinity (EA) is essentially zero. This large energy deficit has to be counterbalanced, but at a distance $\sim a_0$ the overlap between the electrons clouds is already very large, leading to a strong repulsion.

The potential in this case can be written as (note that for ground states the potential $E_s(R)$ is replaced by the more familiar notation $V(R)$)

$$V(R) = A_r e^{-aR} - \frac{C_6}{R^6}, \tag{16.24}$$

where A_r and a are constants characterizing the repulsion at short distances, and C_6 is the van der Waals coefficient. (The notation is simplified here by using atomic units.) As first derived by London [167], the C_6 coefficient of the interaction of two atoms A and B can be approximated by

$$C_6 = -\frac{3}{2} \frac{(\text{IP})_A (\text{IP})_B \alpha_A \alpha_B}{(\text{IP})_A + (\text{IP})_B} = -\frac{3}{4} (\text{IP}) \alpha^2 \quad (A = B), \tag{16.25}$$

where α_i is the polarizability of atom i, and the last equality holds for identical atoms. The coefficients A_r and a are determined by the overlap of the electron clouds at short distances, and this is complicated to determine *a priori*. An

		He	Ne	Ar	Kr	Xe
	IP (eV)	24.59	21.56	15.76	14.00	12.13
	α (Å^3)	0.2051	0.395	1.64	2.48	4.04
He	A_r (keV)	1.657	0.909	1.307	0.717	1.665
	a (Å^{-1})	5.05	4.28	3.79	3.49	3.44
	C_6 (eV Å^6)	0.879	2.8	9.5	13.1	20.4
Ne	A_r (keV)		4.583	6.648	3.604	5.865
	a (Å^{-1})		4.61	4.12	3.82	3.76
	C_6 (eV Å^6)		5.4	18.3	25.4	39.5
Ar	A_r (keV)			8.575	5.224	8.521
	a (Å^{-1})			3.62	3.33	3.27
	C_6 (eV Å^6)			62.0	85.8	134
Kr	A_r (keV)				2.854	4.630
	a (Å^{-1})				3.03	2.98
	C_6 (eV Å^6)				119	185
Xe	A_r (keV)					7.555
	a (Å^{-1})					2.92
	C_6 (eV Å^6)					287

Table 16.1 Parameters C_6, a, and A_r for noble-gas atom pairs. Using these parameters the binding energy D_{eq} and equilibrium distance R_{eq} can be calculated [168].

alternative way to estimate these parameters is by using the equilibrium distance R_{eq} and the vibrational "spring constant" k_{eq} to fix these parameters:

$$\left.\frac{dV(R)}{dR}\right|_{R=R_{eq}} = 0 \quad \text{and} \quad \left.\frac{d^2V(R)}{dR^2}\right|_{R=R_{eq}} = k_{eq} \tag{16.26}$$

In combination with Eq. (16.24) this leads to:

$$a = \frac{k_{eq}R_{eq}^8 + 42C_6}{6C_6 R_{eq}} \quad \text{and} \quad A_r = \frac{6C_6 e^{aR_{eq}}}{aR_{eq}^7}. \tag{16.27}$$

In this way the interaction potential is fully determined, and the resulting dissociation energy D_{eq} agrees rather well with the experimental determined values. A list of parameters for noble-gas atom pairs is given in Tab. 16.1.

16.4.2 Covalent bonding

The situation is completely different for atoms with partially filled outer shells, the so-called open-shell atoms. Suppose that the number of electrons in the outer shells of the two atoms exceeds the number needed to fill one of the shells. Then the atoms can share some of the electrons, giving each of them a pseudo-filled shell, and form a covalent bond. For example, two halogen atoms can each share one of the electrons from the outer shell of its "partner" atom. In the case of two chlorine atoms forming Cl_2, for example, each atom keeps six of its original seven outer electrons while sharing the last one. The result is two quasi-filled shells of eight electrons each. For oxygen, each atom would share two electrons, for nitrogen three, etc. Thus the two nuclei could approach each other without the promotion of electrons, and each would have a "filled" outer shell after the two atoms are united. Valence electrons can thus be shared between the two nuclei resulting in a much stronger attraction than the van der Waals interaction. This covalent bonding occurs in the H_2 molecule, but because this is discussed in considerable detail in Chap. 17, it is only introduced here.

16.4.3 Ionic bonding

The case of ionic bonding that dominates for the alkali halide molecules such as NaCl results from the strong electron affinity of the halogen atoms and the low ionization potential of the alkali-metal atoms. This causes the transfer of the weakly bound valence electron of an alkali-metal atom to a halogen atom where it fills the outer shell. The interaction between the two resulting, oppositely charged ions becomes a Coulomb attraction that is much stronger than the covalent or van der Waals bonding and leads to a very stable molecule.

Because the ionic bond derives from a Coulomb attraction between two ions, this interaction is strong at short distances. It compensates the energy deficit between the ionization potential of one atom and the electron affinity of the other atom. Consider the bond in NaCl. Here, Na has an IP = 5.14 eV and Cl has an EA = −3.65 eV. Thus the Coulomb interaction has to compensate for a deficit of 1.49 eV or 0.055 au, and this is achieved at a distance as large as $18a_0$. Because of the increase in Coulomb attraction for shorter distances, the bonding for NaCl is strong.

Rittner [169] has developed an electrostatic model for all the alkali halide molecules. This model takes into account the short-range repulsion, the long-range van der Waals interaction, and the Coulomb attraction between the ions, and it also accounts for the polarization of each ion by the other ion and the induced dipole–induced dipole term. The induced polarization in this model is given by

$$\mu_{ind} = \frac{e(\alpha_+ + \alpha_-)}{R^2} + \frac{4e\alpha_+\alpha_-}{R^5}, \qquad (16.28)$$

and the energy shift caused by this induced polarization is written $-\mu_{\text{ind}}/2R^2$. The potential in this model is thus given by

$$V(R) = A_r e^{-aR} - \frac{C_6}{R^6} - \frac{C_8}{R^8} - \frac{e^2}{R}$$
$$- \frac{e^2(\alpha_+ + \alpha_-)}{2R^4} - \frac{2e^2\alpha_+\alpha_-}{R^7} + (\text{IP}) + (\text{EA}). \tag{16.29}$$

Table 16.2 gives values for the IP and polarizability of the positive alkali ions along with the EA and polarizability of the negative halogen ions. For each pair, the C_6, C_8, R_{eq} and ω_N values are also given.

The short-range coefficients A_r and a can be derived using the procedure described in Sec. 16.4.1 that yield in this model:

$$a = \frac{k_e - V''_{\text{LR}}(R_{\text{eq}})}{V'_{\text{LR}}(R_{\text{eq}})} \quad \text{and} \quad A_r = \frac{V'_{\text{LR}}(R_{\text{eq}})e^{aR}}{a}, \tag{16.30}$$

where $V_{\text{LR}}(R)$ is the long-range interaction in the Rittner model and it contains all the terms on the right-hand side of Eq. (16.29), apart from the short-range interaction $A_r \exp(-aR)$. From these parameters the dissociation energy D_{eq} can be calculated, and the values agree rather well with the experimental values listed in Tab. 16.2. The polarizabilities calculated using Eq. (16.28) are somewhat lower than the experimental values, which shows that at R_{eq} the overlap between the electron clouds is already so strong that the long-range polarizability model is no longer adequate.

16.5 Electronic states of molecules

Perhaps the most important characteristics of the electronic properties of molecules derive from the presence of more than one nucleus, and therefore more than one center of force. Since the nuclear electric field is not spherically symmetric, its effect on electronic motion does not derive from a central force. Therefore it exerts a torque on the electrons so that their orbital angular momentum is not conserved. Total angular momentum is always conserved, of course, and the difference is made up by the nuclear motion, by coupling of the electronic orbital motion to the molecular rotation (see Sec. 16.3.2).

Although the nuclear Coulomb force is not central, its direction always passes through the $\hat{z}$-axis (chosen along the internuclear axis) so that the component of electronic orbital angular momentum along this $\hat{z}$-direction is indeed conserved (the field is cylindrically symmetric). This $\hat{z}$-component is denoted $\hbar\Lambda$, corresponding to the L of atoms. It is quantized in units of $\hbar$ and is labeled by the Greek letters Σ (for $\Lambda = 0$), Π (for $\Lambda = 1$), Δ (for $\Lambda = 2$), Φ, ..., etc. corresponding to the S,

16.5 Electronic states of molecules

		F	Cl	Br	I
	EA (eV)	3.400	3.616	3.363	3.069
	α_- (Å^3)	0.759	2.974	4.130	6.199
Li	C_6 (eV Å^6)	0.5	1.2	1.6	2.1
IP = 5.390 eV	C_8 (eV Å^8)	0.4	1.5	2.1	3.3
$\alpha_+ = 0.0286$ Å^3	R_{eq} (Å)	1.56	2.02	2.17	2.39
	ω_N (cm^{-1})	910	641	563	498
	μ_{ind} (D)	6.3248	7.1289	7.268	6.25
	D_{eq} (eV)	5.99	4.85	4.36	3.66
Na	C_6 (eV Å^6)	2.8	7.0	8.7	11.9
IP = 5.138 eV	C_8 (eV Å^8)	2.4	8.7	12	19
$\alpha_+ = 0.255$ Å^3	R_{eq} (Å)	1.93	2.36	2.50	2.71
	ω_N (cm^{-1})	536	365	299	259
	μ_{ind} (D)	8.1558	9.0020	9.1183	9.2357
	D_{eq} (eV)	4.94	4.22	3.74	3.43
K	C_6 (eV Å^6)	12.2	39	52	76
IP = 4.339 eV	C_8 (eV Å^8)	13	46	62	97
$\alpha_+ = 1.201$ Å^3	R_{eq} (Å)	2.17	2.66	2.82	3.04
	ω_N (cm^{-1})	426	279	219	187
	μ_{ind} (D)	8.5926	10.289	10.628	11.05
	D_{eq} (eV)	5.08	4.37	3.92	3.45
Rb	C_6 (eV Å^6)	19	49	62	84
IP = 4.176 eV	C_8 (eV Å^8)	25	84	112	175
$\alpha_+ = 1.797$ Å^3	R_{eq} (Å)	2.27	2.79	2.94	3.17
	ω_N (cm^{-1})	373	223	169	139
	μ_{ind} (D)	8.5465	10.515		
	D_{eq} (eV)	5.02	4.31	3.89	3.31
Cs	C_6 (eV Å^6)	32	80.5	102	140
IP = 3.893 eV	C_8 (eV Å^8)	49	156	212	324
$\alpha_+ = 3.137$ Å^3	R_{eq} (Å)	2.35	2.91	3.07	3.32
	ω_N (cm^{-1})	353	214	150	119
	μ_{ind} (D)	7.8839	10.387		12.1
	D_{eq} (eV)	5.17	4.59	4.19	3.57

Table 16.2 Potential parameters C_6, C_8, R_{eq} and ω_N for the alkali halide molecules. From these parameters the induced dipole moment μ_{ind} can be calculated using Eq. (16.28), and the values are somewhat smaller than the experimental values given in the table. The parameters can also be used to calculate the short-range parameters A_r and a that fully determine the potential. The dissociation energy D_{eq} of this potential agrees rather well with the experimental value, as given in the table. Data are from Ref. [168].

P, D, F, ... of the atomic labels (lower-case letters are used for a single electron). Thus

$$L_z\phi_s = \pm\Lambda\hbar\phi_s, \quad \Lambda = |M_L| = 0, 1, 2, 3, \ldots \tag{16.31}$$

The projection Λ can be either positive or negative, but the two cases have the same energy for symmetry reasons.

The degeneracy results because the Hamiltonian for a diatomic molecule is unchanged upon reflection through any plane containing the internuclear axis. For example, the operator that reflects the coordinates in the plane containing the x- and z-axes is denoted as $\mathcal{R}_y$ and transforms $y_i \to -y_i$. It is easy to show that this operator commutes with $\mathcal{H}$, and that its anti-commutator with L_z is zero: $\mathcal{R}_y L_z = -L_z \mathcal{R}_y$. Using the eigenfunctions of L_z of Eq. (16.31), it can be shown that $\mathcal{R}_y \phi_s$ is also an eigenfunction of L_z for $\Lambda \neq 0$, but with eigenvalue $-\Lambda\hbar$. Since $\mathcal{R}_y$ does not change the Hamiltonian, the states with $\pm\Lambda$ are degenerate for $\Lambda \neq 0$ so that only the absolute value is specified.

The coupling between rotational and electronic orbital motion can lead to a splitting of the degenerate states and is called Λ-doubling. The terms $J_x L_x + J_y L_y$ in Eq. (16.12) can be written as $(J_+L_- + J_-L_+)/2$, and this term leads to coupling of states with different Λ. This coupling removes the degeneracy in second-order perturbation theory, as shown in Ref. [170]. States with $\Lambda = 0$ are non-degenerate. Since applying $\mathcal{R}_y$ twice leads to the same wavefunction, the eigenvalue of $\mathcal{R}_y$ is either $+1$ or -1. This is indicated by a superscript on the terms, so Σ^+ states are unchanged upon reflection but Σ^- states change sign.

For homonuclear diatomic molecules there is an additional symmetry operation that leaves the Hamiltonian unchanged, namely an inversion of all coordinates through the midpoint of the internuclear axis. If the wavefunction remains unchanged through this operation, the state is called *gerade*, and otherwise it changes sign and is called *ungerade*. Thus it has non-degenerate states Σ_g^+, Σ_g^-, Σ_u^+, Σ_u^-, Π_g, Π_u, ...

As in the atomic case, the intrinsic magnetic moment of the electron interacts with the magnetic field component along $\hat{z}$ produced by the orbital motion, and is also called the spin–orbit interaction. The spin $\vec{S}$ of the electron is not conserved but its component along $\hat{z}$ is conserved, and this component is designated as Σ (not to be confused with the $\Lambda = 0$ state). The spectroscopic labels are also directly analogous to the atomic labels, e.g. $^3\Pi_g$ is for a *gerade* state whose spin is one unit of $\hbar$ (triplet state) and whose component of orbital angular momentum along the axis is $\hbar$. For most diatomic molecules the ground state has no orbital or spin angular momentum and thus most ground states are $^1\Sigma_{g,u}$. In the next most common case, the orbital and spin angular momenta couple to form a total along $\hat{z}$ given by $\Omega = \Lambda + \Sigma$ in analogy to the atomic $\vec{J}$, and $\vec{\Omega}$ is necessarily parallel to the internuclear axis.

Molecular angular momentum schemes are further enriched by the contribution from the nuclear rotation. Although the circulating positive charges rotate much more slowly than the electron orbits, the consequent reduction of their magnetic field is partially compensated because the nuclear charges are $Z_{A,B}$ times larger. The important feature is that this rotational magnetic moment, and its associated angular momentum $\vec{N}$, are perpendicular to the internuclear axis for diatomic molecules, and this results in a variety of ways to couple these quantities. The differences arise from the relative strengths of the interactions.

The different coupling schemes are attributed to Hund [171] who distinguished five of them by small letters. A proper description of all these cases is given for diatomic molecules by Herzberg [158]. In Hund's case (a) there is a strong coupling between $\vec{L}$ and $\vec{S}$ so that the spin is also coupled to the internuclear axis. Their projections Λ and Σ combine to make $\Omega = |\Lambda + \Sigma|$ where $\vec{\Omega} = \hbar\Omega\hat{z}$ as shown in Fig. 16.5a. Then $\vec{\Omega}$ couples to $\vec{N}$ to make the total angular momentum $\vec{J} = \vec{\Omega} + \vec{N}$. Hund's case (b) usually applies when the spin is zero or weakly coupled to $\vec{L}$ so that it is decoupled from the internuclear axis. In that case, Σ is not a good quantum number and $\vec{L}$ is coupled with $\vec{N}$ to form $\vec{K} = \vec{\Lambda} + \vec{N}$ with $\vec{\Lambda} = \Lambda\hbar\hat{z}$. Next, this angular momentum $\vec{K}$ is coupled with the total spin to form $\vec{J} = \vec{K} + \vec{S}$ as shown in Fig. 16.5b. Even though Hund's case (a) may offer a good description for some lower rotational states, the situation for higher rotational states may be different and for these states the spin can uncouple from the internuclear axis, leading to Hund's case (b). In all cases, the total angular momentum is given by

$$\vec{J} = \vec{L} + \vec{S} + \vec{N}, \tag{16.32}$$

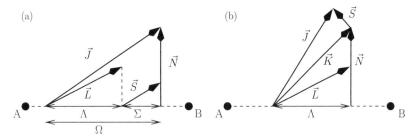

Figure 16.5 Two different coupling schemes for angular momenta of a molecule. In Hund's case (a) in part (a) the spin $\vec{S}$ of the electrons is strongly coupled to the orbital angular momentum $\vec{L}$ and is thus coupled to the internuclear axis. In Hund's case (b) in part (b) the spin is uncoupled. Note that the rotation $\vec{N}$ of the molecule is always perpendicular to the internuclear axis.

where the rotational angular momentum $\vec{N}$ is perpendicular to the internuclear axis. The other Hund's cases are not discussed here.

The intrinsic magnetic moment of the nuclei is unchanged when they are bound into a molecule by their atomic electrons, and these moments couple to the total electronic magnetic moment associated with $\vec{\Omega}$ to produce molecular hyperfine structure (hfs). The consequence is a splitting of the levels similar to the atomic case, and the magnitude is similar to atomic hfs and typically smaller than E_{rot} by about a factor of 10. The Hamiltonian for molecular hfs is derived similarly to the atomic case, but is complicated by the multiple magnetic dipoles of the nuclei. The total electronic angular momentum, including the hfs, is not required to be parallel to the internuclear axis. The hfs energies will not be discussed further here.

Even though the hfs energies are small and difficult to observe, the molecular states of diatomic molecules may be profoundly influenced by the nuclear spins of the atoms. For homonuclear diatomic molecules the total wavefunction has to be symmetric or anti-symmetric with respect to the exchange of the two nuclei, depending on whether the nuclei are bosons or fermions respectively. For example, in molecular hydrogen the nuclei are fermions ($I = 1/2$). The total nuclear spin can be either be singlet ($I_{\text{tot}} = 0$) or triplet ($I_{\text{tot}} = 1$). The first case is called para-hydrogen and the second case is called ortho-hydrogen, and these two cases cannot be coupled by ordinary optical transitions. Moreover, the total wavefunction of the ground state of hydrogen is anti-symmetric so that only even rotational levels exist for para-hydrogen whereas only odd rotational levels exist for ortho-hydrogen.

16.6 Optical transitions in molecules

16.6.1 Introduction

The optical spectrum of molecules is much more complex than that of atoms since molecular transitions can involve changes of electronic, vibrational, and rotational states. This leads to a dense spectrum as shown for N_2 in Fig. 16.1. The separation between rotational states is of the order of hundreds of GHz (mm waves), the separation between vibrational states is of the order of a few THz (far infrared), and the separation between electronic states is of the order of a few PHz (optical), just as in atoms. Since the mm-wave and far infrared spectral regions have historically proven difficult to work in for purely technological regions, most of the knowledge about molecular transitions has been obtained from the structure of spectra in the optical region for electronic transitions. Up to now the molecular electrons have been treated primarily as forming a potential in which the nuclei move, except for some discussion of symmetry and angular momentum of unpaired electrons in Sec. 16.5. This section will describe molecular electronic excitations.

16.6 Optical transitions in molecules

Transitions between a particular pair of vibrational levels $v \to v'$ can involve closely spaced transitions between many different rotational states, and so they form a band of spectral lines. Transitions between a particular pair of electronic states $s \to s'$ can involve transitions between multiple different vibrational states and thus form a band system. The number of distinct spectral lines in a band and in a band system depends on the number of populated states in the lower and upper electronic states. In this section the emphasis is on the electronic transitions using the electric dipole approximation, since they can best be compared to the transitions of atoms. The focus will first be on the transition strength, followed by a discussion on the frequency of the transitions.

The strength of transitions between the different nuclear motions in molecules (rotations and/or vibrations) is small and can be neglected in most cases, so they are not discussed here. Also, in this section spin-dependent interactions and the effects of the nuclear spin are not taken into account. In heteronuclear molecules, pure rotational or vibrational transitions can be observed, whereas in homonuclear molecules they are forbidden since a homonuclear molecule has no permanent electric dipole moment. For pure rovibrational transitions at microwave and infrared frequencies the student is referred to textbooks in molecular physics.

The nomenclature of molecular states is not as straightforward as that for atoms. Usually the lowest-lying electronic state (molecular ground state) corresponds to both atoms in their respective ground states, and its energy lies even lower than the sum of the atomic binding energies because of the potential that binds them (e.g. lower curve of Fig. 16.6). Traditionally this state is labeled with a capital X preceding the angular momentum designator Σ, Π, Δ, etc., of Sec. 16.5. For a homonuclear molecule there could also be a subscript g or u. Higher-lying electronic states are designated by capital letters A, B, C, etc., so there could be states such as $X^2\Sigma_u$ or $B^3\Pi_g$. There are also many cases where lower-case letters are used, sometimes for triplet states when singlet states are also allowed such as in alkali-metal dimers, sometimes for newly discovered states, and multiple conventions can be found in the literature. When a new level is discovered, those lying above it have not been renamed to maintain this ordering, so often additional labels using lower-case letters do not strictly follow this energy-ordered scheme.

16.6.2 Transition strength

Electric dipole transitions between two stationary states of atoms can be driven by resonant light, and the rate is proportional to the square of the electric dipole moment between them, as shown for the alkali-metal atoms in Sec. 10.5. In the case of alkali-metal atoms, the wavefunctions of the states involved can be calculated with high accuracy, and the dipole operator $\vec{\mu} = -e\vec{r}$ contains only the radial

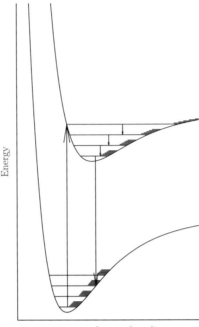

Figure 16.6 The interatomic potentials are for two states that can each make molecular bound states, but the upper state potential has a larger range because the active electron is in a higher excited state. Their energy separation is just the atomic energy at large distances, but it depends on the details of the potentials at shorter distances.

coordinate of the electron. For molecules the description is very similar but the additional degrees of freedom make the description of molecular interactions with light more complicated. In molecular transitions, the electronic, vibrational, and rotational states can change.

The electronic wavefunctions of the molecule are given in the molecular frame, whereas the polarization of the laser light is specified with respect to the laboratory frame. Furthermore, the dipole operator is a complicated function of both the electron and nuclear coordinates. It is necessary to calculate a transition dipole moment of the form $\langle \mu \rangle_{\alpha\alpha'} = \langle \Psi_\alpha | \vec{\mu} \cdot \hat{\varepsilon} | \Psi_{\alpha'} \rangle$ where Ψ_α is of the form in Eq. (16.5), $\hat{\varepsilon}$ is the polarization vector of the light and $\vec{\mu}$ is a molecular dipole operator discussed below.

Figure 16.6 shows molecular potential curves for two states. At large R their energy separation is just the difference of the sum of their atomic energies. The two curves have different shapes and different values of R_{eq}, but the same general features. They each contain a plethora of rovibrational states as given by Eq. (16.20), and two of very many possible electronic transitions are shown. Clearly each of the

16.6 Optical transitions in molecules

bands of Fig. 16.1 corresponds to a different vibrational transition, and the closely spaced transitions between the multiple rotational states of a single coupled pair of vibrational states give rise to the band spectrum.

Calculation of the rate of such molecular transitions was originally described in terms of semi-classical notions by Franck [172] and Condon [173]. The premise is that the electrons' motion is much faster than the nuclear motion so that during the transition the nuclear separation R does not change significantly. Consequently the transition is called "vertical", as shown by the vertical lines in Fig. 16.6. The Franck–Condon principle allows prediction of the internuclear distance R where the transition occurs in a classical sense [173].

The quantum mechanical description of the principle is more complicated because it is contingent upon the dependence of the electric dipole moment on R. Since the electronic wavefunctions themselves depend parametrically on R, the dipole moment calculated from this integral does as well. However, if this R-dependence is very weak, the dipole moment part of the integrand can be approximated as a constant over the range of R so that the integrand has only the vibrational wavefunctions remaining. The resulting overlap of the wavefunctions of the two states is largest for R-values where the spatial variation of the two states are nearly the same. Since the momentum depends on the gradient of the wavefunction, large positional overlap results in large momentum overlap, and this is comforting because the optical transition cannot make a significant change in the nuclear momenta. These consequences substantiate the classical notion of vertical transitions.

The total wavefunction of a diatomic molecule is given by (see Eqs. (16.5) and (16.14))

$$\Psi_\alpha(\vec{R}, \vec{r}_1 \ldots \vec{r}_\sigma) = \frac{1}{R} \mathcal{F}^{(s)}_{v,J}(R) \, \mathcal{Y}^{(\Lambda)}_{JM_J}(\Theta, \Phi) \, \phi_s(\vec{R}; \vec{r}_1 \ldots \vec{r}_\sigma), \tag{16.33}$$

where $\alpha = (s, v, J, M_J, \Lambda)$ is an abbreviation for the quantum numbers specifying the state, s is the electronic state, v is the vibrational quantum number, J is the total angular momentum of the molecule, and Λ, M_J specify the projections of L and J on the molecular and laboratory axis, respectively.

The electric dipole operator for the molecule is given by

$$\vec{\mu} = e \left(\sum_{i=A,B} Z_i \vec{R}_i - \sum_{j=1}^{\sigma} \vec{r}_j \right), \tag{16.34}$$

where the first sum is over the two nuclei and the second sum is over the σ electrons of the molecule. In this vertical transition approximation, $\mathcal{F}$ is calculated for a fixed R so $\langle \mu \rangle_{\alpha'\alpha}$ becomes

$$\langle\mu\rangle_{\alpha'\alpha} = \langle\Psi_{\alpha'}|\vec{\mu}\cdot\hat{\varepsilon}|\Psi_{\alpha}\rangle \qquad (16.35)$$

$$= e\int d\vec{R}\frac{1}{R^2}\int d\vec{r}_1\ldots d\vec{r}_\sigma \mathcal{F}^{(s')}_{v',J'}(R)\, \mathcal{Y}^{(\Lambda')}_{J'M'_J}(\Theta,\Phi)^*\, \phi_{s'}(\vec{R};\vec{r}_1\ldots\vec{r}_\sigma)$$

$$\times\hat{\varepsilon}\cdot\left(\sum_{i=A,B}^{2} Z_i\vec{R}_i - \sum_{j=1}^{\sigma}\vec{r}_j\right)\mathcal{F}^{(s)}_{v,J}(R)\, \mathcal{Y}^{(\Lambda)}_{JM_J}(\Theta,\Phi)\, \phi_s(\vec{R};\vec{r}_1\ldots\vec{r}_\sigma).$$

The second line contains $\Psi_{\alpha'}$ and the bottom line Ψ_α. The first sum in the parenthesis does not depend on the electronic coordinates $(\vec{r}_1,\ldots,\vec{r}_\sigma)$, and since the wavefunctions ϕ_s and $\phi_{s'}$ are orthogonal, the integral over the electronic coordinates of each term in this sum vanishes. However, in the case $s = s'$ there may be purely vibrational and/or rotational transitions that are allowed.

Just as in the atomic case, the electric dipole moment $\vec{\mu}$ of the molecule has three spherical components μ_q with $q = 0, \pm 1$. This dipole moment is specified in the molecular frame, since the orbital angular momentum of the electrons is given with respect to the internuclear axis. However, the excitation takes place in the laboratory frame and is thus specified with respect to this frame. Since the dipole moment is a vector, its components can be rotated from the laboratory frame to the molecular frame with the help of rotation matrices that depend on the Euler angles (Φ, Θ, Υ):

$$\mu_q^{LF} = \sum_{q'} D^1_{qq'}(\Phi,\Theta,\Upsilon)^* \mu^{MF}_{q'}, \qquad (16.36)$$

with $D^1(\Phi,\Theta,\Upsilon)$ the rotation matrix for a tensor with rank 1. Since a diatomic molecule has only two rotational degrees of freedom, three angles are redundant and the angle Υ can be chosen freely. Here the choice is made that $\Upsilon = 0$. Furthermore, for linearly polarized light $\hat{\varepsilon}$ can be chosen along the z-axis in the laboratory frame without loss of generality and thus $q = 0$. In that case the elements of the rotation matrix become

$$D^1_{0q'}(\Phi,\Theta,0)^* = \left(\frac{4\pi}{3}\right)^{1/2} Y_{1q'}(\Theta,\Phi). \qquad (16.37)$$

Then Eq. (16.35) can be written as a product of three terms so that the transition dipole moment becomes

$$\langle\mu\rangle_{\alpha'\alpha} = \sum_{q'} \mu_{el}(R)\, S_{v',v}\, \mathcal{A}^{q'}_{J'M'_J\Lambda';JM_J\Lambda}, \qquad (16.38a)$$

Because the electron wavefunctions ϕ_s depend only weakly on R, the Franck–Condon principle has been used to remove them from the integral over the nuclear

16.6 Optical transitions in molecules

coordinates $d\vec{R}$ in Eq. (16.35). The first factor is the electronic part of the matrix element:

$$\mu_{el}(R) = -e \sum_{j=1}^{n} \int d\vec{r}_1 \ldots d\vec{r}_n \phi_{s'}^*(\vec{R}; \vec{r}_1 \ldots \vec{r}_\sigma) r_j \phi_s(\vec{R}; \vec{r}_1 \ldots \vec{r}_\sigma). \tag{16.38b}$$

To evaluate this integral one chooses a value of R using the R-centroid approximation [158].

The second factor of Eq. (16.38a) is the overlap between the radial wavefunctions of the vibrational states:

$$S_{v',v} = \int dR \, \mathcal{F}_{v',J'}^{(s')}(R)^* \mathcal{F}_{v,J}^{(s)}(R) \tag{16.38c}$$

where the R^2 factor of $d\vec{R}$ cancels the $1/R^2$ in Eq. (16.35). The strength of the allowed transitions is then proportional to the Franck–Condon factor $S_{v',v}^2$. It leads to effects in molecular spectra that are discussed in the next section.

The third factor, usually called the Hönl–London factor, describes the overlap of the angular parts of the nuclear wavefunctions:

$$\mathcal{A}_{J'M'_J\Lambda';JM_J\Lambda}^{q'} = \left(\frac{4\pi}{3}\right)^{1/2} \int \sin\Theta d\Theta d\Phi \, \mathcal{Y}_{J'M'_J}^{(\Lambda')}(\Theta, \Phi)^* Y_{1q'}(\Theta, \Phi) \, \mathcal{Y}_{JM_J}^{(\Lambda)}(\Theta, \Phi). \tag{16.38d}$$

This Hönl–London factor leads to molecular selection rules, as described for atoms in Sec. 3.5. For $\Lambda = \Lambda' = 0$ the rotational wavefunctions $\mathcal{Y}_{JM_J}^{(\Lambda)}(\Theta, \Phi)$ reduce to $Y_{JM_J}(\Theta, \Phi)$ and similarly for the excited state, so the matrix element reduces to the well-known result for atomic transitions, but with the electronic angular wavefunctions replaced by those of the nuclei.

For optical transitions, evaluation of the the Hönl–London factors $\mathcal{A}$ leads to selection rules that determine which transitions are allowed just as in atoms. These selection rules are presented in Tab. 16.3 without proof, but details can be found in Refs. [158] and [170]. Most of these rules can be inferred from their atomic counterparts (see Tab. 4.1). Transitions lead not only to the exchange of energy, but

	$g \to u$,	$g \not\to g$,	$u \not\to u$
$\Sigma^+ \to \Sigma^+$,	$\Sigma^- \to \Sigma^-$,	$\Sigma^+ \not\to \Sigma^-$,	$\Sigma^- \not\to \Sigma^+$
$\Delta\Lambda = 0, \pm 1$	$\Delta\Omega = 0, \pm 1$	$\Delta S = 0$	$\Delta\Sigma = 0$
$\Delta J = 0, \pm 1$,	but	$J = 0 \not\to J = 0$	
$\Omega = 0 \to \Omega = 0$,	only for	$\Delta J \neq 0$.	

Table 16.3 Each row gives molecular selection rules for E1 transitions in molecules as given by Ref. [158]. In cases where the angular momenta of the electrons $\vec{L}$ are zero, the angular momentum $\vec{J}$ can be replaced by $\vec{N}$.

also to changes of the electronic angular momentum by $\pm\hbar$. This angular momentum change can lead to changes in the electronic orbital momentum Λ, and also to changes in the rotational angular momentum N of the nuclei (see Sec. 16.6.3 below).

16.6.3 Vibrational effects in molecular spectra

In electronic transitions in atoms and molecules, the interaction with light primarily affects the motion of the electrons and has little effect on nuclear motion. Thus the factor $S_{v',v}$ of Eq. (16.38a) depends on only the nuclear wavefunctions and the light plays no role in it. In general, the radial parts of the nuclear wavefunctions are not orthogonal since the vibrational states belong to different electronic configurations. Thus the spatial overlap of the wavefunctions $\mathcal{F}_{v,J}^{(s)}(R)$ results in overlap integrals of Eq. (16.38c) called Franck–Condon factors.

Transitions between the ground state and higher excited electronic states are referred to as fundamental progressions. Usually the equilibrium nuclear separation of higher excited states is larger than R_{eq} because electrons in such states have a larger orbital radius so they start to overlap at larger distances. The effect on the Franck–Condon factors is shown in Fig. 16.7 where the position of the minimum of the excited-state potential has been increased from that of the ground state by $0.4a_0$.

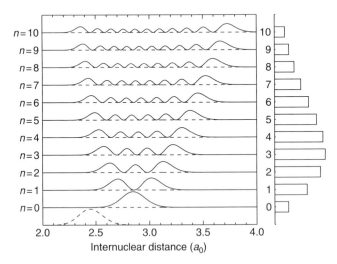

Figure 16.7 Vibrational wavefunctions of the excited states for $v = 0, \ldots, 10$. The lowest wavefunction (dashed line) is for the $v = 0$ ground-state wavefunction. which is shifted inwards with respect to the excited state by $0.4a_0$. The column on the right shows the Franck–Condon factors for the transition from the ground state to the excited states.

The Franck–Condon factors are given by the overlap of the ground state as indicated at the bottom of the figure with the excited state. This overlap is shown at the right of the figure. Clearly, the Franck–Condon factor peaks at $v = 3$, but is appreciable over a large range of v values. The excitation from the ground state with $v = 0$ leads to excitation of higher vibrational states, but collisions can cause these states to decay non-radiatively to the lowest vibrational state of the excited electronic state faster than radiative decay out of it.

16.6.4 Rotational effects in molecular spectra

Molecular electrons are bound by a non-central force as discussed in Sec. 16.5 so their orbital angular momentum is not conserved. However, their motion in a plane perpendicular to the molecular axis does conserve angular momentum, and its magnitude is $\Lambda\hbar$. Conservation of total molecular angular momentum is mediated by coupling between the component of electron motion containing the molecular axis and the molecular rotation. This results in the possibility of a change of N caused by an electronic transition even though the light does not act directly on the nuclear motion.

The transition frequency $v_{\alpha\alpha'}$ in electronic transitions between a state $\alpha = (s, v, J)$ and a state $\alpha' = (s', v', J')$ is given by the difference in energy between the two states involved divided by h:

$$v_{\alpha\alpha'} = \frac{E_{s',v',J'} - E_{s,v,J}}{h}, \tag{16.39}$$

where the energies of the molecular states are given by Eq. (16.22). Here s, s' denote the electronic states, v, v' the vibrational states, and J, J' the rotational states. Within a band the difference between the transition frequencies is determined by the rotational quantum numbers J and J' involved.

Under the restriction that $\Delta L = \pm 1$, the change of Λ can still be zero if there is a change of N. The selection rule for rotational quantum number J in electric-dipole transitions is $\Delta J = J' - J = 0, \pm 1$, as given in Tab. 16.3. For $\Delta J = -1$ the transitions form the P-branch, for $\Delta J = 0$ they form the Q-branch, and for $\Delta J = +1$ they form the R-branch. Neglecting the coupling between rotation and vibration, the frequency difference within a band is given by the difference in rotational energy, or,

$$\begin{aligned}\Delta E_{\text{rot}} &= B'(J + \Delta J)(J + 1 + \Delta J) - BJ(J + 1), \\ &= (B' - B)J^2 + (B' - B + 2\Delta J B')J + B'(\Delta J^2 + \Delta J),\end{aligned} \tag{16.40}$$

where $B = \hbar^2/(2\mu_N R_{eq}^2)$ is the rotational constant for molecular state α and B' for state α'. The equilibrium nuclear separation R_{eq} of higher excited states is usually larger than that of the lower states, so B' is usually smaller than B. If $B' = B$, the difference ΔE_{rot} is linear in J, but for $B' < B$ the difference is instead quadratic in J.

A plot of Eq. (16.40) is shown in Fig. 16.8, where it is assumed that $\Delta B = B' - B < 0$. For the P-branch the transition frequency decreases for increasing J at low J, but starts to increase for higher values of J. For the Q- and R-branches the transition frequency is always increasing for increasing J. The minimum for the P-branch occurs for J_{min} given by

$$J_{min} = \frac{1}{2} + \frac{B'}{\Delta B}, \qquad (16.41)$$

which yields $J_{min} = 9$–10 for a typical value $\Delta B = 0.1B$. The lower part of Fig. 16.8 shows the frequencies of the transitions for the different branches, and the minimum frequency is referred to as the band head. Figure 16.9 shows the rotational lines in a band of the Cs_2 molecule, and the band head is easily identified. Separation of a band structure into its different branches allows for a determination of the difference ΔB of the rotational constants.

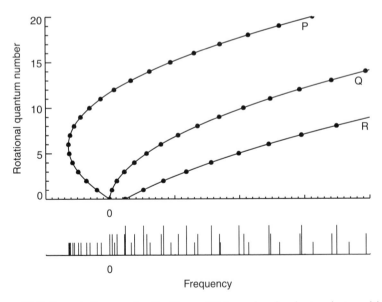

Figure 16.8 Fortrat diagram for P-, Q-, and R-branches in electronic transitions, where $\Delta B = B' - B = 0.1875B$. The transition frequencies are indicated with filled circles for the different rotational quantum numbers J. The lower part of the figure shows the frequencies, where the P-branch is indicated with a short marker, the Q-branch with an intermediate marker and the R-branch with a long marker.

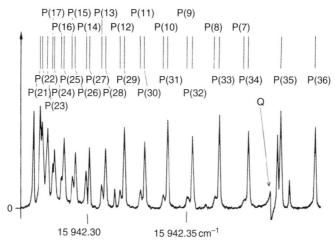

Figure 16.9 Sub-Doppler spectroscopy for the $v = 14 \rightarrow v' = 9$ vibrational transition of the $X^1\Sigma_g^+ \rightarrow C^1\Pi_u$ electronic transition in the Cs_2 molecule. (Courtesy of Prof. W. Demtröder.)

Appendix

16.A Morse potential

Molecular potentials come in all kinds of shapes and forms. In order to obtain the vibrational frequency of the lowest states, the minimum of the potential can be approximated by a harmonic function. However, this is only valid close to the minimum and does not account for the anharmonicity of the potential. A better approximation is to use the Morse potential [166] given by

$$V(R) = D_{eq}\left(e^{-2\alpha(R-R_{eq})} - 2e^{-\alpha(R-R_{eq})}\right). \tag{16.42}$$

The function is repulsive for $\alpha > 0$ at short distances and for large R connects asymptotically to $V(R) = 0$. Not only does the Morse potential "look" like most molecular potentials, its parameters have physical meaning. The potential minimum is at $R = R_{eq}$, the minimum is given by $V(R_{eq}) = -D_{eq}$, and α is proportional to the effective charge. Moreover, Morse has shown that the vibrational levels can be well approximated by Eq. (16.21). The second-order coefficient of the Taylor expansion of the potential around its minimum is given by $k_{eq} = 2D_{eq}\alpha^2$. Finally, the classical inner turning point R_c (see Fig. 16.2) is given by

$$R_c = R_{eq} - \frac{\ln 2}{\alpha}. \tag{16.43}$$

When one uses the Morse potential, there is thus a direct relation between its parameters and its physical shape.

With only a finite number of potential energy values for a given interaction, a Morse potential can easily be constructed as follows. Using the potential at three points $R_0 - \Delta R$, R_0, and $R_0 + \Delta R$ with values E_{-1}, E_0, and E_{+1}, respectively, substitution of these three values into Eq. (16.42) leads to a cubic relation:

$$B^3 E_{-1} - B^2 E_0 - BE_0 + E_1 = 0, \quad (16.44)$$

where the parameter $B = \exp(-\alpha \Delta R)$. Since α needs to be positive, one takes the solution of Eq. (16.44) with $0 \le B \le 1$. Taking $C = \exp[-\alpha(R_0 - R_{eq})]$ leads to

$$C = \frac{2B(BE_{-1} - E_0)}{B^2 E_{-1} - E_0}, \quad (16.45)$$

which can be used to extract R_{eq}. Finally, the dissociation energy is given by

$$D_{eq} = \frac{E_0}{C(C-2)}. \quad (16.46)$$

These three relations are sufficient to fully specify the Morse potential.

Exercises

16.1 Transitions between molecular vibrational states are forbidden to lowest order because the electric dipole moment operator's matrix elements are zero between different nuclear vibrational states. How do such transitions occur? Give some examples. What are Franck–Condon factors and how do they enter this topic?

16.2 The assertion in Sec. 16.2 that the atomic and molecular "spring constants" are comparable can be further supported as follows. Find the position of the minimum of the potential of Eq. (7.9b) and evaluate the spring constant for the binding of an electron in its orbit. Show that it is of the order of R_∞/a_0^2. Then use the approximation that $R_c \approx R_{eq}/2$ suggested by Fig. 16.2, Eq. (16.43), and the expression for k_{eq} just above it to find the molecular "spring constant" in the Morse potential. Show that $k_{eq} \approx 4D_{eq}/R_{eq}^2$ and compare this with the atomic value for reasonable values of D_{eq} and R_{eq}.

16.3 Draw the Morse potential for the ground state of H_2 using molecular constant from Herzberg [158].

16.4 Use your favorite program to generate lowest-order charge distributions for the four molecular orbitals in the separate atom limit starting from the 2p state of H. Your answers should look something like the images below. Label the charge distributions using the united and neutral schemes. Indicate which orbitals are bonding and which are anti-bonding. (Figure from Ref. [174].)

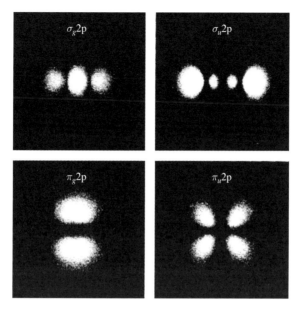

16.5 The moment of inertia of $H^{79}Br$ is 3.30×10^{-47} kg m². Calculate the energies of the first rotational levels of the molecule in eV and the corresponding wavenumbers ($\tilde{\nu} = E/hc$) in units of cm⁻¹. Find the internuclear distance in atomic units and in angstroms.

16.6 The wavenumber $\tilde{\nu}_0$ of the fundamental vibrational motion of the molecule $H^{79}Br$ is 2,650 cm⁻¹. Calculate
(a) the energy of the lowest and first excited states in eV;
(b) the corresponding periodic times;
(c) the force constant in SI units.

16.7 Find the energy of dissociation D_0 of the deuterium molecule D_2, given that the energy of dissociation of H_2 is 4.48 eV and that the energy of the lowest vibrational level of H_2 is 0.26 eV.

16.8 Different electronic states may have different values of R_{eq} resulting in rotational constants B and B' (see Eq. (16.19)).
(a) Show that the rotational contribution to the energy difference between states of J and $J + 1$ is given by $J^2(B' - B) + J(B' - B \pm 2B) + (1 \pm 1)B$
(b) Show that the case of $B' - B \ll |2B|$ still has equally spaced lines at small J, but as J gets larger, the spacing becomes smaller and eventually changes sign. Plot the energies of such a system.
(c) Sketch such a spectrum, noting that it folds back upon itself. The turning point is called a band head, and these are visible in Fig. 16.1. These energy differences

can be plotted differently in a Fortrat diagram [158]. Look this up and draw such a diagram.

16.9 Consider an electronic transition between two states of a diatomic molecule that have rotational constants B and B' respectively. Show that there are three distinct bands corresponding to $\Delta J = 0, +1$, and -1, and that only one of them can have a band head of rotational lines. Draw a Fortrat diagram for the case $B - B' = B/10$.

17

Binding in the hydrogen molecule

For chemists, molecular hydrogen plays the paradigm role that atomic hydrogen plays for physicists. Because it is relatively simple, a detailed study is possible [161, 175–185]. For the hydrogen molecule, a method for building up the electronic wavefunctions by linear combination of atomic orbitals (LCAO) is often done. The simplest configuration to start with is actually the hydrogen molecular ion H_2^+, where one electron has been removed from the H_2 molecule. In this case the interaction between the single electron and nuclei within the Born–Oppenheimer approximation is discussed in Sec. 17.1. Then the neutral H_2 molecule can be treated by taking a combination of these one-electron states using the resulting molecular orbitals, and this molecular orbital (MO) approach is discussed in Sec. 17.2. Here the interaction between the two electrons plays a key role, similar to helium. An alternative to this approach is the valence bond (VB) approach, where it is explicitly assumed that the valency of the electrons is equally shared between the two nuclei, as discussed in Sec. 17.3. Finally, the methods explored for H_2 are used to discuss the bonding of this diatomic molecule in more detail. Note that the methods discussed in this chapter use the Born–Oppenheimer approximation, as derived in Sec. 16.3.1.

17.1 The hydrogen molecular ion

In the Born–Oppenheimer approximation, the Hamiltonian for the electron in the hydrogen molecular ion is given by

$$\mathcal{H} = -\frac{\hbar^2}{2m}\nabla^2 + \frac{e^2}{4\pi\varepsilon_0}\left(\frac{1}{R} - \frac{1}{r_A} - \frac{1}{r_B}\right), \quad (17.1)$$

where R denotes the distance between the nuclei A and B, and $r_{A,B}$ denote the distance from the electron to nucleus A or B, respectively. The vectors $\vec{r}_A$ and $\vec{r}_B$ are given by $\vec{r}_A = \vec{r} + \vec{R}/2$ and $\vec{r}_B = \vec{r} - \vec{R}/2$, where $\vec{r}$ is the position vector of the electron to the midpoint of the internuclear axis. Note that all terms in Eq. (17.1) depend only on the distance of the electron to the nuclei, and not on the orientation. In this chapter atomic units will be used for the radius and energy (see App. B.1). In atomic units $e^2/4\pi\varepsilon_0 \equiv 1$, thereby reducing the length of the formulas considerably.

To use the LCAO method, start with a large internuclear distance R where the electron is expected to be attached to one of the nuclei. For the ground state of H_2^+, the electron will be in the 1s orbital of H at large R, either at nucleus A or B. However, the Pauli symmetrization principle applied to the nuclei requires that the spatial wavefunction is either symmetric or anti-symmetric with respect to the exchange of A and B. So the most appropriate wavefunction of H_2^+ for the electron in the field of the two protons for all internuclear distances R is

$$\Psi_{g,u}(\vec{R};\vec{r}) = \frac{1}{2}\sqrt{2}\,[\psi_{1s}(r_A) \pm \psi_{1s}(r_B)], \tag{17.2}$$

where the labels g, u refer to gerade(+)/ungerade(−), respectively. These signs are defined by applying the exchange operator P_{AB} of Chap. 14 to the wavefunctions $\Psi_{g,u}$ and the results are that the *gerade* wavefunction does not change sign, whereas the *ungerade* does. The states can then be labeled $^2\Sigma_g$ and $^2\Sigma_u$. The exchange of electrons does not apply here because there is only one of them in H_2^+. This expression can be used as a trial wavefunction in a variational expression to obtain the potential energies $E_{g,u}(R)$:

$$E_{g,u}(R) = \frac{1}{\mathcal{N}}\int d\vec{r}\,\Psi^*_{g,u}(\vec{R};\vec{r})\,\mathcal{H}\,\Psi_{g,u}(\vec{R};\vec{r}), \tag{17.3}$$

where the normalization $\mathcal{N}$ has to be taken into account explicitly because of the overlap between the electron wavefunctions at nuclei A and B, even though each individual wavefunction is properly normalized at each nucleus. Thus $\mathcal{N}(R) = 1 \pm \mathcal{I}(R)$, where the overlap integral $\mathcal{I}(R)$ is given by

$$\mathcal{I}(R) = \int \psi_{1s}(r_A)\psi_{1s}(r_B)d\vec{r}. \tag{17.4}$$

This is a one-electron, two-center integral, since it involves the integration of two wavefunctions at different centers.

It is instructive at this point to introduce the confocal elliptical coordinates ξ, η, and ϕ (see also App. 17.A). Then it can be shown that the solution of the Schrödinger equation can be written as $\Phi(\xi, \eta, \phi) = F(\xi)G(\eta)e^{im\phi}$, where $F(\xi)$ and $G(\eta)$ are solutions of differential equations that depend on only ξ or η, respectively. These equations can be solved numerically for each internuclear distance R and the resulting energies plotted. In this respect the H_2^+ molecular ion allows for numerical solutions to any degree of precision, as the H atom allows for exact analytic solutions in the atomic case. However, at this point, it is more interesting to use the H_2^+ ion to illustrate the LCAO-method. As is shown in App. 17.B, the integrals of Eqs. (17.3) and (17.4) can be solved efficiently using confocal elliptical coordinates.

The energy of Eq. (17.3), including Eq. (17.4), can now be written as

$$E_{g,u}(R) = \frac{\mathcal{H}_{AA}(R) \pm \mathcal{H}_{AB}(R)}{1 \pm I(R)} \quad (17.5a)$$

where the terms $\mathcal{H}_{AA}(R)$ and $\mathcal{H}_{AB}(R)$ are given by

$$\mathcal{H}_{AA}(R) = \int d\vec{r}\, \psi_{1s}(r_A)\, \mathcal{H}\, \psi_{1s}(r_A) \quad (17.5b)$$

and

$$\mathcal{H}_{AB}(R) = \int d\vec{r}\, \psi_{1s}(r_A)\, \mathcal{H}\, \psi_{1s}(r_B), \quad (17.5c)$$

where Eq. (17.2) is used for the wavefunctions. Since the wavefunction $\psi_{1s}(r_A)$ is for the H(1s) orbital of the electron at nucleus A, the first and third terms of the Hamiltonian in Eq. (17.1) yield the energy E_{1s} for $\mathcal{H}_{AA}$ of Eq. (17.5)b. The second term of Eq. (17.1) yields $1/R$. However, the last term is more complicated and is denoted by the one-electron, two-center integral $\mathcal{J}_1(R)$, which is written in atomic units as

$$\mathcal{J}_1(R) = \int \psi_{1s}(r_A) \frac{1}{r_B} \psi_{1s}(r_A) d\vec{r}. \quad (17.6)$$

There are several different integrals of this type that will be denoted by an index $n = 1, 2, 3$ using the notation $\mathcal{J}$ and $\mathcal{K}$ as introduced for helium in Eq. (14.17). This does not follow the older molecular literature where these symbols are often exchanged. Since the integral involves only wavefunctions at one nucleus, it is called a direct integral, similar to the case of helium. The same analysis for the $\mathcal{H}_{AB}$ term of Eq. (17.5c) also yields four terms, where the $1/r_A$ term leads to the exchange integral $\mathcal{K}_1(R)$ given by

$$\mathcal{K}_1(R) = \int \psi_{1s}(r_A) \frac{1}{r_A} \psi_{1s}(r_B) d\vec{r}, \quad (17.7)$$

which involves integration of wavefunctions at different nuclei. As shown in App. 17.B, both integrals can be solved analytically using confocal elliptical coordinates.

The energy of the states then becomes

$$E_{g,u}(R) = E_{1s} + \frac{1}{R} - \frac{\mathcal{J}_1(R) \pm \mathcal{K}_1(R)}{1 \pm I(R)}, \quad (17.8)$$

which is plotted in Fig. 17.1. The results are compared with the "exact" numerical result, which can be obtained from numerical integration of the differential

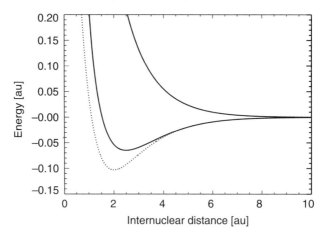

Figure 17.1 Molecular potential curves for the H_2^+ molecule (solid lines) as calculated using the H(1s) state as trial state. The *gerade* state (lower curve) is binding and the *ungerade* state (upper curve) is anti-binding. The dotted line is the "exact" potential, which shows a larger binding energy at a shorter distance R.

equations for the wavefunctions $F(\xi)$ and $G(\eta)$ in confocal elliptical coordinates. Clearly, the *gerade* function is binding and has a minimum of $E_g = -0.0648$ au at $R = 2.49a_0$. This has to be compared with the "exact" results of -0.1025 eV at $R = 2.01a_0$ (see Tab. 17.1). The *ungerade* function is repulsive over the whole internuclear range. Although the trial wavefunction predicts the position of the minimum in the potential reasonably well, it underestimates the binding energy by more than 0.035 au or 1 eV (~50%).

The striking difference between the form of the energies $E_{g,u}(R)$ of the *gerade* and *ungerade* states, namely the bonding and anti-bonding, can be understood by looking at the electronic distributions around the nuclei. In Fig. 17.2 the electron distribution $\rho_g(\vec{r}) = e|\Psi_g(R;\vec{r})|^2$ is compared with its *ungerade* counterpart. As the figure shows, the electron distribution between the two nuclei is appreciable in the *gerade* case, and this leads to an effective screening of the repulsion of the two nuclei. The *ungerade* wavefunction is small in this region and is even zero in the midpoint. The repulsion of the two nuclei remains unscreened and this leads to a repulsive state for each R. The issue of bonding of molecules will be discussed in more detail in Sec. 17.5.

17.2 The molecular orbital approach to H_2

Knowing the electronic wavefunction of the H_2^+ molecule as discussed in Sec. 17.1 allows the use of the molecular orbital (MO) approach to construct the electronic wavefunctions for the H_2 molecule. In the same spirit as the Aufbau principle

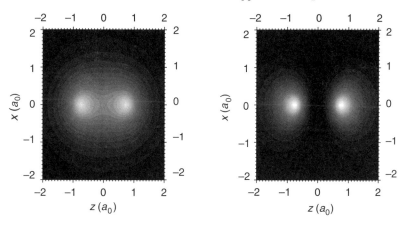

Figure 17.2 Charge distribution of the *gerade* (left) and *ungerade* wavefunctions (right) for the H_2^+ molecule. Note that the *gerade* distribution shows a large electron density between the two nuclei, whereas this density is strongly suppressed for the *ungerade* case.

for atoms, the electronic wavefunction for the molecular system comprising many electrons can be constructed from one-electron wavefunctions. This method was pioneered by Hund and Mulliken for the H_2 molecule, and the method is often referred to by their names [69].

For H_2 one has to combine two one-electron wavefunctions $\Psi_{g,u}(\vec{R};\vec{r})$. By choosing the electrons at each nucleus to be either in the *gerade* or *ungerade* state, one can construct four wavefunctions. However, the product of a *gerade* and *ungerade* wavefunction does not have definite symmetry with respect to the exchange of electrons and thus does not obey the Pauli symmetrization principle. In a similar way as in done in Chap. 14, one can construct linear combinations with proper symmetry. Together with the proper linear combinations of the wavefunctions for the electron spins as given in Sec. 14.2.3, the following wavefunctions for the two electrons of the H_2 molecule can be constructed:

$$\Psi_A^{MO}(1,2) = \Psi_g(\vec{r}_1)\Psi_g(\vec{r}_2) \, {}^1\chi(1,2) \tag{17.9}$$
$$\Psi_B^{MO}(1,2) = \Psi_u(\vec{r}_1)\Psi_u(\vec{r}_2) \, {}^1\chi(1,2)$$
$$\Psi_C^{MO}(1,2) = (\Psi_g(\vec{r}_1)\Psi_u(\vec{r}_2) + \Psi_u(\vec{r}_1)\Psi_g(\vec{r}_2)) \, {}^1\chi(1,2)$$
$$\Psi_D^{MO}(1,2) = (\Psi_g(\vec{r}_1)\Psi_u(\vec{r}_2) - \Psi_u(\vec{r}_1)\Psi_g(\vec{r}_2)) \, {}^3\chi(1,2),$$

where the index on the spin wavefunction denotes either its anti-symmetric singlet (${}^1\chi$) or symmetric triplet (${}^3\chi$) character. The first two wavefunctions $\Psi_{A,B}^{MO}(1,2)$ are ${}^1\Sigma_g$ states, the third wavefunction $\Psi_C^{MO}(1,2)$ is a ${}^1\Sigma_u$ state, whereas the last wavefunction $\Psi_D^{MO}(1,2)$ is a ${}^3\Sigma_u$ state and is three-fold degenerate because of the

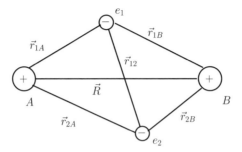

Figure 17.3 Coordinates in the H_2 molecule, where R denotes the internuclear distance. Here r_{ij} is the distance between two particles $i, j = 1, 2, A$, and B, where the nuclei are indicated by A and B and the electrons by 1 and 2.

spin. It is clear by inspection that the total wavefunction is always anti-symmetric with respect to the exchange of the two electrons, as required for fermions.

The Hamiltonian for the H_2 molecule can be conveniently written in atomic units as

$$\mathcal{H} = \mathcal{H}_1 + \mathcal{H}_2 + \frac{1}{r_{12}} + \frac{1}{R}, \tag{17.10}$$

where $\mathcal{H}_i$ denotes the one-electron Hamiltonian of Eq. (17.1) for electron $i = 1, 2$ excluding the $1/R$-term (see Fig. 17.3 for the different coordinates). Similar to the case of helium, the most problematic term to evaluate in Eq. (17.10) is the $1/r_{12}$ term. Concentrating on $\Psi_A^{MO}(1, 2)$ which combines attractive states of the H_2^+ ion and thus should have the lowest energy, the energy is given by

$$E_A^{MO}(R) = \frac{1}{N} \int d\vec{r}\, \Psi_A^*(\vec{R}; \vec{r})\, \mathcal{H}\, \Psi_A(\vec{R}; \vec{r}), \tag{17.11}$$

which contains 64 terms, both in the numerator and the denominator N (see Eq. (17.10)). Because $\Psi_g(\vec{R}; \vec{r}_i)$ satifies

$$\mathcal{H}_i \Psi_g(\vec{R}; \vec{r}_i) = \left(E_g - \frac{1}{R}\right) \Psi_g(\vec{R}; \vec{r}_i), \tag{17.12}$$

this lowest energy becomes

$$E_A^{MO}(R) = 2E_g(R) - \frac{1}{R} + \int d\vec{r}_1 d\vec{r}_2 \frac{1}{r_{12}} |\Psi_g(\vec{r}_1)\Psi_g(\vec{r}_2)|^2. \tag{17.13}$$

The last term is called a two-electron, two-center integral, since it involves both electronic coordinates located at each of the two nuclei. The number of terms to be evaluated is 16.

Expanding the spatial electronic wavefunction more explicitly, disregarding the spin wavefunctions since spin-dependent interactions are not included explicitly in the Hamiltonian, yields

17.2 The molecular orbital approach to H_2

$$\Psi_A^{MO}(\vec{r}_1, \vec{r}_2) \equiv \Psi_a + \Psi_b + \Psi_c + \Psi_d$$
$$= \psi_{1s}(\vec{r}_{A1})\psi_{1s}(\vec{r}_{B2}) + \psi_{1s}(\vec{r}_{B1})\psi_{1s}(\vec{r}_{A2})$$
$$+ \psi_{1s}(\vec{r}_{A1})\psi_{1s}(\vec{r}_{A2}) + \psi_{1s}(\vec{r}_{B1})\psi_{1s}(\vec{r}_{B2}). \quad (17.14)$$

The electrons are on opposite nuclei in the first two terms, so they are commonly called covalent states. The electrons are both at the same nucleus in the last two terms, so they are called ionic states. Denoting the electron–electron interaction by $V = 1/r_{12}$, the four wavefunctions $\Psi_{a,b,c,d}$ lead to 16 terms $\mathcal{V}_{ij} = \langle \Psi_i | V | \Psi_j \rangle$ with $i, j = a, b, c, d$, which can be written as

$$\mathcal{V}_{aa} = \mathcal{V}_{bb} \equiv \mathcal{J}_2(R) = \int d\vec{r}_1 d\vec{r}_2 \frac{|\psi_{1s}(\vec{r}_{A1})\psi_{1s}(\vec{r}_{B2})|^2}{r_{12}},$$

$$\mathcal{V}_{cc} = \mathcal{V}_{dd} \equiv \mathcal{J}_3(R) = \int d\vec{r}_1 d\vec{r}_2 \frac{|\psi_{1s}(\vec{r}_{A1})\psi_{1s}(\vec{r}_{A2})|^2}{r_{12}},$$

$$\mathcal{V}_{ab} = \mathcal{V}_{ba} = \mathcal{V}_{cd} = \mathcal{V}_{dc} \equiv \mathcal{K}_2(R)$$
$$= \int d\vec{r}_1 d\vec{r}_2 \frac{\psi_{1s}(\vec{r}_{A1})\psi_{1s}(\vec{r}_{A2})\psi_{1s}(\vec{r}_{B1})\psi_{1s}(\vec{r}_{B2})}{r_{12}},$$

$$\mathcal{V}_{a,b\ c,d} = \mathcal{V}_{c,d\ a,b} \equiv \mathcal{K}_3(R)$$
$$= \int d\vec{r}_1 d\vec{r}_2 \frac{|\psi_{1s}(\vec{r}_{A1})|^2 \psi_{1s}(\vec{r}_{A2})\psi_{1s}(\vec{r}_{B2})}{r_{12}}. \quad (17.15)$$

Note that the integrals $\mathcal{J}_2$, $\mathcal{J}_3$, $\mathcal{K}_2$, and $\mathcal{K}_3$ bear a strong resemblance to the integrals for helium in Chap. 14. For the term $\mathcal{J}_2$ the two electrons are on different nuclei, whereas for the term $\mathcal{J}_3$ the two electrons are on the same nucleus. For the term $\mathcal{K}_2$ both electrons are exchanged, whereas for the term $\mathcal{K}_3$ one electron is located at one nucleus and the other electron is exchanged. The terms all have to be evaluated, and there are many erroneous results in the literature and in textbooks.

The evaluation of the terms is given in App. 17.C and for the energy $E_A(R)$ is given by

$$E_A^{MO}(R) = 2E_g(R) - \frac{1}{R} + \frac{\mathcal{J}_2(R) + \mathcal{J}_3(R) + 2\mathcal{K}_2(R) + 4\mathcal{K}_3(R)}{2(1 + \mathcal{I}(R))^2}. \quad (17.16)$$

The result is plotted in Fig. 17.4 along with the results for $E_{B,C,D}^{MO}(R)$. As expected $E_A^{MO}(R)$ has the lowest energy at short internuclear distances. The binding energy is 0.0991 au or 2.70 eV, which has to be compared to the experimental value of 4.747 eV (see Tab. 17.1). The minimum distance is at $1.601 a_0$, as compared with the experimental value of $1.4011 a_0$. However, since this state also contains an ionic

		α	R_{eq}		E_{min}	
		(1/a_0)	(a_0)	(Å)	(au)	(eV)
H_2^+	non-optiomal	1.000	2.493	1.319	−0.0648	−1.764
	optimal	1.238	2.014	1.066	−0.0865	−2.354
	"exact"		2.00	1.06	−0.103	−2.79
H_2	VB-method	1.000	1.642	0.869	−0.1160	−3.156
	optimal	1.166	1.409	0.745	−0.1391	−3.784
	MO-method	1.000	1.601	0.847	−0.0991	−2.696
	optimal	1.190	1.346	0.712	−0.1277	−3.476
	"exact"		1.4011		−0.1744	−4.7471
	experimental		1.4006	0.742	−0.1745	−4.74759

Table 17.1 Potential parameters for H_2^+ and H_2 using different methods. The first lines in each case are the results using the trial wavefunction H(1s). The optimal values are for the case of the H(1s)-wavefunction after optimizing the effective charge α. The "exact" values are theoretical values, where the variational solution has fully converged.

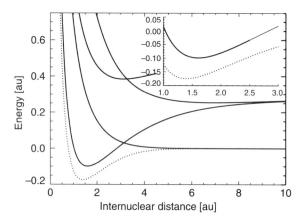

Figure 17.4 Potential curves for H_2 in the MO-approach. Three of the four curves are attractive and connect higher excited states. The repulsive state connects to the limit of two separated H atoms in the ground state. The "exact" potential curve for the ground state of H_2 is indicated with a dotted line. The inset shows an enlargement of the ground-state potential around the equilibrium distance.

part, the state does not asymptotically connect to the limit of two separated H atoms in the ground state at very large distances. Two of the three other states are also attractive at long range, but they asymptotically connect to higher-lying states. The fourth state is purely repulsive, but it asymptotically connects to the limit of two H atoms, each in the ground state.

17.3 The valence bond approach to H$_2$

An alternative method for H$_2$ described by Heitler and London [175] is called the valence bond (VB) method. It explicitly assumes that the valency of the electrons is equally shared by the two nuclei, so in the ground state the wavefunctions are

$$\Psi^{VB}_{g,u}(\vec{r}_1, \vec{r}_2) = [\psi_{1s}(\vec{r}_{A1})\psi_{1s}(\vec{r}_{B2}) \pm \psi_{1s}(\vec{r}_{B1})\psi_{1s}(\vec{r}_{A2})]^{1,3}\chi. \quad (17.17)$$

Again the total wavefunction changes sign when exchanging the two electrons, but the spatial part of the *gerade* wavefunction does not change sign in such an exchange.

Using the wavefunction of Eq. (17.17) and the results of the integrals $\mathcal{J}_1$, $\mathcal{J}_2$, $\mathcal{K}_1$, and $\mathcal{K}_2$ of the previous section leads directly to

$$E^{VB}_{g,u}(R) = 2E_{1s} + \frac{1}{R} + \frac{[\mathcal{J}_2(R) - 2\mathcal{J}_1(R)] \pm [\mathcal{K}_2(R) - 2\mathcal{K}_1(R)\mathcal{I}(R)]}{1 + \mathcal{I}(R)^2}, \quad (17.18)$$

and the energy curves are plotted in Fig. 17.5. The binding energy (0.116 au) and the minimum distance (1.64a_0) are comparable to the results of the MO-approach. However, both states connect to the asymptote of two H atoms in the ground state at large range, as they should.

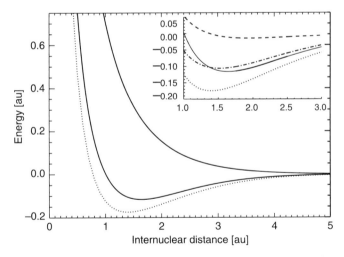

Figure 17.5 Potential curves (solid lines) for H$_2$ in the VB-approach. The *gerade* state (lower curve) is attractive, whereas the *ungerade* state (upper curve) is repulsive. The "exact" potential curve for the ground state of H$_2$ is indicated with a dotted line. The inset shows an enlargement of the ground state potential around the equilibrium distance. In addition to the two curves in the main figure, the dashed curve represents the "direct" part of the energy of the *gerade* state, whereas the dashed-dotted curve represents the "exchange" part. The curves show that most of the binding energy is caused by the exchange integral.

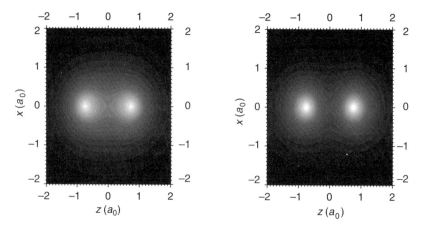

Figure 17.6 Charge distribution of the *gerade* (left) and *ungerade* (right) wavefunction in the VB approach for the H_2 molecule. Note that the *gerade* distribution shows a large electron density between the two nuclei, whereas this density is strongly suppressed for the *ungerade* case, although the reduction is not as strong as in the H_2^+ case.

Similar to the electron density plots in Fig. 17.2 for the H_2^+ molecule, the electron distribution for the VB wavefunctions of the H_2 molecule are plotted in Fig. 17.6. In the VB method, the distributions are given by

$$\rho_\pm(R, \vec{r}) = \frac{e}{\pi(1 \pm \mathcal{I}(R))} \left[e^{-2r_A} + e^{-2r_B} \pm 2\mathcal{I}(R) e^{-(r_A + r_B)} \right]. \tag{17.19}$$

Again the *gerade* state shows an appreciable electron density between the two nuclei, whereas for the *ungerade* state the density is considerably less. Thus the shielding of the nuclear repulsion is more effective for the *gerade* state than for the *ungerade* state; and the *gerade* state is binding, whereas the *ungerade* state is anti-binding. The binding is discussed in more detail in Sec. 17.5.

17.4 Improving the methods

The results of both the MO and VB methods disagree with the experimental values for the binding energy of H_2. One possible source of the discrepancy may be the choice of the hydrogenic (1s) orbital as the trial wavefunction. There are many reports in the literature of better choices for the wavefunction. For example, using a variational approach allows the choice of a trial wavefunction with one or more free parameters to minimize the energy for each internuclear distance.

An obvious approach is to scale the exponent of the (1s) orbital and thus effectively use the charge of the nucleus as an optimization parameter:

$$\psi(\vec{r}) = \frac{\alpha^{3/2}}{\sqrt{\pi}} e^{-\alpha r}, \tag{17.20}$$

where $\alpha = 1$ returns the original wavefunction. By finding the position R_{min} that minimizes the binding energy for each value of α one obtains the binding energy $E(R_{min}, \alpha)$. If the electron coordinates and the internuclear distance R are scaled with α, or $\tilde{r} = \alpha r$ and $\tilde{R} = \alpha R$ respectively, the kinetic energy $\propto d^2/dr^2$ is scaled by α^2, whereas the potential energy $\propto 1/R$ is scaled with α. Thus for different values of α the minimum is always located at $R_{min} = \tilde{R}_{min}/\alpha$, where the value of $\tilde{R}_{min}$ can easily be found by minimizing the energy for $\alpha = 1$. The energy can then be written as a function of α as

$$E_g(\alpha) = p\alpha^2 + q\alpha, \tag{17.21}$$

where for the VB method the coefficients p for the kinetic energy and q for the potential energy are given by

$$p = -1 + \frac{2 + 2k_1(\rho_{min})\mathcal{I}(\rho_{min})}{1 + \mathcal{I}^2(\rho_{min})}, \tag{17.22}$$

$$q = \frac{1}{\rho_{min}} - \frac{2 + 2j_1(\rho_{min}) + 4k_1(\rho_{min}) - \mathcal{I}(\rho_{min})(j_2(\rho_{min}) + k_2(\rho_{min}))}{1 + \mathcal{I}^2(\rho_{min})},$$

and the parameters $j_{1,2} = \mathcal{J}_{1,2}/\alpha$ and $k_{1,2} = \mathcal{K}_{1,2}/\alpha$ are independent of α. The energy can be minimized using $dE_g(\alpha)/d\alpha = 0$ and the minimum becomes $E_g(\alpha_{min}) = -q^2/4p$ at $\alpha_{min} = -q/2p$.

The results for the various methods are collected in Tab. 17.1. Since the increase in the binding energy for both the MO and VB methods is of the order of 0.02–0.03 au, which is still a lot smaller than the difference between the theoretical and experimental values, it is clear that an accurate description of the electronic wavefunction of the H_2 molecule requires more flexibility.

Another way to implement more flexibility in the variational methods for the ground state of the H_2 molecule is to use a trial wavefunction:

$$\Psi(\vec{r}_1, \vec{r}_2) = \sum_i c_i \Phi_i(\vec{r}_1, \vec{r}_2), \tag{17.23}$$

where the basis wavefunctions $\Phi_i(\vec{r}_1, \vec{r}_2)$ can be selected at will. For instance, the wavefunctions in the VB method have only a covalent character, whereas the wavefunctions in the MO method all have an equal mixture of covalent and ionic character. In both cases the wavefunctions have a fixed ratio between the two characters, but one can use a variable mixture of ionic and covalent.

For optimization one takes c_1 for the amplitude of the covalent wavefunction and c_2 for the amplitude of the ionic wavefunction and finds the minimum energy as a function of these two parameters. Defining matrix elements $\mathcal{H}_{ij}$ for the Hamiltonian as

$$\mathcal{H}_{ij} = \int d\vec{r}_1 d\vec{r}_2 \; \Phi_i(\vec{r}_1, \vec{r}_2) \, \mathcal{H} \, \Phi_j(\vec{r}_1, \vec{r}_2) \tag{17.24}$$

and the elements S_{ij} for the overlap matrix as

$$S_{ij} = \int d\vec{r}_1 d\vec{r}_2 \; \Phi_i(\vec{r}_1, \vec{r}_2) \, \Phi_j(\vec{r}_1, \vec{r}_2), \tag{17.25}$$

the resulting set of equations for minimizing the energy as a function of c_i has a non-trivial solution only when the secular determinant is zero:

$$\det(\mathcal{H} - ES) = 0. \tag{17.26}$$

There are two solutions for two coefficients c_i and the minimum is for the lower state. For the linear combination of covalent and ionic wavefunctions, the binding energy increases by a few tenths of eV, as can be seen from Tab. 17.1. The optimization leads to a ratio of $c_2/c_1 = 0.16$, which shows that the wavefunction for H_2 at its equilibrium distance has mainly covalent character, and only a little ionic character.

To facilitate the calculations, trial wavefunctions can be chosen of the form

$$\Phi_i(\vec{r}_1, \vec{r}_2) = e^{-\alpha(\xi_1 + \xi_2)} r_{12}^{\mu_i} \left(\xi_1^{p_i} \eta_1^{q_i} \xi_2^{r_i} \eta_2^{s_i} + \xi_1^{r_i} \eta_1^{s_i} \xi_2^{p_i} \eta_2^{q_i} \right), \tag{17.27}$$

where p_i, q_i, r_i, s_i, and μ_i are integers and ξ and η the confocal elliptical coordinates. The variational parameters are α and $c_1, c_2, \ldots$, and for the wavefunction to be symmetrical under the exchange of nucleus A and B, $q_i + s_i$ must be even. It is interesting to follow the increased computational power from the results obtained with this wavefunction for H_2. In 1933 James and Coolidge [180] used 13 wavefunctions to obtain $V = -0.1735$ au at $R_{eq} = 1.40a_0$. In 1968 Kolos and Wolniewicz [184] used 100 wavefunctions to obtain $V = -0.174\,474\,983$ au at $R_{eq} = 1.4011a_0$. Recently, this value was improved by Pachucki [161] in 2010 to $-0.174\,475\,931\,400\,215\,99$ au at the same distance using 22,000 wavefunctions. The latest results are very impressive and the accuracy of the results is claimed to be of the order 10^{-16}. It is thus generally believed by chemists that quantum mechanics provides a complete description of chemical phenomena, although the actual calculation might be too complex to be carried out.

17.5 Nature of the H_2 bond

The virial theorem can provide insight into the binding of the H_2 molecule. For an atom the theorem is given by $\langle T \rangle = -\langle V \rangle / 2$, which shows that for all the states in hydrogen the kinetic energy is equal to the binding energy of the electron, whereas

17.5 Nature of the H₂ bond

the potential energy is twice the binding. For a diatomic molecule the virial theorem becomes [186]

$$\langle T \rangle = -1/2 \langle V \rangle - 1/2 R \frac{dE}{dR}, \quad (17.28)$$

which shows that the virial theorem for atoms is reproduced for $R = R_{\min}$, where dE/dR vanishes. Note that the generalized theorem for diatomic molecules cannot be applied to all wavefunctions, since it requires that the variational approach is used to optimize its energy. Combined with the relation $E = \langle T \rangle + \langle V \rangle$, Eq. (17.28) can be used to calculate the kinetic and potential energy at each internuclear distance R:

$$\langle T \rangle = -\left(E + R\frac{dE}{dR}\right) \quad \text{and} \quad \langle V \rangle = 2E + R\frac{dE}{dR}. \quad (17.29)$$

In this way the energy $E(R)$ is sufficient to find the kinetic and potential energy of the electrons. The results for E, $\langle T \rangle$, and $\langle V \rangle$ are plotted in Fig. 17.7. At large internuclear distance ($R \geq 3a_0$) the electron distribution is extended along the internuclear axis. This lowers the kinetic energy consistent with the uncertainty principle, since the wavefunction is extended over a larger region. The electron distribution is slightly shifted from the region close to the nuclei to the region between the nuclei, and this increases the potential energy somewhat. Overall, the total energy E is lowered. For shorter distances ($1a_0 \leq R \leq 3a_0$) the region between the nuclei is smaller and this increases the kinetic energy, but the potential energy is strongly decreased because of the lowering of the potential between the nuclei.

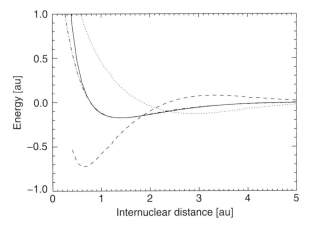

Figure 17.7 Kinetic energy (dotted line), potential energy (dashed line), and total energy (solid line) for the H₂ molecule calculated using the potential of Ref. [161]. The binding at long range results from the decrease of kinetic energy, whereas the binding at short range is caused by the decrease of potential energy. The total energy using a Morse potential adaptation is shown by the dashed-dotted line.

At very short distance ($R \leq 1a_0$) the repulsion between the nuclei becomes very strong and this increases the potential energy. The balance between the kinetic and potential energy leads to the binding of the molecule at large R and the repulsion at short R. Note that the electron distribution between the two nuclei is proportional to the overlap $I(R)$ (see Eq. (17.19)), and thus a strong overlap between the wavefunctions leads to a strong binding.

Appendices

17.A Confocal elliptical coordinates

The confocal elliptical coordinates for an electron in a diatomic molecule are defined by

$$\xi = \frac{r_A + r_B}{R}, \quad \eta = \frac{r_A - r_B}{R}, \quad \text{and} \quad \phi, \tag{17.30}$$

where $r_{A,B}$ is the distance from the electron to nucleus A, B, respectively, R the internuclear distance, and ϕ the azimuthal angle with respect to the internuclear axis. The limits for the coordinates are $1 \leq \xi \leq \infty$, $-1 \leq \eta \leq +1$, and $0 \leq \phi \leq 2\pi$. The foci are located $R/2$ from the origin and the surfaces of constant ξ are ellipsoids of revolution with respect to these foci, whereas the surfaces of constant η are hyperboloids of revolution having the same foci, as shown in Fig. 17.8. The Cartesian coordinates x, y, and z are given by

$$x = 1/2 R \sqrt{\xi^2 - 1} \sqrt{1 - \eta^2} \cos\phi,$$
$$y = 1/2 R \sqrt{\xi^2 - 1} \sqrt{1 - \eta^2} \sin\phi,$$
$$z = 1/2 R \, \xi \, \eta. \tag{17.31}$$

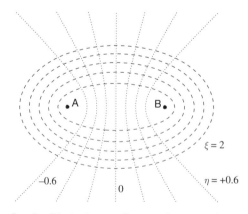

Figure 17.8 Confocal elliptical coordinates for an electron in a diatomic molecule. The contour lines for ξ are from 1.2 to 2 with steps of 0.2, and the contour lines for η are from −0.6 to 0.6 with steps of 0.2.

Using these expressions it is easy to show that the volume element $d\vec{r}$ of the integral over the electronic coordinates is given by $d\vec{r} = 1/8 R^3 (\xi^2 - \eta^2) d\xi d\eta d\phi$.

17.B One-electron, two-center integrals

One-electron, two-center integrals can be solved used confocal elliptical coordinates, as introduced in App. 17.A. Consider the more general integral $Q(p, q)$

$$Q(p, q) = \int \frac{e^{-pr_A} e^{-qr_B}}{r_A r_B} d\vec{r}, \qquad (17.32)$$

where p and q are constants. Using the confocal elliptical coordinates one arrives at

$$Q(p, q) = \frac{R}{2} \int_1^\infty d\xi \int_{-1}^{+1} d\eta \int_0^{2\pi} d\phi \exp\left[-\frac{R}{2}((p+q)\xi - (p-q)\eta)\right]$$

$$= \frac{4\pi}{(p^2 - q^2)R} \left(e^{-qR} - e^{-pR}\right). \qquad (17.33)$$

Differentiating with respect to q leads to

$$Q'(p, q) \equiv -\frac{\partial Q(p, q)}{\partial q} = \int \frac{e^{-pr_A} e^{-qr_B}}{r_A} d\vec{r}$$

$$= \frac{4\pi}{R} \left[\frac{R}{p^2 - q^2} e^{-qR} + \frac{2q}{(p^2 - q^2)^2} \left(e^{-pR} - e^{-qR}\right) \right]. \qquad (17.34)$$

Finally, differentiating $Q'(p, q)$ with respect to p leads to

$$Q''(p, q) \equiv -\frac{\partial Q'(p, q)}{\partial p} = \int e^{-pr_A} e^{-qr_B} d\vec{r}$$

$$= \frac{8\pi}{R(p^2 - q^2)^2} \left[R\left(pe^{-pR} + qe^{-qR}\right)\right.$$

$$\left. + \frac{4pq}{p^2 - q^2} \left(e^{-pR} - e^{-qR}\right)\right]. \qquad (17.35)$$

These expressions can now be used to find relations for the integrals $I(R)$, $\mathcal{J}_1(R)$, and $\mathcal{K}_1(R)$.

The wavefunction of the H(1s) orbital is given in atomic units by

$$\psi(r) = \frac{1}{\sqrt{\pi}} e^{-r}. \qquad (17.36)$$

Using the above relations one obtains

$$I(R) \equiv \frac{1}{\pi} Q''(1, 1) = (1 + R + \tfrac{1}{3}R^2) e^{-R},$$

$$\mathcal{J}_1(R) \equiv \frac{1}{\pi} Q'(2, 0) = \frac{1}{R} - \frac{1}{R}(1 + R) e^{-2R},$$

$$\mathcal{K}_1(R) \equiv \frac{1}{\pi} Q'(1, 1) = (1 + R) e^{-R}. \qquad (17.37)$$

Using these expressions in combination with Eq. (17.8) yields the energy of the *gerade* and *ungerade* states of H_2^+ molecular ion.

17.C Electron–electron interaction in molecular hydrogen

The electron–electron interaction in H_2 is difficult to calculate, and in the literature many erroneous results are reported. Here the evaluation will be shown in detail for the $\mathcal{J}_2(R)$-integral, which can be applied to the other integral as well. The $\mathcal{J}_2(R)$-integral is

$$\mathcal{J}_2(R) = \int d\vec{r}_1 d\vec{r}_2 \frac{|\psi_{1s}(\vec{r}_{A1})\psi_{1s}(\vec{r}_{B2})|^2}{r_{12}}$$

$$= \frac{1}{\pi^2} \int d\vec{r}_1 d\vec{r}_2 \frac{e^{-2(r_{A1}+r_{B2})}}{r_{12}}. \tag{17.38}$$

The term $1/r_{12}$ can be evaluated using the expansion in Legendre polynomials of Eq. (14.26). They are given in terms of spherical coordinates, whereas confocal elliptical coordinates are usually used to evaluate two-center integral. Using the Legendre expansion around nucleus A the two-electron integral $\mathcal{J}_2(R)$ depends on only r_{A1} and not on r_{B1}. This allows for the integration over $\vec{r}_1$ in spherical coordinates. Integration of the (θ_1, ϕ_1)-coordinates of the spherical harmonic $Y_{kq}(\theta_1, \phi_1)$ in Eq. (14.26) yields

$$\int Y_{kq}(\theta_1, \phi_1) \sin\theta_1 d\theta_1 d\phi_1 = \sqrt{4\pi}\delta_{k0}\delta_{q0}, \tag{17.39}$$

and thus

$$\int \frac{1}{r_{12}} \sin\theta_1 d\theta_1 d\phi_1 = \frac{4\pi}{r_>}, \tag{17.40}$$

where $r_> = \max(r_{A1}, r_{A2})$. The integral over r_{A1} can now be carried out, yielding

$$\mathcal{J}_2(R) = \frac{4}{\pi} \int dr_{A1} d\vec{r}_2 r_{A1}^2 \frac{e^{-2(r_{A1}+r_{B2})}}{r_{12}}$$

$$= \frac{4}{\pi} \int \int_{r_{A2}}^{\infty} dr_{A1} d\vec{r}_2 r_{A1}^2 \frac{e^{-2(r_{A1}+r_{B2})}}{r_{A1}}$$

$$+ \frac{4}{\pi} \int \int_0^{r_{A2}} dr_{A1} d\vec{r}_2 r_{A1}^2 \frac{e^{-2(r_{A1}+r_{B2})}}{r_{A2}}$$

$$= \frac{1}{\pi} \int d\vec{r}_2 e^{-2(r_{A2}+r_{B2})}(1+2r_{A2}) + \frac{1}{\pi} \int d\vec{r}_2 e^{-2r_{B2}}$$

$$- \frac{1}{\pi} \int d\vec{r}_2 \frac{e^{-2(r_{A2}+r_{B2})}}{r_{A2}}\left(1+2r_{A2}+2r_{A2}^2\right). \tag{17.41}$$

The second term can be identified as $\mathcal{J}_1(R)$, but the other terms can also be evaluated using confocal elliptical coordinates. The result becomes

$$\mathcal{J}_2(R) = \frac{1}{R}\left[1 - e^{-2R}\left(1 + \tfrac{11}{8}R + \tfrac{3}{4}R^2 + \tfrac{1}{6}R^3\right)\right]. \tag{17.42}$$

The same procedure can also be applied to $\mathcal{K}_3(R)$, and the result is

$$\mathcal{K}_3(R) = \int d\vec{r}_1 d\vec{r}_2 \frac{|\psi_{1s}(\vec{r}_{A1})|^2 \psi_{1s}(\vec{r}_{A2})\psi_{1s}(\vec{r}_{B2})}{r_{12}}$$

$$= \frac{1}{R}\left[e^{-R}\left(\tfrac{5}{16} + \tfrac{1}{8}R + R^2\right) - e^{-3R}\left(\tfrac{5}{16} + \tfrac{1}{8}R\right)\right]. \tag{17.43}$$

The integral $\mathcal{J}_3(R)$ has been evaluated for helium, and the result is $\mathcal{J}_3(R) = 5/8$, independent of R.

Finally, the most complicated integral is $\mathcal{K}_2(R)$, where the atomic wavefunctions depend on all four coordinates r_{A1}, r_{A2}, r_{B1}, and r_{B2}. Thus the approach above does not work for this integral. The integral was first evaluated by Sugiura [176]. The result is

$$\mathcal{K}_2(R) = e^{-2R}\left(\tfrac{5}{8} - \tfrac{23}{20}R - \tfrac{3}{5}R^2 - \tfrac{1}{15}R^3\right) + \frac{6}{5R}\left[\mathcal{I}(R)^2(\gamma + \log R)\right.$$

$$\left. + \mathcal{I}'(R)^2 E_i(-4R) - 2\mathcal{I}(R)\mathcal{I}'(R) E_i(-2R)\right], \tag{17.44}$$

where $\gamma = 0.5772\ldots$ is the Euler constant, E_i the integral logarithm, and $\mathcal{I}'(R) \equiv \mathcal{I}(-R)$. All these factors have to be evaluated in order to find the energy for the MO- and VB-methods.

Exercises

17.1 For the molecular ion H_2^+, the wavefunction is given by Eq. (17.2).
(a) Find the energy of the molecule by using Eq. (17.3) and express your result in terms of $\mathcal{H}_{AA}(R)$, $\mathcal{H}_{AB}(R)$, and $\mathcal{I}(R)$, as defined in Eqs. (17.4) and (17.5).
(b) Express $\mathcal{H}_{AA}(R)$ and $\mathcal{H}_{AB}(R)$ in terms of $\mathcal{J}_1(R)$ and $\mathcal{K}_1(R)$.
(c) Plot your results as a function of R using the relations of Eq. (17.37) and determine the minimum and the equilibrium position in the ground state. Compare your results with the well-known values for the H_2^+ molecule (see Tab. 17.1).

17.2 For the molecular ion H_2^+, Sec. 17.1 uses the atomic hydrogen wavefunctions. In this problem the variational principle will be used by taking the one-electron wavefunction as

$$\psi_{1s}(r_A) = N e^{-\alpha r_A} \qquad \psi_{1s}(r_B) = N e^{-\alpha r_B},$$

with the normalization condition $N = \sqrt{\alpha^3/\pi}$. The Hamiltonian is given by Eq. (17.1) and the wavefunction is given by Eq. (17.2), where for the *gerade* ground state the sign in the brackets is +.

320   Binding in the hydrogen molecule

(a) Determine the action of the kinetic operator $-\hbar^2 \nabla^2/2m$ on this wavefunction and compare your result for $\alpha = 1$ with the well-known result for atomic hydrogen.
(b) Determine the terms $\mathcal{H}_{AA}(R)$ and $\mathcal{H}_{AB}(R)$ and express your results in terms of $\mathcal{J}_1(R)$ and $\mathcal{K}_1(R)$.
(c) Use the results to determine the energy $E_g(R)$ of the *gerade* ground state as a function of α.
(d) Minimize this energy using the relations of Eq. (17.37) with respect to α and compare your result to the well-known values for the H_2^+ molecule (see Tab. 17.1).

17.3 Calculate the overlap integral $I(R)$ of Eq. (17.4) directly by substitution of the hydrogen wavefunction $\psi(r_i) = \exp(-r_i)/\sqrt{\pi}$ with $i = A, B$ and transformation to confocal elliptical coordinates of Eq. (17.30). Compare your result with Eq. (17.37).

17.4 There is a discussion in Sec. 16.2 about Bohr's model for molecular hydrogen based on his famous atomic hydrogen model. He assumed that the two electrons orbit in a circle whose plane is perpendicular to the internuclear axis at fixed positions z_i on the internuclear axis. He also assumed that the two electrons satisfied the quantization condition $mvr = n\hbar$ as discussed in Sec. 1.5 with $n = 1$.
(a) Using Bohr's assumption for the quantization of the electron, find the kinetic energy T of the two electrons as a function of their orbital radius r_i with $i = 1, 2$. The potential energy V of the system is given by

$$V = \frac{e^2}{4\pi\varepsilon_0}\left(-\frac{Z}{r_{1A}} - \frac{Z}{r_{1B}} - \frac{Z}{r_{2A}} - \frac{Z}{r_{2B}} + \frac{1}{r_{12}} + \frac{Z^2}{R}\right),$$

where the distances between the different particles are shown in Fig. 17.3.
(b) Determine the distances r_{iA}, r_{iB}, and r_{12} as a function of r_i, z_i, and ϕ, where ϕ is the difference between the azimuth angles of electron 1 and 2.
Bohr considered four different configurations: (1) $z_1 = z_2 = 0$, $\phi = \pi$, (2) $z_1 = -z_2$, $\phi = \pi$, (3) $z_1 = z_2$, $\phi = \pi$, and (4) $z_1 = -z_2$, $\phi = 0$.
(c) Determine the symmetry of the configurations and relate these states to the $^1\Sigma_g^+$ or $^3\Sigma_u^+$ state of hydrogen.
(d) Show to which asymptote these four configurations are related for large R, namely $H + H$, $H^+ + H^-$, or $2 H^+ + 2e$.
(e) Minimize numerically the total energy $E = T + V$ at a fixed internuclear distance R as a function of r_1, r_2, z_1, and z_2 for the four different configurations. Plot your results as a function of R and compare your results with Fig. 17.4 for the MO approach and Fig. 17.5 for the VB approach.

17.5 Show that the wavefunction $\Psi_u^{VB}(\vec{r}_1, \vec{r}_2)$ of Eq. (17.17) in the VB approach is identical to the wavefunction $\Psi_D^{MO}(1, 2)$ of Eq. (17.9) in the MO approach.

18
Ultra-cold chemistry

18.1 Introduction

Interactions between separated atoms are complicated by the large number of degrees of freedom of a many-atom system, and are very difficult to describe quantitatively. Although the Hamiltonian for such systems can be written down easily, finding the eigenfunctions requires a sophisticated effort and large computer resources. This area of quantum chemistry, with all its intricacies, is a complex subfield of science that cannot be adequately described in a single book chapter. Therefore the student is referred to specialized books in this field for further reading [165, 170].

In the 1990s a new field of chemistry emerged that considered such interactions between atoms at ultra-cold temperatures. The field focuses strongly on atomic interactions at large internuclear distances where the atoms are well separated and their electronic clouds do not strongly overlap. In such circumstances the interactions are dominated by the properties of the states of the atoms and they can be understood from an atomic point of view. In this chapter the basic features of this field will be described.

The interaction energy (electronic potential energy) between two colliding partners separated by a large distance is called the dissociation limit. Light interactions with the particles can occur with the total energy above or below this limit. The relative kinetic energy does not contribute because it is so very small at such low temperatures. In the presence of a light field tuned below this limit for a pair that has one of the atoms in an excited state, two unbound atoms can be associated into a bound molecular state (for example, an S–P collision of two alkali atoms).

Such photo-association is more likely at a large interatomic separation because there is good radial overlap between the incoming, low-energy scattering state and the outer turning point of the bound excited state, so it usually produces a molecule in a high vibrational state. Studying these states provides important information about interatomic interactions near the dissociation limit. The long-range part of the potential is dominated by atomic parameters and in Sec. 18.2 it is shown how these potentials can be constructed. The LeRoy–Bernstein method, which was developed long before the advent of cold-atom photo-association, is described in Sec. 18.3 and can be used to extract these parameters from measurements of the bound states.

For the case of atomic interactions above the dissociation limit, scattering theory that was developed more than a century ago, and is discussed in many textbooks, can be applied to describe its features. For ultra-cold atoms the interactions are dominated by scattering of the lowest partial waves (Sec. 18.4), and in most cases can be described accurately by the s-wave scattering length. A short description of scattering theory is given in Sec. 18.4, followed by a longer description of the scattering length that plays an important role in ultra-cold chemistry in Sec. 18.5. Finally a simple model of Feshbach resonances is given in Sec. 18.6. These play an important role in ultra-cold chemistry for the production of ultra-cold molecules as described in Sec. 22.4.

18.2 Long-range molecular potentials

The interatomic potentials for diatomic molecules are very complicated. To narrow the discussion, the focus here is on the dimers of alkali-metal atoms. Even so, the atomic interactions at short range ($<30a_0$) are especially complicated so reliable potentials can only be obtained by sophisticated quantum chemistry calculations. However, at long range ($>30a_0$) the interactions are dominated by the electrostatic forces between the atomic valence electrons that depend on their particular atomic states.

For instance, for two atoms interacting in their S+P states respectively, the long-range interaction is fully determined by the dipole–dipole interaction. The interaction potential is of the form C_3/R^3, and the long-range coefficient C_3 can be calculated from first principles. In the case of the alkali-metals, those coefficients have been calculated by Refs. [187, 188]. At ultra-cold temperatures the potentials are experimentally probed only at long range, and the analysis of the results can be carried out with little knowledge of the short-range potentials. For the alkali-metal atoms, fine structure and even hyperfine structure may also play important roles. As is shown below, those effects can be incorporated in the analysis of the long-range dipole–dipole interactions.

For a pseudo two-electron system the long-range interaction can be written as (see Eq. (16.3)):

$$V_{\text{int}} = \frac{e^2}{4\pi\varepsilon_0}\left(\frac{1}{|\vec{R}|} + \frac{1}{|\vec{R}+\vec{r}_b-\vec{r}_a|} - \frac{1}{|\vec{R}+\vec{r}_b|} - \frac{1}{|\vec{R}-\vec{r}_a|}\right), \qquad (18.1)$$

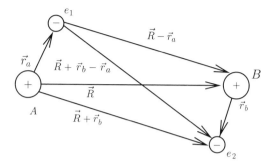

Figure 18.1 Coordinates in the long-range interaction in a diatomic molecule, where R denotes the internuclear distance. Here r_a is the position of electron 1 with respect to the nucleus A and r_b the position of electron 2 with respect to the nucleus B.

with $\vec{R} \equiv \vec{R}_A - \vec{R}_B$ position of nucleus A with respect to B, $\vec{r}_a$ the position of electron 1 with respect to nucleus A and $\vec{r}_b$ the position of electron 2 with respect to nucleus B, as shown in Fig. 18.1. The interaction between each electron and its own nucleus is included in the atomic structure.

Each of the terms of Eq. (18.1) can be written as a multipole expansion of Legendre polynomials or spherical harmonics. The resulting sums can be combined for notational convenience resulting in [189]:

$$V_{\text{int}} = \sum_{j,k=1}^{\infty} U_{jk}(\vec{R}). \tag{18.2}$$

Here

$$U_{jk}(\vec{R}) = (-1)^k C_{jk} \frac{e^2 r_a{}^j r_b{}^k}{4\pi\varepsilon_0 R^{j+k+1}}$$

$$\times \sum_{p=-j}^{j} \sum_{q=-k}^{k} Q_{pq} Y_{jp}(\hat{r}_a) Y_{kq}(\hat{r}_b) Y^*_{j+k, p+q}(\hat{R}), \tag{18.3}$$

where C_{jk} and Q_{pq} are dimensionless numerical parameters of order unity and independent of R.

Now choose a combination of one-electron wavefunctions as an electron basis set that are designated as

$$|\alpha\rangle = |n_a \ell_a m_a; n_b \ell_b m_b\rangle \tag{18.4}$$

where a and b refer to the individual atoms. The wavefunctions are not properly symmetrized with respect to electron exchange, but the two electron clouds are non-overlapping and the energy does not depend on the symmetry of the wavefunctions. Then the matrix elements of Eq. (18.2) are given by

$$\langle\alpha|V_{\text{int}}|\alpha'\rangle = \sum_{j,k=1}^{\infty} \frac{e^2 D_{\alpha\alpha'}(j,k)}{4\pi\varepsilon_0 R^{j+k+1}} Y^*_{j+k,\,p+q}(\hat{R}), \tag{18.5}$$

with $p = m_a - m'_a$, $q = m_b - m'_b$ and

$$D_{\alpha\alpha'}(j,k) = (-1)^k C_{jk} \langle n_a\ell_a|r_a^j|n'_a\ell'_a\rangle \langle n_b\ell_b|r_b^k|n'_b\ell'_b\rangle$$
$$\times Q_{pq} \langle\ell_a m_a|Y_{jp}|\ell'_a m'_a\rangle \langle\ell_b m_b|Y_{kq}|\ell'_b m'_b\rangle. \tag{18.6}$$

Here the primed quantum numbers refer to state α'. From the properties of the spherical harmonics $Y_{\ell m}$ (see Eq. (C.23) in the end-of-book appendix) one can immediately see that $D_{\alpha\alpha'}(j,k) = 0$ unless

$$|p| \leq j; \quad |q| \leq k; \quad |\ell_a - \ell'_a| \leq j \leq (\ell_a + \ell'_a); \quad |\ell_b - \ell'_b| \leq k \leq (\ell_b + \ell'_b). \tag{18.7}$$

Choosing the z-direction along the internuclear axis one finds

$$Y^*_{j+k\,p+q}(\hat{R}) = \sqrt{\frac{2(j+k)+1}{4\pi}} \delta_{p+q,0}. \tag{18.8}$$

Therefore the z-component of the total orbital angular momentum of the two electrons is conserved, or $m_a + m_b = m'_a + m'_b$. Finding the interaction potential relies on the identification of the states that are coupled.

For the Coulomb interaction between two charged atoms (ions) both $j = 0$ and $k = 0$, so the lowest-order interaction reduces to C_1/R with $C_1 = e^2/4\pi\varepsilon_0$, as expected. But for the case of two neutral atoms interacting in electronic states that differ by $\Delta\ell = \pm 1$, the states are coupled via the electric dipole interaction and the matrix elements are proportional to $\langle|r|\rangle$ or $j = k = 1$. The interaction thus becomes C_3/R^3 and is called the dipole–dipole interaction. Table 18.1 summarizes the different types of interactions and identifies the order of the interaction. The van der Waals interaction is a special case, since it derives from the second order of the dipole–dipole interaction, as will be shown later.

For the S–P asymptote there are couplings between the states $|\ell_a m_a; \ell_b m_b\rangle = |s0; pm\rangle$ and $|pm; s0\rangle$, with $m=0, \pm 1$. Note that $m_a + m_b = m$ in this case and since $m_a + m_b$ is conserved, one can consider the different values of m separately. These two states are coupled by the electric dipole interaction and the first non-zero element is $\langle n_a\ell_a|r_a|n'_a\ell'_a\rangle$. Therefore the minimum value for j and k is 1 and the lowest-order interaction is the dipole–dipole interaction. Equation (18.5) shows that the diagonal elements are zero and the non-diagonal elements are given by

$$\langle\alpha|V_{\text{int}}|\alpha'\rangle = -\frac{2}{3}\frac{\mu_d^2}{4\pi\varepsilon_0 R^3}, \quad m = 0; \text{ and } \langle\alpha|V_{\text{int}}|\alpha'\rangle = \frac{1}{3}\frac{\mu_d^2}{4\pi\varepsilon_0 R^3}, \quad |m| = 1 \tag{18.9}$$

18.2 Long-range molecular potentials 325

Atoms	j	k	n	Interaction
2 charged atoms	0	0	1	monopole–monopole
1 charged atom and 1 atom with permanent dipole moment	0	1	2	monopole–dipole
2 atoms with permanent dipole moment	1	1	3	dipole–dipole
2 atoms in states with $\Delta \ell = \pm 1$	1	1	3	resonant dipole–dipole
1 charged atom and 1 neutral atom	0	3	4	monopole–induced dipole
1 atom with permanent dipole moment and 1 atom with permanent quadrupole moment	1	2	4	dipole–quadrupole
2 atoms with permanent quadrupole moments	2	2	5	quadrupole–quadrupole

Table 18.1 Lowest-order interaction for two atoms in states defined in the first column. The lowest order is given by $n = j+k+1$, where j and k are the coefficients determined by the individual atoms.

with $\mu_d = -e\langle s||r||p\rangle$ the dipole moment for the S ↔ P transition. Diagonalizing the matrix is simple since the diagonal elements are zero and yields

$$\epsilon_{1,2} = \pm \frac{2}{3} \frac{C_3}{R^3}, \quad \Sigma\text{-state}; \quad \text{and} \quad \epsilon_{1,2} = \pm \frac{1}{3} \frac{C_3}{R^3}, \quad \Pi\text{-state}, \quad (18.10)$$

where the generic dispersion coefficient for dipole–dipole interaction $C_3 = \mu_d^2/4\pi\epsilon_0$ has been introduced.

To introduce the fine structure into the calculation of the potential curves, one can choose the basis functions including the fine structure of the atoms as

$$|\alpha\rangle = |n_a(\ell_a s_a) j_a m_a; n_b(\ell_b s_b) j_b m_b\rangle, \quad (18.11)$$

where $m_{a,b}$ is the projection of the total angular momentum $j_{a,b}$ on the internuclear axis. Since the interaction term V_{int} depends on $\ell_{a,b}$ and not on $j_{a,b}$, the basis functions can be expanded in terms of the eigenstates $|n_a \ell_a m_a^\ell\rangle$:

$$|n_a(\ell_a s_a) j_a m_a\rangle = \sum_{\ell_a, m_a^\ell, s_a, m_a^s} |n_a \ell_a m_a^\ell; s_a m_a^s\rangle \langle \ell_a m_a^\ell; s_a m_a^s | j_a m_a\rangle \quad (18.12)$$

and a similar relation for the electron at nucleus b. By using the Clebsch–Gordan coefficients and summing over all angular momenta $\ell_{a,b}$ and $s_{a,b}$, the coupling terms can be calculated. The diagonal elements are given by the energy of the state, i.e. the energy of fine-structure states.

Even though the wavefunctions are not properly symmetrized, the results are symmetrized wavefunctions since the Hamiltonian is symmetric. Furthermore, in the analysis there is no difference between electron 1 at nucleus a and electron 2 at nucleus b, versus electron 1 at nucleus b and electron 2 at nucleus a, since the two electron clouds do not overlap. Therefore the exchange symmetry for the spatial wavefunction cannot be determined from the result. However, the Pauli principle requires the total wavefunction to be anti-symmetric with respect to the exchange of electrons and thus the exchange symmetry of the electron spin wavefunction determines the exchange symmetry for the spatial part.

In Fig. 18.2 the potentials connected to the 3S + 3P asymptote of Na_2 are shown. In Fig. 18.2a the two asymptotes $3S_{1/2} + 3P_{1/2}$ and $3S_{1/2} + 3P_{3/2}$ are shown, and they are split by the 515 GHz fine structure of the 3P state. In Fig. 18.2b the region around the $3S_{1/2} + 3P_{3/2}$ asymptote is enlarged, and one clearly observes

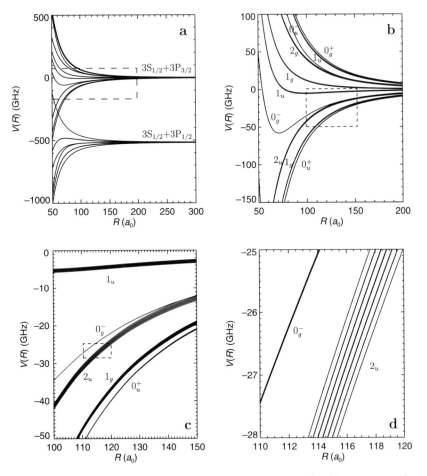

Figure 18.2 Calculated Hund's case (c) potential curves for Na_2 connected to the $3S_{1/2} + 3P_{1/2,3/2}$ dissociation limit shown in two different panels. In (a) all the molecular potentials are shown including fine structure (the fine-structure splitting is $\Delta_{fs} = 515.52$ GHz), (b) shows all the curves connecting to the $3S_{1/2} + 3P_{3/2}$ dissociation limit and (c) shows only the attractive curves connected to this threshold. Finally, in (d) the hyperfine structure of the 2_u state can also be distinguished. (Figure from Ref. [190].)

18.2 Long-range molecular potentials

the characteristic C_3/R^3 dependence of the potentials combined with the avoided crossings of some of the potentials caused by the fine-structure interaction. Zooming in more closely on some of the potentials, in Fig. 18.2c,d one observes the splitting of each of those potentials into 10 different ones caused by the hyperfine interactions. Although the number of states involved becomes large and the detailed shape of the potential becomes complicated, all those features can be calculated by relying on a relatively simple model description of the long-range forces.

In experiments the probe laser is tuned to the red (blue) side of the $3S_{1/2} \rightarrow 3P_{3/2}$ ($3P_{1/2}$) atomic transition frequency. Therefore the only molecular potentials that need to considered are attractive and asymptotically connecting to the $3S_{1/2} + 3P_{3/2}$ dissociation limit. Repulsive potentials connecting to the $3S_{1/2} + 3P_{1/2}$ dissociation limit cannot sustain bound states and are therefore of no relevance to the experiments discussed in Sec. 22.5. Figure 18.2b shows that five molecular symmetries fulfill the necessary requirements: 1_u, 0_g^-, 2_u, 1_g, and 0_u^+. Here the molecular symmetry is specified in the Hund's case (b) notation (see Sec. 16.5), since the spin of the electrons is not coupled to the internuclear axis because there is little overlap between the two electronic clouds. The states are identified as $\Lambda_{u,g}^\pm$, where $\Lambda\hbar$ is the projection of the orbital angular momentum of the electrons on the internuclear axis, u, g specifies the inversion symmetry with respect to the midpoint of the internuclear axis and $\pm$ is the reflection symmetry for $\Lambda = 0$ states with respect to a plane containing the internuclear axis (see Sec. 16.5). The 1_u potential is called a purely long-range potential with a well depth of only ~5 GHz, while the 0_g^- potential is also purely long range with a well depth of $\approx$ 57 GHz [191–193]. Although these two states are attractive at long range, they are repulsive at distances of 50–100a_0 (see Fig. 18.2b) because of avoided crossings with repulsive states connecting to the $3S_{1/2} + 3P_{1/2}$ dissociation limit. Bound levels in these potentials have inner turning points at large internuclear distances and their properties can be calculated to high precision from the atomic properties of Na. The 2_u, 1_g, and 0_u^+ potentials remain attractive to much shorter internuclear distances. Transitions from the ground state to the 2_u state are dipole forbidden.

For the van der Waals interaction the states to consider are $|s0; s0\rangle$ and $|p+m; p-m\rangle$ with $m = 0, \pm 1$. Thus there are four states coupled, and for the coupling elements one finds (see Eq. (18.9))

$$\langle p+m; p-m|V_{\text{int}}|s0; s0\rangle = \begin{cases} -\dfrac{2C_3}{3R^3} & m = 0; \\ \dfrac{C_3}{3R^3} & |m| = 1. \end{cases} \quad (18.13)$$

Note that the state $|s0; s0\rangle$ has a different energy than the states $|p+m; p-m\rangle$, where the shift is given by twice the atomic energy shift $\Delta E = E_p - E_s$. Diagonalizing the matrix leads to a shift of the ground state given by

$$E_g = -\frac{2\mu_d^4}{6\,\Delta E\,(4\pi\varepsilon_0)R^6} \equiv -\frac{C_6}{R^6}. \qquad (18.14)$$

This shows the characteristic dependence of the van der Waals interaction ($n = 6$) on the internuclear separation. Since the matrix elements are given by the square of the dipole coupling in each atom, it is also referred as the second-order dipole–dipole interaction. For Na, the coefficient becomes $C_6 = 1,521$ au, which has to be compared with the theoretical value $C_6 = 1,472$ au. Considering the fact that only four states are taken into account in this simple model calculation, the agreement between the two values is rather good.

18.3 LeRoy–Bernstein method

In general it is difficult to find the eigenstates for a particular potential $V(R)$ if the potential is not very well known. The energies of the states depend strongly on the detailed shape of the whole potential. At short range, where the electronic wavefunctions of the two atoms overlap significantly, the potential is especially difficult to determine. Ultra-cold collisions can provide experimental knowledge of only the highest bound states so it would be convenient to be able to predict the energies using only the long-range part of the potential. Such a method was discussed in Ref. [194] before the advent of laser cooling and ultra-cold molecules.

In their method the WKB approximation was used to find a solution for the wavefunction in a certain potential. This wavefunction is formally the solution of the Schrödinger equation given by Eq. (18.31). In this section it is assumed that the rotational potential is included in the potential $V(R)$. Then a local wavenumber for the nuclei can be defined in a similar way as Eq. (10.2) for an electron as

$$k(R) = \frac{\sqrt{2\mu_N(E - V(R))}}{\hbar} \qquad (18.15)$$

and if this wavenumber does not depend strongly on R the wavefunction can be written as $u(R) \propto \exp(\pm ikR)$. However, $k(R)$ depends on the internuclear distance R and in the WKB approximation a trial wavefunction $u_0(R) \propto \exp(\pm i\phi_0(r))$ can be used with

$$\phi_0 = \int k(R)\,\mathrm{d}R + C_0, \qquad (18.16)$$

which can be substituted in Eq. (18.31). This leads to the next-order approximation, which can be written as

$$u_1(R) \propto \frac{1}{\sqrt{|k(R)|}} \exp\left(\pm i \int k(R)\,\mathrm{d}R\right). \qquad (18.17)$$

18.3 LeRoy–Bernstein method

Although higher-order approximations can be found, for the present method this will be sufficient.

One of the requirements for this method to work is that $k(R)$ does not change significantly within one oscillation of the wavefunction, or

$$\left|\frac{\mathrm{d}k(R)}{\mathrm{d}R}\right| \ll |k(R)|^2. \tag{18.18}$$

Although this is easily fulfilled in the center of the potential well and also far in the forbidden region, this is not true near the classical turning points, where $|k(R)|$ becomes small. At the classical turning points, $k(R) = 0$. To accommodate for this exception the wavefunction can be integrated analytically, making certain assumptions about the potential near the classical turning points. This leads to so-called connection formulas that relate the wavefunctions in the inner region to the wavefunctions in the classically forbidden region. These formulas show that in the classically forbidden region the wavefunction picks up a phase of $\pi/4$.

For a stable bound state the phase during one round trip of the molecule should change by $2v\pi$ with v an integer, otherwise the returning wavefunction interferes destructively with the original wavefunction. The phase accumulated in the inner region between the classical turning points is given by

$$\phi_{\mathrm{inner}} \equiv \frac{\sqrt{2\mu_{\mathrm{N}}}}{\hbar}\int_{R_1}^{R_2}\sqrt{E(v)-V(R)}\,\mathrm{d}R = \frac{2v\pi - 4\pi/4}{2} = (v+\tfrac{1}{2})\pi, \tag{18.19}$$

where R_1 and R_2 are the classical turning points. Here the factor 4 in front of $\pi/4$ in the numerator accounts for the fact that the system goes in and out of the inner and outer forbidden region, and the factor 2 in the denominator accounts for the fact that in one round trip the phase in the inner region has been accumulated twice. The leads to the well-known WKB condition for bound levels in a potential well,

$$v + \tfrac{1}{2} = \frac{\sqrt{2\mu_{\mathrm{N}}}}{\pi\hbar}\int_{R_1}^{R_2}\sqrt{E(v)-V(R)}\,\mathrm{d}R, \tag{18.20}$$

with $E(v)$ the energy of the vibrational state v. The condition allows one to solve an integral instead of solving the wave equation Eq. (18.31).

For the energy levels close to the dissociation level LeRoy and Bernstein [194] employ the WKB condition of Eq. (18.20), but what is needed is a functional form for the position of the resonances $E(v)$. Differentiating both sides with respect to $E(v)$ leads to

$$\frac{\mathrm{d}v}{\mathrm{d}E(v)} = \frac{\sqrt{\mu_{\mathrm{N}}}}{\sqrt{2}\pi\hbar}\int_{R_1}^{R_2}\frac{1}{\sqrt{E(v)-V(R)}}\mathrm{d}R. \tag{18.21}$$

For the long range of the potential $V(R)$ one uses

$$V(R) = D - \frac{C_n}{R^n}, \tag{18.22}$$

where D is the dissociation limit. Note that at $R = R_2$ the binding energy $E(v)$ equals the potential $V(R_2)$, so the factor under the square-root of Eq. (18.21) becomes

$$E(v) - V(R) = \frac{C_n}{R^n} - \frac{C_n}{R_2^n} \equiv \frac{C_n}{R_2^n}(x^n - 1), \qquad x = \frac{R_2}{R}. \tag{18.23}$$

Then the integral of Eq. (18.21) can be written as

$$\frac{dv}{dE(v)} = \frac{\sqrt{\mu_N}}{\sqrt{2\pi}\hbar} \frac{C_n^{1/n}}{[D - E(v)]^{1/2+1/n}} \int_1^{R_2/R_1} \frac{1}{x^2\sqrt{x^n - 1}} dx. \tag{18.24}$$

If the upper limit is taken to be infinite, this is a standard integral. Using the value of this standard integral in Eq. (18.24) leads to

$$\frac{dv}{dE(v)} = A_n [D - E(v)]^{-(n+2)/2n}, \qquad A_n = \sqrt{\frac{\mu_N}{2\pi\hbar^2}} \frac{\Gamma(1/2 + 1/n) C_n^{1/n}}{\Gamma(1 + 1/n) n}. \tag{18.25}$$

Taking the factor $dE(v)$ to the right-hand side and integrating both sides leads to

$$v - v_D = B_n [D - E(v)]^{(n-2)/2n}, \tag{18.26}$$

where v_D is an integration constant and $B_n = 2nA_n/n - 2$. Finally, one obtains the binding energy of the state $D - E(v)$ as a function of the vibrational quantum number $v_D - v$:

$$D - E(v) = \left(\frac{v_D - v}{B_n}\right)^{2n/(n-2)}. \tag{18.27}$$

This is the final result in the LeRoy–Bernstein analysis [194].

The integration constant v_D plays an important role. Since the number of bound states is not known, its absolute value is unimportant. However, since the states are counted from the top, its fractional value determines the position of the resonances. If the fractional value is zero, this indicates that there is a bound state at the dissociation limit. If the value is small, it indicates that there is a resonance close to the threshold. If the value is close to 1, this indicates that the last bound level is deeply bound and that the next higher vibrational state is just into the continuum. Figure 18.3 shows the result of Eq. (18.27) for the measurement of the binding energy of the vibrational states of the Hund's case (c) 1_g potential of Na_2 near the dissociation limit $Na(3s)$–$Na(3p)$ [195]. In this case the potential has the dipole–dipole C_3/R^3 form and thus the exponent in Eq. (18.27) is 6. Therefore the discrepancy between the fit and the experimental values increases strongly as $v_D - v$ becomes large.

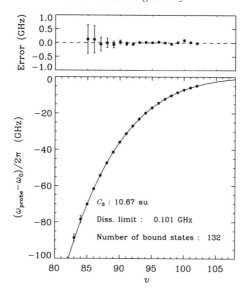

Figure 18.3 Binding energies of vibrational states in the Hund's case (c) 1_g dipole–dipole C_3/R^3 potential of Na$_2$. The experimental values are fitted to Eq. (18.27) to obtain the C_3 coefficient and the dissociation limit D_{eq}. (Figure from Ref. [195].)

18.4 Scattering theory

In the twentieth century, atomic and molecular physics consisted of two subfields: spectroscopy and collisional physics. Both aimed at gaining understanding of the structure of atoms and molecules and their interaction with light, electrons, and one another. In spectroscopy the wavelengths of the resonances in atoms and molecules yielded information about the energies of the states involved. In collision physics the profile of the scattered particles and the emission of light and electrons during scattering provided information about the interaction potential between the scatterers. The enormous increase in resolution and sensitivity that arose from the use of lasers in the latter part of the twentieth century enabled spectroscopy to flourish and dominate over scattering experiments, and has become the primary source of knowledge of atomic and molecular structure. However, atoms and molecules still interact with one another, and information about these interactions is of vital importance in this field.

Interactions play a major role for ultra-cold atoms. In the widely used process of evaporative cooling, elastic collisions of atoms enable the rethermalization process that is essential for the extension of the optical cooling process, and are therefore called "good collisions". By contrast, inelastic collisions, where the internal states of the atoms change or molecules are formed, are considered as "bad collisions", since they lead to a loss of atoms from the cold cloud. Thus the ratio of good-to-bad collisions should be large in order to have efficient evaporative cooling. In this section a short overview of collisions will be given with the emphasis on ultra-cold interactions.

The interaction of two particles can be described in the frame of their center-of-mass since their total momentum is a conserved quantity. This allows the description of such collisions as a scattering of one particle with reduced mass $\mu_N = M_A M_B/(M_A + M_B)$ in a potential $V(\vec{R})$ depending on the relative position $\vec{R} = \vec{R}_A - \vec{R}_B$. For two ultra-cold atoms the internal structures of the colliding partners plays essentially no role, and the potential depends only on the distance R between them.

Far away from the origin of the scattering, the wavefunction can be written as

$$\Psi(\vec{R}) \propto e^{i\vec{k}\cdot\vec{R}} + f(\theta,\phi)\frac{e^{ikR}}{R}. \tag{18.28}$$

The physical significance of this wavefunction is as follows. The particle comes in as a plane wave with a wavevector $\vec{k}$. After the scattering the scattered wave goes out as a spherical wave in the direction (θ,ϕ), which decreases as $1/R$ in order to conserve probability. The scattering amplitude $f(\theta,\phi)$ determines the angular distribution of the scattering amplitude, and the differential cross-section for scattering in a direction Ω is given by $d\sigma/d\Omega = |f(\theta,\phi)|^2$. Here only elastic collisions are considered, and the wavevector's magnitude k is unchanged by the collision. The result of scattering theory shows that the scattering amplitude can be found from the interaction potential in the following way:

$$f(\theta,\phi) = -\frac{\mu_N}{2\pi\hbar^2} \int e^{-i\vec{k}'\cdot\vec{R}'} V(\vec{R}')\Psi(\vec{R}') \, d\vec{R}'. \tag{18.29}$$

One of the main tasks of scattering theory is to find approximations where Eq. (18.29) can be solved.

For a spherically symmetric potential $V(R)$ the scattering amplitude can be expanded as partial waves:

$$\Psi(\vec{R}) = \sum_{\ell=0}^{\infty} \frac{u_\ell(R)}{R} P_\ell(\cos\theta), \tag{18.30}$$

where the spherically symmetric potential results in no azimuthal dependence. The wavefunctions $u_\ell(R)$ are solutions of the radial wave equation:

$$-\frac{\hbar^2}{2\mu_N}\frac{d^2 u_\ell(R)}{dR^2} + \left(V(R) + \frac{\ell(\ell+1)\hbar^2}{2\mu_N R^2} - E\right)u_\ell(R) = 0, \tag{18.31}$$

and E is the total energy of the collision. The term proportional to $\ell(\ell+1)$ is a centrifugal barrier called the rotational energy, and for $\ell \neq 0$ it effectively repels the wavefunction from the small R region, depending on the incoming energy E.

18.4 Scattering theory

To find the scattering amplitude one needs to determine the change of the solution $u_\ell(R)$ caused by the potential $V(R)$. At long range, where $V(R)$ becomes small, the solutions of $u_\ell(R) \propto \exp(ikR)$ integrated outwards with or without $V(R)$ included in Eq. (18.31) differ only by a phase shift $\delta_\ell(k)$. These phase shifts $\delta_\ell(k)$ contain all the information about the scattering process. In terms of these phase shifts, the differential cross-section is given by

$$f(\theta) = \frac{1}{2ik} \sum_{\ell=0}^{\infty} (2\ell + 1) P_\ell(\cos\theta) \left(e^{2i\delta_\ell(k)} - 1 \right). \tag{18.32}$$

Once the phase shifts $\delta_\ell(k)$ are determined by solving Eq. (18.31) with and without $V(R)$, the differential cross-section can be determined. The total cross-section is then

$$\sigma(k) = \frac{4\pi}{k^2} \sum_{\ell=0}^{\infty} (2\ell + 1) \sin^2 \delta_\ell(k). \tag{18.33}$$

So far the symmetry of the colliding particles has not been taken into account. The scattering over an angle θ is the same as a scattering over $\pi - \theta$, since the two particles are indistinguishable. Since $P_\ell(\cos(\pi - \theta)) = (-1)^\ell P_\ell(\cos\theta)$, this leads to the condition that ℓ should be even for bosons. Furthermore, since one cannot distinguish between a scattering in the direction θ from $\pi - \theta$, the amplitudes in the two directions have to be added, leading to an additional factor of 2 in the amplitude. The cross-section for identical bosons is thus given by

$$\sigma(k) = \frac{8\pi}{k^2} \sum_{\ell=0,\ell \text{ even}}^{\infty} (2\ell + 1) \sin^2 \delta_\ell(k). \tag{18.34}$$

Note that for large wavevectors k a large number of partial waves ℓ can contribute to the cross-section $\sigma(k)$ and approximating $\sin^2 \delta_\ell(k) \approx 1/2$ yields the same cross-section for distinguishable (Eq. (18.33)) and indistinguishable particles (Eq. (18.34)). This shows that the bosonic nature of the particles becomes important only for low energies.

For small values of k or low energies E, the number of partial waves that can contribute becomes rather small. Consider the case where the interaction potential at long range is given by $V(R) = -C_n/R^n$, where n is of the range 3–6 for typical atom–atom scattering (see Tab. 18.1). At long range the rotational potential $\ell(\ell+1)\hbar^2/2\mu_N R^2$ dominates, whereas at short distances the potential $V(R) = -C_n/R^n$ becomes larger. This leads to a maximum in the potential energy given by

$$V_{\max}(\ell) = \left(\frac{n-2}{2}\right)\left(\frac{\hbar^2 l(l+1)}{n\mu_N}\right)^{(n/n-2)} C_n^{(-2/n-2)} \quad \text{at} \quad R_{\max}(\ell) = \left(\frac{n\mu_N C_n}{\hbar^2 l(l+1)}\right)^{(1/n-2)}.$$

$$\tag{18.35}$$

If the incoming energy E is smaller than this rotational barrier $V_{\max}(\ell)$ for a certain partial wave ℓ, the particle cannot penetrate to small internuclear distances where the potential $V(R)$ becomes important. As such, the phase shift $\delta_\ell(k)$ becomes zero and this value of ℓ does not have to be included in the sum of Eq. (18.34). For energies E below $V_{\max}(\ell = 2)$ the sum reduces to just one term, namely for $\ell = 0$. This regime is called s-wave scattering and is obtained for the low temperatures reached in laser and evaporative cooling. Defining a scattering length a given by

$$a = -\lim_{k \to 0} \frac{\tan \delta_0(k)}{k}, \tag{18.36}$$

the cross-section for elastic scattering at these extremely low energies becomes equal to $\sigma_s = 8\pi a^2$ independent of the incoming energy. Note that s-wave scattering means that the differential cross-section does not depend on the scattering angle θ, since $P_0(\theta)$ is independent of θ.

18.5 The scattering length

In most cases the scattering length a is the only interaction parameter that determines the properties of binary interactions at low energies. The cross-section for elastic scattering $\sigma_s = 8\pi a^2$ becomes independent of energy and solely dependent on a. The stability of a degenerate Bose gas depends on the sign of the scattering length, as discussed in Sec. 21.4. In this section the scattering length will be inspected in more detail.

The simplest case to consider for elastic scattering is for a potential that is simply a square well, since it allows for simple analytical expressions. The potential, as depicted in the inset of Fig. 18.4, is given by

$$V_{\text{int}} = \begin{cases} -V_0; & R < R_0 \\ 0; & R \geq R_0 \end{cases} \tag{18.37}$$

The wavefunctions in the inner ($R < R_0$) and outer ($R > R_0$) region are simply given in terms of the wavevectors $k_< = \sqrt{2\mu_N V_0}/\hbar$ and $k_> = \sqrt{2\mu_N E}/\hbar$, assuming that the total energy E is much smaller than the depth of the potential V_0:

$$u_<(R) = \sin k_< R; \qquad u_>(R) = C \sin(k_> R + \delta_0). \tag{18.38}$$

Note that in the inner region only the sine-function is a regular solution at the origin. Matching the wavefunction and its derivative at $R = R_0$ yields two relations that fully determine the amplitude C and phase shift δ_0. For δ_0 it can be shown that

$$\delta_0 = -k_> R_0 + \arctan\left(\frac{k_>}{k_<} \tan k_< R_0\right). \tag{18.39}$$

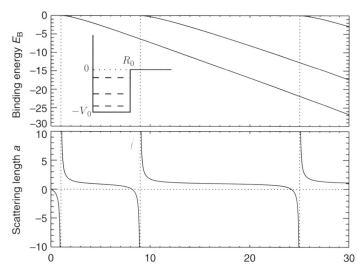

Figure 18.4 Potential scattering for a square well. The lower panel shows the scattering length a in units of R_0 as a function of the well depth V_0 in units of $\pi^2\hbar^2/8\mu_N R_0^2$. The upper panel shows the energy of the bound states E_B in units of $\pi^2\hbar^2/8\mu_N R_0^2$ as a function of V_0. At these values of V_0, where the scattering length diverges, an extra bound state appears in the potential well.

Using Eq. (18.36) this leads to

$$a = R_0\left(1 - \frac{\tan k_< R_0}{k_< R_0}\right), \qquad (18.40)$$

so the scattering length is proportional to the width of the potential R_0. In the lower panel of Fig. 18.4 the result is shown, where V_0 is scaled by $\pi^2\hbar^2/8\mu_N R_0^2$ and a by R_0. For small values of V_0 the scattering length a is finite and negative, and a goes to zero as V_0 goes to zero. However, for the value of V_0, where $k_< R_0 = \pi/2$, the phase shift δ_0 becomes $\pi/2$ and the scattering length diverges to negative infinity. For larger values of V_0 the scattering length decreases from positive infinity to approximately R_0, but for the value of V_0 where $k_< R_0 = 3\pi/2$, it diverges again. So the value of a depends strongly on the exact value of the depth V_0 of the potential.

Apart from the low-energy scattering state discussed above, there can also be bound states that are supported by the potential well. For a bound state the wavefunction in the inner region is given by the left side of Eq. (18.38), where in this case the wavevector is given by $k'_< = \sqrt{2\mu_N(V_0 - E_B)}/\hbar$ and E_B is the energy of the bound state ($E_B < 0$). For a bound state the outer region becomes a classically forbidden region, and the wavefunction should damp exponentially to zero,

$u_> \propto \exp(-\kappa_> R)$ with $\kappa = \sqrt{-2\mu_N E_B}/\hbar$. Matching the wavefunction and its derivative at $R = R_0$ leads to the relation

$$k'_< R_0 \cot k'_< R_0 = -\kappa R_0, \qquad (18.41)$$

which has to be fulfilled for a bound state to exist. In the upper panel of Fig. 18.4 the energies of the bound states are shown. At values where $k_< = (2n + 1)\pi/2$ with $n = 0, 1, 2, \ldots$, the number of bound states increases by one and the last bound state is at threshold ($E_B = 0$). It can be shown that this leads to a relation between the scattering length a and the energy of the last bound state E_B:

$$E_B \approx -\frac{\hbar^2}{2\mu_N a^2}, \qquad (18.42)$$

which shows that E_B becomes zero when a diverges.

Large scattering lengths are not merely of academic interest, but indeed occur in nature. Consider the classical example of the helium dimer $^4\text{He}_2$ molecule. For a long time it was uncertain whether this system supports a bound state. Careful experimental measurements show that there is one bound state whose radial probability density is given in Fig. 18.5. It has a binding energy of $E_B = -4.2 \times 10^{-9}$ au, which is extremely small, and a classical outer turning point of $27a_0$. The wavefunction in the outer region is given by $u_>(R) \propto \exp(-R/a)$ and is thus fully determined by a. Using this wavefunction, the average internuclear distance is given by $\langle R \rangle = a/2$ and in this case it predicts $\langle R \rangle = 5.2$ nm, in good agreement with the experimental value $\langle R \rangle = 5.2(4)$ nm derived from scattering experiments.

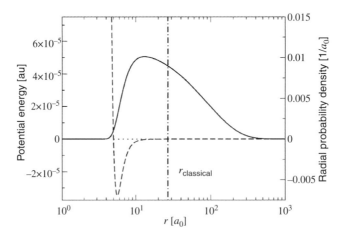

Figure 18.5 Radial probablity density (solid line) of the $^4\text{He}_2$ molecule together with the interaction potential (dashed line). The figure shows that more than 80% of the probability is located outside the classical turning point (note the logaritmic internuclear distance scale). Figure adapted from Ref. [196].)

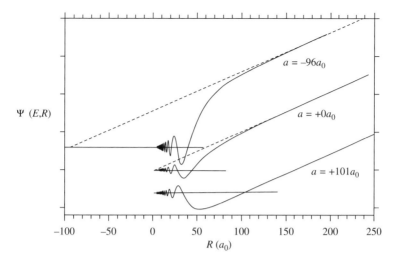

Figure 18.6 Wavefunctions for different scattering length. Although the scattering length in the three examples changes considerably, the wavefunctions in the inner range are nearly unaffected, since at short range the interaction energy is much larger than the incoming energy. Changes in the potential are only reflected in a phase shift of the wavefunction in the outer range. The scattering length is thus a sensitive parameter to determine the interaction potential. (Figure adapted from Ref. [197].)

The fact that the wavefunction does not depend on the details of the interaction forces but only on the scattering length is referred to as universality.

To illustrate the meaning of the scattering length, Fig. 18.6 shows three wavefunctions calculated for very low energy, where in each case the potential $V(R)$ has changed slightly. The wavefunctions in the inner region are nearly not affected, since the change of the potential is much smaller than the depth of the potential. Thus the local wavevector $k(R)$ of Eq. (18.15) is nearly identical in the three cases. At long range all wavefunctions are similar, since the wavefunctions are proportional to $\sin[k(R - a)]$, where $k = \sqrt{2\mu_N E}/\hbar$. The wavefunctions only differ in the accumulated phase $\phi = ka$. The scattering length can be found be extrapolating the wavefunction to $R = 0$ and the cut-off of this line on the internuclear distance is the scattering length a. The difference between an effective repulsive or attractive potential is whether this cut-off is for negative or positive R, respectively. So although the wavefunctions in both cases are very similar, the details of the potential $V(R)$ determine the value of the scattering length.

18.6 Feshbach molecules

Consider scattering in a situation as depicted in Fig. 18.7. The scattering starts in the lower channel, which is called the open channel. The total energy of the

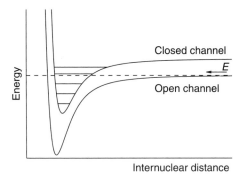

Figure 18.7 Mechanism for a Feshbach resonance [198]. Scattering starts in the lower channel, which is open, since the incoming energy is above its threshold. The upper channel is closed, since the total energy is below its threshold. However, if the incoming energy is close to a bound state in the closed channel, the coupling between the two states can change the scattering in the open channel considerably.

system is too low to reach the upper channel, which is called the closed channel. However, if the total energy is such that it coincides with a bound state in the closed channel, this changes the scattering in the lower channel considerably. For ultra-low collision energies this changes the scattering length, which is the only relevant parameter at low energy. In fact, adjusting the bound state in the closed channel to coincide with the threshold of the open channel makes the scattering length approach infinity at ultra-low collision energies. An infinite scattering length tunes the interactions in the system to become much larger than the range of the potential.

To bring out the essentials of the Feshbach resonance, the following simple model is considered (Fig. 18.8). Both the closed and open channel are described by square well potentials, where the thresholds of the two channels are shifted by an amount ΔV. To further simplify the model, it is assumed that the two states are coupled only at $R = R_0$, but for smaller and larger distances the two states are uncoupled. The wavefunction $u(r)$ of the state can be described by a vector that is given for $R \leq R_0$ by

$$u_{\leq}(R) = \begin{pmatrix} C_1 \sin k_1 R \\ C_2 \sin k_2 R \end{pmatrix}. \tag{18.43}$$

Here $C_{1,2}$ are coefficients to be determined for the open and closed channel, respectively. The wavevectors $k_{1,2}$ are given by $k_1 = \sqrt{2\mu_N V_0}/\hbar$ for the open channel, and $k_2 = \sqrt{2\mu_N(V_0 + \Delta V)}/\hbar$ for the closed channel. Here μ_N is the reduced mass, and it is assumed that the incoming energy E is much smaller than the potential difference ΔV. Note that the choice of wavefunctions causes the wavefunctions at $R = 0$

18.6 Feshbach molecules

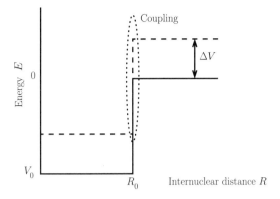

Figure 18.8 Model for the Feshbach resonance. Both the open and closed channel are approximated by square well potentials with depth V_0, where the asymptote of the closed channel is shifted up by an energy ΔV with respect to the open channel. At some distance R_0 the two states are coupled.

to remain finite. In the region $R \geq R_0$ the closed-channel wavefunction decreases exponentially with R, whereas it is an oscillating function in the open channel:

$$u_{\geq}(R) = \begin{pmatrix} C_3 \sin(k_3 R) + C_4 \cos(k_4 R) \\ \exp(-\kappa R) \end{pmatrix}. \tag{18.44}$$

Here $C_{3,4}$ are again coefficients to be determined, the wavevector k_3 is given by $k_3 = \sqrt{2mE}/\hbar$, and $\kappa = \sqrt{2m\Delta V}/\hbar$. The coefficient of the closed channel is taken to be unity without loss of generality. Since the coupling takes place only at $R = R_0$, the wavefunctions at that point are related by $u_{\leq}(R_0) = \mathcal{U} u_{\geq}(R_0)$, where the unitary transformation matrix can simply be expressed in terms of one parameter ϕ:

$$\mathcal{U} = \begin{pmatrix} \cos\phi & \sin\phi \\ -\sin\phi & \cos\phi \end{pmatrix}. \tag{18.45}$$

These ingredients are sufficient to capture the main features of the Feshbach resonance.

To determine the coefficients C_{1-4} one requires that the wavefunction and its derivative are continuous at $R = R_0$. For sufficiently low energy E this allows for the determination of the scattering length, given by Eq. (18.36). In Fig. 18.9a the result is shown as a function of the potential difference ΔV. For increasing ΔV the scattering length starts to increase, until at a certain value of $\Delta V_{\rm res}$ it diverges to plus infinity. This markes the resonance. Above the resonance the scattering length starts to increase from minus infinity to its value before the resonance. Calculating the bound states in the closed channel without the coupling indicates that at $\Delta V_{\rm res}$ a resonance in the closed channel becomes equal in energy with the energy of the incoming channel. Calculating the wavefunction in the closed channel for

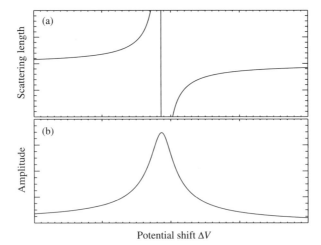

Figure 18.9 (a) Scattering length as a function of potential difference ΔV between the closed and open channel. Owing to the Feshbach resonance the scattering length varies strongly as a function of the potential difference ΔV and becomes infinite at the resonance. (b) Amplitude of the wavefunction in the closed channel. The Feshbach resonance increases the coupling between the open and closed channel causing a strong increase in the amplitude of the closed channel.

$\Delta V \approx \Delta V_{\text{res}}$ (see Fig. 18.9b), shows that its amplitude at the resonance is strongly increased. So in the scattering process the state in the closed channel becomes strongly coupled to the incoming scattering state, which leads to a strong change in the scattering length for the open channel. Just like in the one-channel model discussed in Sec. 18.4, a bound state at threshold leads to an infinite scattering length. Since the coupling described by ϕ is finite, the resonance has a finite width ΔE. This finite coupling also causes the resonance not to appear at the position of the bound state, but shifted upwards.

To change the potentials with respect to each other in an experiment, the Zeeman effect can be used. If the magnetic dipole moment of two states involved is different, the magnetic field can be used to shift one of the states in the closed channel into resonance with the incoming open channel. Figure 18.10b shows the energy of a bound state, which (depending on molecular symmetry) shifts up in energy for increasing field. At some point the energy coincides with the energy of the open channel and this causes a strong change in the scattering length (see Fig. 18.10a). Far away from the resonances the scattering length becomes equal to the zero-field, background scattering length, but close to the Feshbach resonance the effect on the scattering length is clearly visible. In most cases the scattering length a can be described by

$$a = a_b \left(1 - \frac{\Delta B}{B - B_{\text{res}}}\right), \tag{18.46}$$

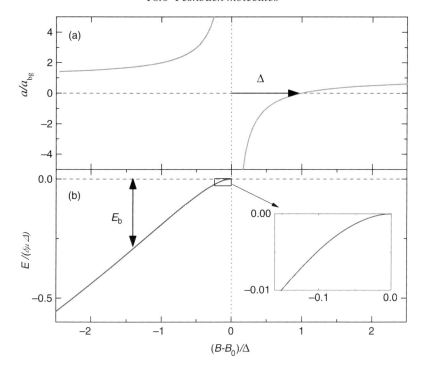

Figure 18.10 Magnetically induced Feshbach resonance. (a) By tuning the magnetic field the scattering length can be increased by many orders over the background scattering length. At the resonance the scattering length diverges. (b) By increasing the magnetic field the energy of the last bound state is tuned towards the dissociation limit, where the state crosses with the unbound state of two atoms. This causes a strong coupling between the open and bound channel leading to the increase of the scattering length. bg indicates background, b indicates binding. (Figure from Ref. [198].)

where ΔB is the width of the resonance and B_{res} the resonance position. The position of the resonance B_{res} depends strongly on the detailed potential of the closed channel, and the observation of the position of many Feshbach resonances allows for a complete determination of the interaction potential for the closed channel.

The Feshbach resonance can also be used to produce cold molecules out of two ultra-cold atoms. The crossing between the open and closed channel is not a real crossing, but because of the coupling between the two states, it forms an anti-crossing. If the magnetic field is quickly ramped over the anti-crossing, it will follow the state diabatically and remain in the open channel. However, if subsequently the magnetic field is slowly ramped down over the anti-crossing, the system will follow the potential curves adiabatically and end up in the closed channel, i.e. the molecular state. In this way molecules can be produced that are translationally very cold, since in the process no translational energy is added to the molecules. However, the molecules are in a highly excited vibrational state. But since there is no rethermalization, the molecules thus produced can have a long lifetime.

Exercises

18.1 Use the Hamiltonian of Eq. (18.2) and the wavefunctions of Eq. (18.4) to derive the matrix elements of Eq. (18.5). Split your result into two parts: one for the radial part of the electrons and one for their angular part. Show that the angular part leads directly to the "selection" rules Eq. (18.7). Why is $m_a + m_b$ conserved only if one takes the internuclear distance along the z-direction?

18.2 Consider the long-range dipole–dipole interaction for the S–P asymptote. Determine the states which should be used in Eq. (18.4). Using the relations $C_{11} = (4\pi)^{3/2}/3\sqrt{5}$ and $Q_{m,-m} = 2, 1$ for $m = 0$ and $|m| = 1$, respectively, calculate the relevant matrix elements of Eq. (18.5). Why can you treat the ($m = 0$)-case and ($|m| = 1$)-case independently? Why does the ($m = 0$)-case corresponds to the Σ-state, and the ($|m| = 1$)-case to the Π-state? Show why it is easy to do the diagonalization for these two cases and derive the results of Eq. (18.10).

18.3 The van der Waals interaction in the ground state of two alkali-metal atoms can be treated as a second-order dipole–dipole interaction. For that case the relevant states are $|n00; n00\rangle$, $|n10; n10\rangle$, $|n1-1; n1+1\rangle$, and $|n1+1; n1-1\rangle$, with n the principal quantum number of the valence electron.
(a) Show that for this problem $j = 1$ and $k = 1$ in Eq. (18.5).
(b) Using your result of Prob. 18.2 show that the relevant matrix elements $\langle \alpha|V_{\text{int}}|\alpha'\rangle$ are $-2\mu_d^2/3(4\pi\varepsilon_0)R^3$ for $m = 0$ and $-\mu_d^2/3(4\pi\varepsilon_0)R^3$ for $|m| = 1$.
(c) Show that one can construct a linear combination of the excited states $|n1+m; n1-m\rangle$ with $m = 0, \pm 1$ that is coupled to the ground state, and two linear combinations that are not coupled.
(d) Calculate the coupling between the ground state and the coupled excited state.
(e) Diagonalize the interaction matrix V_{int} for these two states including the asymptote energies of the ground and excited state.
(f) Show that for a coupling element much smaller than the energy difference between the asymptotes, the shift of the ground state can be written as $-C_6/R^6$, and find an expression for C_6 in terms of the coupling element and the energy difference.
(g) Calculate for your favorite alkali-metal element the C_6 coefficient using μ_d of Tab. 10.3 and compare your result with the calculated value of Ref. [188].

18.4 Using the effective potential

$$V_{\text{eff}}(R) = -\frac{C_n}{R^n} + \frac{\ell(\ell+1)\hbar^2}{2\mu_N R^2},$$

show that the maximum of the effective potential V_{max} is given by Eq. (18.35). At which internuclear distance R_{max} does this maximum occurs. Calculate V_{max} and R_{max} for two Na atoms in the ground state and $\ell = 2$ using $C_6 = 1,472$ au. Calculate using this result the temperature, for which only s-wave scattering takes place in this potential. Are those temperatures accessible for laser-cooled Na atoms?

Part III
Applications

19
Optical forces and laser cooling

The usual form of electromagnetic forces is given by $\vec{F} = q(\vec{\mathcal{E}} + \vec{v} \times \vec{B})$, but for neutral atoms $q = 0$. The next order of force is the dipole term, but this also vanishes because neutral atoms have no inherent dipole moment. However, a dipole moment can be induced by a field, and this is most efficient if the field is alternating near the atomic resonance frequency. Since these frequencies are typically in the optical range, dipole moments are efficiently induced by shining nearly resonant light on the atoms.

Laser cooling and trapping rely on the interaction between laser light and atoms to exert a controllable force on the atoms, and many sophisticated schemes have been developed using the special properties of the interaction. This chapter begins with the simplest schemes for exerting optical forces on atoms, namely a single-frequency light field interacting with a two-level atom. Although this is the simplest possible scheme, it is pedagogically valuable because it shows many of the features that will be encountered further on. The ultimate temperature using such a scheme is limited, but lower temperatures can be achieved with multilevel atoms. This topic is addressed later in the chapter.

19.1 Two kinds of optical forces

When an atom absorbs nearly resonant light, it makes a transition to the excited state, and its subsequent return to the ground state can be either by spontaneous or by stimulated emission. The nature of the optical force that arises from these two different processes is quite different, and will be described separately [1]. The spontaneous emission case is different from the familiar quantum mechanical calculations using state vectors to describe the state of the system, because spontaneous emission causes the state of the system to evolve from a pure state into a mixed state (see Exercise 6.9). Spontaneous emission is an essential ingredient for the dissipative nature of the optical forces in monochromatic light, and the density matrix is needed to describe it (see Chap. 6).

When an atom absorbs light, its energy $\hbar\omega$ excites the atom and its angular momentum changes the electron's orbit (recall $\Delta\ell = \pm 1$). Both of these are internal atomic properties. However, the linear momentum of the light, $\hbar\omega/c = \hbar k$, can change only the external, translational motion of the atom. The absorption leaves the atom in its excited state, and if the light intensity is low enough so that it is much more likely to return to the ground state by spontaneous emission than by stimulated emission, the resulting fluorescent light carries off momentum $\hbar k$ in a random direction. The momentum exchange from the fluorescence averages zero, so the net total force is given by $\hbar k \gamma_p$ (see Eq. (6.18)). Such momentum exchange is called the light pressure force, radiation pressure force, or dissipative force, since it relies on the scattering of light out of the laser beam.

When the detuning $\delta \gg \gamma$, spontaneous emission may be much less frequent than stimulated emission. In this case, absorption is most often followed by stimulated emission, and might appear to produce zero momentum transfer because the stimulated light has the same momentum as the exciting light. However, if the optical field has beams with at least two different $\vec{k}$-vectors present, such as in counterpropagating beams, absorption from one beam followed by stimulated emission into the other indeed produces a non-zero momentum exchange. This is called the dipole force, reactive force, gradient force, or redistribution force. It has the same origin as the force of an inhomogeneous dc electric field on a classical dipole, but relies on the redistribution of light from one laser beam to the other.

It needs to be emphasized that these are two fundamentally different kinds of forces. Dipole forces can be made large by using high-intensity light because they do not saturate. Even though they cannot be used to cool a sample of atoms, they can be combined with the dissipative scattering force to enhance cooling in several different ways, as described below. By contrast, scattering forces are always limited by the rate of spontaneous emission γ and cannot be made arbitrarily strong.

19.2 Low-intensity laser light pressure

The philosophy of the correspondence principle requires a smooth transition between quantum and classical mechanics. Clearly the orbits of the planets can be described with arbitrary accuracy using classical mechanics, but just as clearly, they must conform to the rules of quantum mechanics. The quantum version of Newton's laws is embodied in the Ehrenfest theorem [199], a simple statement that the expectation value of an operator must correspond to the behavior of its classical counterpart.

In this section the semi-classical description of the interaction of a light field with a two-level atom is used to derive the laser light pressure on an atom. The force F on an atom is defined as the expectation value of the quantum mechanical

19.2 Low-intensity laser light pressure

force operator $\mathcal{F}$, as defined by $F = \langle \mathcal{F} \rangle = \mathrm{d}\langle p \rangle/\mathrm{d}t$. The time evolution of the expectation value of a time-independent quantum mechanical operator $\mathcal{A}$ is given by [69] $(\mathrm{d}/\mathrm{d}t)\langle A \rangle = (i/\hbar)\langle[\mathcal{H}, \mathcal{A}]\rangle$. The commutator of $\mathcal{H}$ and p is given by $[\mathcal{H}, p] = i\hbar(\partial \mathcal{H}/\partial z)$, where the operator p has been replaced by $-i\hbar(\partial/\partial z)$. The force on an atom is thus given by $F = -\langle \partial \mathcal{H}/\partial z \rangle$. This relation is a specific example of the Ehrenfest theorem and forms the quantum mechanical analog of the classical expression that the force is the negative gradient of the potential.

Discussion of the force on atoms caused by light fields begins with the relevant part of the Hamiltonian of the system, $\mathcal{H}'(t)$ [1]. Using the electric dipole approximation, i.e. neglecting the spatial variation of the electric field over the size of an atom, allows the interchange of the gradient with the expectation value, and gives $F = e(\partial/\partial z)\langle \vec{\mathcal{E}}(\vec{r}, t) \cdot \vec{r} \rangle$, whose matrix has only off-diagonal entries. The expectation value can be found using the definition of the Rabi frequency of Sec. 2.3 and the expectation value $\langle \mathcal{A} \rangle = \mathrm{Tr}(\rho \mathcal{A})$, resulting in

$$F = \hbar \left(\frac{\partial \Omega}{\partial z} \rho_{eg}^* + \frac{\partial \Omega^*}{\partial z} \rho_{eg} \right). \tag{19.1}$$

Deriving this result requires the RWA (see Sec. 3.2.3), which neglects terms oscillating with the laser frequency. Note that the force depends on the state of the atom, and in particular on the optical coherence between the ground and excited states, ρ_{eg}.

Although it may seem a bit artificial, it is instructive to split $\partial \Omega/\partial z$ into its real and imaginary parts (the matrix element that defines Ω in Sec. 2.3 can certainly be complex): $(\partial \Omega/\partial z) = (q_r + iq_i)\Omega$. Here $q_r + iq_i$ is the logarithmic derivative of Ω. In general, for a field $\mathcal{E}(z) = \mathcal{E}_0(z)\exp(i\phi(z)) + \mathrm{c.c.}$ the real part of the logarithmic derivative corresponds to a gradient of the amplitude $\mathcal{E}_0(z)$ and the imaginary part to a gradient of the phase $\phi(z)$. Then the expression for the force becomes

$$F = \hbar q_r (\Omega \rho_{eg}^* + \Omega^* \rho_{eg}) + i\hbar q_i (\Omega \rho_{eg}^* - \Omega^* \rho_{eg}). \tag{19.2}$$

Equation (19.2) is a very general result that can be used to find the force for any particular situation as long as the optical Bloch equations (OBE) for ρ_{eg} can be solved (see Sec. 6.3). In spite of the chosen complex expression for Ω, it is important to note that the force itself is real, and that first term of the force is proportional to the real part of $\Omega \rho_{eg}^*$, whereas the second term is proportional to the imaginary part.

In the presence of low-intensity, nearly resonant light, the force on atoms is dominated by the spontaneous emission part, and so the contribution from stimulated emission will be neglected. The steady-state solutions of the OBE for an atom at rest are given in Sec. 6.4. Substituting the solution for ρ_{eg} of Eq. (6.15) into Eq. (19.2) gives

$$F = \frac{\hbar s}{1+s}\left(-\delta q_{\rm r} + \frac{1}{2}\gamma q_{\rm i}\right). \tag{19.3}$$

Note that the first term is proportional to the detuning δ, whereas the second term is proportional to the decay rate γ. For zero detuning, the force becomes $F = (\hbar k\gamma/2)[s_0/(s_0+1)]$, a very satisfying result because it is simply the momentum per scattering event $\hbar k$, times the scattering rate $\gamma_{\rm p}$.

It is instructive to identify the origin of both of the terms in Eq. (19.3) [1]. Absorption of light leads to the transfer of momentum from the optical field to the atoms. If the atoms decay by spontaneous emission, the recoil associated with the spontaneous fluorescence is in a random direction, so its average over many emission events results in zero net effect on the atomic momentum. Thus the force from absorption followed by spontaneous emission can be written as $F_{\rm sp} = \hbar k\,\gamma\,\rho_{ee}$, where the first factor is the momentum transfer for each scattering event, the second factor is the rate for the process, and the last factor is the probability for the atoms to be in the excited state. Although it may seem natural for this expression to depend on the ground-state population ρ_{gg} and not the excited-state population ρ_{ee}, using ρ_{ee} simply builds in the dependence of absorption on detuning and intensity, including saturation. Using Eq. (6.18), the force resulting from absorption followed by spontaneous emission becomes

$$F_{\rm sp} = \frac{\hbar k s_0 \gamma/2}{1 + s_0 + (2\delta/\gamma)^2} = \hbar k \gamma_{\rm p}, \tag{19.4}$$

which saturates at high intensity as a result of the term s_0 in the denominator. Increasing the rate of absorption by increasing the intensity does not increase the force without limit, since that would only increase the rate of stimulated emission, where the transfer of momentum is opposite in direction to the absorption. Thus the force saturates to a maximum value of $\hbar k\gamma/2$, because ρ_{ee} has a maximum value of 1/2 (as discussed in Sec. 6.4).

On the other hand, the dipole force can be understood intuitively from an energy picture. Atomic energy levels are shifted by the light field as discussed in Sec. 2.3.4. The energy levels are given by Eq. (2.10) and the resulting light shifts are $\omega_{\rm ls} = [\sqrt{\Omega^2 + \delta^2} - \delta]/2$. For sufficiently large detuning $\delta \gg \Omega$, these light shifts are approximately $\omega_{\rm ls} \approx \Omega^2/4\delta = \gamma^2 s_0/8\delta$.

In a standing wave in 1D with $\delta \gg \Omega$, the light shift $\omega_{\rm ls}$ varies sinusoidally from node to antinode. When δ is sufficiently large, the spontaneous emission rate may be negligible compared with that of stimulated emission so that $\hbar\omega_{\rm ls}$ may be treated as a potential U. The resulting dipole force is

$$\vec{F} = -\vec{\nabla}U = -\frac{\hbar\gamma^2}{8\delta I_{\rm s}}\vec{\nabla}I, \tag{19.5}$$

where I is the total intensity distribution of the standing-wave light field of period $\lambda/2$, and I_s is the saturation intensity. For such a standing wave, the optical electric field (and the Rabi frequency) at the antinodes is double that of each traveling wave that composes it, and so the total intensity $I_{\max}$ at the antinodes is four times that of the single traveling wave.

19.2.1 A two-level atom at rest

The electric field of a traveling wave is given by $\mathcal{E}(z) = \mathcal{E}_0(e^{i(kz-\omega t)} + \text{c.c.})/2$. In calculating the Rabi frequency from this, the RWA causes the positive frequency component of $\mathcal{E}(z)$ to drop out. Then the gradient of the Rabi frequency becomes proportional to the gradient of the surviving negative frequency component, so that $q_r = 0$ and $q_i = k$. For such a traveling wave the amplitude is constant but the phase is not, and this leads to the non-zero value of q_i.

Examination of Eq. (19.4) shows that it clearly corresponds to the second term of Eq. (19.2). It vanishes for an atom at rest in a standing wave where $q_i = 0$, and this can be understood because atoms can absorb light from either of the two counterpropagating beams that make up the standing wave, and the average momentum transfer then vanishes. This force is dissipative because the reverse of spontaneous emission is not possible.

Just the opposite is true for a standing wave, composed of two counterpropagating traveling waves for which the electric field is given by $\mathcal{E}(z) = \mathcal{E}_0 \cos(kz)(e^{-i\omega t} + \text{c.c.})$, so that $q_r = -k\tan(kz)$ and $q_i = 0$. Again, only the negative frequency part survives the RWA, but the gradient does not depend on it. Thus a standing wave has an amplitude gradient, but not a phase gradient. The singularity in q_r from the tangent function for a standing wave does not lead to problems, since it occurs at the node of the field where the Rabi frequency Ω is zero.

It is clear that the first term in Eq. (19.2) derives from the light shifts of the ground and excited states, described in Sec. 2.3.4. Such light shifts depend on the strength of the optical electric field. A standing wave is composed of two counterpropagating laser beams, and their interference produces an amplitude gradient that is not present in a traveling wave. The resulting spatially modulated light shift produces a force that is proportional to the gradient of the light shift, and Eq. (2.10) can be used to find the force on ground-state atoms in low-intensity light:

$$F_{\text{dip}} = -\frac{\partial(\Delta E_g)}{\partial z} = \frac{\hbar\Omega}{2\delta}\frac{\partial\Omega}{\partial z}. \tag{19.6}$$

For an amplitude-gradient light field such as a standing wave, $\partial\Omega/\partial z = q_r\Omega$, and this force corresponds to the first term in Eq. (19.2) in the limit of low saturation ($s \ll 1$).

For the case of a standing wave, Eq. (19.3) becomes

$$F_{dip} = \frac{2\hbar k \delta s_0 \sin 2kz}{1 + 4s_0 \cos^2 kz + (2\delta/\gamma)^2}, \quad (19.7)$$

where s_0 is the saturation parameter of each of the two beams that form the standing wave. For $\delta < 0$ the force drives the atoms to positions where the intensity has a maximum, whereas for $\delta > 0$ the atoms are attracted to the intensity minima. The force is conservative and can be written for an atom at rest as the gradient of a potential.

19.3 Atomic beam slowing and collimation

Among the earliest laser-cooling experiments was deceleration of atoms in a beam [200]. The authors exploited the Doppler shift to make the momentum exchange (hence the force) velocity dependent. It worked by directing a laser beam opposite to an atomic beam so the atoms could absorb light, and hence momentum $\hbar k$, very many times along their paths through the apparatus [200, 201]. Of course, excited-state atoms cannot absorb light efficiently from the laser that excited them, so between absorptions they must return to the ground state by spontaneous decay, accompanied by emission of fluorescent light. The spatial symmetry of the emitted fluorescence results in an average of zero net momentum transfer from many such fluorescence events so the net force on the atoms is in the direction of the laser beam.

The maximum deceleration is limited by the spontaneous emission rate γ. Since the maximum deceleration $\vec{a}_{max} = \hbar \vec{k} \gamma/2M$ is fixed by atomic parameters, it is straightforward to calculate the minimum stopping length L_{min} and time t_{min} for the rms velocity of atoms $\bar{v} = 2\sqrt{k_B T/M}$ at the source temperature. The result is $L_{min} = \bar{v}^2/2a_{max}$ and $t_{min} = \bar{v}/a_{max}$.

When an atomic beam crosses a 1D optical molasses (see Sec. 19.4), the transverse motion of the atoms is quickly damped while the longitudinal component is essentially unchanged. This transverse cooling of an atomic beam is an example of a method that can actually increase its brightness (atoms/sec-sr-cm^2) because such active collimation uses dissipative forces to compress the phase space volume occupied by the atoms. By contrast, the usual realm of beam focusing or collimation techniques for light beams, and most particle beams, is restricted to selection by apertures or conservative forces that preserve the phase space density of atoms in the beam. Clearly optical techniques can create atomic beams 10^6 or more times as intense as ordinary thermal beams, and also many orders of magnitude brighter.

19.4 Optical molasses

19.4.1 Two-level atoms moving in a standing wave

Laser cooling requires velocity-dependent forces that cannot derive from the gradient of a potential. Instead, it depends upon dissipative forces that are velocity dependent. Including the velocity of the atoms in the OBE is usually done as Doppler shifts, but the resulting equations are generally too hard to solve analytically.

Instead, the procedure will be to treat the velocity of the atoms as a small perturbation, and make first-order corrections to the solutions of the OBE obtained for atoms at rest [202]. It begins by adding drift terms in the expressions for the relevant quantities. For example, the Rabi frequency satisfies

$$\frac{d\Omega}{dt} = \frac{\partial \Omega}{\partial t} + v\frac{\partial \Omega}{\partial z} = \frac{\partial \Omega}{\partial t} + v(q_\mathrm{r} + iq_\mathrm{i})\Omega, \tag{19.8}$$

where the gradient of Ω has been separated into real and imaginary parts as above. In the same way, differentiating Eq. (6.15)a leads to

$$\frac{dw}{dt} = \frac{\partial w}{\partial t} + v\frac{\partial w}{\partial z} = \frac{\partial w}{\partial t} - \frac{2vq_\mathrm{r}s}{(1+s)^2}, \tag{19.9a}$$

since $s_0 = 2|\Omega|^2/\gamma^2$ and Ω depends on z. Similarly, differentiating Eq. (6.15b) leads to

$$\frac{d\rho_{eg}}{dt} = \frac{\partial \rho_{eg}}{\partial t} + v\frac{\partial \rho_{eg}}{\partial z} = \frac{\partial \rho_{eg}}{\partial t} - \frac{iv\Omega}{2(\gamma/2 - i\delta)(1+s)}\left[q_\mathrm{r}\left(\frac{1-s}{1+s}\right) + iq_\mathrm{i}\right]. \tag{19.9b}$$

In both of these calculations it must be remembered that Ω is complex, so differentiating s_0 results in two terms that give $\partial s_0/\partial z = 2q_\mathrm{r} s_0$. In Eqs. (19.9) the value of w in $\partial w/\partial z$ has been taken from its steady-state value given by Eq. (6.15a), and similarly for ρ_{eg}. Since neither w nor ρ_{eg} is explicitly time dependent, both $\partial w/\partial t$ and $\partial \rho_{eg}/\partial t$ vanish.

The solution for the special case of a standing wave, $q_\mathrm{i} = 0$, provides some new insight. The resulting coupled equations for w and ρ_{eg} can be separated and substituted into Eq. (19.2) for the force. In the limit of $s \ll 1$, this force is

$$F = \hbar k \frac{s_0 \delta \gamma^2}{2(\delta^2 + \gamma^2/4)}\left(\sin 2kz + kv\frac{\gamma}{(\delta^2 + \gamma^2/4)}(1 - \cos 2kz)\right). \tag{19.10}$$

Here s_0 is the saturation parameter of each of the two beams that compose the standing wave. The first term is the velocity-independent part of Eq. (19.3) and is sinusoidal in space, with a period of $\lambda/2$. Thus its spatial average vanishes. The force remaining after such averaging is $F_{av} = -\beta v$, where the damping coefficient β is given by

$$\beta = -\hbar k^2 \frac{8 s_0(\delta/\gamma)}{(1 + (2\delta/\gamma)^2)^2}. \tag{19.11}$$

This is a true damping force because atoms are slowed toward $v = 0$ independent of their initial velocities. Note that this expression for β is valid only for $s \ll 1$ because it depends on spontaneous emission to return excited atoms to their ground state.

19.4.2 Intuitive discussion of optical molasses

There is an appealing description of the mechanism for this kind of cooling in a standing wave. With light detuned below resonance, atoms traveling toward one laser beam see it Doppler shifted upward, closer to resonance. Since such atoms are traveling away from the other laser beam, they see its light Doppler shifted further downward, hence further out of resonance. Atoms therefore scatter more light from the beam counterpropagating to their velocity, and thus their velocity is lowered. This is the damping mechanism called optical molasses (OM) and is one of the most important tools of laser cooling. This is illustrated in Fig. 19.1.

It is straightforward to estimate the force on atoms in OM from Eq. (19.4). The discussion here is limited to the case where the light intensity is low enough that stimulated emission is not important. This eliminates consideration of excitation of

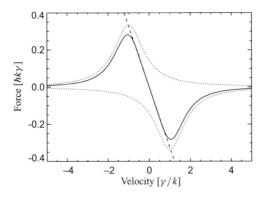

Figure 19.1 Velocity dependence of the optical damping forces for one-dimensional optical molasses. The two dotted traces show the force from each beam, and the solid curve is their sum. The straight line shows how this force mimics a pure damping force over a restricted velocity range. These are calculated for $s_0 = 2$ and $\delta = -\gamma$ so there is some power broadening evident (see Sec. 6.4). (Figure from Ref. [1].)

an atom by light from one beam and stimulated emission by light from the other, a sequence that can lead to very large, velocity-independent changes in the atom's speed. In this low-intensity case the forces from the two light beams are simply added to give $\vec{F}_{OM} = \vec{F}_+ + \vec{F}_-$, where

$$\vec{F}_\pm = \pm \frac{\hbar \vec{k} \gamma}{2} \frac{s_0}{1 + s_0 + \left[2(\delta \mp |\vec{k} \cdot \vec{v}|)/\gamma\right]^2} \tag{19.12}$$

from Eq. (19.4) with δ replaced by $\delta \mp |\vec{k} \cdot \vec{v}|$. Then the sum of the two forces is

$$\vec{F}_{OM} \cong \frac{8\hbar k^2 \delta s_0 \vec{v}}{\gamma(1 + s_0 + (2\delta/\gamma)^2)^2} \equiv -\beta \vec{v}, \tag{19.13}$$

where terms of order $(kv/\gamma)^4$ and higher have been neglected (see Eq. (19.11)).

For $\delta < 0$, this force opposes the velocity and therefore viscously damps the atomic motion. $\vec{F}_{OM}$ has maxima near $v = \pm(\gamma'/2k)(X/\sqrt{3})$ and decreases rapidly for larger velocities. Here $\gamma' \equiv \gamma \sqrt{s_0 + 1}$ is the power-broadened linewidth (see Eq. (6.19b)), X is the numerical factor given by $\sqrt{x - 1 + 2\sqrt{x^2 + x + 1}}$, and $x \equiv (2\delta/\gamma')^2$. For $x \gg 1$ these maxima appear at $v = \pm \delta/k$ as expected, but for the usual realm of OM, $x \sim 1$ and $X \sim \sqrt{3}$.

The slowing force is proportional to velocity for small enough velocities, resulting in viscous damping [203, 204] as shown in Fig. 19.1 that gives this technique the name "optical molasses". By using three intersecting orthogonal pairs of oppositely directed beams, the movement of atoms in the intersection region can be severely restricted in all three dimensions, and many atoms can thereby be collected and cooled in a small volume. Note that OM is not a trap for neutral atoms because there is no restoring force on atoms that have been displaced from the center. Still, the detainment times of atoms caught in OM of several mm diameter can be remarkably long.

19.5 Temperature limits

If there were no other influence on the atomic motion, all atoms in OM would quickly decelerate to $v = 0$ and the sample would reach $T = 0$, a clearly unphysical result. There is also some heating caused by the light beams that must be considered, and it derives from the discrete size of the momentum steps the atoms undergo with each emission or absorption. Since the atomic momentum changes by $\hbar k$, the kinetic energy changes on the average by at least the recoil energy $E_r = \hbar^2 k^2 / 2M = \hbar \omega_r$, where ω_r is called the recoil frequency. This means that the average frequency of each absorption is $\omega_{abs} = \omega_{eg} + \omega_r$ and the average frequency of each emission is $\omega_{emit} = \omega_{eg} - \omega_r$. Thus the light field loses an average energy of $\hbar(\omega_{abs} - \omega_{emit}) = 2\hbar\omega_r$ for each scattering. This loss occurs at a rate $2\gamma_p$ (two beams), and the energy becomes atomic kinetic energy because the atoms recoil from each event. The atomic sample is thereby heated because these recoils are in random directions.

The competition between this heating with the damping force of Eq. (19.13), results in a non-zero kinetic energy in steady state. At steady state, the rates of heating and cooling for atoms in OM must be equal. Equating the cooling rate, $\vec{F} \cdot \vec{v}$, to the heating rate, $4\hbar\omega_r\gamma_p$, the steady-state kinetic energy is found to be $(\hbar\gamma/8)(2|\delta|/\gamma + \gamma/2|\delta|)$. This result is dependent on $|\delta|$, and it has a minimum at $2|\delta|/\gamma = 1$, whence $\delta = -\gamma/2$. The temperature found from the kinetic energy is then $T_D = \hbar\gamma/2k_B$, where k_B is Boltzmann's constant and T_D is called the Doppler temperature or the Doppler cooling limit. For ordinary atomic transitions T_D is typically below 1 mK.

Another instructive way to determine T_D is to note that the average momentum transfer of many spontaneous emissions is zero, but the rms scatter of these about zero is finite. One can imagine these decays as causing a random walk in momentum space with step size $\hbar k$ and step frequency $2\gamma_p$, where the factor of 2 arises because of the two beams. The random walk results in diffusion in momentum space with diffusion coefficient $D_0 \equiv 2(\Delta p)^2/\Delta t = 4\gamma_p(\hbar k)^2$. Then Brownian motion theory gives the steady-state temperature in terms of the damping coefficient β to be $k_B T = D_0/\beta$. This turns out to be $\hbar\gamma/2$ as above for the case $s_0 \ll 1$ when $\delta = -\gamma/2$. There are many other independent ways to derive this remarkable result that predicts that the final temperature of atoms in OM is independent of the optical wavelength, atomic mass, and laser intensity (as long as it is not too large).

19.6 Experiments in three-dimensional optical molasses

Optical molasses experiments can also work in three dimensions at the intersection of three mutually orthogonal pairs of opposing laser beams (see Ref. [205] and Fig. 19.2). Even though atoms can be collected and cooled in the intersection region, it is important to stress again that this is *not* a trap. That is, atoms that wander away from the center experience no force directing them back. They are allowed to diffuse freely and even escape, as long as there is enough time for their very slow diffusive movement to allow them to reach the edge of the region of the intersection of the laser beams. Because the atomic velocities are randomized during the damping time $1/\omega_r$, atoms execute a random walk with a step size of a few μm [1]. To diffuse a distance of 1 cm requires about 10^7 steps or about 30 s [206, 207].

Three-dimensional OM was first observed in 1985 [204]. Preliminary measurements of the average kinetic energy of the atoms were done by blinking off the

19.6 Experiments in three-dimensional optical molasses

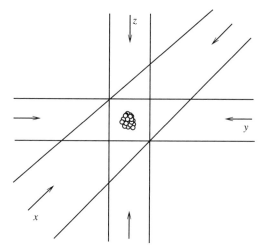

Figure 19.2 Schematic diagram of the arrangement of laser beams for 3D optical molasses. Three mutually perpendicular standing waves are formed by reflecting three laser beams from mirrors. Because of the red-detuned laser light, atoms experience a friction force in all directions and are therefore confined in a viscous medium, the optical molasses. (Figure from Ref. [1].)

laser beams for a fixed interval. Comparison of the brightness of the fluorescence before and after the turnoff was used to calculate the fraction of atoms that left the region while it was in the dark. The dependence of this fraction on the duration of the dark interval was used to estimate the velocity distribution and hence the temperature. The result was not inconsistent with the two-level atom theory described in Sec. 19.5.

Soon, other laboratories had produced 3D OM. The photograph in Fig. 19.3 shows OM in Na at the laboratory in the National Bureau of Standards (now NIST) in Gaithersburg. The phenomenon is readily visible to the unaided eye, and the photograph was made under ordinary snapshot conditions. The three mutually perpendicular pairs of laser beams appear as a star because they are viewed along a diagonal.

This NIST group developed a more accurate ballistic method to measure the velocity distribution of atoms in OM [209]. The limitation of the first measurements was determined by the size of the OM region and the unknown spatial distribution of atoms [204]. The new method at NIST used a separate measuring region composed of a 1D OM about 2 cm below the 3D region, thereby reducing the effect of this limitation. When the laser beams forming the 3D OM were shut off, the atoms dropped because of gravity into the 1D region, and the time-of-arrival distribution was measured. This was compared with calculated distributions for T_D and 40 µK. Using a series of such data sets it was possible to determine the

Figure 19.3 Photograph of optical molasses in Na taken under ordinary snapshot conditions in the lab at NIST. The upper horizontal streak is from the slowing laser while the three beams that cross at the center are on mutually orthogonal axes viewed from the (111) direction. Atoms in the optical molasses glow brightly at the center. (Figure from Ref. [208].)

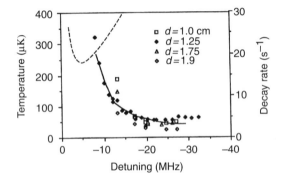

Figure 19.4 Temperature vs detuning determined from time-of-flight data for various separations d between the optical molasses and the probe laser (data points). The solid curve represents the measured molasses decay rate; it is not a fit to the temperature data points, but its scale (shown at right) was chosen to emphasize its proportionality to the temperature data. The dashed line shows the temperature expected on the basis of the two-level atom theory of Sec. 19.4. (Figure from Ref. [209].)

dependence of temperature on detuning, and that is shown in Fig. 19.4, along with the theoretical calculations for a two-level atom, as given in Sec. 19.5.

It was an enormous surprise to observe that the ballistically measured temperature of the Na atoms was as much as 10 times lower than $T_D = 240\mu K$ [209], the temperature minimum calculated from the theory. This breaching of the Doppler limit forced the development of an entirely new picture of OM that accounts for the fact that in 3D, a two-level picture of atomic structure is inadequate. The multi-level structure of atomic states, and optical pumping among these sublevels, must be considered in the description of 3D OM [1].

These experiments also found that OM was less sensitive to perturbations and more tolerant of alignment errors than was predicted by the 1D, two-level atom theory. For example, if the intensities of the two counterpropagating laser beams forming an OM were unequal, then the force on atoms at rest would not vanish, but the force on atoms with some non-zero drift velocity *would* vanish. This drift velocity can be easily calculated by using Eq. (19.12) with unequal intensities s_{0+} and s_{0-}, and following the derivation of Eq. (19.13). Thus atoms would drift out of an OM, and the calculated rate would be much faster than observed by deliberately unbalancing the beams in the experiments [210].

Section 19.7 describes the startling new view of OM that emerged in the late 1980s as a result of these surprising measurements. The need for a new theoretical description resulting from incontrovertible measurements provides an excellent pedagogical example of how physics is truly an experimental science, depending on the interactions between observations and theory, and always prepared to discard oversimplified descriptions as soon as it is shown that they are inadequate.

19.7 Cooling below the Doppler temperature

Section 19.6 showed that the model theories used to describe laser cooling had failed, and because physics is an experimental science, these models were deemed inadequate. In the place of the two-level atom there arose the need for considering the internal structure, including the hyperfine and Zeeman sublevels presented in Chaps. 9 and 11. To complete the picture it was necessary to consider the light shifts of Chap. 2, extended for the multilevel atom case using the selection rules of Chap. 3. It becomes clear that this subject does not consist of a collection of isolated bits of information, but instead is a unified whole, with each of the subtopics having a vital role to play.

In response to the surprising measurements of temperatures below T_D shown in Fig. 19.4, two groups developed a model of laser cooling that could explain the lower temperatures [211, 212]. The key feature of this model that distinguishes it from the earlier picture is the inclusion of the multiplicity of sublevels that make up an atomic state (e.g. Zeeman and hfs). The dynamics of optically pumping atoms among these sublevels provides the new mechanism for producing the ultra-low temperatures [208].

For atoms moving in a light field that varies in space, optical pumping acts to adjust the atomic orientation to the changing conditions of the light field. In a weak pumping process, the orientation of moving atoms always lags behind the orientation that would exist for stationary atoms. It is this phenomenon of non-adiabatic following that is the essential feature of the new cooling process.

Production of spatially dependent optical pumping processes can be achieved in several different ways. As an example, consider two counterpropagating laser beams that have orthogonal polarizations, as discussed in Ref. [1]. The superposition of the two beams results in a light field having a polarization that varies on the wavelength scale along the direction of the laser beams. Laser cooling by such a light field is called polarization gradient cooling. In a three-dimensional optical molasses, the transverse wave character of light requires that the light field always has polarization gradients.

The cooling process that derives from this non-adiabatic following is effective over a limited range of atomic velocities. The force is maximum for atoms that travel a distance $\lambda/4$ during one optical pumping process. If atoms travel at a lower velocity, they will not have reached a very different part of the optical field before a spontaneous emission causes the pumping process to occur; if atoms travel faster, they will already go beyond the largest change in the field before being pumped toward another sublevel [202]. Of course the velocity at which this force is effective scales with the characteristic distance over which the optical field changes. Although this is typically $\lambda/4$, it can be much larger for two light waves with oblique $\vec{k}$-vectors.

The nature of this cooling process is fundamentally different from the Doppler laser-cooling process discussed in Sec. 19.4. In that case, the differential absorption from the laser beams was caused by the Doppler shift of the laser frequency, and the process is therefore known as Doppler cooling. In the cooling process described in this section, the force is still caused by differential absorption of light from the two laser beams, but the velocity-dependent differential rates, and hence the cooling, rely on the non-adiabaticity of the optical pumping process. Since lower temperatures can usually be obtained with this cooling process, it is called sub-Doppler laser cooling [208, 209, 213].

19.7.1 Polarization and interference

Consider the light field of two counterpropagating plane-wave laser beams with the same frequency ω. If the polarizations of the two laser beams are identical, then the polarization of the resulting light field is everywhere the same as that of the incoming laser beams. However, the two plane waves interfere and produce a

19.7 Cooling below the Doppler temperature

standing wave. The resulting electric field for a linear polarization $\hat{\varepsilon}$ can be written as $\vec{\mathcal{E}} = \mathcal{E}_0\, \hat{\varepsilon}\, \cos(\omega t - kz) + \mathcal{E}_0\, \hat{\varepsilon}\, \cos(\omega t + kz) = 2\mathcal{E}_0\, \hat{\varepsilon} \cos kz\, \cos \omega t$. The intensity of the light field has a $\cos^2 kz$ spatial dependence with a period of $\lambda/2$. This situation of a standing wave is very common in laser cooling, and it will reappear in the discussion of optical traps and lattices.

If the polarization of the laser beams is not identical, then the situation becomes rather complicated. Only one of the two special cases that play important roles in laser cooling will be considered here, namely where the two counterpropagating laser beams are both linearly polarized, but their $\hat{\varepsilon}$-vectors are perpendicular (which is called lin ⊥ lin or lin-perp-lin). Then the total field is the sum of the two counterpropagating beams given by

$$\vec{\mathcal{E}} = \mathcal{E}_0\, \hat{x}\, \cos(\omega t - kz) + \mathcal{E}_0\, \hat{y}\, \cos(\omega t + kz)$$
$$= \mathcal{E}_0\, [(\hat{x} + \hat{y}) \cos \omega t\, \cos kz + (\hat{x} - \hat{y}) \sin \omega t\, \sin kz]. \qquad (19.14)$$

At the origin, where $z = 0$, this becomes $\vec{\mathcal{E}} = \mathcal{E}_0(\hat{x} + \hat{y}) \cos \omega t$, which corresponds to linearly polarized light at an angle $+\pi/4$ to the x-axis. The amplitude of this field is $\sqrt{2}\mathcal{E}_0$. Similarly, for $z = \lambda/4$, where $kz = \pi/2$, the field is also linearly polarized but at an angle $-\pi/4$ to the x-axis.

Between these two points, at $z = \lambda/8$, the total field is $\vec{\mathcal{E}} = \mathcal{E}_0[\hat{x}\, \sin(\omega t + \pi/4) - \hat{y}\, \cos(\omega t + \pi/4)]$ because $kz = \pi/4$. Since the $\hat{x}$- and $\hat{y}$-components have sine and cosine dependence, they are $\pi/4$ out of phase, and so this represents circularly polarized light rotating about the z-axis in the negative sense. Similarly, at $z = 3\lambda/8$ where $kz = 3\pi/4$, the polarization is circular but in the positive sense. Thus in this lin ⊥ lin scheme the polarization cycles from linear to circular to orthogonal linear to opposite circular in the space of only half a wavelength of light, as shown in Fig. 19.5. It truly has a very strong polarization gradient. To study the effects of this polarization gradient on the cooling process, the authors of

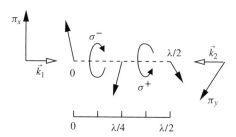

Figure 19.5 Polarization gradient field for the lin ⊥ lin configuration. (Figure from Ref. [1].)

Ref. [211] considered a $J_g = 1/2$ to $J_e = 3/2$ transition. This is one of the simplest transitions that shows sub-Doppler laser cooling.

19.7.2 Lin-perp-lin polarization gradient cooling

In the place where the light field is purely σ^+, the pumping process drives the ground-state population to the $M_g = +1/2$ sublevel. This optical pumping occurs because absorption always produces $\Delta M = +1$ transitions, whereas the subsequent spontaneous emission produces $\Delta M = \pm 1, 0$ (see discussion near the end of Sec. 3.5). Thus the average $\Delta M \geq 0$ for each scattering event. For σ^--light the population will be pumped toward the $M_g = -1/2$ sublevel. Thus in traveling through a half wavelength in the light field, atoms have to readjust their population completely from $M_g = +1/2$ to $M_g = -1/2$ and back again.

The interaction between nearly resonant light and atoms not only drives transitions between atomic energy levels, but also shifts their energies as given in Eq. (2.10). This light shift plays a crucial role in this scheme of sub-Doppler laser cooling, and the changing polarization has a strong influence on the light shifts. In the low-intensity limit of two laser beams each of intensity $s_0 I_s$, the light shifts ΔE_g of the ground magnetic sub-states are given by

$$\Delta E_g = \frac{\hbar \delta s_0 C_{ge}^2}{1 + (2\delta/\gamma)^2}, \tag{19.15}$$

where C_{ge} is the Clebsch–Gordan coefficient that describes the coupling between the atom and the light field (Sec. 10.5.2). This relation has to be compared with the result obtained in Eq. (2.10) for a two-level atom in a traveling wave. First, the light shift is twice as large, since there are two traveling waves. Second, the coupling has been modified because of the multiplicity of the ground state, which is expressed by the coefficients C_{eg}^2. Finally, the semi-classical analysis of Sec. 2.3.4 did not take into account the spontaneous emission process, and a more careful analysis [211] leads to the result of Eq. (19.15). Since C_{ge} depends on the magnetic quantum numbers and on the polarization of the light field, the light shifts are different for different magnetic sublevels. The ground-state light shift is negative for a laser tuning below resonance ($\delta < 0$) and positive for $\delta > 0$ (see Eq. (2.10)).

In the present case of orthogonal linear polarizations and $J = 1/2 \to 3/2$, the light shift for the magnetic sub-state $M_g = 1/2$ is three times larger than that of the $M_g = -1/2$ sub-state when the light field is completely σ^+. On the other hand, when the light field becomes σ^-, the shift of $M_g = -1/2$ is three times larger. So in this case the optical pumping discussed above causes there to be a larger population in the state with the larger light shift. This is generally true for any transition J_g to $J_e = J_g + 1$. A schematic diagram showing the transitions and light shifts for this particular case of negative detuning is shown in Fig. 19.6.

19.7 Cooling below the Doppler temperature

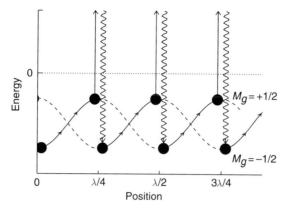

Figure 19.6 The spatial dependence of the light shifts of the ground-state sublevels of the $J = 1/2 \Leftrightarrow 3/2$ transition for the case of the lin $\perp$ lin polarization configuration. The arrows show the path followed by atoms being cooled in this arrangement. Atoms starting at $z = 0$ in the $M_g = +1/2$ sublevel must climb the potential hill as they approach the $z = \lambda/4$ point where the light becomes σ^--polarized, and there they are optically pumped to the $M_g = -1/2$ sublevel. Then they must begin climbing another hill toward the $z = \lambda/2$ point where the light is σ^+-polarized and they are optically pumped back to the $M_g = +1/2$ sublevel. The process repeats until the atomic kinetic energy is too small to climb the next hill. Each optical pumping event results in absorption of light at a lower frequency than emission, thus dissipating energy to the radiation field. (Figure from Ref. [1].)

19.7.3 The damping force and the temperature limit

To discuss the origin of the cooling process in this polarization gradient scheme, consider atoms with a velocity v at a position where the light is σ^+-polarized, as shown at the lower left of Fig. 19.6. The light optically pumps such atoms to the strongly negative light-shifted $M_g = +1/2$ state. In moving through the light field, atoms must increase their potential energy (climb a hill) because the polarization of the light is changing and the state $M_g = 1/2$ becomes less strongly coupled to the light field. After traveling a distance $\lambda/4$, atoms arrive at a position where the light field is σ^--polarized, and are optically pumped to $M_g = -1/2$, which is now lower than the $M_g = 1/2$ state. Again the moving atoms are at the bottom of a hill and start to climb. In climbing the hills, the kinetic energy is converted to potential energy, and in the optical pumping process, the potential energy is radiated away because the spontaneous emission is at a higher frequency than the absorption (see Fig. 19.6). Thus atoms seem to be always climbing hills and losing energy in the process. This process brings to mind a Greek myth, and is thus called "Sisyphus laser cooling".

The cooling process described above is effective over a limited range of atomic velocities. The damping is maximum for atoms that undergo one optical pumping process while traveling over a distance $\lambda/4$. Slower atoms will not reach the hilltop before the pumping process occurs, and faster atoms will already be descending the hill before being pumped toward the other sublevel. In both cases the energy loss is smaller and therefore the cooling process less efficient.

The damping force $F = -\beta v$ can be estimated from the distance dependence of the energy loss. Denote the optical pumping time by $\tau_p \equiv 1/\gamma_p$ and then the optimum speed is $v_c \cong \gamma_p/k$. The force at this velocity v_c is $F = \Delta W/\Delta z \cong k\Delta E \equiv -\beta v_c$. To find the order of magnitude of the friction coefficient β, both the light shift ΔE and the pumping rate γ_p need to be estimated. For a detuning $|\delta| \gg \gamma$, Eq. (6.18) yields the pumping rate $\gamma_p = s_0 \gamma^3/4\delta^2$ (the 4 instead of an 8 accounts for the presence of two laser beams), and then choosing $C_{ge}^2 = 1$ in Eq. (19.15) yields $\Delta E = \hbar \gamma^2 s_0/4\delta$. Then the damping rate $\beta/M = \hbar k^2 \delta/2M\gamma = \omega_r \delta/\gamma$. The velocity-dependent force has the same order of magnitude as the Doppler force, but its velocity range is γ_p/k. The best result are often obtained with $|\delta| \gg \gamma/2$, so usually this is much smaller than γ/k and therefore β is much larger.

This result is of particular significance because it shows that the friction coefficient for this sub-Doppler process is larger by a factor $(2|\delta|/\gamma)$ than the maximum friction coefficient for Doppler laser cooling. It can be shown that the momentum diffusion coefficient of this process is of the same order of magnitude as that of Doppler cooling, so that the temperature will be smaller than the Doppler temperature by the same factor. Furthermore, it shows that the friction coefficient for this case is independent of intensity, since both ΔE and γ_p are proportional to the intensity.

Calculation of the temperature limit for such Sisyphus cooling is much more subtle than for Doppler cooling [1]. Intuitively, one might expect that the last spontaneous emission in a cooling process would leave atoms with a residual momentum of the order of $\hbar k$, since there is no control over its direction. Thus the randomness associated with this would put a lower limit on such cooling of $v_{min} \sim v_r$, where $v_r = \hbar k/M$ is the recoil velocity and M is the atomic mass. Second, the polarization gradient cooling mechanism described above in Sec. 19.7.3 requires that atoms be localizable within the scale of $\sim \lambda/2\pi$ in order to be subject to only a single polarization in the spatially inhomogeneous light field. The uncertainty principle then requires that these atoms have a momentum spread of at least $\hbar k$.

Exercises

19.1 Each scattering event of visible light changes the speed of an atom by a "few" cm/s.
(a) Calculate how much is a "few" for the Na 3s→3p transition.
(b) For Na atoms coming from a source heated to temperature 400 °C, what is the average thermal speed v_T of the atoms?

Exercises

(c) How many scattering events are required to stop atoms with this average speed?

(d) Suppose such a slowing experiment is done with a laser tuned to frequency $\omega = \omega_{eg} - kv_T$ to compensate the Doppler shift. How many scattering events will occur before such an atom is slowed enough to be Doppler shifted out of resonance with the laser?

(e) Calculate the minimum length required to decelerate Na atoms at the average velocity from a thermal source as described above.

19.2 Use the formula for β given in Eq. (19.11) and the expression $\dot{E}_{cool} = \vec{F} \cdot \vec{v}$ to calculate the rate of energy loss of an atom in optical molasses at low intensity. Then assume that the rate of energy gain $\dot{E}_{heat}$ is twice the recoil $E_r \equiv \hbar^2 k^2/2M$ per scattering event (why?), and that the scattering rate for each of the two beams is γ_p (total scattering rate is $2\gamma_p$). Set these rates equal (steady state) and calculate the expected final kinetic energy of such laser-cooled atoms as a function of detuning. Plot your results.

19.3 The force on an atom in one dimension from counterpropagating light beams is given by $F = \beta v$, which is a damping force for $\beta < 0$. Give the formula for β, and calculate the damping time for Na driven on the $3^2S \to 3^2P$ transition near $\lambda = 589$ nm. The excited 3^2P state has a lifetime of 16 ns. Calculate the natural width γ, and from that, the range of velocities for which the approximation required to derive $F = \beta v$, namely $(kv/\gamma)^4 \ll 1$, is valid. Discuss the meaning of this velocity range, what "temperature" it corresponds to, and why such a limit arises. For an atom moving at $v = \gamma/k$, what is the characteristic damping distance (the atom's travel) in this optical molasses?

19.4 As shown in Eq. (19.1), the force depends only on the coherences ρ_{eg} and ρ_{ge}, and not on the populations ρ_{gg} and ρ_{ee}. Equivalently, use the effective Hamiltonian of Eq. (2.17) and the Ehrenfest theorem to show that the force depends only on the Pauli matrices σ_x and σ_y and not on the Pauli matrix σ_z. Show why these two statements are equivalent.

19.5 In laser cooling, the polarization of light plays an important role. Different configurations of polarization result in different cooling mechanisms. Consider two beams that are counterpropagating along the z-axis.

(a) Suppose the two beams have opposite circular polarizations. Here, the sense of circularity is determined in the laboratory coordinates and has no relation to the direction of travel. Write an expression for the total electric field $\vec{\mathcal{E}}_{total}(z)$. Show that the intensity, proportional to $|\vec{\mathcal{E}}_{total}(z)|^2$, is independent of z. Describe the nature of the field. This configuration is normally called $\sigma^+ - \sigma^-$.

(b) Suppose the two beams are linearly polarized with $\vec{\mathcal{E}}_1(z)$ perpendicular to $\vec{\mathcal{E}}_2(z)$ (e.g. one beam has $\hat{x}$-polarization and the other $\hat{y}$). Write an expression for the total electric field $\vec{\mathcal{E}}_{\text{total}}(z)$. Describe its spatial variation, and calculate the intensity as a function of z. This case is normally called lin-perp-lin or LPL.

(c) Repeat Exercise 19.5b for the case where the angle between $\vec{\mathcal{E}}_1(z)$ and $\vec{\mathcal{E}}_2(z)$ is not $\pi/2$, but is some arbitrary angle $\theta < \pi/2$. What implications does this θ-dependence have for an optical lattice? Show that $\theta = 0$ gives an ordinary standing wave. This case is called lin-angle-lin or LAL.

19.6 An atom moving with velocity V along the axis of a standing wave can undergo absorption followed by stimulated emission in the opposite direction, thereby changing its momentum by $2\hbar k$. Since the energy of the light field is unchanged by this sequence, so must be the kinetic energy of the atom. Thus $MV^2/2 = M(V \pm 2\hbar k/M)^2/2$. How can this be? Note that the direction of this change depends on the relative phase of the standing wave at the position of the atom, and the exchange can therefore be repeated many times in the same direction, thus producing a large force (the dipole force).

19.7 In sub-Doppler cooling, the velocity dependence of the force has a similar shape as in Fig. 19.1. What is the range of this force, and how do these range boundaries arise?

20

Confinement of neutral atoms

In order to confine any object, it is necessary to exchange kinetic for potential energy in the trapping field, and in neutral atom traps, the potential energy must be stored as internal atomic energy. Thus practical traps for ground-state neutral atoms are necessarily very shallow compared with thermal energy because the energy level shifts that result from conveniently sized fields are typically considerably smaller than $k_B T$, even for $T = 1$ K. Neutral atom trapping therefore depends on substantial cooling of a thermal atomic sample, and is often connected with the cooling process. In most practical cases, atoms are loaded from magneto-optical traps, where they have been efficiently accumulated and cooled to mK temperatures (see Sec. 20.3), or from optical molasses, where they have been optically cooled to μK temperatures (see Sec. 19.4).

The small depth of typical neutral atom traps dictates stringent vacuum requirements, because an atom cannot remain trapped after a collision with a thermal energy background gas molecule. Since these atoms are vulnerable targets for thermal energy background gas, the mean free time between collisions *must* exceed the desired trapping time. The cross-section for destructive collisions is quite large because even a gentle collision (i.e. large impact parameter) can impart enough energy to eject an atom from a trap. At pressure P sufficiently low to be of practical interest, the trapping time is $\sim(10^{-8}/P)$ s, where P is in torr.

This chapter begins with a discussion of ordinary, single-center traps, but also considers arrays of traps. Such "optical lattices" are playing an increasingly important role in research in atomic physics, so discussion of their properties is an appropriate part of the education of a student of the subject.

20.1 Dipole force optical traps

20.1.1 Single-beam optical traps for two-level atoms

The simplest imaginable optical trap consists of a single, strongly focused Gaussian laser beam (see Fig. 20.1) [214, 215] whose intensity at the focus varies

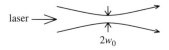

Figure 20.1 A single focused laser beam produces the simplest type of optical trap. (Figure from Ref. [1].)

transversely with r as $I(r) = I_0 e^{-2r^2/w_0^2}$ where w_0 is the beam waist size. Such a trap has a well-studied and important macroscopic classical analog in a phenomenon called optical tweezers (see App. 1.C) [216–218].

With the laser light tuned below resonance ($\delta < 0$), the ground-state light shift is everywhere negative, but largest at the center of the Gaussian beam waist. Ground-state atoms therefore experience a force attracting them toward this center given by the gradient of the light shift which is found from Eq. (2.10), and for $\delta/\gamma \gg s_0$ is given by Eq. (19.5). For a Gaussian beam, this transverse force at the waist is harmonic for small r and is given by

$$F \simeq \frac{\hbar \gamma^2}{4\delta} \frac{I_0}{I_s} \frac{r}{w_0^2} e^{-2r^2/w_0^2}. \tag{20.1}$$

In the longitudinal direction there is also an attractive force, but it is more complicated and depends on the details of the focusing. Thus this trap produces an attractive force on atoms in three dimensions.

Although it may appear that the trap does not confine atoms longitudinally because of the radiation pressure along the laser beam direction, careful choice of the laser parameters can indeed produce trapping in 3D. This can be accomplished because the radiation pressure force decreases as $1/\delta^2$ (see Eq. (19.4)), but by contrast, the light shift and hence the dipole force only decreases as $1/\delta$ for $\delta \gg |\Omega|$ (see Eq. (2.10)). If $|\delta|$ is chosen to be sufficiently large, atoms spend very little time in the untrapped (actually repelled) excited state because its population is proportional to $1/\delta^2$. Thus a sufficiently large value of $|\delta|$ both produces longitudinal confinement and maintains the atomic population primarily in the trapped ground state.

The first optical trap was demonstrated in Na with light detuned below the D-lines [215]. With 220 mW of dye laser light tuned about 650 GHz below the atomic transition and focused to a waist of $\sim$10 µm, the trap depth was about $15\hbar\gamma$ corresponding to 7 mK.

20.1.2 Blue detuned optical traps

One of the principal disadvantages of the optical traps discussed above is that the negative detuning attracts atoms to the region of highest light intensity. This may

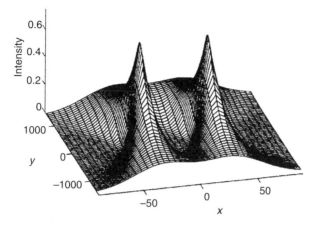

Figure 20.2 The light intensity experienced by an atom located in a plane 30 μm above the beam waists of two quasi-focused sheets of light traveling parallel and arranged to form a V-shaped trough. The x- and y-dimensions are in μm. (Figure from Ref. [222].)

result in significant spontaneous emission such as in the Nd:YAG laser trap [219] or the CO_2 laser trap [220], unless the detuning is a large fraction of the optical frequency. More important in some cases is that the trap relies on Stark shifting of the atomic energy levels by an amount equal to the trap depth, and this severely compromises the capabilities for precision spectroscopy in a trap [221].

Attracting atoms to the region of *lowest* intensity would ameliorate both of these concerns, but such a trap requires positive detuning (blue), and an optical configuration having a dark central region. One of the first experimental efforts at a blue detuned trap used the repulsive dipole force to support Na atoms that were otherwise confined by gravity in an optical "cup" [222]. Two rather flat, parallel beams detuned by 25% of the atomic resonance frequency were directed horizontally and oriented to form a V-shaped trough. Their Gaussian beam waists formed a region about 1 mm long where the potential was deepest, and hence provided confinement along their propagation direction as shown in Fig. 20.2. The beams were the $\lambda = 514$ nm and $\lambda = 488$ nm from an argon laser, and the choice of two frequencies was not simply to exploit the full power of the multi-line Ar laser, but also to avoid the spatial interference that would result from a single frequency.

Obviously a hollow laser beam would also satisfy the requirement for a blue-detuned trap, but conventional textbook wisdom shows that such a beam is not an eigenmode of a laser resonator [223]. Some lasers can make hollow beams, but these are illusions because they consist of rapid oscillations between the TEM_{01} and TEM_{10} modes of the cavity. Nevertheless, Maxwell's equations permit the propagation of such beams, and in the recent past there have been studies of the LaGuerre–Gaussian modes that constitute them [224–226].

20.2 Magnetic traps

20.2.1 Introduction

Magnetic trapping of neutral atoms is well suited for use in very many areas, including high-resolution spectroscopy, collision studies, Bose–Einstein condensation, and atom optics. Although ion trapping, laser cooling of trapped ions, and trapped ion spectroscopy were known for many years [227], it was only in 1985 that neutral atoms were first trapped [228]. Such experiments offer the capability of the spectroscopic ideal of an isolated atom at rest, in the dark, available for interaction with electromagnetic field probes. Because trapping requires the exchange of kinetic for potential energy, the atomic energy levels will necessarily shift as the atoms move in the trap. These shifts can severely affect the precision of spectroscopic measurements. Since one of the potential applications of trapped atoms is in high-resolution spectroscopy, such inevitable shifts must be carefully considered.

20.2.2 Magnetic confinement

The Stern–Gerlach experiment in 1924 first demonstrated the mechanical action of inhomogeneous magnetic fields on neutral atoms having magnetic moments, and the basic phenomenon was subsequently developed and refined. An atom with a magnetic moment $\vec{\mu}$ can be confined by an inhomogeneous magnetic field because of an interaction between the moment and the field. This produces a force given by $\vec{F} = \vec{\nabla}(\vec{\mu} \cdot \vec{B})$, where $\vec{F} = -\vec{\nabla}(U)$ and $U = -\vec{\mu} \cdot \vec{B}$. Several different magnetic traps with varying geometries that exploit this force have been studied in some detail in the literature. The general features of the magnetic fields of a large class of possible traps have been presented [229].

Wolfgang Paul originally suggested a quadrupole trap comprising two identical coils carrying opposite currents (see Fig. 20.3). This trap clearly has a single center where the field is zero, and is the simplest of all possible magnetic traps. When the coils are separated by 1.25 times their radius, such a trap has equal depth in the radial (x–y plane) and longitudinal (z-axis) directions [229]. Its experimental simplicity makes it most attractive, both because of ease of construction and because of optical access to the interior. Such a trap was used in the first neutral atom trapping experiments [228].

The magnitude of the field is zero at the center of this trap, and increases in all directions as $B = A\sqrt{\rho^2 + 4z^2}$, where $\rho^2 \equiv x^2 + y^2$, and A is the field gradient (see

20.2 Magnetic traps

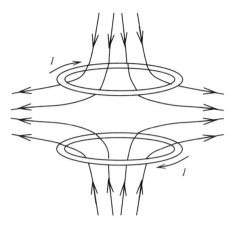

Figure 20.3 Schematic diagram of the coil configuration used in the quadrupole trap and the resultant magnetic field lines. Because the currents in the two coils are in opposite directions, there is a $|\vec{B}| = 0$ point at the center. (Figure from Ref. [1].)

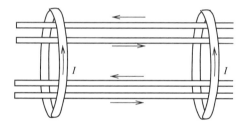

Figure 20.4 The Ioffe trap has four straight current elements that form a linear quadrupole field. The axial confinement is accomplished with end coils as shown. These fields can be achieved with many different current configurations as long as the geometry is preserved. (Figure from Ref. [1].)

Ref. [229]). The field gradient is constant along any line through the origin, but has different values in different polar directions because of the 4 in the formula above. Therefore the magnetic force that confines the atoms in the trap is neither harmonic nor central, and orbital angular momentum is not conserved.

The requisite field for the quadrupole trap can also be provided in two dimensions by four straight currents as indicated in Fig. 20.4. The field is translationally invariant along the direction parallel to the currents, so a trap cannot be made this way without additional fields. These are provided by end coils that close the trap, as shown.

Although there are very many different kinds of magnetic traps for neutral particles, this particular one has played a special role. There are certain conditions required for trapped atoms not to be ejected in a region of zero field such as occurs at the center of a quadrupole trap (see Sec. 20.2.3). This problem is not easily cured, so the Ioffe trap has been used in many of the BEC experiments because it has $|\vec{B}| \neq 0$ everywhere.

20.2.3 Motion of atoms in a quadrupole trap

Because of the dependence of the trapping force on the angle between the field and the atomic moment, the orientation of the magnetic moment with respect to the field must be preserved as the atoms move about in the trap. Otherwise the atoms may be ejected instead of confined by the fields of the trap. The precession frequency ω_L must be large enough that the magnetic moment can follow the changing field seen by the moving atom. This requires velocities low enough to ensure that the interaction between the atomic moment $\vec{\mu}$ and the field $\vec{B}$ is adiabatic, especially when the atom's path passes through a region where the field magnitude is small and therefore the energy separation between the trapping and non-trapping states is small. This is especially critical at the low temperatures of the BEC experiments. Therefore energy considerations that focus only on the trap depth are not sufficient to determine the stability of a neutral atom trap: orbit and/or quantum state calculations and their consequences must also be considered.

For the two-coil quadrupole trap, the adiabaticity condition can be easily calculated. Using $v = \sqrt{\rho_1 a}$ for circular orbits of radius ρ_1 in the $z = 0$ plane, the adiabatic condition for a practical trap ($\nabla B \sim 1$ T/m) requires $\rho_1 \gg (\hbar^2/M^2 a)^{1/3} \sim 1$ μm as well as $v \gg (\hbar a/M)^{1/3} \sim 1$ cm/s, where $a \equiv \mu \nabla B/M$ and M is the whole atom mass. Note that violation of these conditions (i.e. $v \sim 1$ cm/s in a trap with $\nabla B \sim 1$ T/m) results in the onset of quantum dynamics for the motion (de Broglie wavelength ≈ orbit size).

Since laser and evaporative cooling have the capability to cool atoms to energies where their de Broglie wavelengths are on the micron scale, the motional dynamics must be described in terms of quantum mechanical variables and suitable wavefunctions. Studying the behavior of extremely slow (cold) atoms in the two-coil quadrupole trap begins with a heuristic quantization of the orbital angular momentum using $Mr^2\omega_T = n\hbar$ for circular orbits. The energy levels are then given by $E_n = (3/2)E_1 n^{2/3}$ where $E_1 = (Ma^2\hbar^2)^{1/3}$, which corresponds to about 5 kHz. For velocities of optically cooled atoms of a few cm/s, $n \sim 10$–100. By contrast, evaporative cooling can produce velocities ~ 1 mm/s resulting in $n \sim 1$.

It is readily found that $\omega_Z = n\omega_T$, so that the adiabatic condition is satisfied only for $n \gg 1$. The lower-lying (small-n) bound states, whose orbits are confined to a region near the origin where the field is small, are strongly coupled to unbound states as a result of the motion (dynamic coupling), and these are rapidly ejected from the trap [229, 230]. On the other hand, the large-n bound states are less coupled because they spend most of their time in a stronger field, and thus satisfy the condition of adiabaticity of the orbital motion relative to the Larmor precession. In this case the separation of the rapid precession from the slower orbital motion is reminiscent of the Born–Oppenheimer approximation for molecules.

20.3 Magneto-optical traps

20.3.1 Introduction

The most widely used trap for neutral atoms is a hybrid, employing both optical and magnetic fields to make a magneto-optical trap (MOT), first demonstrated in 1987 [231]. The operation of a MOT depends on both inhomogeneous magnetic fields and radiative selection rules to exploit both optical pumping and the strong radiative force [231, 232]. The radiative interaction provides cooling that helps in loading the trap, and enables very easy operation. The MOT is a very robust trap that does not depend on precise balancing of the counterpropagating laser beams or on a very high degree of polarization.

The magnetic field gradients are modest and have the convenient feature that the configuration is the same as the quadrupole magnetic traps discussed in Sec. 20.2.2. Appropriate fields can readily be achieved with simple, air-cooled coils. The trap is easy to construct because it can be operated with a room-temperature cell where alkali atoms are captured from the vapor. Furthermore, low-cost diode lasers can be used to produce the light appropriate for many atoms, so the MOT has become one of the least expensive ways to make atomic samples with temperatures below 1 mK.

Trapping in a MOT works by optical pumping of slowly moving atoms in a linearly inhomogeneous magnetic field $B = B(z)$, such as that formed by a magnetic quadrupole field. Atomic transitions with the simple scheme of $J_g = 0 \to J_e = 1$ have three Zeeman components in a magnetic field, excited by each of three polarizations, whose frequencies tune with field (and therefore with position) as shown in Fig. 20.5 for 1D. Two counterpropagating laser beams of opposite circular polarization, each detuned below the zero field atomic resonance by δ, are incident as shown.

Because of the Zeeman shift, the excited state $M_e = +1$ is shifted up for $B > 0$, whereas the state with $M_e = -1$ is shifted down. At position z' in Fig. 20.5 the magnetic field therefore tunes the $\Delta M = -1$ transition closer to resonance and the $\Delta M = +1$ transition further out of resonance. If the polarization of the laser beam incident from the right is chosen to be σ^- and correspondingly σ^+ for the other beam, then more light is scattered from the σ^- beam than from the σ^+ beam. Thus the atoms are driven toward the center of the trap where the magnetic field is zero. On the other side of the center of the trap, the roles of the $M_e = \pm 1$ states

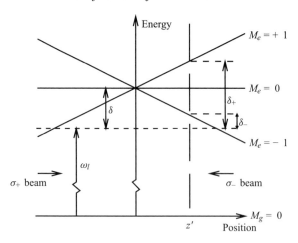

Figure 20.5 Arrangement for MOT in 1D. The horizontal dashed line represents the laser frequency seen by an atom at rest in the center of the trap. Because of the Zeeman shifts of the atomic transition frequencies in the inhomogeneous magnetic field, atoms at $z = z'$ are closer to resonance with the σ^- laser beam than with the σ^+ beam, and are therefore driven toward the center of the trap. (Figure from Ref. [1].)

are reversed and now more light is scattered from the σ^+ beam, again driving the atoms towards the center.

So far the discussion has been limited to the motion of atoms in 1D. However, the MOT scheme can easily be extended to 3D by using six instead of two laser beams. Furthermore, even though very few atomic species have transitions as simple as $J_g = 0 \to J_e = 1$, the scheme works for any $J_g \to J_e = J_g + 1$ transition. Atoms that scatter mainly from the σ^+ laser beam will be optically pumped toward the $M_g = +J_g$ sub-state, which forms a closed system with the $M_e = +J_e$ sub-state.

20.3.2 Cooling and compressing atoms in a MOT

For a description of the motion of the atoms in a MOT, consider the radiative force in the low-intensity limit (see Eq. (19.4)). The total force on the atoms is given by $\vec{F} = \vec{F}_+ + \vec{F}_-$, where $\vec{F}_\pm$ is given in Eq. (19.13), and the detuning $\delta_\pm$ for each laser beam is given by $\delta_\pm = \delta \mp \vec{k} \cdot \vec{v} \pm \mu' B/\hbar$. Here $\mu' \equiv (g_e M_e - g_g M_g)\mu_B$ is the effective magnetic moment for the transition used. Note that the Doppler shift $\omega_D \equiv -\vec{k} \cdot \vec{v}$ and the relative Zeeman shift $\omega_L = \mu' B/\hbar$ both have opposite signs for opposite beams.

The situation is analogous to the velocity damping in optical molasses from the Doppler effect as discussed in Sec. 19.4.2, but here the effect operates in position space, whereas for molasses it operates in velocity space. Since the laser light is detuned below the atomic resonance in both cases, compression and cooling of the atoms is obtained simultaneously in a MOT.

When both the Doppler and Zeeman shifts are small compared with the detuning δ, the denominator of the force can be expanded as for Eq. (19.13) and the result becomes

$$\vec{F} = -\beta\vec{v} - \kappa\vec{r}, \tag{20.2}$$

where the damping coefficient β is defined in Eq. (19.13). The spring constant κ arises from the similar dependence of $\vec{F}$ on the Doppler and Zeeman shifts, and is given by $\kappa = \mu'\nabla B\beta/\hbar k$.

The force of Eq. (20.2) leads to damped harmonic motion of the atoms, where the damping rate is given by $\Gamma_{MOT} = \beta/M$ and the oscillation frequency $\omega_{MOT} = \sqrt{\kappa/M}$. For magnetic field gradients $\nabla B \approx 10$ G/cm, the oscillation frequency is typically a few kHz, and this is much smaller than the damping rate that is typically a few hundred kHz. Thus the motion is overdamped, with a characteristic restoring time to the center of the trap of $2\Gamma_{MOT}/\omega_{MOT}^2 \approx$ several ms for typical values of the detuning and intensity of the lasers.

20.3.3 Capturing atoms in a MOT

Although the approximations that lead to Eq. (20.2) for the force hold for slow atoms near the origin, they do not apply for the capture of fast atoms far from the origin. In the capture process, the Doppler and Zeeman shifts are no longer small compared with the detuning, so the effects of the position and velocity can no longer be disentangled. However, the full expression for the force still applies, and the trajectories of the atoms can be calculated by numerical integration of the equation of motion [195].

The capture velocity of a MOT is serendipitously enhanced because atoms traveling across it experience a decreasing magnetic field just as in beam deceleration [200]. This enables resonance over an extended distance and velocity range because the changing Doppler shift of decelerating atoms can be compensated by the changing Zeeman shift as atoms move in the inhomogeneous magnetic field. Of course, it will only work this way if the field gradient ∇B does not demand an acceleration larger than the maximum acceleration a_{max}. Thus atoms are subject to the optical force over a distance that can be as long as the trap size, and can therefore be slowed considerably.

The very large velocity capture range v_{cap} of a MOT can be estimated by using $F_{max} = \hbar k \gamma/2$ and choosing a maximum size of a few cm for the beam diameters. Thus the energy change can be as large as a few K, corresponding to $v_{cap} \sim 100$ m/s [232]. The number of atoms in a vapor with velocities below v_{cap} in the

Boltzmann distribution scales as v_{cap}^4, and there are enough slow atoms to fall within the large MOT capture range even at room temperature, because a few K includes one in 10^4 of the atoms.

20.4 Optical lattices

20.4.1 Quantum states of motion

As the techniques of laser cooling advanced from a laboratory curiosity to a tool for new problems, the emphasis shifted from attaining the lowest possible steady-state temperatures to the study of elementary processes, especially the quantum mechanical description of the atomic motion. In the completely classical description of laser cooling, atoms were assumed to have well-defined position and momentum that could be known simultaneously with arbitrary precision. However, when atoms are moving sufficiently slowly that their de Broglie wavelength precludes their localization to less than $\lambda/2\pi$, these descriptions fail and a quantum mechanical description is required. Such exotic behavior for the motion of whole atoms, as opposed to electrons in the atoms, had not been considered before the advent of laser cooling simply because it is too far out of the range of ordinary experiments. A series of experiments in the early 1990s provided striking evidence for these new quantum states of motion of neutral atoms, and led to the debut of de Broglie wave atom optics.

The quantum description of atomic motion requires that the energy of such motion be included in the Hamiltonian. The total Hamiltonian for atoms moving in a light field would then be given by

$$\mathcal{H} = \mathcal{H}_{\text{atom}} + \mathcal{H}_{\text{rad}} + \mathcal{H}_{\text{int}} + \mathcal{H}_{\text{kin}}, \qquad (20.3)$$

where $\mathcal{H}_{\text{atom}}$ describes the motion of the atomic electrons and gives the internal atomic energy levels, $\mathcal{H}_{\text{rad}}$ is the energy of the radiation field and is of no concern here because the field is not quantized, $\mathcal{H}_{\text{int}}$ describes the excitation of atoms by the light field and the concomitant light shifts, and $\mathcal{H}_{\text{kin}}$ is the kinetic energy operator of the motion of the atoms' center of mass. This Hamiltonian has eigenstates of not only the internal energy levels and the atom–laser interaction that connects them, but also of the kinetic energy operator $\mathcal{H}_{\text{kin}} \equiv P^2/2M$. These eigenstates will therefore be labeled by quantum numbers of the atomic states as well as the center of mass momentum p. An atom in the ground state $\Psi_{g,p}$ has an energy $E_g + (\hbar k)^2/2M$, which can take on a range of values.

In 1968, Letokhov [233] suggested that it is possible to confine atoms in the wavelength-size regions of a standing wave by means of the dipole force that arises from the light shift. This was first accomplished in 1987 in one dimension with an

20.4 Optical lattices

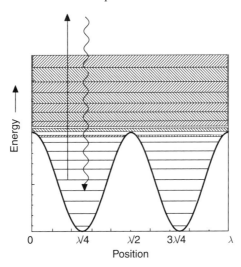

Figure 20.6 Energy levels of atoms moving in the periodic potential of the light shift in a standing wave. There are discrete bound states deep in the wells that broaden at higher energy, and become bands separated by forbidden energies above the tops of the wells. Under conditions appropriate to laser cooling, optical pumping among these states favors populating the lowest ones as indicated schematically by the arrows. (Figure from Ref. [1].)

atomic beam traversing an intense standing wave [234]. Since then, the study of atoms confined in wavelength-size potential wells has become an important topic in optical control of atomic motion because it opens up configurations previously accessible only in condensed matter physics using crystals.

The limits of laser cooling suggest that atomic momenta can be reduced to a "few" times $\hbar k$. This means that their de Broglie wavelengths are equal to the optical wavelengths divided by a "few". If the depth of the optical potential wells is high enough to contain such very slow atoms, then their motion in potential wells of size $\lambda/2$ must be described quantum mechanically, since they are confined to a space of size comparable to their de Broglie wavelengths. Thus they do not oscillate in the sinusoidal wells as classical localizable particles, but instead occupy discrete, quantum mechanical bound states [235], as shown in the lower part of Fig. 20.6.

The basic ideas of the quantum mechanical motion of particles in a periodic potential were laid out in the 1930s with the Kronig–Penney model and Bloch's theorem, and optical lattices offer important opportunities for their study. For example, these lattices can be made essentially free of defects with only moderate care in spatially filtering the laser beams to assure a single transverse mode structure. Furthermore, the shape of the potential is exactly known, and does not depend on the effect of the crystal field or the ionic energy level scheme. Finally, the laser parameters can be varied to modify the depth of the potential wells without changing the lattice vectors, and the lattice vectors can be changed independently by redirecting the laser beams. The simplest optical lattice to consider is a 1D pair of counterpropagating beams of the same polarization, as was used in the first experiment [234]. Of course, such tiny traps are usually very shallow, so loading them requires cooling to the μK regime.

Atoms trapped in wavelength-sized spaces occupy vibrational levels similar to those of molecules (see Chap. 16). The optical spectrum can show Raman-like sidebands that result from transitions among the quantized vibrational levels [236, 237] as shown in Fig. 20.8. These quantum states of atomic motion can also be observed by stimulated emission [236, 238] and by direct rf spectroscopy [239, 240].

20.4.2 Properties of 3D lattices

The name "optical lattice" is used rather than optical crystal because the filling fraction of the lattice sites is often limited to only a few percent. The limit arises because the loading of atoms into the lattice is typically done from a sample of trapped and cooled atoms, such as a MOT for atom collection, followed by an optical molasses for laser cooling. The atomic density in such experiments is limited to a few times $10^{11}/cm^3$ by collisions and multiple light scattering. Since the density of lattice sites of size $\lambda/2$ is a few times $10^{13}/cm^3$, the filling fraction is necessarily small. With the advent of experiments that load atoms directly into a lattice from a BEC, the filling factor can be increased to 100%, and in some cases it may be possible to load more than one atom per lattice site [241, 242].

Because of the transverse nature of light, any mixture of beams with different $\vec{k}$-vectors necessarily produces a spatially periodic, inhomogeneous light field. The importance of the "egg-crate" array of potential wells arises because the associated atomic light shifts can easily be comparable to the very low average atomic kinetic energy of laser-cooled atoms. A typical example projected against two dimensions is shown in Fig. 20.7.

20.4.3 Spectroscopy in 3D lattices

The NIST group studied atoms loaded into an optical lattice using Bragg diffraction of laser light from the spatially ordered array [243]. They cut off the laser beams that formed the lattice, and before the atoms had time to move away from their positions, the NIST group applied a pulsed probe laser beam at the Bragg angle appropriate for one of the sets of lattice planes. Not only did the Bragg diffraction enhance the reflection of the probe beam by a factor of 10^5, but by varying the time between the shut-off of the lattice and turn-on of the probe, the researchers could measure the "temperature" of the atoms in the lattice. The reduction of the

20.4 Optical lattices

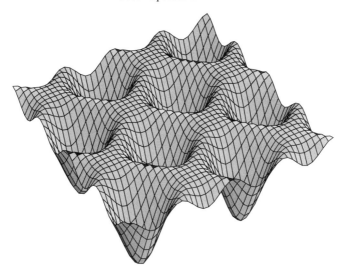

Figure 20.7 The "egg-crate" potential of an optical lattice shown in two dimensions. The potential wells are separated by $\lambda/2$. (Figure from Ref. [1].)

amplitude of the Bragg scattered beam with time provided some measure of the diffusion of the atoms away from the lattice sites, much like the Debye–Waller factor in X-ray diffraction.

The group at NIST also developed a new method that superposed a weak probe beam of light directly from the laser upon some of the fluorescent light from the atoms in a 3D optical molasses, and directed the light from these combined sources on a fast photodetector [244]. The resulting beat signal carried information about the Doppler shifts of the atoms in the optical lattices [237]. These Doppler shifts were expected to be in the sub-MHz range for atoms with the previously measured 50 µK temperatures. The observed features confirmed the quantum nature of the motion of atoms in the wavelength-size potential wells (see Fig. 20.8) [209].

20.4.4 Quantum transport in optical lattices

In the 1930s Bloch realized that applying a uniform force to a particle in a periodic potential would not accelerate it beyond a certain speed, but instead would result in Bragg reflection when its de Broglie wavelength became equal to the lattice period. Thus an electric field applied to a conductor could not accelerate electrons to a speed faster than that corresponding to the edge of a Brillouin zone, and at longer times the particles would execute oscillatory motion. Ever since then, experimentalists have tried to observe these Bloch oscillations in increasingly pure and/or defect-free crystals.

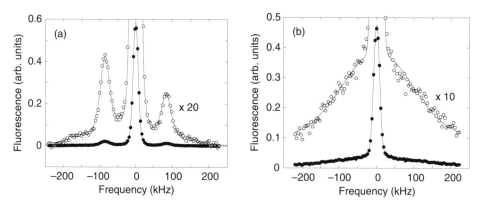

Figure 20.8 (a) Fluorescence spectrum in a 1D lin ⊥ lin optical molasses. Atoms are first captured and cooled in a MOT, then the MOT light beams are switched off leaving a pair of lin ⊥ lin beams. Then the measurements are made with $\delta = -4\gamma$ at low intensity. The open symbols are scaled up to emphasize the sidebands by a factor of 20 compared with the original data indicated by the filled symbols. The center peak is due to spontaneous emission of the atoms to the same vibrational state from which they are excited, whereas the sideband on the left (right) is due to spontaneous emission to a vibrational state with one vibrational quantum number lower (higher) (see Fig. 20.6). The presence of these sidebands is a direct proof of the existence of the band structure. (b) Same as (a) except the 1D molasses is σ^+-σ^- which has no spatially dependent light shift and hence no vibrational states. (Figures from Ref. [237].)

Atoms moving in optical lattices are ideally suited for such an experiment, as was beautifully demonstrated in 1996 [245]. The authors loaded a 1D lattice with atoms from a 3D molasses, further narrowed the velocity distribution, and then instead of applying a constant force, simply changed the frequency of one of the beams of the 1D lattice with respect to the other in a controlled way, thereby creating an accelerating lattice. Seen from the atomic reference frame, this was the equivalent of a constant force trying to accelerate them. After a variable time t_a the 1D lattice beams were shut off, and the measured atomic velocity distribution showed beautiful Bloch oscillations as a function of t_a. The centroid of the very narrow velocity distribution was seen to shift in velocity space at a constant rate until it reached $v_r = \hbar k/M$, and then it vanished and reappeared at $-v_r$ as shown in Fig. 20.9. The shape of the "dispersion curve" allowed measurement of the "effective mass" of the atoms bound in the lattice.

Exercises

20.1 In a quadrupole magnetic trap, the magnitude of the B field is in cylindrical coordinates given by $|\vec{B}| = A\sqrt{\rho^2 + 4z^2}$, where A is a measure of the field gradient.

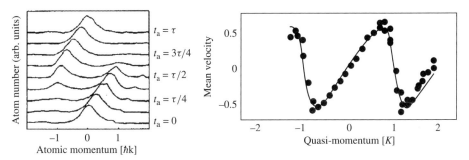

Figure 20.9 Plot of the measured velocity distribution vs time in the accelerated 1D lattice. (a) Atoms in a 1D lattice are accelerated for a fixed potential depth for a certain time t_a and the momentum of the atoms after the acceleration is measured. The atoms accelerate only to the edge of the Brillouin zone where the velocity is $+v_r$, and then the velocity distribution appears at $-v_r$. (b) Mean velocity of the atoms as a function of the quasi-momentum, i.e.,, the force times the acceleration time. (Figures from Ref. [245].)

Consider a classical circular orbit in a plane of constant z and calculate the adiabatic condition. Show that the adiabatic condition is manifest in terms of ω_T which is a measure of the orbital frequency in the trap. That is, what is the limit on the orbital parameters that satisfy $\omega_L \gg \omega_T$, where ω_L is the Larmor precession frequency (Zeeman shift) about the local field?

20.2 For very cold atoms in a quadrupole magnetic trap, a heuristic quantization of the atomic motion can be made by requiring the orbital angular momentum to have a magnitude $n\hbar$, where n is an integer. What are the conditions on such orbits that satisfy the adiabatic condition above? Prove your result.

20.3 The complete expressions for the force on an atom in optical molasses and in a MOT are very similar. The difference is that the detuning in molasses derives only from the Doppler shift, but in a MOT there is **also** a Zeeman shift. In order to get the simplified expression for the MOT force, $F = \beta v$ where β could be negative, the velocity had to be limited by the approximation $(kv/\gamma)^4 \ll 1$ (Eq. (19.13)). Write the complete expression for the force on an atom in a MOT, and find the corresponding approximation. What limits does this place on MOT design? On MOT properties, such as capture range?

21

Bose–Einstein condensation

21.1 Introduction

Bose–Einstein condensation (BEC) is one of the most intriguing phenomena in physics. The basic idea goes back to the 1920s, when Einstein made predictions for the behavior of atoms at sufficiently low temperatures and high densities. However, it took seven decades for experimental physicists to realize this phase under the low-density conditions envisioned by Einstein. In the intermediate period there were many experiments in which aspects of BEC played important roles. Therefore, phenomena such as superfluidity and superconductivity were connected to BEC long before its 1995 realization in ultra-cold, low-density atomic vapors. Since BEC can be discussed with very many different backgrounds and applications in mind, the subject can sometimes be overwhelming for interested students.

There is a very good reason why a chapter on BEC does not belong in this book. In all the other chapters here, the processes described involve one or at most two atoms, and the interactions that play a role are between the particles that constitute the atom, namely between the nucleus (or nuclei) and their electrons. The interactions are described quantum mechanically by their Hamiltonian that can always be written exactly in non-relativistic terms, and in some cases can even be solved exactly. This makes atoms and molecules a branch of physics where comparisons between theory and experiments become very meaningful since theoretical predictions are often confirmed to very high accuracy.

In BEC, it is the interactions among many atoms that are the key for understanding the phenomenon. Although in principle one could write a Hamiltonian for a Bose condensed gas, its large number of particles makes this not very practical. The starting point for the description of a BEC is in the area of many-body physics that belongs to the expertise of condensed-matter physicists. The interactions between $\sim 10^{23}$ electrons in a solid state with a lattice formed by $\sim 10^{23}$ ions resemble the kind of interactions in a BEC much more closely than the simple few-body interactions in an atom or molecule. Thus all the theoretical models that have been described up to now are not applicable for the contents of this chapter.

On the other hand, there are many reasons why a chapter on BEC does belong to this book. First, there is a very large community in atomic, molecular, and optical (AMO) physics that is currently active in this field. Although atomic physics is not very important for the description of a BEC, atomic properties play a crucial role in the creation and observation of BECs, and thus are required knowledge for working in such an experimental group. Second, the field of BEC physics is a excellent demonstration of the cross-fertilization of one field of physics (atomic physics) with many other fields, for example condensed matter, high-energy, and statistical physics. Moreover, the unique feature of a BEC of ultra-cold atoms that permits many of its properties to be described using exact models allows theoreticians to cast their theories on a solid basis, making the outcomes of their theories highly testable. Third, ultra-cold atomic systems allow physicists to engineer their underlying models to a high degree, exploring many different experimental realizations using very similar setups.

The field of BEC encompasses many different areas that have been recently explored with ultra-cold atoms and have all generated their own interest. The range of topics is so vast and diverse that it is not possible to cover it adequately in just a single book chapter. Thus choices have to be made, and in this chapter only a few of these topics are discussed. After a brief history, the chapter covers the initial description of BEC by Bose and Einstein in Sec. 21.3 since this elegantly shows how relatively simple statistical notions can lead to the concept. To understand the crucial role of the quantum statistics of the atoms (bosonic or fermionic), the distribution functions for bosons and fermions are discussed separately in App. 21.A. Next, the properties of a BEC are discussed using a mean-field approach because it gives a very good description of the shape of the BEC and the lowest-order excitations of it. The quantum mechanical nature of a BEC requires the assignment of a phase that should be observable in an interference experiment, and in Sec. 21.5 such an experiment is described, along with its theoretical description which outlines the unusual aspects of such an experiment.

Since a real BEC always consist of two parts for finite temperatures, namely the condensed and thermal particles, it is an ideal system for the study of quantum hydrodynamics as initially conceived by Landau. This is because the interactions between the atoms in the condensed (degenerate) part are weak and can be modeled nearly exactly. The quantum hydrodynamical behavior is discussed for a particular case, namely that of second sound, whereas very many other examples could have been given. Finally, it is shown how BECs can be used to study strongly interacting systems in which optical lattices and/or Feshbach resonances produce interactions that are so strong that a mean-field description is no longer adequate.

21.2 The road to BEC

The road to BEC in ultra-cold atomic systems has been marked by many breakthroughs during a period of two decades. First was the demonstration that laser fields can change the velocity distribution of an atomic ensemble and the exploitation of it to cool and trap atoms using electro-magnetic fields. The large tunability of lasers enabled atoms to be cooled to extremely low temperatures, one of the requirements for observation of BEC. Although the temperatures achieved, usually in the range of 0.01–1 mK, are generally too high to achieve BEC for the usual density and mass of atoms, laser cooling nevertheless enabled temperature reduction of the atoms far below that of liquid helium, which had been the standard for low temperatures for the preceding decades. These accomplishments were reflected in the physics Nobel Prize of 1997 for the development of methods to cool and trap atoms with laser light.

To reduce the temperature even further, a new technique had to be developed and it is the well-known, classical phenomenon of evaporative cooling. It is based on the removal of those atoms with the highest energy, followed by a rethermalization of the remaining sample to lower temperatures caused by elastic collisions. Although the technique is very well known and has been applied to many different situations, its application to ultra-cold atoms in a magnetic trap paved the way to achieve BEC in 1995. Although evaporative cooling results in the loss of particles, it turned out that it actually is very effective in increasing the phase-space density of the gas, since the decrease of temperature of the trapped atoms leads to a smaller occupied volume and thus an increase of their density. Eventually the phase-space density increases to values above the minimum for BEC predicted by Einstein, and thus the vapor undergoes the transition to BEC. The physics Nobel Prize in 2001 was for the achievement of BEC in dilute gases of trapped alkali atoms, and for early fundamental studies of the properties of the condensates.

It is remarkable that BEC in ultra-cold atomic gases had been envisioned by Hecht as early as 1959 [246]. He estimated that the attractive interaction between two spin-polarized hydrogen atoms could not overcome their magnetic repulsion, so there could be no bound state, and then such atoms would be stable against molecule formation. These ideas were revived in 1976 by Stwalley and Nosanow [247] who argued that the system would be both a BEC and also a superfluid. Spin-polarized hydrogen was first stabilized by Silvera and Walraven [248] and later trapped magnetically by the group led by Greytak and Kleppner [249]. They developed evaporative cooling for ultra-cold atoms [250] and showed that it is a promising technique for increasing the phase-space density. However, the experimental challenge of achieving BEC for spin-polarized hydrogen turned out to be more difficult than for the alkali-metal atoms, partly because laser cooling of atomic hydrogen is more difficult because of the need for VUV light, and partly because three-body collisions at the achievable density and temperature could reorient the atomic magnetic moments and thus destabilize the gas [251]. BEC in spin-polarized hydrogen was finally achieved in 1998 [252].

Up to now, more than 20 different atoms of the periodic table have undergone Bose condensation since the first realization in Rb. Contrary to the initial suggestions, the existence of bound molecular states for two interacting atoms that are cooled to degeneracy does not preclude the formation of an atomic BEC. First, the densities of the atoms can be sufficiently small that three-body interactions necessary for molecule formation are unlikely. Second, the time for atoms to evolve to a solid state is exceedingly long. Thus a BEC is effectively a metastable state, but its metastability lifetime is much longer than the time it can be maintained in a magnetic or optical trap.

Moreover, the bound molecular states of some of these atoms can also be condensed using techniques such as photo-association spectroscopy or Feshbach resonance to form the cold molecules from cold atomic vapors (see Sec. 22.4). In most cases the translationally cold molecules are in a highly excited vibrational state, but there are some cases where the molecules are in their internal ground states. This has led to exploration of a new rich field of Bose condensation in molecules for which novel applications have been predicted [253].

21.3 Quantum statistics

In many cases, atoms can be considered as point particles because their size is much smaller than the typical range of the interactions between them. For the interaction between atoms and light in the first part of this book, the typical interaction range is the wavelength of the light λ. This is orders of magnitude larger than the atomic size, which is comparable to the Bohr radius $a_0 \approx \alpha^2 \lambda$.

For a gas at a temperature T the atomic thermal wavelength is

$$\lambda_T(T) = \frac{h}{\sqrt{2\pi M k_B T}}. \quad (21.1)$$

This thermal wavelength is approximately equal to the de Broglie wavelength of matter as introduced in Sec. 1.5. At room temperature, λ_T is even smaller than a_0 and the wave properties of atoms can be safely neglected. It is also much smaller than the interparticle distance $d = 1/n^{1/3}$ at atmospheric pressure (here

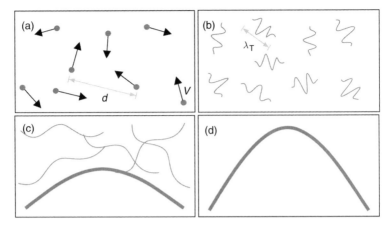

Figure 21.1 The principle of Bose–Einstein condensation. (a) At room temperature, atoms can be considered as point particles, which collide as "billiard balls". (b) At low temperatures, the de Broglie wavelength λ_T becomes larger than the atomic size, and atoms have to be described as waves. (c) At even lower temperatures, below the critical temperature T_c, λ_T becomes larger than the interparticle distance d and the atoms start to Bose condense. (d) At $T = 0$, all the atoms are in the condensate, which becomes a giant matter wave. (Figure adapted from Ref. [254].)

n is their number density) so the gas can be considered as composed of classical point particles (see Fig. 21.1a).

By contrast, at very low temperatures the thermal wavelength is much larger, as shown in Fig. 21.1b, and therefore limits n to be $n \leq 1/\lambda_T(T)^3$. At temperatures of a few tens of μK, it becomes of the same order as the interparticle distance. Lowering the temperature even more (Fig. 21.1c) causes the waves of the different atoms to start to overlap, and the behavior of the gas depends on the quantum statistical properties of the gas. The thermal wavelength becomes equal to the interparticle distance at a temperature T_c given by

$$n\lambda_T^3(T_c) \approx 1, \qquad (21.2)$$

and T_c is called the condensation or critical temperature. For typical experimental conditions T_c is of the order of 0.1–1 μK.

For bosons, the atoms tend to accumulate in identical states at very low $\lambda_T(T)$ because their scattering favors multiple occupation of the lowest energy state, and the gas is called a Bose–Einstein condensate. For temperatures near absolute zero (Fig. 21.1d), nearly all the atoms are in the condensate and its properties have to be described quantum mechanically.

For a more accurate determination of T_c consider the microcanonical ensemble of bosonic atoms, as can be found in any textbook on statistical mechanics. As

21.3 Quantum statistics

shown in App. 21.A, the quantum statistics of bosons lead to an energy-dependent mean occupation number $f_{BE}(\varepsilon)$ given by

$$f_{BE}(\varepsilon) = \frac{1}{e^{(\varepsilon-\mu)/k_B T} - 1}, \quad (21.3)$$

where μ is the chemical potential. Physically acceptable values of f_{BE} (≥ 0) require $\mu < \varepsilon$. With the states of the system enumerated by index i, the total number N of atoms is $N = \sum_i f_{BE}(\varepsilon_i)$. As μ approaches the lowest energy ε_0 the occupation of the ground state starts to diverge, and this is the essence of Bose–Einstein condensation: a macroscopic occupation of the ground state. If the sum for N is evaluated at $\mu = \varepsilon_0$ for states $i = 1\ldots\infty$, its value is finite and for a certain temperature it will fall below the number of atoms in the system. This is the condensation temperature T_c.

For the evaluation of the sum for N excluding the ground state, consider a gas of many atoms contained in a box with dimensions much larger than the interparticle distance. As shown in App. 21.B, the sum can then be replaced by an integral,

$$N_{ex} = \sum_i f_{BE}(\varepsilon_i) \approx \int d\varepsilon \, f_{BE}(\varepsilon) \rho(\varepsilon), \quad (21.4)$$

where N_{ex} is the number of atoms outside the ground state. Here $\rho(\varepsilon)$ is the density of states that is given for an uniform gas (see App. 21.B) by

$$\rho(\varepsilon) = \frac{V}{\pi^2 \hbar^3} \sqrt{\frac{M^3 \varepsilon}{2}}, \quad (21.5)$$

where M is the mass of each atom and V is the volume of the box. The resulting integral Eq. (21.4) can be evaluated in terms of the Bose function $g_n(x) = \sum_{k=1}^{\infty} x^k/k^n$ and yields (choosing $\varepsilon_0 = 0$)

$$N_{ex} = \frac{V g_{3/2}\left(e^{\mu/k_B T}\right)}{\lambda_T^3(T)}. \quad (21.6)$$

At condensation, $\mu = 0$, and one obtains $g_{3/2}(1) \approx 2.612$. Using $n = N/V$, it leads to

$$n\lambda_T^3(T_c) = 2.612, \quad (21.7)$$

a more accurate condition for condensation than Eq. (21.2). Below T_c the ground state, namely the condensate, has a macroscopic occupation number N_0, and the total number of particles is given by

$$N = N_0 + N_{ex}. \quad (21.8)$$

Below T_c this leads to

$$\frac{N_0}{N} = 1 - \left(\frac{T}{T_c}\right)^{3/2}, \quad (21.9)$$

showing that on lowering the temperature the fraction of the atoms in the ground state increases. At the experimentally unattainable temperature $T = 0$ all atoms are in the ground state.

In typical BEC experiments the atoms are trapped in an external potential, and this has some consequences for the results above. Although many trapping potentials exist, the analysis here is restricted to a harmonic potential

$$V(r) = \tfrac{1}{2}M\left(\omega_x^2 x^2 + \omega_y^2 y^2 + \omega_z^2 z^2\right) \quad (21.10)$$

where ω_i is the trap frequency in direction $i = x, y, z$. The density of states is modified to become

$$\rho(\varepsilon) = \frac{\varepsilon^2}{2\hbar^3 \bar{\omega}^3}, \quad (21.11)$$

where $\bar{\omega}$ is the geometrical mean of the trap frequencies in the three directions: $\bar{\omega}^3 = \omega_x \omega_y \omega_z$. The condition for condensation remains $N = N_{\text{ex}}$, where N_{ex} is given by Eq. (21.4). This yields

$$kT_c = \frac{\hbar \bar{\omega} N^{1/3}}{g_3(1)^{1/3}}, \quad (21.12)$$

where $g_3(1) \approx 1.202$. Note that in the center of the trap the condition of Eq. (21.7) is still fulfilled.

There are several kinds of small corrections to Eq. (21.12), two of which are discussed here. To begin, it is valid only in the thermodynamic limit $N \to \infty$. For a finite number of atoms there are corrections to T_c, but $\delta T_c/T_c \approx -N^{-1/3}$, which is of the order of 1% for a million atoms in the trap. Another correction that also shifts the critical temperature is caused by the interactions between the atoms and has so far been neglected. This shift is given by $\delta T_c/T_c \approx aN^{1/6}/a_{\text{ho}}$, where a is the scattering length and $a_{\text{ho}} = \sqrt{\hbar/M\bar{\omega}}$ is the characteristic trap length. Since typically $a \ll a_{\text{ho}}$, the shift is the order of a few % for a million atoms.

21.4 Mean-field description of the condensate

In Sec. 21.3 the description is restricted to the thermal particles, and their thermodynamic properties are derived. The atoms that cannot be accommodated in the thermal cloud accumulate in the ground state: the condensate. As introduced in Sec. 21.1, they have to be described by quantum mechanics since they are all in an identical quantum state. The techniques used to describe such a macroscopic

21.4 Mean-field description of the condensate

occupation of a single state are best treated within the framework of quantum-field theory. In that case the atoms are described by creation and annihilation operators just as in App. 5.B for light. Since quantum-field theory is beyond the scope of this book, a slightly simpler approach is taken.

Since ultra-cold atoms mainly occupy the ground state at zero temperature, the creation operator can be replaced by its expectation value, and fluctuations around this expectation value are neglected in first order. This allows the introduction of the condensate wavefunction $\Phi(\vec{r}, t)$, which is also called the order parameter. It is quite different from the usual wavefunctions in quantum mechanics because it is not normalized to unity, but instead it yields the total number of condensate atoms when integrated over space. Thus

$$N_0 = \int d\vec{r} \, |\Phi(\vec{r}, t)|^2. \tag{21.13}$$

The density n_0 of the condensate is given by $n_0(\vec{r}, t) = |\Phi(\vec{r}, t)|^2$, the square of the wavefunction. The evolution of the condensate wavefunction can be derived from quantum-field theory and is given by the non-linear Schrödinger or Gross–Pitaevskii equation:

$$i\hbar \frac{\partial}{\partial t} \Phi(\vec{r}, t) = \left(-\frac{\hbar^2}{2M} \nabla^2 + V(\vec{r}) + U_0 |\Phi(\vec{r}, t)|^2 \right) \Phi(\vec{r}, t) \tag{21.14}$$

The first two terms on the right side are the usual terms for the evolution of the wavefunction for an atom moving in an external potential $V(\vec{r})$, but the last term is unusual. It describes the effect of the interparticle interactions on the evolution, where U_0 is the interaction term given by

$$U_0 = \frac{4\pi \hbar^2 a}{M}. \tag{21.15}$$

Here a is the scattering length of the interparticle interaction and is the only parameter needed to describe the effects of interparticle interaction at these low temperatures.

For stationary situations the *Ansatz* $\Phi(\vec{r}, t) = \phi(\vec{r}) \exp(-i\mu t/\hbar)$ can be used, where μ is the chemical potential. The Gross–Pitaevskii equation reduces to

$$\left(-\frac{\hbar^2}{2M} \nabla^2 + V(\vec{r}) + U_0 |\phi(\vec{r})|^2 \right) \phi(\vec{r}) = \mu \phi(\vec{r}), \tag{21.16}$$

where the first term on the left side is the kinetic energy, the second term is the potential energy in the trap, and the last term is the interaction energy. This equation can easily be solved in the Thomas–Fermi limit, where the kinetic energy is neglected with respect to the interaction energy. In that limit the solution is

$$n(\vec{r}) = |\phi(\vec{r})|^2 = \max\left(\frac{\mu - V(\vec{r})}{U_0}, 0\right), \tag{21.17}$$

where the maximum assures that the density is non-negative everywhere. In experiments the external potential is often given by the harmonic potential of Eq. (21.10). In that case the density profile can be described by the inverted parabola

$$n_0(\vec{r}) = n_0 \max\left(1 - \frac{x^2}{\sigma_x^2} - \frac{y^2}{\sigma_y^2} - \frac{z^2}{\sigma_z^2}, 0\right), \tag{21.18}$$

with the central density $n_0 = \mu/U_0$, and the width of the condensate in the direction i is $\sigma_i = \sqrt{2\mu/M\omega_i^2}$. The Thomas–Fermi profile is frequently used to describe the shape of the condensate, and the width of the condensate is a good measure of the chemical potential $\mu = \frac{1}{2}M\omega_i^2\sigma_i^2$. Integration of the density profile over space then yields the total number N_0 of the condensate:

$$N_0 = \frac{8\pi}{15}\left(\frac{2\mu}{M\bar{\omega}^2}\right)^{3/2}\frac{\mu}{U_0}. \tag{21.19}$$

Since the scattering length a is well known, measuring the width of the condensate provides a good measure of the total number of atoms.

To include fluctuations in the description of the condensate, it is necessary to go beyond this mean-field description. In first-order this can be done using the Bogoliubov approximation, which introduces the two functions $u(\vec{r})$ and $v^*(\vec{r})$ to describe the amplitudes of the creation and annihilation operators, respectively. In the language of quantum-field theory, $u(\vec{r})$ describes the particles, whereas $v^*(\vec{r})$ describes the holes. In the present treatment this method is equivalent to using the Ansatz

$$\Phi(\vec{r}, t) = e^{-i\mu t/\hbar}\left(\phi(\vec{r}) + u(\vec{r})e^{-i\omega t} + v^*(\vec{r})e^{+i\omega t}\right). \tag{21.20}$$

in Eq. (21.14). Then the approximation that $u(\vec{r})$ and $v^*(\vec{r})$ are small compared with $\phi(\vec{r})$ yields the following equations for the coupled amplitude functions:

$$\hbar\omega u(\vec{r}) = \left(-\frac{\hbar^2}{2M}\nabla^2 + V(\vec{r}) - \mu + 2U_0\phi(\vec{r})^2\right)u(\vec{r}) + U_0\phi(\vec{r})^2 v(\vec{r}). \tag{21.21}$$

and

$$\hbar\omega v(\vec{r}) = \left(-\frac{\hbar^2}{2M}\nabla^2 + V(\vec{r}) - \mu + 2U_0\phi(\vec{r})^2\right)v(\vec{r}) + U_0\phi(\vec{r})^2 u(\vec{r}). \tag{21.22}$$

In a uniform gas the solutions of these two equations lead to the Bogoliubov dispersion relation, where the frequency ω_q of an excitation with wavevector $\vec{m}$ and energy $\varepsilon_q = \hbar^2 q^2/2M$ is given by

$$\hbar\omega_q = \sqrt{\varepsilon_q(\varepsilon_q + 2U_0 n_0)}. \tag{21.23}$$

For low momenta $\vec{m}$ with $\varepsilon_q \ll 2U_0 n_0$ this reduces to

$$\omega_q = c_B q \qquad c_B = \sqrt{\frac{U_0 n_0}{M}} \tag{21.24}$$

and the collective excitations are *sound*-like, where c_B is the speed of sound. For high momenta with $\varepsilon_q \gg 2U_0 n_0$, Eq. (21.23) reduces to

$$\hbar\omega_q = \varepsilon_q. \tag{21.25}$$

In this case, atoms are excited individually and the excitations are now *particle*-like.

In many experiments the Gross–Pitaevskii equation is an excellent and sufficient description of the evolution of the condensate. There are quantum corrections, but they are small (order few percent). Furthermore, its description is not exact for finite temperatures, a condition that is usually difficult to incorporate. In some experiments the interactions between the atoms are quite strong, and in that case the Gross–Pitaevskii equation is not sufficient. The physics beyond the Gross–Pitaevskii equation, the strongly correlated regime, is the subject of Sec. 21.7.

21.5 Interference of two condensates

Atomic motion in the condensate is described by a single wavefunction, as discussed in Sec. 21.4. One of the striking aspects of such a common wavefunction is that it has a single, well-defined phase so that the condensate is very similar to a coherent radiation field. Bose–Einstein condensation makes it possible to perform experiments similar to those with laser light, but now with massive particles. One of the easiest way to visualize this phase coherence is by allowing interference of two coherent sources and measuring the interference pattern. Since these experiments for condensates are performed with massive particles, it is necessary to re-examine the description of such an interference experiment.

Consider two phase-coherent laser beams incident on a beam splitter as in Fig. 21.2. If the laser beams remain phase-coherent during the experiment, the signal on one detector is proportional to $\cos^2 \phi_0$ and the signal on the other detector is proportional to $\sin^2 \phi_0$, where $\phi_0 = (\phi_A - \phi_B)/2$ is the phase difference between the two beams. Thus the ratio of the two signals is $\tan^2 \phi_0$, and this allows measurement of ϕ_0. However, a similar experiment using two condensates as phase-coherent sources forces one to rethink the outcome of the experiment.

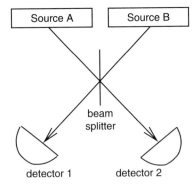

Figure 21.2 Interference experiment on a 50:50 beam splitter between two coherent sources A and B, where the signals are detected by two detectors D_1 and D_2.

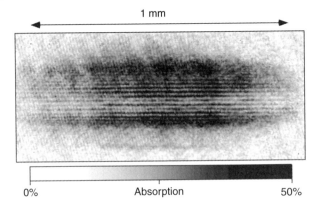

Figure 21.3 Matter-wave interference. A condensate is split in two by a laser-light sheet. After the light sheet is switched off, the two condensates are allowed to expand by switching off the trap. After a time-of-flight of 40 ms, an interference pattern between the two condensates is observed. (Figure from Ref. [255].)

Since condensates consist of massive particles, where the number of atoms is a conserved quantity, the phase of a condensate is completely undetermined because of the phase–number uncertainty relation for quantum systems. It would appear that, since the phase difference between the two condensates is random, no interference between them can be observed, so the ratio of the two signals is simply $I_A/I_B = 1$.

One of the first experiments to detect the interference between two condensates is described in Ref. [255]. The condensate produced in a magnetic trap was split in two by a far-detuned laser beam, and after some time both the magnetic trap and the far-detuned laser beam were switched off. This allowed the two clouds to expand, to overlap and, if possible, to interfere. The result in Fig. 21.3 shows the two overlapping clouds after $\tau = 40$ ms time-of-flight, and a clear interference pattern is

21.5 Interference of two condensates

observed. The fringe period equals the de Broglie wavelength $\lambda_{\text{deB}} = h\tau/(Md)$, where d is the initial separation between the two clouds and thus Md/τ is the relative momentum of the interfering particles. The interference pattern is a remarkable manifestation of the wave character of the condensed atoms.

To explain why such an interference pattern arises, Ref. [256] exploited the special properties of bosons with the following *Gedanken* experiment, depicted in Fig. 21.2. If both condensates are in a coherent state with an arbitrary phase between them, the ratio of the two detection rates allows the determination of this relative phase. But if the number of atoms in both condensates is constant and thus their phase undetermined, the two detection rates would seem to be equal. In particular, the detection of an atom in one channel at a certain time t would seem to be independent of the outcome of the detections that have been done in either channel before time t. However, Ref. [256] shows why this is not the case.

To understand this, suppose for simplicity that the number of atoms N is initially equally split between the two condensates with $N/2$ atoms in each one. Each independent condensate is thus in a Fock state, and the total system has to be described by the wavefunction $|\Psi\rangle$:

$$|\Psi\rangle = |N/2, N/2\rangle. \tag{21.26}$$

Write the detection operator for detector D_1 as $\frac{1}{2}\sqrt{2}(\hat{a}+\hat{b})$ with $\hat{a}$ and $\hat{b}$ the annihilation operators of condensate A and B, respectively. After the detection of the first atom in channel D_1, the wavefunction is changed to

$$|\Psi'\rangle = \frac{1}{2}\sqrt{2}(\hat{a}+\hat{b})|\Psi\rangle \propto |N/2-1, N/2\rangle + |N/2, N/2-1\rangle. \tag{21.27}$$

After the detection of the second atom in channel D_1 the system becomes

$$|\Psi''\rangle = \frac{1}{2}\sqrt{2}(\hat{a}+\hat{b})|\Psi'\rangle \tag{21.28}$$

$$\propto \sqrt{N/2-1}\,|N/2-2, N/2\rangle + \sqrt{N/2-1}\,|N/2, N/2-2\rangle + 2\sqrt{N/2}\,|N/2-1, N/2-1\rangle$$

and after a detection of the second atom in channel D_2 it becomes

$$|\Psi''\rangle = \frac{1}{2}\sqrt{2}(\hat{a}-\hat{b})|\Psi'\rangle \propto \sqrt{N/2-1}\,|N/2-2, N/2\rangle + \sqrt{N/2-1}\,|N/2, N/2-2\rangle. \tag{21.29}$$

Thus the probability ratio between the two channels is 3:1, which leads to a strong increase of a detection of the second atom in the channel D_1, where the first atom has also been detected. Reference [256] shows that the rate of detection of a large number of atoms in one channel and no atoms in the other channel is much larger than one can expect from a random detection probability in the two channels. Moreover, Ref. [256] shows that the outcomes for an initial Fock state or initial coherent state with random phase are equivalent.

To relate such an analysis to the interference experiment described in Ref. [255], the authors of Ref. [256] used phase states $|\phi\rangle_N$ to describe the system:

$$|\phi\rangle_N = \frac{1}{\sqrt{2^N N!}} \left(\hat{a}^\dagger e^{+i\phi} + \hat{b}^\dagger e^{-i\phi}\right)^N |0\rangle, \tag{21.30}$$

with $|0\rangle$ the vacuum state. Applying the operator $\left(\hat{a}^\dagger e^{+i\phi} + \hat{b}^\dagger e^{-i\phi}\right)$ on a state creates a new state with one more atom in the system, but with a well-defined phase difference 2ϕ between the two condensates. The initial state $|\Psi\rangle$ can be expanded in a set of phase states:

$$|\Psi\rangle = \frac{1}{\pi} \int_{-\pi/2}^{+\pi/2} d\phi\, c(\phi) |\phi\rangle_N, \tag{21.31}$$

with phase amplitude $c(\phi)$. For the initial state with $|\Psi\rangle = |N/2, N/2\rangle$ with an equal number of atoms $N/2$ in each condensate, the initial phase amplitudes $c_i(\phi)$ are independent of ϕ.

If n atoms are detected in channel D_1 and m atoms are detected in channel D_2 the state has evolved to $|\Psi(n,m)\rangle$ given by (for details see Ref. [256])

$$|\Psi(m,n)\rangle \propto (\hat{a}+\hat{b})^n (\hat{a}-\hat{b})^m |N/2, N/2\rangle \tag{21.32}$$

$$\propto \int_{-\pi/2}^{+\pi/2} d\phi\, \cos^n\phi \sin^m\phi\, |\phi\rangle_{N-n-m}$$

$$\propto \int_{-\pi/2}^{+\pi/2} d\phi \left(e^{-(n+m)(\phi-\phi_0)^2} + (-1)^m e^{-(n+m)(\phi+\phi_0)^2}\right) |\phi\rangle_{N-n-m}.$$

So from an initial random phase the state has evolved to a phase distribution as a double Gaussian centered at ϕ_0 and $-\phi_0$, where the ambiguity of the sign of ϕ_0 is similar to that of the experiment using laser beams. Note that the number of detections $n+m$ can be small with respect to the total number of atoms N, and also that the sequence in which the atoms are detected in the two detectors is irrelevant. So the results for a Fock state, as described in this section where the initial phase is not determined, and the results for a coherent state with a pre-existing, random phase are identical. The experimental results cannot distinguish between these two very different situations.

21.6 Quantum hydrodynamics

In the 1940s, Landau [257] and Tisza [258] developed a theory to describe two-fluid hydrodynamics. At that time, only liquid helium was known to form a superfluid, but their theory is general and applies to all systems where a normal fluid and superfluid can be identified.

21.6 Quantum hydrodynamics

In this section the two-fluid hydrodynamics of ultra-cold atoms will be described, accompanied by an experimental realization where the two-fluid character is shown. In the special case of ultra-cold atoms, the thermal atoms act as a normal fluid whereas the condensed atoms form a superfluid. The identification of the superfluid is straightforward with ultra-cold atoms since the superfluid is formed by the condensate. This has to be compared with the situation of liquid helium, where all atoms can be in the superfluid even though only about 10% are condensed. In this section the experimental proof of superfluidity is second sound, which was first reported in a BEC in Ref. [259]. Here results from Ref. [260] will be shown.

In hydrodynamics it is assumed that the constituent particles undergo many collisions, and that there is consequently a thermal equilibrium. This is normally not achieved in ultra-cold gases since the densities are usually low and the interactions are weak. The behavior is then normally described by the collisionless regime, where the atoms oscillate many times through the trap before they collide once. However, increasing the number of atoms eventually leads to a situation where the thermal atoms can be brought into thermal equilibrium. Producing a Bose–Einstein condensate in such a sample brings the system into the regime of two-fluid hydrodynamics. Since the condensate is quantum, this is also called quantum hydrodynamics.

Hydrodynamics is governed by continuity equations. In the usual hydrodynamics the three relations for continuity are for the number of particles, the flux, and the entropy. These relations can be written as

$$\frac{\partial n}{\partial t} = -\vec{\nabla}(n_n \vec{v}_n + n_s \vec{v}_s), \quad \text{(number)}$$

$$\frac{\partial (n_n \vec{v}_n + n_s \vec{v}_s)}{\partial t} = \frac{-\vec{\nabla} P}{M}, \quad \text{(flux)} \qquad (21.33)$$

$$\frac{\partial ns}{\partial t} = -\vec{\nabla}(ns\vec{v}_n), \quad \text{(entropy)}$$

where the indices indicate the normal fluid (n) and the superfluid (s). Here the total number density is given by $n = n_s + n_n$, $\vec{v}$ is the velocity, P is the pressure and s is the entropy per particle. The first equation shows that the total number of particles is conserved. The second equation shows that the total flux of superfluid and normal particles is caused by a gradient in the pressure. Finally, the third equation shows that the entropy of the system resides only in the normal fluid, whereas the superfluid carries no entropy. This is one of the basic assumptions of Landau and Tisza [257, 258] and discriminates the normal from the superfluid.

For quantum hydrodynamics the only additional relation concerns the superfluid velocity, given by the gradient of the chemical potential μ in the model by Landau and Tisza [257, 258]:

$$\frac{\partial \vec{v}_s}{\partial t} = \frac{-\vec{\nabla}\mu}{M}. \tag{21.34}$$

Adding this relation to the hydrodynamic relations of Eq. (21.33) fully determines the system. Using the Gibbs–Dunheim relation, the time-derivative of the superfluid velocity can be written in terms of the gradients of the pressure and temperature:

$$\frac{\partial \vec{v}_s}{\partial t} = \frac{-1}{Mn}\vec{\nabla}P + \frac{s}{M}\vec{\nabla}T. \tag{21.35}$$

Thus the unknown variables are n, P, s and T. The object is now to solve such relations under the assumption that the perturbations with respect to equilibrium are small.

To do this, assume that all thermodynamic variables are close to their equilibrium values with a small perturbation. So for instance, for the density one can write $n = n_0 + \delta n$, and similarly for the other thermodynamic variables. By keeping only the terms linear in the perturbations, the following two relations are obtained:

$$\frac{\partial^2 \delta s}{\partial t^2} = \frac{n_s}{Mn_n} s^2 \nabla^2 \delta T \quad \text{and} \quad \frac{\partial^2 \delta n}{\partial t^2} = \frac{1}{M} \nabla^2 \delta P. \tag{21.36}$$

The first equation relates the second-order time derivative of the entropy to the spatial dependence of the temperature, whereas the second one relates the second-order time derivative of the density to the spatial dependence of the pressure. Note that both equations resemble wave equations.

To find solutions to these equation, eliminate two unknown variables by expressing them in terms of two variables n and T:

$$\delta s = \left(\frac{\partial s}{\partial n}\right)_T \delta n + \left(\frac{\partial s}{\partial T}\right)_n \delta T, \tag{21.37}$$

and similarly for δP. Next is to make a plane wave expansion of the perturbation in n and T: $\delta n = \delta n_0 \exp\left(i(\vec{k}\cdot\vec{r} - \omega t)\right)$, and similarly for T. Here the speed of the wave is given by $c = \omega/k$. This leads to the following two coupled equations:

$$c^2\left(\frac{\partial s}{\partial n}\right)_T \delta n + \left(c^2\left(\frac{\partial s}{\partial T}\right)_n - \frac{n_s}{Mn_n}s^2\right)\delta T = 0 \tag{21.38}$$

and

$$\left(c^2 - \frac{1}{M}\left(\frac{\partial P}{\partial n}\right)_T\right)\delta n - \frac{1}{M}\left(\frac{\partial P}{\partial T}\right)_n \delta T = 0. \tag{21.39}$$

These equations yield non-trivial solutions only for

$$2c_\pm^2 = \left(c_1^2 + c_2^2 + c_3^2\right) \pm \sqrt{\left(c_1^2 + c_2^2 + c_3^2\right)^2 - 4c_1^2 c_2^2}, \tag{21.40}$$

21.6 Quantum hydrodynamics

where

$$c_1^2 = \frac{1}{M}\left(\frac{\partial P}{\partial n}\right)_T, \quad c_2^2 = \frac{n_s s^2 T}{M n_n C_V} \quad \text{and} \quad c_3^2 = \frac{T}{M^2 n^2 C_V}\left(\frac{\partial P}{\partial T}\right)_n. \quad (21.41)$$

It can be shown [262] that

$$c_1^2 + c_3^2 = \frac{1}{M}\left(\frac{\partial P}{\partial n}\right)_S, \quad (21.42)$$

where S is the system entropy, so whereas c_1 corresponds to the relation for isothermal sound for a thermal gas, $\sqrt{c_1^2 + c_3^2}$ is its proper relation for sound.

For $c_3 \ll c_1, c_2$ Eq. (21.40) reduces to $c_+ \approx c_1$ and $c_- \approx c_2$. Usually, c_1 is called first sound and c_2 is called second sound. One can show [263, 264] that c_3 is proportional to $(C_P - C_V)/C_V$ and thus proportional to the difference between the heat capacity at constant pressure and the heat capacity at constant volume. For liquid helium C_P and C_V are nearly equal so that the coupling term is small. In that case the sound velocity is either c_1 or c_2, depending on the sound mode. For liquid helium these sound modes are uncoupled, but for ultra-cold atoms they are coupled.

The difference between first is second sound is displayed graphically in Fig. 21.4. For first sound, the density oscillates in space as a wave, where the normal and superfluid components are in phase. Since these two components are in phase, the temperature is constant because the ratio between normal and superfluid components is determined by the temperature. This is the kind of sound one is used to in normal life. For second sound, the density is constant, but the temperature oscillates in space. Thus the normal and superfluid are out of phase, but the total density is constant. This sound mode can exist only if there are two components in the fluid, and this is a signature of superfluidity.

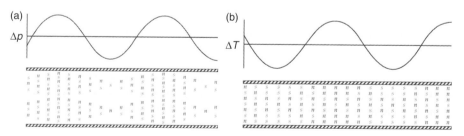

Figure 21.4 In two-fluid hydrodynamics the system consist of a normal (n) fluid and a superfluid (s). Therefore two sound modes exist. (a) "First sound", where the density is modulated periodically, and density of the normal and superfluid are in phase. (b) "Second sound", where the density is constant, but the temperature becomes modulated periodically. In that case, the normal and superfluid are out of phase. (Figure adapted from Ref. [261].)

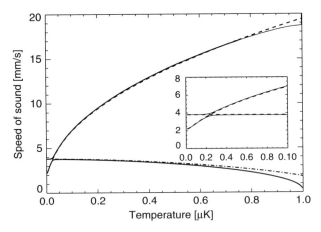

Figure 21.5 Speed of first and second sound for an uniform gas of Na atoms at a density of 5.5×10^{19} atoms/m^3. The solid lines are the result of Eq. (21.40), whereas for the dashed lines the coupling between the two modes is neglected. The inset shows an enlargement of the region of the crossing.

To determine the velocities of first and second sound, one has to find expressions for $(\partial P/\partial n)_T$, $(\partial P/\partial T)_n$ and $(\partial s/\partial T)_n$, and these depend on the equation of state of the system. For ultra-cold atoms, a description is the so-called Hartree–Fock equation of state. It allows for an analytical evaluation of these quantities, and the results for first and second sound are shown in Fig. 21.5. The figure shows that under practical conditions the first sound velocity is of the order of 1–20 mm/s, whereas the second sound velocity is about 1–4 mm/s. Since the condensate of Ref. [260] is of the order of a few mm, the propagation of sound can be observed in a period over 1 second. The coupling between the two sound modes can be observed, as indicated by the difference between the solid (coupled) and dashed (uncoupled) lines. A very special feature shows up at low temperatures, as shown in the inset, where the two sound velocities form an anti-crossing and thus switch their character. Note that the dashed line indicates that for the uncoupled case the anti-crossing becomes a crossing, as expected.

To experimentally investigate sound in a ultra-cold gas, a method has to be designed to induce sound in such systems. The easiest way to arrange this is to create a hole in the center of the condensate by shining in a repulsive laser beam, and shut off the laser beam at a certain instant. The condensate will replenish the hole by moving atoms from the side of the hole to the center, and this creates dips in the density moving out to the left and the right. Figure 21.6 shows the difference between the condensate density with and without the repulsive dipole laser and shows, for some time after the switching of the laser, the motion of two

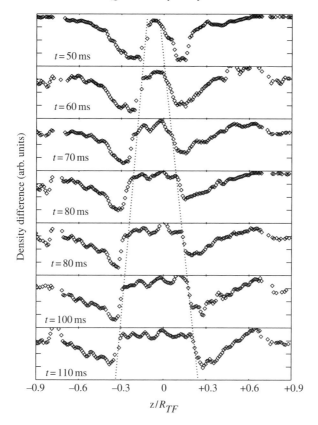

Figure 21.6 Density profiles of the condensate as a function of the propagation time after the excitation of a sound mode. The profiles show the difference between the perturbed and unperturbed ones. The Thomas–Fermi radius R_{TF} = 1.4 mm in this experiment. (Figure from Ref. [260].)

waves: one to the right and one to the left. Determining the speed of sound from such experiments shows that the speed of the dip coincides with speed of second sound. The analysis above predicts that second sound is primarily a density wave in the condensate, and this is what is observed. Modulation of the density of the normal fluid, which is an indication for sound, has not been observed systematically.

Figure 21.7 shows the measured speed of sound compared with the Bogoliubov speed of sound. In absence of coupling between first and second sound, the speed of sound is determined by the Bogoliubov dispersion relation Eq. (21.24), given by $c_B = \sqrt{U_0 n_0 / M}$, and this corresponds to c_2 under those conditions. The figure shows that the speed of second sound is about 10% smaller than c_B, which is a clear indication that the coupling between the normal and superfluid plays an important role in the experiment.

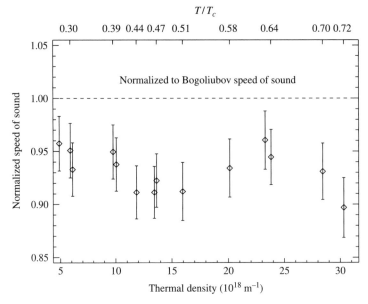

Figure 21.7 The normalized speed of sound, i.e. the ratio between measured speed of sound and the Bogoliubov speed of sound c_B, as a function of the thermal density. For a large range of thermal densities, the speed of sound is 5–10% smaller than c_B, owing to the interactions between the normal and superfluid. (Figure from Ref. [260].)

21.7 The superfluid–Mott insulator transition

A few years after the experimental realization of BEC in ultra-cold atoms, it was realized that the system allows ideal tests of many of the conventional models of condensed matter physics [265]. Ultra-cold atoms can be contained in an optical lattice (see Sec. 20.4), and such a system has a strong resemblance to conductance electrons in the ionic lattice of many typical solids.

The scales of the two systems are very different in many respects. First, the periodic potential of the ionic lattice has been replaced by an optical lattice produced by the interference of laser beams and thus the spacing between lattice sites is $\sim 1/2\,\mu$m, much larger than the few angstroms typical of ionic lattices. Second, the "conducting particles" in optical lattices are neutral atoms with a much larger mass and a much reduced velocity than those of the electrons in an ionic lattice. Third, the temperature of the working sample is typically less than one μK, compared with room temperature or a few K characteristic of condensed matter experiments. Fourth, unlike real ionic lattices, which always have very many defects and dislocations, optical lattices have neither, as long as their laser beams are spatially filtered. The resulting structural purity allows direct tests of the condensed matter physics

21.7 The superfluid–Mott insulator transition

models derived since the 1930s, because the measurements are not obscured by the effects of lattice defects that are usually difficult to include in the models.

One such case is the Bose–Hubbard model described for condensed matter physics in Ref. [266]. For ultra-cold atoms, a mean-field description does not suffice since the model requires quantum correlations. Assuming that the atoms are only in the lowest Bloch band, the operators $\hat{c}_i^\dagger$ and $\hat{c}_i$ can be introduced for the creation and annihilation of an atom at site i. These obey the standard commutation relation $\left[\hat{c}_i, \hat{c}_j^\dagger\right] = \delta_{ij}$ and $n_i = \hat{c}_i^\dagger \hat{c}_i$ is the number operator. The Hamiltonian can then be written as [265]

$$H = -t \sum_{\langle i,j \rangle} \hat{c}_i^\dagger \hat{c}_j + \frac{1}{2} U \sum_i \hat{c}_i^\dagger \hat{c}_i^\dagger \hat{c}_i \hat{c}_i - \mu \sum_i \hat{c}_i^\dagger \hat{c}_i, \qquad (21.43)$$

where the bracketed $\langle i, j \rangle$ means that only nearest-neighbor interactions are included. Here t is the hopping parameter from one site to another, U is the on-site interaction strength, and μ is the chemical potential.

When every lattice site i contains the same number of atoms (can be more than one) the lattice is exactly filled. For $t \ll U$, the interaction between the particles makes it energetically unfavorable for a particle to move from one site to another, and the gas is in the Mott-insulating phase. If an additional particle is added to such a system, there is only a small energy cost to move it from one site to another because its interaction energy is the same on each site. By contrast, an incompletely filled lattice is in a superfluid phase at zero-temperature.

Thus the gas can be in either an insulator or a superfluid phase. These circumstances are illustrated in Fig. 21.8, where in the insulator state the atoms are equally spread out across the lattice, forming a perfectly filled atomic lattice. However, since the number of atoms per lattice site is fixed, the complementarity between number and phase dictates that the phase difference between atoms at different sites is random. For the superfluid phase, the atoms are non-uniformly distributed over the lattice, and the number of atoms per site fluctuates. Since the atoms are allowed to jump easily between different lattice sites, the superfluid has a well-defined phase and the phase difference between atoms at different sites is zero.

To describe the zero-temperature phase transition from the superfluid to the Mott-insulating phase, a superfluid order parameter $\psi = \sqrt{n_i} = \langle \hat{c}_i^\dagger \rangle = \langle \hat{c}_i \rangle$ can be introduced. The Hamiltonian of Eq. (21.43) becomes diagonal in the site index i and can be recast as an effective onsite Hamiltonian [265, 268]:

$$H_{\text{eff}} = \frac{1}{2} U \hat{n} (\hat{n} - 1) - \mu \hat{n} - \psi z t \left(\hat{c}^\dagger + \hat{c} \right) + z t \psi^2, \qquad (21.44)$$

where z is the number of nearest-neighbor sites. The ground state can be found by variationally minimizing the total energy as a function of ψ for given values of t

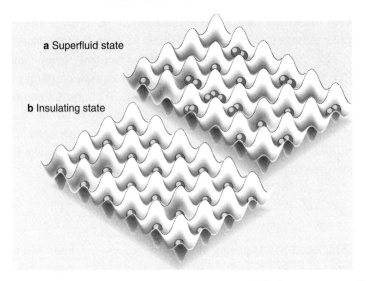

Figure 21.8 Atoms in an optical lattice. In the superfluid phase corresponding to a weak optical lattice, the atoms readily tunnel from one site to another, leading to a non-uniform distribution of the atoms over the sites. On increasing the lattice potential, the system makes a quantum phase transition to a insulator phase, where all atoms are distributed uniformly over the whole lattice. (Figure from Ref. [267].)

and U, and the result depends only on the ratio t/U. It is important to realize that both parameters t and U can be related directly to the microscopic parameters. For $\psi \neq 0$ the system is in the superfluid phase, whereas for $\psi = 0$ it is in the Mott-insulating phase. Note that this zero-temperature phase transition, also referred to as a quantum phase transition, can be driven by changing the ratio t/U and is not driven by temperature as are those occurring in nature.

Suppose atoms are confined in a 3D optical lattice constructed with four laser beams in tetrahedral symmetry that intersect pairwise at angles 2θ as described in Ref. [262]. Each of the two pairs of laser beams lie in orthogonal planes (e.g. xz and yz), and all four beams intersect at a common point. The result has a tetragonal structure with a periodicity of $\lambda/\sin\theta$ in one direction and $\lambda/\cos\theta$ in the other direction. The lattice sites have alternately right-handed or left-handed circular polarization, and the lattice potential can be written as [268]

$$V_{\pm} = \frac{1}{3}\hbar\delta s \left(\cos^2(k_{\perp}x) + \cos^2(k_{\perp}y) \pm \cos(k_{\perp}x)\cos(k_{\perp}y)\cos(2k_{//}z)\right). \quad (21.45)$$

where $k_{\perp}$ and $k_{//}$ refer to the z-axis, the factor $\frac{1}{3}$ arises from the Clebsch–Gordan coefficients for $J = 1/2 \to J = 3/2$ transitions, and s is the off-resonance saturation parameter.

21.7 The superfluid–Mott insulator transition

Using a Gaussian wavefunction for the atoms as in the case of a 1D lattice

$$\Psi(r) = \left(\frac{1}{\pi\beta^2}\right)^{3/4} e^{-r^2/2\beta^2}, \quad (21.46)$$

the energy is minimal for $\beta = (\hbar^2/M\kappa)^{1/4}$. Here κ is the harmonicity of the potential $\kappa = -4\hbar\delta s k^2/3$ and defines the oscillation frequency $\omega = \sqrt{\kappa/M}$. From this wavefunction one can derive the on-site interaction strength U:

$$U = \int d\mathbf{r} \int d\mathbf{r}' \Psi^*(\mathbf{r})\Psi^*(\mathbf{r}')V_{\text{int}}(\mathbf{r}-\mathbf{r}')\Psi(\mathbf{r})\Psi(\mathbf{r}') = \frac{2\hbar\omega}{\sqrt{2\pi}}\left(\frac{a_s}{\beta}\right), \quad (21.47)$$

where a_s is the triplet s-wave scattering length. The hopping parameter t is given by [268]

$$t = -\int d r^3 \Psi^*(\vec{r})\left(\frac{p^2}{2m} + V_\pm\right)\Psi(\vec{r} + a\hat{e}_j) = \frac{\hbar\omega}{8}\left[1 - \left(\frac{2}{\pi}\right)^2\right]\left(\frac{a}{\beta}\right)^2 e^{-\frac{1}{4}(a/\beta)^2}. \quad (21.48)$$

These parameters allow a quantitative comparison between the predicted and observed phase transition point.

Figure 21.9 shows the phase diagram for this system, where the chemical potential μ is plotted as a function of t, and both energies are normalized using the on-site interaction U. Inside the lobes indicated by $n = 1, 2$, and 3, the system is in the Mott-insulating phase with a fixed number of atoms per lattice site. Outside these lobes, the system is in the superfluid state, since the order parameter

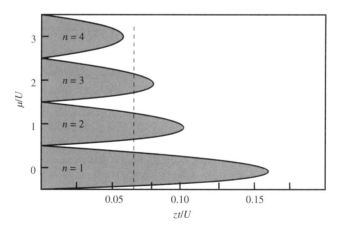

Figure 21.9 Phase diagram for ultra-cold atoms in an optical lattice. Here the quantum phase is indicated for reduced chemical potential μ/U as a function of the reduced hopping rate t/U, where U is the on-site interaction energy. Within the lobes the number of atoms per site is fixed and the system is in the insulator state. Outside the lobes the system is in a superfluid state. (Figure adapted from Ref. [268].)

is non-zero. If the chemical potential is increased from zero as indicated by the dashed line, the system starts in the superfluid phase and eventually reaches the Mott-insulating phase with $n = 1$. For larger μ the system stays in that state, until it becomes superfluid again for $\mu/U \approx 1$. Increasing the chemical potential even more causes the system to become insulating again, but now with $n = 2$. However, once t/U becomes larger than a critical value, the system always remains a superfluid, irrespective of μ.

The phase transition occurs for the first time for $V_{\text{trap}}/E_{\text{r}} = -2\hbar\delta s/3E_{\text{r}} \approx 10$ with E_{r} the recoil energy that depends slightly on the dimensionality of the lattice. This parameter can easily be realized in the case of the alkali-metal atoms. Note that the line signifies the result of going from the edge of the lattice to the center of the lattice, where the magnetic trapping potential causes the cloud to be inhomogeneous, with the chemical potential being largest in the center and decreasing near the edges.

In a pioneering experiment by Mandel and co-workers [269], Rb atoms were loaded into an optical lattice formed by six intersecting lasers detuned by more than 50 nm from resonance. A pair of laser beams in one direction formed a standing wave for the atoms in that direction and was detuned by tens of MHz with respect to the pairs in the other directions. This caused the interference between the standing waves to oscillate at this detuning frequency and was too fast for the atoms to follow, thus creating a cubic optical lattice.

The quantum phase transition was observed by using absorption imaging of the atoms after time-of-flight of 15 ms along the line of site of the imaging beam. The images for various values of the potential depth are shown in Fig. 21.10. In the absence of an optical lattice ($V_0 = 0$), as shown in Fig. 21.10a, the usual expansion of a superfluid is shown, where the width of the peaks reflects the chemical potential μ. However, for a finite depth of the optical potential, atoms expanding from one site of the lattice can interfere with atoms expanding from another site of the lattice leading to so-called Bragg peaks at finite momenta $\pm\hbar k$. Whether or not interference is observed depends on the existence of a phase relation between atoms at different lattice sites. For an insulator state, atoms at different sites are isolated from one another and do not have a well-defined phase relationship. As shown in Fig. 21.10, after the appearance of Bragg peaks for finite potential depth in panels b–f, the peaks disappear for larger potential depths in panels g and h, signaling the transition from a superfluid to a Mott insulator. The potential depths for which this transition occurs is in good agreement with the predictions of theory.

It is remarkable that the experimentalists succeeded in making a reversible transition from the superfluid to Mott-insulating phase and back. This was achieved by first slowly ramping up the potential to $V_0 = 22E_{\text{r}}$, where the insulator state is

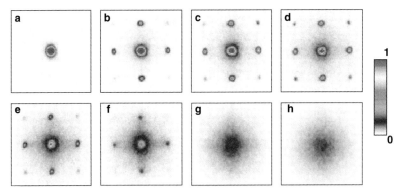

Figure 21.10 Absorption images of the cloud after release and expansion for different values of the depth of the optical lattice V_0/E_r, where E_r is the recoil energy: (a) 0, (b) 3, (c) 7, (d) 10, (e) 13, (f) 14, (g) 16, and (h) 20. The occurrence of interference peaks in (b)–(f) signals the phase coherence between atoms at different sites. The absence of those peaks in (g) and (h) is an indication that phase coherence is lost and the atoms are in an insulating state. (Figure from Ref. [269].)

formed. After some time the potential was ramped down in a few ms to $V_0 = 9E_r$ and the superfluid phase became visible again in the absorption image. The finite time was attributed to the tunneling time for an atom between two neighboring sites, which was 2 ms for $V_0 = 9E_r$. This was contrasted with the case of atoms in a phase-incoherent state, where such a superfluid state was not reached even after a ramp-down time of 400 ms. It shows that the loss of coherence in the insulator state cannot be described to a dephasing of the condensate wavefunction.

Appendices

21.A Distribution functions

A fundamental postulate of statistical mechanics is that in thermal equilibrium each microstate of a system with the same energy is equally populated. Although two particles may exchange energy in a collision, their total energy is conserved, and in equilibrium the temperature T is a measure of their average kinetic energy. The quantum statistics of the particles determine how the different kinetic energy configurations need to be counted.

In classical mechanics all particles are distinguishable, whereas in quantum mechanics they are not. Furthermore, quantum particles can be either fermions (half-integer spin) or bosons (integer spins), and these behave differently because of the Pauli symmetrization principle. No two fermions can occupy the same state, whereas bosons can. Although these principles are relatively easy to state, they complicate the counting of possible configurations considerably.

Begin the counting process by labeling the distinct energies of the separate states by ε_i. To find the most probable state, count the number of configurations $W(\{n_i\})$ that have the total number of particles N distributed over the states i with n_1 particles in state 1, and n_2 particles in state 2, etc. Since all configurations with equal energy are equally likely to be populated in thermal equilibrium, the system is most likely to be in the state with the largest number of configurations.

Classical particles

Counting the configurations for classical particles assumes that all particles are distinguishable. Picking n balls out of a bucket of N balls can be done in $N(N-1)(N-2)\ldots(N-n+1) = N!/(N-n)!$ ways. Since there is no interest in the order in which the n balls came out of the bucket, the possible ways have to be divided by the $n!$ ways in which this order can be arranged. The number of different ways $w_{c\ell}(N,n)$ thus becomes

$$w_{c\ell}(N, n) = \binom{N}{n} = \frac{N!}{(N-n)!\, n!}, \tag{21.49}$$

where the expression in the middle is the usual binomial coefficient. For selecting n_1 particles out of N particles, followed by n_2 particles out of $N - n_1$ particles, etc., one obtains

$$W_{c\ell}(\{n_i\}) = w_{c\ell}(N, n_1) w_{c\ell}(N-n_1, n_2) w_{c\ell}(N-n_1-n_2, n_3) \ldots = N! \prod_i \frac{1}{n_i!}. \tag{21.50}$$

So far the degeneracy g_i of the states has not been taken into account, but in general some of the sub-states can have the same energy ε_i. Since n_i particles can be distributed in $g_i^{n_i}$ ways over the g_i sub-states, the final result becomes

$$W_{c\ell}(\{n_i\}) = \prod_i \frac{g_i^{n_i}}{n_i!}. \tag{21.51}$$

Here the $N!$ factorial in front of the product is taken out to allow for correct Boltzmann counting (see, for instance, Ref. [270]). Since each state is equally likely in thermal equilibrium, the configuration $\{n_i\}$ that can be obtained in the largest number of different ways is the most probable, and in this configuration the number of ways $W(\{n_i\})$ is a maximum.

In both classical and quantum cases, the total number of particles $\sum_i n_i$ and the total energy $\sum_i n_i \varepsilon_i$ need to be fixed, so one requires additional constraints on the optimization. Thus

$$N = \sum_i n_i, \qquad E = \sum_i n_i \varepsilon_i. \tag{21.52}$$

These constraints can be taken into account in the optimization process using Lagrange multipliers. This is done by defining a new function $f(\{n_i\})$

$$f(\{n_i\}) = \ln W(\{n_i\}) + \alpha \left(N - \sum_i n_i \right) + \beta \left(E - \sum_i n_i \varepsilon_i \right), \tag{21.53}$$

where α and β are the Lagrange multipliers and the logarithm of W is used for calculation to allow the product to be written as a summation. Then the maximum for W is found by setting the following derivatives of f to zero:

$$\frac{\partial f}{\partial n_i} = 0, \qquad \frac{\partial f}{\partial \alpha} = 0, \qquad \frac{\partial f}{\partial \beta} = 0. \tag{21.54}$$

The last two equations above assure that the constraints of Eq. (21.52) are automatically fulfilled.

This procedure for finding the maximum of W is tedious but straightforward using Stirling's approximation for the factorials:

$$\ln N! \approx N \ln N - N, \qquad N \gg 1. \tag{21.55}$$

For the case of classical particles, using $W_{c\ell}$ of Eq. (21.51) leads to

$$\frac{\partial f}{\partial n_i} = \ln g_i - \ln n_i - \alpha - \beta \varepsilon_i, \tag{21.56}$$

and setting the derivative to zero yields

$$n_i = g_i e^{-(\alpha + \beta \varepsilon_i)}. \tag{21.57}$$

The Lagrange multipliers α and β can be found by substituting this relation back into the expression for $W_{c\ell}$ of Eq. (21.51). For $n_i/g_i \ll 1$ this yields

$$\ln W = \alpha N + \beta E. \tag{21.58}$$

The same conclusion arises in the grand-canonical approach even without $n_i/g_i \ll 1$, but the calculation here will continue within the micro-canonical approach.

Differentiating Eq. (21.58) leads to

$$dE = \frac{1}{\beta} d(\ln W_{c\ell}) - \frac{\alpha}{\beta} dN, \tag{21.59}$$

and identifying the result with the second law of thermodynamics

$$dE = T dS + \mu dN, \tag{21.60}$$

gives with the Boltzmann relation $S = k_B \ln W$:

$$\alpha = -\frac{\mu}{k_B T}, \qquad \beta = \frac{1}{k_B T}. \tag{21.61}$$

Here μ is the chemical potential. The distribution for a classical gas thus becomes

$$f_{MB}(\varepsilon_i) \equiv n_i = g_i e^{-(\varepsilon_i - \mu)/k_B T}, \quad (21.62)$$

and this Maxwell–Boltzmann distribution provides a very accurate description of the occupation of the states in a gas when the temperature is sufficiently large.

The chemical potential can be found by setting the sum of n_i over the states i equal to N. For an uniform gas this leads to

$$\mu = k_B T \ln\left(\frac{N \lambda_T^3}{V}\right). \quad (21.63)$$

Note that the condition $\mu = 0$ has been used in Sec. 21.3 to identify the onset of condensation, and it leads to $n = N/V = 1/\lambda_T^3$. The Maxwell-Boltzmann approach can be applied to calculate thermodynamic properties of a gas when the density of the gas is sufficiently low. It has been used in App. 15.A for the description of paramagnetism, but can equally well be used to calculate the heat capacity of a solid for low temperatures, as described by the Einstein model.

Quantum particles: fermions

In the quantum case of fermions all particles are indistinguishable, so it does not matter which particle is in which state. Distributing n particles over g possible sub-states, where each sub-state is occupied not more than once, yields

$$w_f(n, g) = \binom{g}{n} = \frac{g!}{n!(g-n)!}. \quad (21.64)$$

Choosing the factorial of a negative integer to be infinite assures that $w_f(n, g) = 0$ for $n > g$. The number of configurations simply becomes

$$W_f(\{n_i\}) = \prod_i w_f(n_i, g_i) = \prod_i \frac{g_i!}{n_i!(g_i - n_i)!}, \quad (21.65)$$

where the total number of configurations is just the product of all configurations of each state.

The same procedure using Lagrange multipliers can be applied to the relation W_f for fermions of Eq. (21.65). Thus

$$\frac{\partial f}{\partial n_i} = \ln(g_i - n_i) - \ln n_i - \alpha - \beta \varepsilon_i, \quad (21.66)$$

leading to the Fermi–Dirac distribution function for fermions using α and β as given by Eq. (21.61)

$$f_{FD}(\varepsilon) = \frac{1}{e^{(\varepsilon - \mu)/k_B T} + 1}. \quad (21.67)$$

For $(\varepsilon - \mu) \gg k_B T$ the +1 in the denominator can be neglected with respect to the exponential and Eq. (21.67) reduces to the Maxwell–Boltzmann distribution of Eq. (21.62). Since the occupation of each state in that case is quite small, this limit is often called the low-density limit.

For sufficiently low temperatures and $\varepsilon < \mu$, the exponential of Eq. (21.67) is small and $f_{FD} \approx 1$. Thus all the lower-lying states are fully occupied with one particle up to energy $\varepsilon = \mu$. Since the total number of atoms is N, this leads to $\mu \propto N/g$ for states with degeneracy g. As $T \to 0$ the highest occupied state has an energy μ, and this is called the Fermi energy ε_F. If the Fermi energy is related to a temperature $T_F = \varepsilon_F/k_B$, the temperature can be rather large. For instance, for electrons in a metal this temperature is of the order of 1000–10,000 K, and such a large temperature is solely the result of the Pauli symmetrization principle. By contrast, for bosons the particles can all occupy the lowest state and $T_F = 0$.

Quantum particles: bosons

For the quantum case of bosons the counting is rather complicated. The best way is to visualize n particles as n balls and the g sub-states as $g - 1$ partitions, and count the number of ways the $n + g - 1$ elements can be distributed, where more than one particle is allowed to occupy a sub-state. The $n + g - 1$ elements can be distributed in $(n + g - 1)!$ ways but the n balls and $g - 1$ partitions can each be distributed in $n!$ and $(g - 1)!$ ways respectively, and these different ways do not lead to distinct configurations. The number of distinct configurations becomes

$$w_b(n, g) = \binom{n + g - 1}{n} = \frac{(n + g - 1)!}{n!(g - 1)!}, \tag{21.68}$$

and there is no restriction on n. The total number of configurations is just the product of all configurations for each state, resulting in

$$W_b(\{n_i\}) = \prod_i w_b(n_i, g_i) = \prod_i \frac{(n_i + g_i - 1)!}{n_i!(g_i - 1)!}. \tag{21.69}$$

Clearly the number of configurations of bosons is much larger than that of fermions when the number of sub-states g_i becomes comparable to the number of particles n_i. For bosons the procedure for determining the maximum occupied state using W_b of Eq. (21.69) leads to

$$\frac{\partial f}{\partial n_i} = \ln(n_i + g_i - 1) - \ln n_i - \alpha - \beta \varepsilon_i. \tag{21.70}$$

This leads to the Bose–Einstein distribution function using $n_i \gg 1$

$$f_{BE}(\varepsilon) = \frac{1}{e^{(\varepsilon - \mu)/k_B T} - 1}. \tag{21.71}$$

The only difference between the distribution functions for bosons and fermions is the sign in front of the one in the denominator, but its implications are tremendous. For large temperatures the distribution function approximates the Maxwell–Boltzmann function just as in the case of the Fermi–Dirac distribution. Quantum statistics plays an important role only at low temperatures. For low temperatures the occupation of the states with $\varepsilon > 0$ is finite, and, as described in Sec. 21.3, this leads to the formation of the condensate for sufficiently low temperatures. The application of the Bose–Einstein distribution function for atoms with bosonic character can be found in the text of this chapter.

The quantum statistical properties of light

The Bose–Einstein distribution function applies equally well for "photons" that also have integer spin. The difference between photons and atoms with integer spin is that the number of atoms are conserved, whereas the number of photons is not. In the optimization approach this can easily be accomodated by choosing $\mu = 0$, such that the condition on particle conservation is automatically lifted. The distribution function for photons is then given by Eq. (21.71) with $\mu = 0$. Using the density of states for photons, as given by Eq. (21.80), leads immediately to the energy density of light

$$u(\varepsilon)\,d\varepsilon \equiv f_{BE}(\varepsilon)\rho(\varepsilon)\varepsilon\,d\varepsilon = \frac{8\pi V}{(hc)^3}\frac{\varepsilon^3 d\varepsilon}{e^{\varepsilon/k_B T}-1}, \qquad (21.72)$$

and is the Planck radiation formula of Eq. (5.1).

In 1923, Bose arrived at this distribution formula using the methods described in this appendix. However, at the time of his work the "counting" of states considered particles as strictly distinguishable, and it is said that Bose made an error that led to this formula that turned out to be in agreement with experiments. His paper describing the calculation was not accepted for publication until Einstein recognized its significance and applied it not only to photons but also to atoms with bosonic nature. This led to the whole field that is now referred to as Bose–Einstein condensation.

21.B Density of states

Thedensity of states of a system is an important tool to establish a connection between quantum mechanics, where states are enumerated by their quantum numbers, and thermodynamics, where the properties of the system are described by continuous variables. The density of states is the necessary ingredient to convert a sum over a property of the system (number of particles, energy) to an integral over the same property. Or, to express it in a mathematical manner:

21.B Density of states

$$\sum_i Q \rightarrow \int \rho(\varepsilon) Q \, d\varepsilon, \quad (21.73)$$

where the property of the system is denoted by Q. Here, the energy ε of the system is the parameter that is integrated. Clearly, the density of states can work only when the difference between a sum and an integral becomes small, or expressed in another way, when the number of states over which the sum runs is large.

Consider the case of a system enclosed in a cubical box of side L. In quantum mechanics the boundary condition requires that the wavefunction vanishes on the sides of the box. Also, for a light field quantized in such a box, the electromagnetic field on the sides of the box also vanishes. In both cases this leads to a quantization condition for the wavevector $\vec{k} = (k_x, k_y, k_z)$ of the field, namely

$$k_i = \frac{n_i \pi}{L}, \quad i = x, y, z, \quad (21.74)$$

where n_i are positive integers. The states with equal wavevector $|\vec{k}|$ lie on a sphere in k-space.

To determine the number of states, it is convenient first to determine the total number of states in a sphere with radius k (see Fig. 21.11). The total volume in k-space is given by $4/3 \pi k^3$, whereas the volume of one cell in k-space is given by $(\pi/L)^3$. The total number N of cells is thus the volume divided by the volume of one cell, or

$$N = \frac{1}{8} \frac{4 \pi k^3}{3} \frac{L^3}{\pi^3}, \quad (21.75)$$

where the factor $1/8$ stems from the fact that n_i is a positive integer and only states in the uppermost octant count. Differentiating both sides of the equation with respect to k leads to the density of states $\rho(k)$ in k-space:

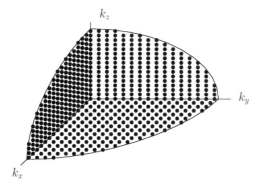

Figure 21.11 Allowed wavevectors $\vec{k} = (k_x, k_y, k_z)$ indicated by a dot for a system quantized in a box. For clarity, only wavevectors in the principal planes are shown.

$$\rho(k)dk = \left(\frac{dN}{dk}\right)dk = \frac{V}{2\pi^2}k^2 dk, \tag{21.76}$$

with the volume $V = L^3$.

In most cases one is not interested in the density of states in k-space, but in ε-space. For particle with mass M the energy ε is given by

$$\varepsilon = \frac{\hbar^2 k^2}{2M} = \frac{\hbar^2 \pi^2}{2ML^2}\left(n_x^2 + n_y^2 + n_z^2\right). \tag{21.77}$$

To transform from k-space to ε-space requires the inclusion of the Jacobian $dk/d\varepsilon$, and the density of states becomes

$$\rho(\varepsilon) = \rho(k)\frac{dk}{d\varepsilon} = \frac{V}{4\pi^2 \hbar^3}\sqrt{8M^3 \varepsilon}. \tag{21.78}$$

This gives the typical square-root dependence on the energy of the density of states in three dimensions. Similar calculations can be done in one and two dimensions, yielding other dependences on the energy.

For a light field in a box the same density of states in k-space applies, although the inclusion of the two polarizations of light leads to an additional factor of 2 for the density of states. Photons do not have mass, but the relationship between the energy and wavenumber is given by

$$\varepsilon = \hbar\omega = \hbar c k, \tag{21.79}$$

with c the speed of light. The density of states in ε-space becomes

$$\rho(\varepsilon)d\varepsilon = \frac{8\pi V \varepsilon^2}{(hc)^3} d\varepsilon, \tag{21.80}$$

which is identical to Eq. (5.10) and is used in App. 21.A to calculate Planck's distribution law.

22
Cold molecules

22.1 Slowing, cooling, and trapping molecules

One of the major achievements in physics during the last two decades of the twentieth century was the ability to cool and trap atoms [1]. Laser cooling can produce previously inconceivable atomic temperatures of less than 1 μK. Atoms with sufficiently low kinetic energy can be trapped in magnetic and optical traps and thus studied for a prolonged time, thereby improving the resolution of spectroscopy by many orders of magnitude as well as reducing Doppler broadening. Furthermore, besides the external degrees of freedom, the internal degrees of freedom of the atoms can also be manipulated, leading to full control of the atoms. One of the holy grails of physics, the achievement of Bose–Einstein condensation in a dilute gas, was achieved mainly because of laser cooling and trapping of atoms.

One of the natural extensions of laser cooling and trapping of atoms is the application of the techniques to molecules. Molecules offer a much richer internal structure than atoms, and thus provide extended possibilities to experimentalists. However, atomic laser cooling is based on the repetitive scattering of light and requires a closed, or at least nearly closed, two-level system. Because of the complicated internal structure of molecules, a closed two-level system cannot usually be found because they have many different electronic, vibrational, and rotational states and in general can decay to many different states after absorption of light (see Chap. 16). This requires a complete rethinking of the schemes that are used to cool atoms.

Still, because of the extended possibilities cold molecules can offer with respect to cold atoms, there have been a large number of techniques developed to slow, cool, and trap them. These techniques can be divided into two classes, namely direct and indirect cooling.

In the first case, molecules at cryogenic or room temperature are directly slowed and cooled. Although many different slowing techniques are used, this chapter focuses on Stark slowing of molecules since it was the first technique developed.

Apart from the Stark shift of the molecules, the Zeeman and light shift of molecules can also be used to slow them. The idea behind Stark slowing is to subject the moving molecules to a continuously increasing electric field so that their kinetic energy is converted into potential energy. Since such an increasing field cannot be sustained over a long distance, the field is switched off suddenly after certain time intervals, the potential energy vanishes, but their kinetic energy is unchanged. This process is reminiscent of Sisyphus laser cooling of atoms, although in this case the switching is done by external means and thus it may not lead to cooling.

Another direct method is buffer gas or sympathetic cooling. Here the molecules are mixed with atoms that are already cold (for instance, the vapor from liquid helium) or are cooled using laser or evaporative cooling (for instance, the alkali-metal atoms). By kinetic energy exchange through collisions, the temperature of the molecules is lowered. This is a very versatile technique, although it requires collisions between the molecules and the atoms in the cold reservoir and thus the method can be rather slow.

In the second case, molecules are produced by assembling them from ultra-cold atoms. The "assembling" can be through either photo-association or Feshbach coupling. In these processes the kinetic energies of the constituents are nearly unchanged, and thus the molecules are already ultra-cold once produced. Up to now, photo-association or Feshbach coupling has worked mainly for the alkali-metal atoms to produce bialkali molecules, either homo- or heteronuclear, and they are in highly excited vibrational states. Efforts to bring these highly excited molecules to their ground states have been successful, but they complicate the experimental arrangement considerably.

So far none of the techniques has dominated the efforts, and a growing number of researchers are improving each of the current techniques or are developing new techniques. In Tab. 22.1 the current state of the field is given in terms of the achieved temperatures and densities. The table shows that the direct methods produce large densities, but the temperatures are rather high, whereas the indirect techniques provide very low temperatures but at a much reduced density. Since molecules have an extensive internal structure, there are still a large number of pathways to be explored in the cooling of molecules, and many substantial improvements are to be expected in the decades to come (for an overview see Refs. [253, 271]).

Although laser cooling of molecules is problematic because of the large number of states that can be populated in the spontaneous emission process, there are certain molecules for which the number of states involved can be small. In a 1D collimation experiment on SrF it was shown that by using three laser beams it is possible to scatter light $\sim$1,000 times without the molecules becoming trapped in an inaccessible state [272]. Magnetic fields are used to eliminate the dark states, since

Technique	Temperature	Density
Stark slowing	10^{-1} K	$10^6/\text{cm}^3$
Buffer gas cooling	1 K	$10^9/\text{cm}^3$
Photo- and magneto-association	10^{-6}–10^{-3} K	$10^4/\text{cm}^3$

Table 22.1 Typical temperatures and densities for ultra-cold molecules. The direct techniques offer large densities, but modest temperatures. The indirect techniques provide very low temperatures, but the densities are low.

the cooling is applied mainly to a $F \to F - 1$ transition. The results show effects of both Doppler and sub-Doppler laser cooling, where their relative contributions can be varied using the magnetic field strength. Since it has been shown [273] that the favorable Franck–Condon overlap for SrF results in the population of only three vibrational states after 10^5 scattering events, these three populated states can be pumped to the excited state by three laser beams. This opens the way for laser cooling of SrF and many other molecules using the techniques of atomic laser cooling and trapping.

Applications of cold molecules are to be found in many fields. First, cold molecules allow the exploration of cold chemistry that is important in astrophysics. The study of the evolution of the universe requires the knowledge of interactions of astrophysically relevant molecules that often take place at low temperatures. Second, molecules form a rich field for precision measurements, in particular for the search for the variation of fundamental constants and for the permanent dipole moment of the electron (see Sec. 24.6). In the last case, the effect of any permanent dipole moment of the electron is enhanced in some molecules by more than four orders of magnitude because of interactions in the molecule, and this makes its search more feasible. Third, molecules can have a permanent electric dipole moment that alters their mutual interactions strongly compared with atoms. Interactions in the ground state of atoms are of the van der Waals type, and they have a short range. However, for molecules the dipolar interactions have a much larger range and thus allow for the creation of strongly interacting dipolar gases. These have novel quantum phases because the dipole interaction is not spherically symmetric [274].

22.2 Stark slowing of molecules

One of the first methods developed for the slowing of molecules is Stark slowing, whose principle is depicted in Fig. 22.1. Molecules with a permanent electric dipole

416   Cold molecules

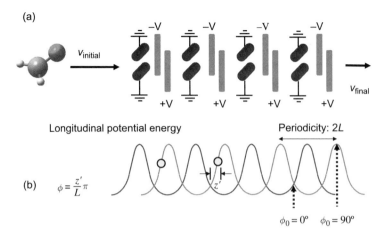

Figure 22.1 Principles of Stark slowing of molecules. (a) Molecules enter the slower, where they are slowed down by an increasing electric field. At certain time intervals the electric field is switched off, the next electrodes are turned on and the slowing starts again. (b) The efficiency of the slower is determined by the phase $\phi = z'\pi/L$, where for $\phi = 0$ no slowing occurs, whereas for $\phi = \pi/2$ the slowing is maximized. (Figure adapted from Ref. [253].)

moment are sent through a set of electrodes that have a voltage of $+10$ kV or -10 kV, or are grounded. As the molecules approach the first set of electrodes that have a high electric field, their energies are Stark-shifted upward, and thus the molecules convert part of their kinetic energy into potential energy. At a certain time the voltage is switched between the electrodes, and the first set is grounded, decreasing the potential energy of the molecules to zero. On approaching the second set of electrodes the process is repeated, and in each step the molecules lose kinetic energy and thus are slowed down. Since the whole process is reversible, the phase-space density is not increased. For the method to be successful it requires supersonic beams with already large phase-space densities. The method is very versatile and in principle can be adopted for all molecules, although they need to have a sufficiently large dipole moment for the method to be effective. Similar ideas were developed for atoms in Rydberg states [275] in the early 1980s but have never been implemented in an experiment.

To quantify the slowing method, consider the "synchronous molecules" that always have the same phase angle called ϕ with respect to the moment of switching. Here the phase angle is $\phi \equiv 2\pi z'/L$, z' is the molecules' position between two sets of the electrodes, and L is the spacing between them. As shown in Fig. 22.1b, a phase $\phi = 0$ indicates that at the moment of switching the molecules are halfway between two electrode sets and have the same Stark shift just before and just after the switching. These molecules are thus unaffected by the switching. A phase

$\phi = 90°$ indicates that the electrodes are switched when the Stark shift has a maximum, and the molecules thus lose the maximum amount of potential energy at each stage. It has been shown that molecules with a slightly different velocity are attracted to the phase of the synchronous molecules, and their velocity oscillates harmonically around the velocity of the synchronous molecules. For $\phi = 90°$ there are few molecules in this range and thus the phase-space density for them is small, whereas the phase-space density for $\phi = 0$ has a maximum. Thus the choice of ϕ is a matter of trade-off between maximum phase-space density and maximum slowing. The whole process is deterministic, and thus using the initial distribution of molecules and the sequence of switching times allows calculation of the final velocity distribution.

The slowing affects only the longitudinal velocities. For the transverse direction the molecules experience a field that has a zero on the axis, and thus the molecules are focused in the direction of the orientation of the electrodes. Therefore alternating sets of electrodes are oriented in orthogonal directions to assure that this focusing occurs in both transverse directions.

In a first proof-of-principle experiment (see Fig. 22.2) it was shown that CO molecules could be slowed from 225 m/s to 98 m/s [276]. The choice of CO molecules with a dipole moment of 1.37 Debye was made since they have a linear Stark shift in their metastable $a^3\Pi_1$ state. The state is prepared by laser excitation and it has a lifetime of 3.7 ms. When they impinge on a gold surface, ionization occurs, and by detecting the time-of-flight between excitation and detection, the velocity of the molecules can be inferred. Figure 22.3 shows the results when the number of deceleration stages is changed from 0 to 63. As the figure shows, there

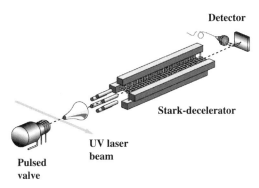

Figure 22.2 Experimental arrangement for Stark slowing. The CO molecules emerging from a pulsed nozzle are excited to a metastable state by a UV laser beam and are subsequently slowed down in a Stark slower of 63 stages. From the arrival time of the molecules on the detector, the final velocities of the molecules can be derived. (Figure from Ref. [276].)

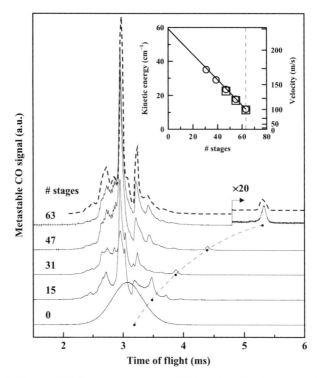

Figure 22.3 Time-of-flight measurements of CO molecules decelerated by a number of stages. The peaks connected by the dashed line show the arrival times of slowed molecules. The inset shows that the energy extracted from the molecules is linearly proportional to the number of stages. (Figure from Ref. [276].)

is a small peak of slow molecules where the arrival time on the detector increases from 3.2 ms for 0 stages to 5.3 ms for 63 stages. The dotted lines show the expected signal and also show excellent agreement between experiment and model. The inset shows that the extracted kinetic energy of the molecules is linearly proportional to the number of stages. The deceleration is about 10^5 m/s^2 and is similar to the deceleration for atoms using laser slowing. Soon after, it was shown that Stark slowing leads to sufficiently slow molecules for trapping with electrostatic fields [277]. The Earnshaw theorem [278] does not apply to dipoles, since the force on them is proportional to the field gradient. This has led to a strong increase in the research activity of molecule trapping using different configurations of electric fields.

22.3 Buffer gas cooling

Another direct method for cooling of molecules relies on the ability to thermalize them through elastic collisions. Copious numbers of laser-cooled atoms are readily

22.3 Buffer gas cooling

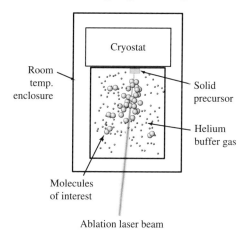

Figure 22.4 Buffer gas cooling of molecules. Molecules ablated by a laser pulse are cooled in a buffer gas above liquid helium and subsequently trapped. (Figure from Ref. [279].)

available, and they are the prime candidate for this method of sympathetic cooling. However, for more than a century large numbers of cryogenically cooled He atoms have also been readily available, and they can cool molecules to temperatures ranging from a few K using ^{4}He to 300 mK using ^{3}He. The choice of He as a buffer gas is suitable because it is structureless and chemically inert, and has a large elastic cross-section for thermalization. In addition, it has an appreciable saturated vapor pressure at temperatures as low as 300 mK.

Cryogenic refrigerators (see Fig. 22.4) often have magnetic traps created by a pair of anti-Helmholtz superconducting coils (see Sec. 20.2). Such traps can be deep enough to trap molecules with magnetic moments of a few Bohr magnetons. This permits the thermalization of the molecules with cold helium vapor for an extended time and thus permits the molecules to become quite cold. Thermalization of the initial room-temperature molecules needs to be sufficiently fast, typically 1 ms, or else the molecules can diffuse to the walls where they normally stick. Since the number of collisions to thermalize is of the order 100, the vapor density of He needs to be larger than $3 \times 10^{14}/\text{cm}^3$. For the molecules to remain trapped during thermalization, the elastic cross-section should be 10^4 times larger than the spin-flip cross-section that would leave them in an untrapped state. Furthermore, about 1,000 He atoms are necessary to cool one molecule because of the large difference of temperature. Since the He atoms are much more abundant than the molecules, the final temperature can be close to that of the He. After the cooling, the helium vapor can be pumped off, and the cold, trapped molecules remain at a temperature of a few K and densities of $10^9/\text{cm}^3$.

420   *Cold molecules*

Soon after the He vapor is pumped out, the trapping time of the molecules increases since reducing the density of the buffer gas reduces the rate of spin-flip collisions to untrapped states that cause ejection of the molecules. However, below a certain density the system is no longer hydrodynamic, and molecules that have obtained a large enough kinetic energy in collisions with the background gas can diffuse away, leading to a reduction of the trapping time. For still lower densities of the buffer gas this "evaporative" cooling of the molecules is no longer dominant; the trapping time increases again, since the molecules are thermally isolated. The crossing of this so-called "valley of death" is experimentally challenging.

Molecules in an apparatus for such collisional cooling with He vapor can be produced using different techniques such as laser ablation from a solid, beam injection, or capillary filling. The first experiment using this technique was carried out by the Doyle group on CaH [280], which has a magnetic moment of 1 μ_B. The molecules were produced by laser ablation of CaH_2, and 10^{10} molecules were produced by each laser pulse. The molecules were detected by either fluorescence or absorption imaging.

Figure 22.5 shows the evolution of the CaH spectrum after the ablation. Strong-field seekers, i.e. molecules that have a negative Zeeman shift, are not trapped. As seen from the figure, those molecules diffuse out of the trap in about 100 ms. However, the weak-field seekers that are trapped remain to be observed even 300 ms after ablation. Moreover, the spectral peak shifts towards resonance, indicating that the temperature of the molecules has decreased because they are closer to the zero-field point at the trap center. A fit of the distributions shows that there are about 10^8 molecules in the trap with a temperature of 400 mK. However, the lifetime of the molecules in the trap is limited to 500 ms as a result of ejection by spin-flip transitions. This first experiment has stimulated further research into buffer gas cooling of molecules, as discussed in Ref. [279], and this has led to the development of buffer gas cooling of molecular beams [281].

22.4 Binding cold atoms into molecules

The production of cold molecules by either photo- or magneto-association of cold atoms is remarkable. The heating of the system by the association process is negligible even though the binding in the potentials of the states involved is 10^{10} times larger than the atomic temperatures. In both cases the molecules are produced in one particular rovibrational state, so that the internal "temperature" of the molecules is also very small. However, the association works only for atoms that can be laser-cooled, and as such the studies so far concentrate on bialkali molecules.

22.4 Binding cold atoms into molecules

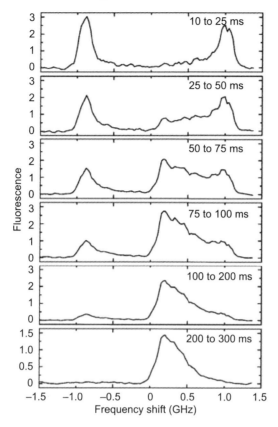

Figure 22.5 Excitation spectra of CaH molecules vs. the time after the laser ablation. The signals for negative frequency shifts are for strong-field seekers that are not trapped. The signals for positive frequency shifts are for weak-field seekers that are trapped. The shift of the spectrum toward the center frequency for the latter is a clear indication of the cooling of the molecules in the buffer gas. (Figure from Ref. [280].)

The resulting molecules are very unstable because of inelastic collisions that quickly lead to their decay. In the case of homonuclear systems the dipole moment of the molecules is zero, and thus these molecules are not of importance for the study of exotic states of matter that rely on dipolar interactions. But for heteronuclear systems the dipole moment depends on the mass difference of the two alkali atoms involved and can be as large as 5 debye.

The two ways of binding cold atoms together discussed below are really variations on similar physical principles. When resonant light is used to couple two atomic states, the dressed atom picture of Sec. 2.3.4 describes the resultant mixing of the states. When the Feshbach resonance of Sec. 18.6 brings an atomic state to degeneracy with a molecular state, the "avoided crossing" picture describes the

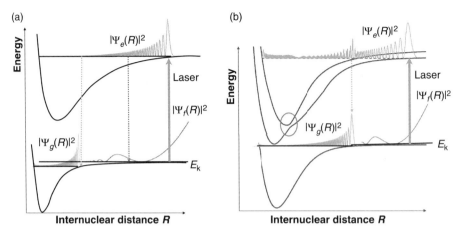

Figure 22.6 Photo-association of ultra-cold atoms. (a) Excitation of two atoms at long–range leads to the formation of a bound molecule that decays back to the ground state leading to two unbound atoms again. (Subscript f indicates "free".) The Franck–Condon overlap with the bound states in the ground potential of the molecule is negligible. (b) With an avoided crossing in the excited state there is some wavefunction of the excited state at short range, leading to the occupation of bound states in the ground potential after spontaneous emission. (Figure from Ref. [253].)

resultant mixing of the states. In both cases, an external field is used to couple two states thereby making "dressed" eigenstates and resulting in a transition from one of the bare states to another.

22.4.1 Photo-association

Photo-association, as depicted in Fig. 22.6, is the process of exciting two colliding atoms with laser light to a bound state of an excited molecule formed by the two atoms [282]. Such a free-bound transition is usually very weak because the wavefunctions of the free and bound states have nearly no overlap. This overlap is expressed in terms of the Franck–Condon factor, as discussed in Sec. 16.6.3. However, for low collision energies that can be attained by laser cooling the atoms before the association, the incoming free wavefunction has about the same periodicity as a weakly bound state in the molecular excited state, so photo-association is more likely. Since the process involves a single step, the recoil causes a negligible increase of the kinetic energy. Thus photo-association seems to provide an ideal means to produce ultra-cold molecules. However, the molecules are in an excited state and quickly decay back to their ground state. For homonuclear molecules the potential with one atom in the excited state is a C_3/R^3 dipole–dipole potential (see Sec. 18.2) so the two atoms are strongly accelerated towards each other. In the

22.4 Binding cold atoms into molecules

process the atoms can pick up sufficient energy that they can no longer be contained in the trap. Thus, photo-association can be studied by observing trap loss induced by the association laser.

To increase the efficiency of producing cold molecules with photo-association, the crossing of different potentials in the excited state can be exploited as shown in Ref. [283] for Cs_2 molecules. Figure 22.6a shows that the Franck–Condon factor for highly excited vibrational states in the excited-state potential, where there are deeply bound vibrational states in the ground-state potential, is small. However, when two excited-state potentials, one the incoming open channel and the other a closed channel, are coupled at some distance, bound states in the two potentials are strongly mixed and the wavefunction in the open channel is strongly modified, leading to an appreciable wavefunction at the outer turning point of the closed channel. This leads to an appreciable Franck–Condon overlap with bound states in the ground potential, and the decay can lead to the population of several bound ground states. This way, molecules can be produced with binding energies of 1 to 50 cm^{-1} and thus to vibrational states ranging from $v = 130$ to 150.

The situation for the production of ultra-cold molecules using photo-association is more favorable for heteronuclear molecules than for homonuclear molecules. In the latter case the interaction in the excited state is the C_3/R^3 potential and thus it is binding at large internuclear distances with nearly no overlap to the ground state, where interaction is the C_6/R^6 van der Waals interaction (see Sec. 18.2). For heteronuclear molecules the interaction in the excited state is also of the van der Waals-type and thus there is a much better overlap with the ground state than for the homonuclear case. However, the excitation has to take place at a much reduced internuclear distance requiring a larger intensity of the photo-association laser. In Ref. [284] it is shown that for the case of RbCs, molecules can be produced through photo-association followed by spontaneous emission.

In a subsequent experiment [285] it was shown that RbCs molecules in these excited vibrational states can be transferred to the ground state by exploiting a "pump–dump" method. The molecules are first excited by one laser pulse, and then a second laser pulse dumps them into the ground vibrational state by stimulated emission. The efficiency of the "pump–dump" process can be as high as 6%, leading to an appreciable production of ground-state $X\,^1\Sigma^+$ molecules, and these have a permanent dipole moment of 1.3 debye. In another experiment with laser-cooled ^{7}Li and ^{133}Cs atoms in a MOT, it was shown that molecules can be formed in the $v = 0$ ground state of the $X\,^1\Sigma^+$ potential with one photo-association pulse followed by spontaneous emission [286].

A similar situation arises for the homonuclear systems by exploiting "vibrational repumping" [287]. Here the molecules of Cs_2 that are produced by photo-association are excited using pulsed laser light and allowed to decay back to the ground state by spontaneous emission. This repumps the molecules over the various vibrational states, but the trick here is to shape the pulse so that there is no light exciting the molecules out of their ground vibrational state. This way the molecules start to accumulate in the ground state, and it is found that 5,000 pulses of laser light are needed to trap about 1,000 molecules in the ground state after 60 μs. The use of photo-association to produce ultra-cold molecules has been discussed in detail in Ref. [288].

22.4.2 Magneto-association

Another method to associate two atoms into a molecule is called magneto-association, using the mechanism of Feshbach resonances that was discussed in Sec. 18.6 [198]. Because the energy of a bound molecular state can be shifted by a magnetic field to the dissociation threshold, two unbound ultra-cold atoms can become bound as depicted in Fig. 22.7. Two ultra-cold atoms in a scattering state remain there if the magnetic field is quickly swept over a Feshbach resonance, but if the magnetic field is slowly reduced through the Feshbach resonance the atoms are transferred from the scattering state to the molecular state. The coupling between the two states causes an avoided crossing at the Feshbach resonance, and the system adiabatically follows the lowest branch of the avoided crossing, leading to the bound state. Transfer efficiencies can be close to 100%, leading to the production of a pure molecular gas at ultra-cold temperatures [289]. One advantage of magneto-association over photo-association is that the molecules are produced in the ground state, albeit in a highly excited vibrational state.

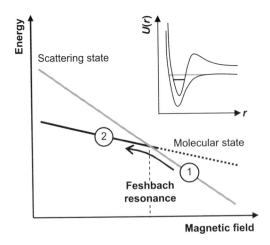

Figure 22.7 Magneto-association of two ultra-cold atoms. On ramping the magnetic field sufficiently slowly over a Feshbach resonance, the two atoms (1) are associated into a molecule (2). (Figure from Ref. [289].)

Magneto-association has been shown to create molecules very efficiently. In the first observation of Feshbach resonances in a Bose–Einstein condensate in Na [290] it was shown that the resonances are able to vary the scattering length by a factor of 10. However, they also give rise to inelastic losses caused by three-body collisions. These can be suppressed considerably by using fermionic instead of bosonic atoms for the association. The fermionic ^{40}K atoms with a total angular momentum $F = 9/2$ in the hyperfine ground state have been magneto-associated into ultracold molecules close to degeneracy [291]. Although the atoms are fermions, the ^{40}K$_2$ molecules are bosons and can thus condense into a BEC. The lifetime of the molecules is surprisingly large (1 ms) because of the Pauli blocking of three-body recombination, making fermions much more attractive than bosons. At the end of 2003, it was shown that fermionic ^{6}Li atoms [292, 293] or fermionic ^{40}K atoms [294] can be magneto-associated with a very high efficiency and that the resulting molecular cloud is below the transition temperature to BEC.

22.4.3 Vibrational state transfer by STIRAP

A very efficient process to transfer atoms coherently between two states is STIRAP, as discussed in Sec. 23.5. But STIRAP can also be used for molecules, and in a proof-of-principle experiment [295] it was shown that Rb$_2$ molecules created by magneto-association could be transferred to a more deeply bound state by applying two laser fields coupling the initial state through an excited state to the final ground state in a counter-intuitive sequence (see Fig. 22.8).

Two problems have to be addressed in the application of STIRAP to molecules. First, although there are many intermediate levels to couple to in principle, many of these intermediate levels are close to other levels and thus can lead to large losses by coupling to these levels. Second, the phase difference between the two laser fields needs to be very stable, and since the two coupled states can have a large energy difference, this can present a problem. In the proof-of-principle experiment this stability was provided by selecting the two coupled states close by, such that the energy difference could be produced by an acousto-optical modulator to shift the frequency of one of the beams.

Stabilizing the phase of two lasers for STIRAP can be achieved by locking them to a frequency comb. Since the comb can span a spectral range of 600–1,100 nm with a mode spacing of 800 MHz, it is ideally suited for the coherent transfer of molecules from a weakly bound state to a strongly bound state. For Cs$_2$, it has been shown that atoms in the $v = 155$ vibrational state of the $a\,^3\Sigma_u^+$ potential can be transferred to the $v = 73$ state in the same potential with a 80% efficiency [296]. For KRb, it has been shown that STIRAP can even produce molecules in lowest vibrational state of the $a\,^3\Sigma$ and $X\,^1\Sigma$ potential [297]. In the triplet potential this

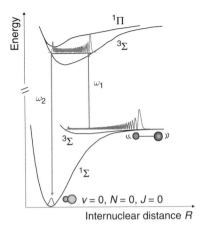

Figure 22.8 STIRAP on molecules that have been magneto-associated. Two laser fields with frequencies ω_1 and ω_2 couple the weakly bound Feshbach molecules to deeply bound states in the ground-state potential. In this case the final state is the ground state: electronic, vibrational, and rotational. The process requires careful selection of the intermediate state because it needs to have sufficient overlap with both the initial and the final states. (Figure from Ref. [253].)

leads to a phase space density of 0.1. Although the electric dipole moment in the triplet state is only 0.05 debye, the dipole moment in the singlet state, which is the true ground state of the molecule, is ten times as large and is measured to be 0.566 debye. In another experiment, STIRAP has been used to fully transfer Rb_2 molecules from the $v = 36$ state of the $a^3\Sigma_u^+$ potential to the $v = 0$ state in the same potential [298]. The two lasers in the STIRAP process are phase-stabilized on the same cavity using Pound–Drever–Hall locking obtaining laser linewidths of the order of 100 kHz, which indicates that the transfer has to take place on a µs timescale in order not to lose coherence in the process. The transfer efficiency is 90%, and the lifetime of the molecules is 200 ms, since they are contained in an optical lattice, which prevents three-body recombination.

22.5 A case study: photo-association spectroscopy

One of the techniques that has been developed to study ultra-cold collisions is photoassociation of ultra-cold atoms, which was introduced in Sec. 22.4.1. In this process a pair of colliding ground-state atoms is associated into an excited molecular state by absorption of light. From photo-association spectroscopy (PAS), it has been possible to obtain information on molecular systems, such as high-precision molecular binding energies, atomic lifetimes, retardation effects, and ground-state scattering lengths [191, 192, 299, 300]. Discussion of how this information is extracted from PAS will be given in this section.

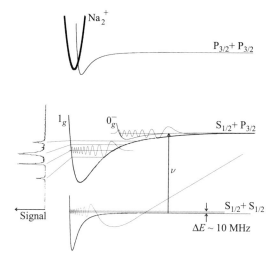

Figure 22.9 Schematic diagram showing the ground, excited, and doubly excited-state potentials of the Na$_2$ molecule that are connected asymptotically to the $3S_{1/2}+3S_{1/2}$, $3S_{1/2}+3P_{3/2}$, and $3P_{3/2}+3P_{3/2}$ dissociation limits, respectively. Two atoms colliding slowly in the ground state absorb red detuned light to produce an electronically excited, translationally cold molecule in a well-defined rovibrational level. The production rate of cold molecules is monitored as a function of laser frequency ν, and the small kinetic energy spread ensures a high resolution. (Figure from Ref. [190].)

The principle of PAS is explained schematically in Fig. 22.9. Although PAS has been performed for many alkali-metal atoms and also for some metastable rare gas atoms, in this section the focus is on the Na$_2$ system. Typically a sample of cold (order 1 mK) atoms is produced by laser cooling and held in a trap. The density of atoms in the trap is large enough to have a significant two-body collision rate between ground-state atoms. A "probe laser" detuned to the red of the atomic resonance ($3S_{1/2} \rightarrow 3P_{3/2}$ for Na) is focused on the sample of cold atoms, and at a certain frequency two slowly colliding atoms absorb light to produce an electronically excited, translationally cold molecule in a well-defined ro-vibrational level. The production rate of cold molecules is monitored as a function of probe laser frequency ν, and a signal is observed only if the frequency is resonant with a transition from the ground state to a singly excited molecular bound state. Note that, contrary to "traditional" molecular spectroscopy, the frequency is measured with respect to the well-known dissociation limit. Since the kinetic energy spread in the atomic sample is very small, the resolution in PAS is of order 10 MHz (for Na), and thus one or two orders of magnitude better than "traditional" molecular spectroscopy. Moreover, PAS has the advantage that it also reveals information about the dynamics of the collisions between slow atoms.

Most of the work on photo-association has focused on the alkali-metal atoms. This section will focus on the case of sodium, but for other atoms see Ref. [282]. Although the principles of photo-association are the same for all the alkali-metal atoms, the case of Na is special because of the possibility of detecting ions produced by two absorptions from the photo-association laser. This not only provides a nearly background-free detection method, but also enables the study of the doubly excited states.

By monitoring the fluorescence of the trapped atomic sample as a function of probe laser frequency, the photo-association process can be detected. When a transition is made to a rovibrational level of an attractive molecular state, the atoms are accelerated towards each other, thereby gaining kinetic energy. After spontaneous emission, the molecule dissociates into two ground-state atoms. According to the Franck–Condon principle, the molecule conserves the kinetic energy gained in the excited state. If the velocity of the atoms gained in the excited state exceeds the capture velocity of the trap, the atoms can escape. This loss from the trap is observed as a decrease in the fluorescence of the trapped atomic sample.

It is also possible to ionize the photo-associated molecules with a second absorption of the same frequency and detect the resulting Na_2^+-ions. For Na, the cold excited molecule is promoted to a doubly excited state that autoionizes at short internuclear distances. If the excitation is to the continuum of the doubly excited states, then the second step adds no structure to the spectrum, and the peaks in the ionization spectrum reflects only the structure in the first photo-association step.

Since there is no background signal present in the ion spectra, the signal-to-noise ratio is much better than for trap loss spectra. Another disadvantage of the trap loss detection method is that atoms that have not gained sufficient energy in the excited state are not lost from the trap. Excitation to the lowest-lying vibrational levels of Na_2^* therefore does not result in an observable trap loss signal. In this section the focus will therefore be on monitoring the photo-association process by detection of Na_2^+ ions.

To be able to study the purely long-range molecules, there is a different ionization mechanism that uses two frequencies. The first laser has to be tuned to the red of atomic resonance for photo-association to occur, and the second laser has to be tuned to the blue of atomic resonance to reach the dissociation limit of the autoionizing doubly excited states. One of these frequencies (ν_{red}) is kept to the red and the other (ν_{blue}) to the blue of atomic resonance. Figure 22.10 shows the measured ion signal as a function of ν_{red}. The spectrum shows two series of regularly spaced ion peaks. Identification of these peaks is based on their vibrational spacing and hyperfine structure [301–303]. The spectrum in this region is dominated by photoassociation at long range ($R \geq 140\ a_0$) to vibrational levels of the singly-excited 1_g and 0_g^- potentials, as indicated in the figure.

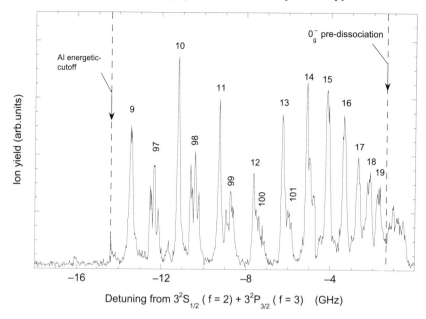

Figure 22.10 Measurement of the Na$_2^+$ signal as a function of ν_{red}. The spectrum shows two series of regularly spaced ion peaks. The series have been identified as the 1_g ($\nu = 97–101$) and the 0_g^- ($\nu = 9–19$) vibrational levels. A cutoff of the ion signal is observed at a detuning of -14.3 GHz. AI, auto-ionization. (Figure from Ref. [301].)

One thing to be noticed in Fig. 22.10 is the sudden disappearance of clearly resolved 0_g^- vibrational peaks for detunings less than 1.5 GHz from atomic resonance. As shown by Stwalley et al. [193], the 0_g^- state supports 40 rotationless vibrational states, of which the levels up to state $\nu = 19$ are clearly resolved. Although the resolution in the experiment is much better than the splitting between the levels, no well-resolved state with $\nu > 19$ is observed. Instead, the ionization signal is at a minimum at the position of the $\nu = 20$ state and increases at smaller detunings without any clear vibrational progression.

This behavior can be fully accounted for by carefully examining the detailed potential curves including the fine- and hyperfine-structure effects. In Fig. 22.11 all potential curves are shown with 0_g^- symmetry connected to the $3S_{1/2} + 3P_{3/2}$ asymptote. The inset of that figure shows a detailed look at the curves around -1.5 GHz. There are 10 curves in total with 0_g^- symmetry that are split by the hyperfine coupling. For internuclear distances smaller than $300a_0$, all the curves seem to connect to the $3S_{1/2}(F_g = 2) + 3P_{3/2}$ asymptote. However, there is an avoided crossing at $320a_0$ with curves connected to the $3S_{1/2}(F_g = 1) + 3P_{3/2}$ asymptote, and 7 of the 10 curves connect to this asymptote. This crossing is just at the position of the $\nu = 20$ vibrational state. Thus excitation of this state leads immediately to

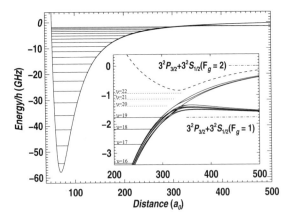

Figure 22.11 Calculated potential energies and bound states for the hyperfine components of the 0_g^-. The inset shows an enlarged view for small detunings. The bound states are indicated by the solid lines. The dotted lines indicate the positions of the hypothetical higher bound states. The dashed line indicates one of the potentials that causes the avoided crossings. Most of the 0_g^- potentials are connected to $3^2S_{1/2}(F_g = 1) + 3^2P_{3/2}$, only three connect to $3^2S_{1/2}(F_g = 2) + 3^2P_{3/2}$ [301]. (Figure from Ref. [301].)

predissociation of the molecule and does not lead to the formation of a molecule in the singly excited state. This is why no ionization of molecules is observed in this state [301]. Although these couplings play a role at very large internuclear distances and are very weak, their effects have a profound influence on the measured spectra.

To study the doubly excited states of Na_2 involved in the process, the frequency of the photo-associating (PA) laser is fixed to drive a transition from colliding ground state atoms to a specific rovibrational level of the 1_g or 0_g^- potential. The frequency of the second, photoionizing, laser is scanned to drive transitions from the chosen intermediate state to levels up to and above the $3P_{3/2}+3P_{3/2}$ threshold. By studying the bound states just slightly below this limit the symmetries of the autoionizing doubly excited states and their properties can be determined. The question of the identity and properties of the autoionizing molecular states connected to the $Na_2(3P+3P)$ asymptote naturally arises in the study of associative ionization in thermal collisions of $Na(3P)$ atoms as well [304–306]. In this case, collisions occur on doubly excited molecular potentials of any symmetry, whereas in the case of cold collisions, by choosing the symmetry of the intermediate state, doubly excited molecular potentials of specific symmetries can be studied separately.

Figure 22.12 shows a measurement with the PA-laser fixed on the 1_g ($v = 63$, $J = 1, 2, 3$, and 4) states. The bound states are identified as having 1_u and 0_u^- symmetry, and from their binding energies it is inferred that their asymptotic

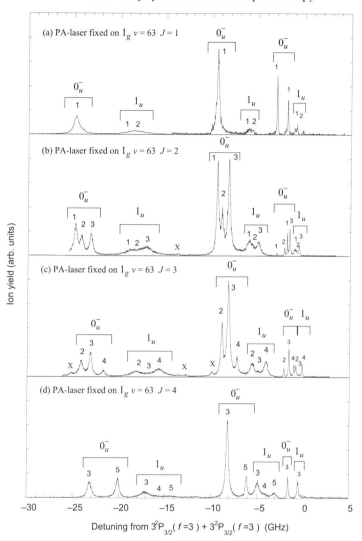

Figure 22.12 Measurement with the PA-laser fixed on the 1_g ($v = 63$, $J = 1, 2, 3$, and 4) states. This vibrational state has an outer turning point at $41a_0$. Fixing the PA-laser on different vibrational levels changes the outer turning point and thus the overlap with the doubly excited states. This is observed as a change in the relative population of the three vibrational levels of the 0_u^- and 1_u potentials. The X marks indicate unidentified lines, possibly from a 2_u potential. (Figure from Ref. [307].)

potentials are well described by a C_5/R^5 interaction with $C_5 \approx -250$ au for both states. The short range ionization probabilities are determined from the widths of the peaks and it is found that the ionization probabilities are 34% for the 1_u state and 15% for the 0_u^- state. These results provide important tests for quantum chemistry calculations [305, 306] that model the thermal associative ionization process. Reproducing these results does not require averaging over thermal distributions and molecular symmetries, and thus permits a more detailed and direct test of the calculations compared to experiments at thermal energies.

It is also possible to determine the ground-state s-wave scattering length using PAS. The scattering length determines the low-energy elastic scattering cross-section as well as the stability of a Bose–Einstein condensate and is therefore an important quantity to measure. Given the form of the asymptotic ground-state wavefunction,

$$\psi_g(R) \propto \frac{\sin[k(R-a)]}{\sqrt{k}}, \qquad (22.1)$$

the scattering length a is approximately determined by the position of the last node of this wavefunction. A node in the ground-state wavefunction is reflected in recorded PAS as a minimum caused by the small Franck–Condon wavefunction overlap at the Condon point $R_C = a$. PAS has been used to determine the s-wave scattering length for Na_2 to be $(85\pm3)a_0$ [192].

The dominant interaction term for the states connecting to the 3S+3P dissociation limit is the C_3/R^3 dipole–dipole interaction. The value of C_3 is proportional to μ^2 with μ the transition dipole moment, $\mu = e\langle 3S|r|3P\rangle$, connecting the 3S and 3P states. Then C_3 depends on the relative orientation of the 3S and the 3P atoms, i.e. on the symmetry of the molecular state. It is possible to determine C_3 for the purely long-range, singly excited 0_g^- potential to great precision by a measurement of the binding energies of vibrational levels in that potential. Using the purely long-range 0_g^- state to determine the transition dipole moment μ avoids the uncertainties associated with the potential in the short range, chemical-binding region. Since C_3 is inversely proportional to the radiative lifetime of the Na(3P) atom, it is possible to measure the atomic Na(3P) lifetime with high accuracy using PAS. The result obtained in Ref. [191] of $\tau(3P_{3/2}) = 16.230(16)$ ns helps to remove a longstanding discrepancy between experiment and theory.

PAS has also been applied to other systems, such as the metastable noble-gas atoms. In that case the spectroscopy can not only provide high-resolution results for the molecular potentials, but also yield insight into the dynamics of the Penning ionization process that is a major loss mechanism for samples of cold, metastable rare gas atoms. PAS can be applied to heteronuclear systems of alkali-metal atoms, and information about the interaction potentials is of crucial importance for the behavior of those systems at ultra-cold temperatures.

23
Three-level systems

23.1 Introduction

In Sec. 4.3.1 the coupling of two states of the same parity by two independent fields (or one field in second order) was considered in the perturbative regime, as initiated by Maria Göppert-Mayer. Needless to say, such electric dipole coupling depended on the presence of at least one additional state having opposite parity. However, laser light (and microwave radiation) can easily induce such three-level transitions where the magnitude of the final-state amplitude is not restricted to be much smaller than 1, so that a perturbative approach is not appropriate. Then there can be complete population exchanges, Rabi oscillations, and other coherent effects associated with significant population in any of the three coupled states. The unexpected new phenomena associated with multiple states result in consequences that are much richer than a pair of two-level transitions might suggest.

Such three-level atomic systems exhibit a wide variety of phenomena that are described carefully in several reference books [46, 308, 309]. Many such phenomena are still under study because they have important applications in precision measurement methods, quantum information storage and retrieval, manipulation of atomic and molecular states, cold quantum gases, and a variety of other current topics. Here the approach to the three-level problem will be treated coherently, in the same non-perturbative manner as Chap. 2 and part of Chap. 3.

The three states are labeled $|g_1\rangle$, $|g_2\rangle$, and $|e\rangle$, and in the spirit of Eq. (2.11) the wavefunction is written as

$$\Psi(t) = c_{g_1}(t)|g_1\rangle + c_{g_2}(t)|g_2\rangle + c_e(t)|e\rangle \qquad (23.1)$$

where the coefficients $c_j(t)$ satisfy $|c_{g_1}(t)|^2 + |c_{g_2}(t)|^2 + |c_e(t)|^2 = 1$ and may be complex. The notation $|e\rangle$ is for "excited", and it has opposite parity from the equal parity states $|g_1\rangle$ and $|g_2\rangle$. Then the Schrödinger equation for $\Psi(t)$ can be written $i\hbar \dot{c}_j(t) = \sum_n c_n(t)\mathcal{H}_{jn}(t)e^{i\omega_{jn}t}$ as in Eq. (2.3).

434　　　　　　　　　　　　　　Three-level systems

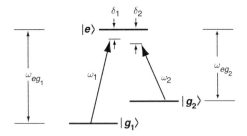

Figure 23.1 Schematic diagram of the three-state energy-level scheme for this chapter. Here ω_1 is the optical frequency that is detuned from the atomic resonance frequency ω_{eg_1} by $\delta_1 \equiv \omega_1 - \omega_{eg_1}$ and similarly for g_2. Both detunings are negative in this picture.

There are different ways to arrange these states energetically, but this chapter will focus on the "Λ" configuration illustrated in Fig. 23.1. If one of the levels $|g_1\rangle$ or $|g_2\rangle$ lies above $|e\rangle$ the arrangement is called the "ladder" or "cascade" configuration (Fig. 4.1c) and if both of them lie above $|e\rangle$ it is called the "vee" configuration (inverted Λ), and $|e\rangle$ has become the ground state. For these other configurations one has to be careful and consistent with the notation and sign choices.

There are electric dipole transitions that can be driven by external fields between $|g_1\rangle$ and $|e\rangle$ at frequency ω_1 having Rabi frequency given by $\Omega_1 \equiv \langle e|e\vec{\mathcal{E}}_1 \cdot \vec{r}|g_1\rangle/\hbar$, and between $|g_2\rangle$ and $|e\rangle$ at frequency ω_2 having Rabi frequency Ω_2 (see Sec. 3.2.1 and Fig. 23.1), but there is no dipole coupling between $|g_1\rangle$ and $|g_2\rangle$ because they have the same parity. The optical frequencies ω_1 and ω_2 are sufficiently different from one another that each one can drive only one transition. Also, $\omega_{eg_1} \equiv \omega_e - \omega_{g_1}$ can be larger or smaller than ω_{eg_2} in the Λ configuration.

The electric field of the light is given by $\vec{\mathcal{E}}_j = \hat{\varepsilon}_j \mathcal{E}_j(t)(e^{i(\omega_j t + \phi_j(t))} + \text{c.c.})/2$ for $j = 1, 2$ where the time dependence of the scalar $\mathcal{E}_j(t)$ allows for a changing amplitude and the time dependence of $\phi_j(t)$ allows for a temporal phase variation, in which case the optical frequency becomes $\omega_j + \dot{\phi}_j(t)$. Both are assumed to be "slowly varying", meaning that they satisfy $|\dot{\phi}_j| \ll \omega_j$ and $(1/\mathcal{E}_j)(d\mathcal{E}_j/dt) \ll \omega_j$. The detunings of the light from atomic resonance are $\delta_1 \equiv \omega_1 - \omega_{eg_1}$ and $\delta_2 \equiv \omega_2 - \omega_{eg_2}$, and in the Λ configuration the two-step transition from $|g_1\rangle$ to $|g_2\rangle$ is resonant when $\delta_1 = \delta_2$. If the atomic population begins in $|g_1\rangle$ the ω_1 and ω_2 fields are sometimes called pump and Stokes, respectively.

After the rotating frame transformation and the rotating wave approximation (Secs. 2.3.2 and 3.2.3) the 3 × 3 Hamiltonian matrix is given by

$$\mathcal{H} = \frac{\hbar}{2}\begin{pmatrix} -2\delta_1 & \Omega_1 & 0 \\ \Omega_1^* & 0 & \Omega_2 \\ 0 & \Omega_2^* & -2\delta_2 \end{pmatrix} \rightarrow \mathcal{H} = \frac{\hbar}{2}\begin{pmatrix} 0 & \Omega_1 & 0 \\ \Omega_1^* & 2\delta & \Omega_2 \\ 0 & \Omega_2^* & 0 \end{pmatrix}, \quad (23.2)$$

where the states are in the order $|g_1\rangle$, $|e\rangle$, and $|g_2\rangle$. For $\delta_1 = \delta_2 \equiv \delta$, shifting the zero of energy by δ leaves the diagonal of Eq. (23.2) with only the (2,2) element non-zero, as shown on the right side of Eq. (23.2). In this case the three differential equations for the coefficients $c_j(t)$ become quite simple (see Eq. (2.3)). Also, it is easily shown that at least one of the eigenvalues is always zero, meaning that there is always a state whose eigenvalue is independent of the Rabi frequencies $|\Omega_i|$, so this state has no light shift and can never be excited.

As can be found from the left side of Eq. (23.2), the eigenvalues for the case of $\delta_1 = \delta_2 \equiv \delta$ are given by

$$E_1 = \hbar\delta \quad \text{and} \quad E_{2,3} = \frac{\hbar}{2}(\delta \pm \Omega') \tag{23.3a}$$

with

$$\Omega' \equiv \sqrt{\delta^2 + |\Omega_1|^2 + |\Omega_2|^2}. \tag{23.3b}$$

Moreover, the eigenvectors can be conveniently expressed as (see App. 23.A)

$$|1\rangle = \begin{pmatrix} \cos\beta \\ 0 \\ -\sin\beta \end{pmatrix}, \quad |2\rangle = \begin{pmatrix} \sin\alpha\sin\beta \\ \cos\alpha \\ \sin\alpha\cos\beta \end{pmatrix}, \quad \text{and} \quad |3\rangle = \begin{pmatrix} \cos\alpha\sin\beta \\ -\sin\alpha \\ \cos\alpha\cos\beta \end{pmatrix}, \tag{23.4}$$

where the mixing angles α and β are given by

$$\tan 2\alpha \equiv \frac{\sqrt{|\Omega_1|^2 + |\Omega_2|^2}}{\delta} \quad \text{and} \quad \tan\beta \equiv \frac{\Omega_1}{\Omega_2}. \tag{23.5}$$

For $\Omega_2 = 0$, the Hamiltonian reduces to that of Eq. (2.9) and $\tan 2\alpha = |\Omega_1|/\delta$ so that $\alpha = \theta$ of Sec. 2.3.4, and conversely for $\Omega_1 = 0$. Then the eigenfunctions are the dressed states of Eq. (2.11). For $\delta_1 \neq \delta_2$ the expressions become very complicated, but the eigenvalues and eigenvectors are given in App. 23.A.

23.2 The spontaneous and stimulated Raman effects

In 1928 Chandrasekhara Raman discovered the effect that bears his name in the spectra of organic liquids. He was awarded the Nobel Prize in 1930, and in his Nobel address, he repeatedly called it an optical analog of the Compton effect. His analogy is clear, but further interpretation in terms of molecular structure came later. As described in Chap. 16, each electronic state of a molecule is composed of multiple rotational and vibrational sublevels, and excitation out of one of these to a corresponding excited sublevel can be followed by spontaneous emission to many sublevels of the lower state as shown in Fig. 4.1e. Thus the fluorescence spectrum can have multiple frequencies, some lower than the excitation frequency (called Stokes components) and some higher (anti-Stokes). In the language of this section, there is excitation of $|g_1\rangle$ to $|e\rangle$ by field $\vec{\mathcal{E}}_1$ and then spontaneous emission of $|e\rangle$ down to $|g_2\rangle$. There is no applied $\vec{\mathcal{E}}_2$ field, and the second transition results from natural decay.

In the perturbative limit of Sec. 4.3.1 the transition rates are given by Eq. (4.13) and the final-state populations are always very much smaller than those of the initial state. However, this chapter focuses on coherent processes for which the populations of initially empty states can become significant, and it is possible for population to slosh back and forth between $|g_1\rangle$ and $|g_2\rangle$ under the influence of the two applied fields. The requirement is that $\delta_1 \approx \delta_2$ and that there be minimal spontaneous decay from $|e\rangle$. The result is called the stimulated Raman effect because both fields are acting, and spontaneous emission is much less important.

For $\delta_1 \sim \delta_2 \gg |\Omega_j|$, the excited state $|e\rangle$ can be adiabatically eliminated from the equations of motion so the Hamiltonian of Eq. (23.2) reduces to that of a two-level atom with an effective Rabi frequency connecting the states $|g_1\rangle$ and $|g_2\rangle$. Using the Hamiltonian matrix on the right side of of Eq. (23.2), the dominant term of the differential equation for $\dot{c}_e(t)$ is $\delta \gg |\Omega_j|$ which means that $c_e(t)$ oscillates very rapidly compared with the time dependence of the coefficients c_g so its time derivative can be neglected. With $\dot{c}_e(t) = 0$ one finds $c_e(t) = -(\Omega_1 c_{g1} + \Omega_2 c_{g2})/2\delta$ and this is substituted into the equations for $\dot{c}_{g1}$ and $\dot{c}_{g2}$. The result is Rabi oscillations between the two states $|g_1\rangle$ and $|g_2\rangle$ just as in the case of a two-level system, and the Rabi frequency is $\Omega_1\Omega_2/4\delta$ for the two-step Raman transition. The two ground states become strongly mixed and can be rewritten as the dressed states of Sec. 2.4.

Such a process allows controlled transfer of atoms or molecules from one ground state to another without the complications of spontaneous emission. In addition, proper sweeping of the phases $\phi_j(t)$ can produce adiabatic rapid passage (see App. 2.D) that also produces controlled population transfers. Such state manipulation constitutes one aspect of "quantum control".

23.3 Coherent population trapping

The three-level system discussed here can provide a demonstration of a stunning example of one of the most bizarre (i.e. non-classical) phenomena of quantum mechanics, namely the existence of a dark state even in the presence of resonant light. In a classical world, a system can be in one state or another (coin showing either heads or tails) and can switch between these states under an external influence. But such a classical system can never be in both states simultaneously. In quantum mechanics the superposition of multiple states is indeed possible, and in fact was dubbed "the heart of the matter" by Schrödinger. The Λ system allows observation of the consequences of a superposition state.

23.3 Coherent population trapping

When atoms are subject to both $\vec{\mathcal{E}}_1$ and $\vec{\mathcal{E}}_2$ that excite them to $|e\rangle$, subsequent spontaneous decay could leave them in either $|g_1\rangle$ or $|g_2\rangle$. Including the coupling caused by the light field in the Hamiltonian results in the new eigenstates that are combinations of $|g_1\rangle$ and $|g_2\rangle$ given in Eq. (23.4). For $\delta_1 = \delta_2 \gg |\Omega_1|$ and $|\Omega_2|$, the normalized eigenstates are called non-coupled ($|1\rangle$) and coupled ($|3\rangle$), and can be found from Eq. (23.4) to be

$$|1\rangle = \cos\beta |g_1\rangle - \sin\beta |g_2\rangle \text{ and } |3\rangle \approx \sin\beta |g_1\rangle + \cos\beta |g_2\rangle \quad (23.6)$$

It is important to note that, for large detuning, neither of these eigenstates has an appreciable component of $|e\rangle$, and even for an arbitrary value of δ, $|1\rangle$ never has any component of $|e\rangle$.

Spontaneous emission can leave an atom in either of these eigenstates, but for the case of $|3\rangle$ there can be repeated cycling. However $|1\rangle$ cannot be excited to $|e\rangle$ since its transition amplitude is given by

$$\langle e|(\vec{\mathcal{E}}_1 + \vec{\mathcal{E}}_2)\cdot e\vec{r} |1\rangle = \cos\beta\langle e|\vec{\mathcal{E}}_1\cdot e\vec{r}|g_1\rangle - \sin\beta\langle e|\vec{\mathcal{E}}_2\cdot e\vec{r}|g_2\rangle$$
$$\propto \Omega_2(\hbar\Omega_1) - \Omega_1(\hbar\Omega_2) = 0, \quad (23.7)$$

so $|1\rangle$ is called a "dark state". The atomic population is trapped there because of an interference between the transition amplitudes of each component of the superposition that makes up the eigenstate, and for this reason the phenomenon is called coherent population trapping (CPT). Its eigenvalue is exactly $\hbar\delta$, which depends only on δ and not on Ω_1 or Ω_2, so it has no light shift because it cannot be coupled to an excited state. After some time, all atoms are optically pumped into $|1\rangle$ and can no longer absorb light. If δ_1 is scanned over a small range through δ_2, then at the $\delta_1 = \delta_2$ point the sample of atoms can neither absorb nor fluoresce light. The consequence is a narrow dark line right in the center of the fluorescence spectrum, first described in Refs. [310, 311] and plotted in Fig. 23.2.

An important extension of dark state physics is the possibility of laser action without population inversion. Normally inversion is required for a laser to work, so that the amplified light will not be absorbed by a high population of atoms in the lower state of the laser transition. But if those atoms are in the dark state, they cannot absorb the light so their population is irrelevant. Thus it is possible to achieve gain even when the lower-state population exceeds that of the upper state.

The phenomenon of CPT can be generalized to the case of any three levels that have an appropriate coupling scheme. It requires one level to be coupled to the other two, but that the other two are not coupled to each other. The couplings need not be E1 transitions as long as there are two terms similar to those of Eq. (23.7). An example is described in Ref. [312] where one coupling is an E1 transition and the other is Zeeman mixing by a dc field as in Chap. 11. For example, one could calculate the eigenfunctions of Eq. (11.15).

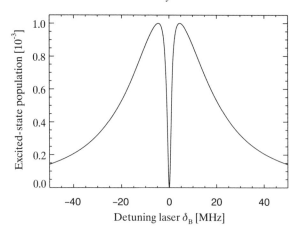

Figure 23.2 Schematic illustration of a spectrum showing a dark resonance. The degeneracy of $|g_1\rangle$ and $|g_2\rangle$ can be lifted by scanning δ_1 in the neighborhood of δ_2 or by using an applied magnetic field to exploit the Zeeman effect. The narrow dip in the center constitutes the region where the non-coupled state is well defined by Eq. (23.6) and cannot absorb light.

23.4 Autler–Townes and EIT

Consider the case where all atoms start in $|g_1\rangle$, for example if $|g_2\rangle$ is also an excited state, and $|\Omega_2| \gg \delta_2 \gg |\Omega_1|$ so that $|g_2\rangle$ and $|e\rangle$ are strongly coupled and $|g_1\rangle$ is so weakly coupled that it is little more than a spectator. Thus there are negligibly small populations in $|e\rangle$ and $|g_2\rangle$ because $|\Omega_1|$ is so small. As shown by Eq. (23.4), the three eigenstates in this case are an almost pure $|g_1\rangle$ and two nearly equal mixtures of $|g_2\rangle$ and $|e\rangle$, each with very little $|g_1\rangle$ (these are the two dressed states of Sec. 2.3.4 and Sec. 2.4). The energies of the strongly mixed pair of dressed eigenstates are approximately $\hbar(\delta \pm \Omega_2')/2$ where $\Omega_2' \equiv \sqrt{\delta^2 + |\Omega_2|^2}$, so they are separated by $\hbar\Omega_2'$.

For $\delta_2 = 0$, a scan of δ_1 over a small range excites atoms out of $|g_1\rangle$ at $\delta_1 \approx \pm \Omega_2'/2$, the light-shifted energies of the dressed states. Excitation can be measured by absorption of $\vec{\mathcal{E}}_1$ or by spontaneous emission, but apart from serving as a detection signal, spontaneous emission plays an insignificant role. The symmetric, double-peaked excitation spectrum is called the Autler–Townes (AT) doublet, and shows essentially no absorption or scattering of $\vec{\mathcal{E}}_1$ at the center where $\delta_1 = 0$.[1] For $\delta_2 \neq 0$ there is a similar but asymmetric double-peaked spectrum.

---

[1] A semi-classical description of this transparency derives from the picture of driving weakly coupled, classical oscillators between their closely spaced normal mode frequencies. One mode is driven above resonance and the other one below so their phase shifts are very nearly $\pm \pi/2$ (opposite signs). Therefore they are nearly π out of phase, and the resulting near-cancelation of their induced oscillating dipole moment precludes significant absorption. The sample is nearly transparent.

23.4 Autler–Townes and EIT

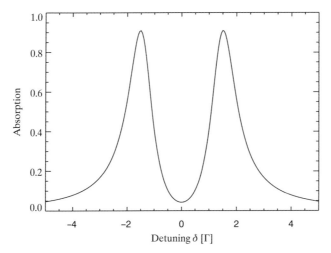

Figure 23.3 Spectrum of an Autler–Townes doublet. The two peaks here represent the excitation of the individual dressed states split by the Rabi frequency, and their separation increases with increasing intensity. By contrast, the two-peaked spectrum of EIT derives from an interference effect similar to CPT, but as the intensity increases the gap between them broadens making it seem as if the peaks are moving apart, although they are not. For more information, see Ref. [313].

There is also a similar double-peaked structure that arises from a quite different effect when spontaneous emission is important, meaning that the decay rate $|e\rangle \to |g_1\rangle$ and $|g_2\rangle$, denoted γ_{eg_1} and γ_{eg_2}, is comparable to $|\Omega_2|$. What is important is not the decay, but that the excited states $|2\rangle$ and $|3\rangle$ have a width that may be comparable to the separation between them $\Omega'_2 \sim \Omega_2$ so that they are not well resolved by scanning δ_1. When the scan of δ_1 brings it to the neighborhood of δ_2, the transition amplitude from $|g_1\rangle$ is a sum of two terms, one for each of the dressed states, and the terms of this sum that connect to the $|e\rangle$ part of the mixture have opposite signs because $\Omega' \gg \delta$ (see Eq. (23.4)). This results in a cancellation of transition amplitudes that leaves the transition probability nearly zero at the center, just as in the AT case. Thus $\vec{\mathcal{E}}_1$ is transmitted because of the mixing of $|e\rangle$ and $|g_2\rangle$ caused by $\vec{\mathcal{E}}_2$.

Unlike the spectroscopic separation of distinct states that characterizes the AT doublet, this is an example of an interference effect arising from the summation of transition amplitudes that must be done before squaring to find the total transition probability. Clearly this is a different phenomenon, and it is called electromagnetically induced transparency (EIT). There are many variations of the EIT level structure, light polarization, optical frequencies, etc., but this is the underlying interference phenomenon.

Since the two quite different phenomena produce a similar double-peaked spectrum when δ_1 is scanned through δ_2, it is important to distinguish between them.

The AT splitting results from two separate absorption spectra, each having a characteristic shape, that are separated by Ω_2'. For different values of $|\Omega_2|$ the peaks change their separation but retain their shape, although they may not be completely resolved so that a convolution is necessary. By contrast, EIT is a transition amplitude interference effect of states broadened by coupling to a continuum, similar to the original description of Fano [314]. The interference arises from the overlap caused by the natural width of $|e\rangle$ so spontaneous emission is necessary, but only for its level broadening effect. The spectral shape is characterized by a sum of two Lorentzians with the same center, one of which is very narrow and inverted, similar to that of Fig. 23.2. For different values of $|\Omega_2|$ the shapes change, and the peaks appear to move apart because the central dip broadens, but the cause is clearly different from that of AT. An objective measure of distinguishing them from one another in the range between extreme AT and EIT has recently been presented [313].

23.5 Stimulated rapid adiabatic passage

Up to now the discussion has been limited to continuous illumination of atoms by incident light beams, but the scalar amplitudes of the fields given at the beginning of this chapter were written as $\mathcal{E}_j(t)$. Carefully timed and shaped pulses of light produce other phenomena in three-level systems, and the best known of these is stimulated rapid adiabatic passage (STIRAP) [46, 308, 315–317]. It exploits the properties of the Hamiltonian of Eq. (23.2) to switch atoms from one state to another with very high efficiency, and is much more robust than the π-pulse method suggested in Sec. 23.2. It is quite analogous to the adiabatic rapid passage of App. 2.D.

It is implemented by applying two overlapping pulses of light, $\mathcal{E}_1(t)$ and $\mathcal{E}_2(t)$, to drive atoms from $|g_1\rangle$ to $|g_2\rangle$ under the condition that $\delta_1 = \delta_2$. However, contrary to intuition, the pulse of $\mathcal{E}_2(t)$ light is applied *before* the pulse of $\mathcal{E}_1(t)$ light, as shown in Fig. 23.4. The process is most easily understood by reference to the eigenstates of Eq. (23.4).

At the start of the pulse pair sequence, $\Omega_1 = 0$ and all the population is in $|g_1\rangle = |1\rangle$ because the $|1\rangle$ eigenstate is pure $|g_1\rangle$. As Fig. 23.4a shows, the Ω_2 light appears first, and couples states $|e\rangle$ and $|g_2\rangle$ even though they have no population. The mixing angle β remains at 0 as shown in region I of Fig. 23.4b, and all the atoms remain in $|g_1\rangle$ as shown in Fig. 23.4c. At the beginning of region II, after the peak of the $\mathcal{E}_2(t)$ pulse, the Ω_1 light intensity begins to increase, the angle β starts to rise, and the nature of eigenstate $|1\rangle$ begins to change because Ω_1 is no longer zero. When the rising $\mathcal{E}_1(t)$ pulse has sufficient intensity relative to that of the falling $\mathcal{E}_2(t)$ pulse, $\Omega_1 = \Omega_2$ and $|1\rangle$ is an equal mixture of $|g_1\rangle$ and $|g_2\rangle$. Of

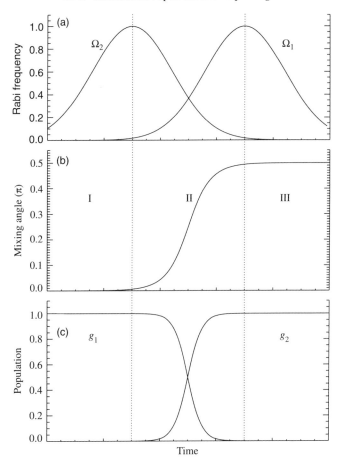

Figure 23.4 Time evolution during the STIRAP process. Part (a) shows the evolution of the Rabi frequencies where Ω_2 precedes Ω_1 even though the population is assumed to start in $|g_1\rangle$ which cannot be excited by this light. Part (b) shows the evolution of the mixing angle β, and part (c) shows how the population of $|g_1\rangle$ is unchanged until Ω_1 begins to rise from zero. Eventually all the atoms finish in $|g_2\rangle$. Note that equal Rabi frequencies in (a) does not necessarily mean equal light intensities.

course, the population remained in $|1\rangle$ so it has evolved to a mixture of $|g_1\rangle$ and $|g_2\rangle$. At all times, $|1\rangle$ has no component of $|e\rangle$. After the $\mathcal{E}_2(t)$ pulse has ended but there is still some light from the $\mathcal{E}_1(t)$ pulse, eigenstate $|1\rangle$ is pure $|g_2\rangle$ and has all the population. Thus the transfer $|g_1\rangle \rightarrow |g_2\rangle$ is complete.

The process requires that the time variation of these light intensities needs to be sufficiently slow that atoms stay in the state $|1\rangle$ where they started when it was pure $|g_1\rangle$. The condition for such adiabatic evolution is $(1/\Omega_j)(d\Omega_j/dt) \ll \Omega_j$, and if it is satisfied and $\delta_1 = \delta_2$, 100% of the population of atoms is transferred from $|g_1\rangle$ to $|g_2\rangle$ because $|1\rangle$ has evolved to it.

The ideal adiabatic process described above, namely the perfect evolution of $|1\rangle$ from $|g_1\rangle$ to $|g_2\rangle$ and completely independent of $|e\rangle$, is deceptive because $|e\rangle$ plays an important role. This is so because perfect adiabaticity is never experimentally possible nor theoretically expected. The origin of the role of $|e\rangle$ arises from two sources: (1) δ_1 is not exactly equal to δ_2 (see App. 23.A), and (2) the time dependence of $\mathcal{H}$ includes the envelopes of the pulsed light fields (in addition to their oscillations at the optical frequencies), and this part of the time dependence survives the rotating frame transformation of Chap. 2, so $\mathcal{H}$ still has some time dependence.

Various schemes to minimize the population of $|e\rangle$ have been developed because its spontaneous emission can compromise the efficiency of the STIRAP process by destroying the coherence between $|g_1\rangle$ and $|g_2\rangle$. The limits of these STIRAP schemes can be discussed by considering the populations of $|e\rangle$ and $|g_2\rangle$ in the actual state vector $\Psi_1(t)$ instead of the ideal adiabatic state $|1\rangle$ [318]. For clarity, note that $|g_1\rangle$, $|g_2\rangle$, and $|e\rangle$ are the time-independent bare states of the atom, the states $|j\rangle$ are the adiabatic states of Eq. (23.4) and are time-dependent through $\alpha(t)$ and $\beta(t)$, and the wavefunctions Ψ_j are the actual wavefunctions for the non-ideal case.

The populations of $|e\rangle$ and $|g_2\rangle$ in Ψ_1 can be conveniently calculated in any basis: using the eigenfunctions is not necessary (see Ref. [317]). Then the desired condition of the closest behavior to perfect adiabaticity requires that both $\Omega_1 \tau \gg \pi$ and $\Omega_2 \tau \gg \pi$, where τ is the pulse overlap time. This means that the adiabatic evolution of Ψ_1 requires the atom to undergo very many Rabi flops during the overlap time of the pulses [318].

23.6 Slow light

There are two important differences between the topics considered above and the subject of "slow light". First, the previous discussion has focused on the response of a three-level atom to two resonant frequencies of incident light, with no discussion of the propagation of the light itself. Second, both light fields under consideration were monochromatic. Slow light refers to the delay of a pulse traveling through a prepared medium, three-level atoms in the case considered here, and a pulse is necessarily polychromatic. The following discussion provides an introduction to pulse propagation.

Monochromatic waves propagate at a speed of (wavelength × frequency) and for light this is written $c = \lambda_0 \nu = \omega/k_0$ where the 0 subscript means vacuum. Of course, such light can carry no information so it has to be modulated to transmit a signal, and then a simple Fourier transform shows that it is no longer monochromatic. Thus there are two speeds to consider: the carrier speed c in vacuum is called

23.6 Slow light

the phase velocity and denoted v_ϕ, and the speed of the modulation that carries the signal that is called the group velocity, v_g.

A vapor of atoms is transparent to light tuned far from atomic resonance, but still the atoms interact with the incident light. The response is primarily forward scattering in accordance with Huygens' principle, but just as in the case of a classical driven oscillator, there is a phase shift of the atomic dipole oscillation. The negative phase shift (lag) is small for light tuned far below resonance, nearly π for tuning far above resonance, and approximately $\pi/2$ near resonance. Light tuned at or near an atomic resonance frequency is not readily transmitted through a sample of two-level atoms because it is absorbed and reradiated as spontaneous emission.

The phase lag appears as a retardation and hence a speed reduction far from resonance where the sample is transparent resulting in $v_\phi = c/n$ where n is called the refractive index. In a vapor or gas such as air at atmospheric pressure, the effect is small, and the refractive index is $n_{air} - 1 \approx 2.9 \times 10^{-4}$. For a transparent solid that is typically 10^3 times denser, $n - 1 \approx 0.3$, 10^3 larger. The lowest absorption frequency of most transparent materials or gases is in the UV, so for visible light the detuning is well below resonance.

In general, $v_g < v_\phi$, and this is most easily seen by considering a pulse sequence formed by the sum of two waves, each of amplitude unity, traveling in the $+\hat{z}$-direction and comprised of two different frequencies, $\omega \pm \delta\omega$. The electric field of such a light beam is written as

$$\cos[(\omega + \delta\omega)t - (k_0 + \delta k)z] + \cos[(\omega - \delta\omega)t - (k_0 - \delta k)z] \quad (23.8)$$
$$= 2\cos[\omega t - k_0 z] \times \cos[(\delta\omega)t - (\delta k)z].$$

The point of constant phase of the first factor on the second line of Eq. (23.8) satisfies $z = (\omega/k_0)t$, meaning that this wave oscillates at the carrier frequency and it travels at speed $v_\phi = \omega/k_0$. By contrast, the point of constant phase of the second factor satisfies $z = (\delta\omega/\delta k)t \to (d\omega/dk)t$ meaning that the "beats" of the modulated wave, namely the pulses, travel at speed $v_g = d\omega/dk$. This result is quite general for pulses with many frequency components.

The group velocity can be found by using $\omega = kc/n(\omega)$ and differentiating to find

$$v_g = \frac{d\omega}{dk} = \frac{c}{n}\left(1 - \frac{k}{n}\frac{dn}{dk}\right) = \frac{c}{n}\left(1 + \frac{\lambda}{n}\frac{dn}{d\lambda}\right) = \frac{c}{n + \omega\frac{dn(\omega)}{d\omega}}, \quad (23.9)$$

although some care is needed to get the last term above. The effective refractive index for the group velocity is thus $n_g = n + dn(\omega)/d\omega > n$ since $dn(\omega)/d\omega$ is usually positive (below resonance) so that $v_g < v_\phi$. This is the background needed to understand slow light propagation.

The sample is prepared with both the atoms in state $|g_1\rangle$ and pump light at frequency ω_2 with sufficiently high intensity that the dressed state picture is appropriate for states $|e\rangle$ and $|g_2\rangle$. The carrier frequency of the "slow" pulse is tuned between the two dressed states so that it is below the resonance for one of them and above it for the other one. Because the splitting between the states $\sim |\Omega_2|$ is very small compared with the optical frequency ω_2, the detuning is small so the phase shifts are large. Nevertheless, the sample is transparent as described in Sec. 23.4.

In this region between the resonance frequencies of the dressed states the refractive index changes strongly. The absolute change is small, perhaps only 10^{-6} amounting to a few percent, but the frequency range is exquisitely small, less than $|\Omega_2|$. Thus $dn(\omega)/d\omega$ in Eq. (23.9) can be very large, resulting in very slow transmission of the pulse. The carrier propagates at $v_\phi \approx c$ and slips through the pulse envelope that is traveling at $v_g \ll v_\phi$.

The frequency components of the pulse must be confined to the region between the dressed states so the pulse duration must be at least as long as $\pi/|\Omega_2|$ which is typically μs. Since the samples are small, the pulse delay is comparable to its duration and therefore hard to measure. The effect can be enhanced using two absorption resonances that are further separated, such as the D-lines of alkali-metal atoms. The anomalous dispersion region between them is ~GHz so that sub-ns pulses can be used and are also delayed by μs. For a sample of atomic vapor of diameter less than 1 mm, a μs transmission time corresponds to a velocity closer to that of sound than of light, less than 1 km/s.

23.7 Observations and measurements

It is clear that three-level systems offer a wide variety of phenomena, many of which seem disjoint as presented in the conventional literature. Here it has been shown that several (but not all) of them are closely related and can be described by the simple Hamiltonian of Eq. (23.2). Moreover, there are others that have not been discussed here. Clearly the differences among these phenomena arise from the relative values of the various laser parameters, and Tab. 23.1 summarizes a few of these.

The three pairs are grouped according to their similarities. The relationship between the two different types of Raman spectroscopy is simple: in the spontaneous case the $|e\rangle \rightarrow |g_2\rangle$ transition is the consequence of spontaneous emission and in the stimulated case it is driven by Ω_2. The cases of CPT and STIRAP are related because both have a dark state, but in STIRAP the objective is to make the transition by keeping the atom always in the dark state and letting it evolve from $|g_1\rangle$ to $|g_2\rangle$ as described in the text, whereas in CPT it is simply observed in the spectrum. For EIT and AT the unpopulated states are strongly coupled into the

Name	Parameter relations	Measured quantity
Spontaneous Raman	$\delta_1 \leq |\Omega_1|$, $\Omega_2 = 0$	Fluorescent light
Stimulated Raman	$\delta_1, \delta_2 \gg |\Omega_1|, |\Omega_2|$	Population transfer
Dark states (CPT)	$|\Omega_1| \sim |\Omega_2|$	Absorption, fluorescence
STIRAP	$|\Omega_1| \sim |\Omega_2|$, pulse timing	Population transfer
EIT	$\gamma_{eg_1} \sim \gamma_{eg_2} \geq |\Omega_2| \gg |\Omega_1|$	Probe transmission
Autler–Townes (AT)	$|\Omega_2| \gg |\Omega_1|$	Splitting of peaks

Table 23.1 Each related pair of phenomena listed here is observed or measured by scanning a parameter associated with $\delta_1 - \delta_2$. Although the entries of Eq. (23.4) are based on $\delta_1 = \delta_2$, the deviations associated with a small scan make only small changes. In the case of EIT and AT, Ω_2 has been chosen as the strong, driving field so that $|g_2\rangle$ and $|e\rangle$ are strongly coupled. Then Ω_1 is a weak beam that serves as a probe, so that $|g_1\rangle$ is often just a spectator state. The converse choice is equally good, but in either case, the weaker frequency or associated level splitting is scanned. In all cases except for EIT, it is assumed that spontaneous emission is of negligible importance. And even for EIT, its role is only to provide some coupling to a reservoir (i.e. provide width to $|e\rangle$ and consequently to $|2\rangle$ and $|3\rangle$).

dressed states by Ω_2 while the system is probed by a weak Ω_1. If the dressed states are well resolved their splitting is the AT doublet, but if they have a width derived from a significant value of γ_{eg_1} making them only partially resolved, the transition amplitudes interfere causing a central dip.

One of the important distinctions among these phenomena is the method of observation. In almost all cases, the experiments are done by scanning one parameter, usually a weak probe that acts on one of the levels that is otherwise not coupled strongly to the other two, and observing the consequences on the probe or some other parameter. Perhaps the most easily visualized example is scanning one of the detunings, say δ_1, in the neighborhood of δ_2 so that they are equal in the center of the scan range. Other possibilities include application of external electric or magnetic fields using the Zeeman or Stark effects to sweep the atomic levels in the presence of fixed optical frequencies. Yet another possible scheme exploits the motion of atoms through inhomogeneous light beams or static fields to use the time-varying light shift to scan the levels.

Appendix

23.A General case for $\delta_1 \neq \delta_2$

In the general case for $\delta_1 \neq \delta_2$ the eigenvalues and eigenvectors become rather complicated [319]. The eigenvalues are the solution of a cubic equation, and in general a cubic equation can have three solutions, of which two are complex in this case. However, in this case one is dealing with a Hermitian matrix, which always has three real solutions. The solutions can be expressed by introducing two frequencies Ω_p and Ω_q:

$$\Omega_p^2 = 4\left(\delta_1^2 - \delta_2\delta_1 + \delta_2^2\right) + 3\left(|\Omega_1|^2 + |\Omega_2|^2\right) \qquad (23.10a)$$

and

$$\Omega_q^3 = 9(2\delta_1 - \delta_2)|\Omega_2|^2 + 9(2\delta_2 - \delta_1)|\Omega_1|^2 + 4(2\delta_1 - \delta_2)(2\delta_2 - \delta_1)(\delta_1 + \delta_2). \qquad (23.10b)$$

In terms of these two frequencies the solutions are $E_i = -\hbar\Delta_i/2$ with Δ_i given by

$$\Delta_1 = -\frac{2}{3}\left(\delta_1 + \delta_2 + \Omega_p \cos\frac{\phi + \pi}{3}\right) \qquad (23.11)$$

$$\Delta_2 = -\frac{2}{3}\left(\delta_1 + \delta_2 + \Omega_p \cos\frac{\phi - \pi}{3}\right)$$

$$\Delta_3 = -\frac{2}{3}\left(\delta_1 + \delta_2 - \Omega_p \cos\frac{\phi}{3}\right),$$

with

$$\tan\phi = \frac{\sqrt{\Omega_p^6 - \Omega_q^6}}{\Omega_q^3}, \qquad (23.12)$$

not to be confused with the phases ϕ_j of the optical frequencies. It can be shown that in case of $\delta_1 = \delta_2$ the values for E_i correspond to the values of (23.3a).

For the ordering of the states $|g_1\rangle$, $|e\rangle$, and $|g_2\rangle$ as in the text, the eigenfunctions are given by

$$|1\rangle = \begin{pmatrix} c\beta\, c\gamma \\ c\beta\, s\gamma \\ -s\beta \end{pmatrix}, \quad |2\rangle = \begin{pmatrix} s\alpha\, s\beta\, c\gamma - c\alpha\, s\gamma \\ c\alpha\, c\gamma + s\alpha\, s\beta\, s\gamma \\ s\alpha\, c\beta \end{pmatrix}, \quad |3\rangle = \begin{pmatrix} c\alpha\, s\beta\, c\gamma + s\alpha\, s\gamma \\ -s\alpha\, c\gamma + c\alpha\, s\beta\, s\gamma \\ c\alpha\, c\beta \end{pmatrix} \qquad (23.13)$$

where sin and cos are abbreviated by s and c, respectively. The mixing angles α, β, and γ are given by

$$\tan\alpha = \sqrt{\frac{(\Delta_3 + 2\delta_2)^2|\Omega_1|^2 + |\Omega_1|^2|\Omega_2|^2 + (\Delta_3^2 + 2\delta_2\Delta_3 - |\Omega_2|^2)^2}{(\Delta_2 + 2\delta_2)^2|\Omega_1|^2 + |\Omega_1|^2|\Omega_2|^2 + (\Delta_2^2 + 2\delta_2\Delta_2 - |\Omega_2|^2)^2}}$$

$$\tan\beta = -\frac{|\Omega_1||\Omega_2|}{\sqrt{(\Delta_1 + 2\delta_2)^2|\Omega_1|^2 + (\Delta_1^2 + 2\delta_2\Delta_1 - |\Omega_2|^2)^2}}$$

$$\tan\gamma = \frac{(\Delta_1 + 2\delta_2)|\Omega_1|}{\Delta_1^2 + 2\delta_2\Delta_1 - |\Omega_2|^2} \qquad (23.14)$$

In case $\delta_1 = \delta_2$ one finds $\Delta_1 = -2\delta_2$, thus $\gamma = 0$, and one retrieves the eigenvectors of Eq. (23.4). Note that the eigenvalues are unchanged by an exchange of the states g_1 and g_2, because the frequencies Ω_p and Ω_q of Eq. (23.10) are unchanged by exchanging 1 and 2. However, the angles α, β, and γ change, since these angles depend on the order of the states in the eigenvectors of Eq. (23.13).

Exercises

23.1 Show by direct substitution that the expressions of Eq. (23.4) are indeed eigenfunctions of the right side of Eq. (23.2).

23.2 Extend the discussion near the end of Sec. 23.5 by adding a small component of $|e\rangle$ into the wavefunction $|g_1\rangle$ so that it reads $|\Psi_1\rangle = |g_1\rangle + \varepsilon |e\rangle$ where $\varepsilon \ll 1$ and the small correction to the normalization is ignored. Then apply the Hamiltonian on the right side of Eq. (23.2) to find an expression for $\mathcal{H}|\Psi_1\rangle$ and equate each of the three components to the appropriate component of $i\hbar\partial|\Psi_1\rangle/\partial t$. Square and add the first and third of these equations, approximate $\dot{\theta} \approx \pi/T$, where T is the pulse overlap time, and show that the condition $\varepsilon \ll 1$ is equivalent to $\Omega'T \gg \pi$, where Ω' is an approximation to the average value of $\sqrt{\Omega_1^2 + \Omega_2^2}$ during T (see Ref. [318]).

23.3 There is a special kind of coherent population trapping that is related to atomic motion and requires the kinetic energy (KE) to be included in the Hamiltonian as $\hat{P}^2/2M$, where $\hat{P}$ is the momentum operator. Then the atomic states need to have momentum among their properties so that a ground-state atom's wavefunction is written $|g, P\rangle$, and the momentum part of the wavefunction is e^{ikz} for $\hat{z}$ motion. Two internally identical ground states having equal but opposite momenta are degenerate.
(a) If an optical standing wave couples both of them to a single excited state they form a three-level Λ system. Show that the condition for this and for one of their $\pm$ combinations to be uncoupled (see Eq. (23.6)) is $k = 2\pi/\lambda$.
(b) For two atomic states with total momentum $P \pm \hbar k$, the KE part of the Hamiltonian is $(P^2 \pm 2P\hbar k + (\hbar k)^2)/2M$. Show that the states $|g, P \pm \hbar k\rangle$ are eigenfunctions of the first and third Hamiltonian terms, but not the middle term. Calculate the resulting off-diagonal term of the Hamiltonian and show that it couples the states $|1\rangle$ and $|3\rangle$ of Eq. (23.6). This coupling term vanishes for $P = 0$ when the momenta are $\pm\hbar k$. Then the state is dark for $P = 0$ and the wavefunction is a de Broglie standing wave. The phenomenon is called velocity-selective CPT (VSCPT, see Refs. [320, 321]).

24

Fundamental physics

Atomic physics has played a vital role in defining the progress of physics since the late nineteenth century. One could say that it began with the observation of discrete spectral lines and Balmer's categorization of their wavelengths in the middle 1880s that culminated in the success of the Bohr model 30 years later. Precise spectral measurements led to the discovery of the Zeeman effect, atomic fine structure, the Lamb shift, the anomalous electron magnetic moment, and a host of other phenomena that made great impacts on the physical description of nature.

Apart from precision measurements, atomic and molecular physics experiments have led to a very long list of important discoveries that have had major impacts on fundamental physics. These include, but are not limited to, the discovery of deuterium and thereby the existence of isotopes, violation of the Bell inequalities, the development of the maser and laser, magnetic resonance imaging, and Bose–Einstein condensation along with its impact on condensed matter physics. Modern commercial aviation, the Internet, and the GPS would not exist without the atomic clocks that synchronize them.

Arguably, the origin and development of quantum mechanics was driven by the inconsistencies between laboratory observations and the classical physics of Newton and Maxwell. Most of what was known about nuclear physics in the 1950s was based on atomic hyperfine studies, and there is currently much information on hadron structure derived from atomic physics [322]. Modern field theories derive from quantum electrodynamics that was founded and established by studies of atomic structure.

More recently there have been attempts to detect an intrinsic electric dipole moment (EDM) of either the electron or the neutron. In the case of the electron, a non-zero EDM would have implications for supersymmetry theories of physics beyond the standard model. Such theories have predictions in the 10^{-26} to 10^{-29} e·cm range (1 e·cm = 1.602 178 $\times 10^{-21}$ C·m), and current limits in the few $\times 10^{-28}$ e·cm range are already casting doubt on some of these. By contrast, the standard model without such additions predicts 10^{-38} e·cm, which is well beyond any current capabilities.

24.1 Precision measurements and QED

Among the dominant contributions of atomic physics is the area of precision measurements. Perhaps the ability to measure frequency to higher and higher precision, eventually leading to the definition of time itself in terms of atomic frequency, has been the major reason why precision measurement has become so important. The extent is so ubiquitous that the American Physical Society has an entire separate topical group devoted to the subject.

24.1.1 The Lamb shift

Detection of the $2S_{1/2}-2P_{1/2}$ splitting in hydrogen that was introduced in Sec. 8.7 derives from the persistence of experimentalists to achieve the highest precision possible in their measurements. The first hint of any spectroscopic discrepancy arose in the study of the measurements of the Balmer-α line of hydrogen and deuterium near $\lambda = 656$ nm in the 1930s. Spectra were taken with the highest-resolution gratings available, the photographic plates were processed with great care for temperature and chemical purity, and exposure traces were made with the best available photogrammetry technology. The consequences shown in Fig. 8.3 were somewhat, not markedly, different from expectations based on the Dirac equation, which was less than a decade old at the time. Nevertheless, the small discrepancy called out for explanation and was actively pursued by careful physicists.

As discussed in Sec. 8.7, the new radio frequency instrumentation that became available after World War II was exploited in an entirely different form of spectroscopy, coupled tightly with new ideas from the emerging notions of quantum mechanics, and resulted in a completely unambiguous demonstration that something more was needed to describe the measurements. What emerged was quantum electrodynamics, and from that other quantum field theories that now dominate theoretical physics.

24.1.2 Anomalous electron magnetic moment

In 1948 there appeared a description of very careful measurements of the ground-state hyperfine structure of hydrogen and deuterium, and comparison of these with calculations similar to those of Sec. 9.3 [323]. The calculated values for the ground state E_{hfs} were less than the experimental values by 0.242% for H and 0.259% for

D, and the discrepancy is 5 times as large as the sum of the published uncertainties of the fundamental constants used to evaluate the formulas. Such high precision and concern with such tiny discrepancies were somewhat unusual for experimental physics at that time, but turned out to be very prophetic.

In a note added in proof of Ref. [323] the authors wrote: "The recent discovery by Kusch and Foley that the magnetic moment of the electron is not μ_0 but $\mu_0(1+0.00118)$ explains the major part of this discrepancy with the theory ... since both moments ... must be multiplied by this factor." (Note that they used the notation μ_0 for the Bohr magneton denoted μ_B in this book and elsewhere.)

Foley and Kusch [324] describe very careful measurements of the Zeeman splittings of various intervals in the ground P states of gallium. Although the interpretation of their measurements leaves some room for alternatives, later measurements confirmed that the intrinsic magnetic moment of the electron is $g_e \mu_B s$, where $g_e \equiv 2(1 + a)$, and a is to be determined experimentally and calculated using the newly established quantum theory of fields (quantum electrodynamics or QED).

In a decade-long series of three major experiments beginning in 1954 led by H. Richard Crane, a was measured to higher and higher precision. Their earlier experiments measured the precession of $\vec{\mu}_e$ during a free electron's flight between two Mott scattering events. The later experiments [325] measured the precession of $\vec{\mu}_e$ of an electron in orbit in a magnetic field (cyclotron motion) arranged so that the orbital and spin precession had opposite directions. This led to a direct measurement of $g - 2$ instead of g itself, resulting in a huge increase in precision because they measured a directly. Their result was $a = 0.001\ 159\ 622 \pm 0.000\ 000\ 027$ [325], and when it is written as $a = \alpha/2\pi - (0.327 \pm 0.005)\alpha^2/\pi^2$, it compares well with the QED value of $a = \alpha/2\pi - 0.328\alpha^2/\pi^2$.

More recent experiments in traps have brought the precision of a for the electron to below one part in 10^{12} [326] in agreement with recent QED calculations. In fact, the QED calculation combined with this measurement of $g - 2$ provides the most accurate value of α (see App. 8.A). Other high-precision measurements have been made of the a-value of the positron (it is the same as the electron to one part in 10^{12}) and $g - 2$ of the muon (it agrees with QED to a few parts in 10^6). The g-factor of various particles remains an active topic of research for many reasons, among them its implications for the standard model.

24.1.3 Atomic clocks

As far back as the 1870s, Maxwell recognized that terrestrial units of time have their deficiencies. He suggested absolute time standards, for example that the period of a particle in gravitational orbit at the surface of a uniform sphere was

24.1 Precision measurements and QED

independent of the size of the sphere, and likewise for the various gravitational vibrational modes of a liquid sphere. In Lord Kelvin's "Treatise on Natural Philosophy" he suggested an atomic time standard based on Na (see Ref. [327]). Lacking the technology for implementation, these suggestions did not receive serious attention until Rabi's 1945 lecture at an American Physical Society meeting, suggesting atomic clocks (see Ref. [328]).

After that, progress was very fast. There were various improvements to the molecular and atomic beam machines (the first version used the inversion transition of NH_3), they were made longer to reduce the Fourier transform limit on the linewidth, the Ramsey method was invented (see App. 2.B), and finally in 1967 the General Conference of Weights and Measures declared the second to be defined in terms of the hfs separation of the Cs ground state.

During this same period, optical pumping was first demonstrated by Kastler (see Sec. 3.5.3). Instead of driving the radio frequency transition between Zeeman levels, one could couple different hfs levels, and thereby detect the microwave clock transitions optically. The application to time standards was immediately exploited because of the simplicity of the apparatus, but collisions between atoms in the vapor cells limited the reproducibility of such atomic clocks. It became clear that atomic clocks would have to operate in collision-free environments, normally that of atomic beams.

Atomic beam machines used a heated oven as a source of atoms, apertures for collimation, magnets for state selection and detection, and microwave regions where the clock transitions were induced. It should be clear from App. 2.B and Fig. 2.8 that the microwave field at zones 1 and 2 must have identical phase in order for the Bloch vector to respond as suggested by Fig. 2.6. Since the beam machines could be nearly 4 m long to extend the time interval between the interaction regions, a single microwave source was used with the power split and fed to two equal length cables connecting to the two regions. However, not all cables are identical, and even with the best efforts, the phase shift problem persisted.

This was recognized as early as 1954 when Zacharias[1] suggested a vertical atomic beam where the interrogation time could be much longer. The idea was that the atoms would fly upward, and then slower atoms would fall back down through the same microwave region (the faster atoms would strike the top), thereby preserving the phase. However, such fountain clocks were not realized until the advent of laser cooling, because collisions caused the loss of all but the fastest atoms in the fountain.

---

[1] Authors often refer to the citation [329], but this is nothing more than the title only of a talk given by Zacharias at a conference. The only known published description of this experiment is in Ramsey's book [18], although there is further information in the proceedings of Zacharias' 61st birthday Festschrift [330].

The dream of Zacharias was fulfilled in 1999 with the construction of a fountain clock that used a laser-cooled sample of Cs atoms. The sample is launched by switching the frequency of one of the beams that produces the vertical optical molasses used for the cooling (Sec. 19.4) by about 10^7 Hz so that the wave is no longer "standing" but moving upward at about 4 m/s. The atoms fly upwards through the microwave region, rise almost 1 m, and then return through the same microwave region almost a full second later. Such clocks routinely achieve stability of 3 parts in 10^{16}.

Detecting the microwave transition between the two hfs levels of the ground state depends on starting with a substantial population difference between them, and up until the 1990s this state selection was done by deflection in an inhomogeneous magnetic field (Stern–Gerlach effect). In 1993 the first United States time and frequency standard using optical pumping for state selection came on line, made possible by advances in laser technology.

The quality of a particular atomic clock is judged by three separate figures of merit: accuracy, reproducibility, and stability. Accuracy is a measure of how well a particular clock agrees with the value determined by the definition of the unit of time. Reproducibility is a measure of how well a particular clock agrees with other similar ones. Stability is a measure of how well a particular clock gives the same result in successive intervals of time, and is characterized by its Allan variance.

The future of atomic time and frequency standards lies in increasing the fundamental frequency of the clock transition, and a huge step is in the making as optical frequency standards based on frequency combs are being developed. Such enormously stable optical frequency sources will have to be referenced to comparably stable atomic transitions, and among the likely candidates are single trapped ions operating on first-order forbidden transitions that are consequently very narrow, as discussed in Sec. 24.2. The present frontier of precision for trapped ions is about 8 parts in 10^{18} [331]. Moreover, there has been recent discussion of using nuclear transitions at still higher frequencies. Typical values for the Allan variance of the current time standard are 10^{-16}, but recent laboratory results are $\sim 10^{-17}$.

24.2 Variation of the constants

Almost every formula in this book, and in all of physics for that matter, contains certain symbols that represent the fundamental physical constants. The most common ones in atomic physics are e, c, $\hbar$, m_e, M_p (the proton mass) and their combination into various secondary constants such as R_∞, μ_B, α, and others. Most of them are dimensioned, meaning that their values are expressed in terms of units chosen for human convenience such as coulombs, meters, joules, seconds, kilograms, etc. As such, they are based on standards that are also established by

24.2 Variation of the constants

humans, unlike mathematical constants as π, e, $\sqrt{2}$, etc. which are believed to be more universal. It is, of course, possible to make combinations of the physical constants that are dimensionless, and therefore thought to be more universal. Perhaps the most well-known of these are α (see App. 8.A) and m_e/M_p.

Spectroscopic measurements of the absorption of light from distant quasars done at the Keck telescope in Hawaii have been interpreted as showing that the Sommerfeld fine-structure constant α was smaller at the time this light passed through interstellar clouds than it is now. If α were different earlier in the universe, we ought to be able to see the evidence in the way distant gas clouds absorb light on its way here from even more distant objects. The discrepancy is tiny, perhaps only one part in 10^5, and is relevant to times of order 10^{10} years ago.

With little else to go on, one might guess among the possibilities that the constants are drifting, that they changed in that era but not currently, that they oscillate, that they vary in different parts of the universe, or that other variations are possible. What is certain is that "variation" of dimensioned numbers can be interpreted in too many ways to be of use, so interest is focused on dimensionless quantities such as α. More recently, theoretical interest in varying constants has been motivated by string theory and other such proposals for going beyond the standard model of particle physics.

The richness of molecular structure provides a plethora of states with splittings that depend on α in different ways, and whose measurements may show the effects of variations. A popular example is the heteronuclear diatomic molecule OH with a Π ground state that has $\Lambda = 1$. Unlike a homonuclear diatomic molecule, there is an electric dipole moment along the internuclear axis that results in an energy difference between the states with opposite projections of Λ on this axis (time reversed orbital motion). The Λ-doubled states have opposite parity so they are connected by an electric dipole transition (see Sec. 16.5).

This Λ doubling is in the microwave region of the spectrum and depends only weakly on α, going as $\alpha^{0.4}$ [332]. On the other hand, the hfs in similar states of OH is of comparable magnitude, and depends on α^4, so variation of their ratio is independent of many experimental artifacts and can be used to detect a variation of α. Measurements can be made both on the emission from interstellar OH masers and in the laboratory, thus spanning time on both the cosmological scale and the human timescale [333].

Another example exploits closely spaced vibrational levels of molecules in states with different binding potentials and hence different dependence on the ratio m_e/M_p. The particular case of Cs_2 studied in Ref. [334] considers the energy spacing between vibrational level $v = 138$ of the ground state and $v = 37$ of an excited state that are spaced by a microwave transition. The sensitivity of the transition frequency to the ratio m_e/M_p is enhanced by the features of the molecular structure and allows laboratory measurements of its time variation on human timescales. In all cases of such precision molecular spectroscopy, a cold dense sample is needed for minimization of Doppler effects and for long interrogation times.

There are also laboratory tests of the variation of the value of fine-structure constant α that depend on the phenomenal precision of present-day atomic spectroscopy. In one case the very complicated structure of atomic dysprosium (Dy) has close-lying excited states whose dependence on α has opposite signs, meaning that their separation would vary in time with α [335]. The sign difference arises because of different relativistic corrections to the energies of these states. Because the two states have opposite parity, there is an electric dipole transition that connects them.

One of these states can be populated by a three-step optical process, and the other one decays by a two-step process, one of which is at the easily detected $\lambda = 564$ nm wavelength of green light. The separation between these pairs is 235 MHz in one isotope and only 3.1 MHz in another. The sensitivity of the separation of each pair is such that a variation of α of about 1 part in 10^{15} would result in a change of 2 Hz in their separations. Measuring such a small frequency change over the period of a year, namely detecting $\Delta\alpha/\alpha \sim 10^{-15}$/year, is indeed a difficult and demanding task, but the authors of Ref. [335, 336] reported small or zero change on this scale over 8 months (one standard deviation).

Perhaps the most accurate determination of the time variation of α was published in 2008 by the NIST ion trapping group [337, 338]. These superb articles describe a comparison of two optical frequency transitions of two separately trapped, single ions in adjacent laboratories. The precision is so good that the height of the different ion traps was considered to result in the gravitational red shift, and was measured on the cm scale.

At first, both frequencies are measured relative to the Cs hfs transition that constitutes the definition of the SI second, maintained at the same NIST laboratory as the ion traps. However, the accuracy of this fountain clock is $\pm 3.3 \times 10^{-16}$ and the stability of the transitions is better than this. The experimenters used a frequency comb based on fs pulsed lasers to compare these two transition frequencies directly to a precision of a few parts in 10^{17} which constrained the variation of α to $\dot{\alpha}/\alpha < (-1.6 \pm 2.3) \times 10^{-17}$ per year. The data were taken over a period of one year.

Quite similar measurements were published in Refs. [339, 340] from two laboratories independently using a single, laser-cooled Yb^+ ion in an ion trap. One used a Cs atomic clock for a standard, and the other used a frequency comb, but both were limited by the definition of the second. In addition to the possible variation of α, which was found to be $\dot{\alpha}/\alpha < 10^{-17}$ per year, their experiments were also sensitive to the dimensionless ratio of the proton to electron mass ($\equiv \mu$). Their results were $\dot{\mu}/\mu < 10^{-16}$/yr. Such measurements appear to rule out a present-day variation of α or μ but not some variation in the past on some cosmological time scale. There

24.3 Exotic atoms and antimatter

The term "exotic atoms" refers to bound states containing one or more charged particles that are not protons, neutrons, or electrons. These particles may be leptonic (e.g. fundamental particles of the standard model such as muons or positrons) or hadronic (e.g. composite particles of the standard model such as pions or antiprotons). Perhaps the most well-known case of more than one replacement particle is anti-hydrogen, consisting of a positron bound to an antiproton. Experiments on such atoms were preceded by the discovery of the exotic particle itself, most often in a high-energy experiment. The naming convention is to add "ium" to the name of the positive particle replacement case, hence "positronium", and also H is often called "protonium". Otherwise add "ic" to the negative particle replaced, hence muonic hydrogen for $p^+\mu^-$.

The first such discovery was the positron, predicted by Dirac from the possible negative sign in the equation that bears his name, and first observed and described by Carl Anderson in 1932. He found cosmic ray tracks in cloud chambers that required them to be made by a particle of positive charge and of mass of at least 100 times less than that of the proton. He called this new particle a positron and was awarded the 1936 Nobel Prize for this discovery. Later there was the discovery of the muon and various mesons that can be bound to form other exotic atoms.

The number of possible discussion topics is enormous, but only five are chosen for this section. Even with this substantially reduced set of exotic atoms, it is hoped that the reader will appreciate the claim that there is very much "high-energy" and fundamental physics to be learned from the spectroscopy of such exotic atoms.

24.3.1 Positronium

Appendix 11.B describes the Zeeman effect and ground-state hfs of the electron–positron bound state called positronium (Ps). Although such a possible bound state had been predicted in the 1930s, it was not until 1951 that Martin Deutsch presented confirming evidence for its observation [341]. It was known that the annihilation rate of these two antiparticles was spin-dependent and that the singlet rate proceeded by emission of two gamma rays with lifetime $\sim 1.3 \times 10^{-10}$ s, whereas conservation laws required that the triplet-state lifetime be $\sim 1.4 \times 10^{-7}$ s by emitting three quanta. He exploited this difference with timing circuits to confirm the production of Ps in various gases.

Since positrons are generally produced from high-energy processes and therefore have energies far above the binding energy for Ps, the formation requires moderating their energy without giving them a chance to annihilate. There is a vast literature on the subject, and at best, the process is quite inefficient. Nevertheless, very clever experiments such as those described in App. 11.B enabled precise measurements of the ground-state hfs splitting. Such a purely leptonic system lends itself to extremely precise theoretical considerations which are not complicated by hadronic interactions or structure. Thus comparisons of such measurements with theory are very important.

Although the excited-state Bohr energies of positronium are similar to those of H, there are several important differences [75]. First, the absence of a massive proton makes the reduced mass $m_e/2$, about half of the value in H, so the Bohr energy levels are given by $E_n = -R_\infty/2n^2$. Second, each particle has a magnetic moment of μ_B so the fine-structure and hyperfine splittings are inseparable. Thus the details of the excited states' energy levels, including the Lamb shift, look quite different from that of H. For example, the equal orbital radii and opposite charge result in cancelation of an orbital magnetic field and thus no "spin–orbit" interaction. Third, the annihilation shifts and broadens many of the levels. Fourth, there are no corrections arising from hadronic structure, e.g. the form factor of the proton. There are also other differences.

After some hotly debated issues in the 1970s, observation of Lyman-α radiation from the 2^3P to the 1^3S transition of Ps was finally confirmed in 1975 [342]. With the advent of tunable lasers there have since been many spectroscopic studies of the excited states, including fine-structure measurements [343] and direct optical excitation of the 2^3S state by a second-order transition (see Sec. 4.3.1) [344]. There have recently been observations of two Ps atoms joining to form molecular positronium [345].

As discussed in App. 11.B, the ground-state splitting of the singlet and triplet forms of Ps has been measured by exploiting the difference in their lifetimes. There is a considerable literature on this topic because there seem to be some discrepancies. The measurements of nearly 30 years ago of 203.389 10(74) GHz still seem to be best [346], and are quite close to the most recent QED values as well.

24.3.2 Muonium

The label muonium belongs to the bound state of an electron and a positive muon (antimuon). Muons are quite hard to produce because they usually derive from the decay of a pi meson, and so experiments are usually done at a high-energy facility such as the Los Alamos Meson Physics Factory (LAMPF).

The mass of the muon is ~206.768 m_e, known to about 1 part in 10^7 or better, and is obtained from the measurement of Zeeman transition frequencies in ground-state muonium. The reduced mass of this two-body system is not very much different from m_e so that the Bohr levels of this purely leptonic atom μ^+e^- are not very different from those of H. There is a good discussion of the history of muonium in [347] even though that paper is quite old.

Also, the muon magnetic moment is quite different from that of the electron or of the proton. It is reduced from the Bohr magneton μ_B by the ratio of the masses, about 207 times smaller. It also has QED corrections that produce an anomalous g-factor similar to that of the electron (see Sec. 24.1.2). It is written the same way, $g_\mu = 2(1 + a_\mu)$ where $a_\mu = \alpha/2\pi + 0.766\, \alpha^2/\pi^2$ + higher orders, similar to that of the electron but having a substantially different α^2-term. Its careful measurement caused quite a stir in the early years of the twenty-first century because of a discrepancy with QED that has since been resolved.

The hfs splitting of the ground state of muonium is ~4.463 302 GHz and is known to about 1 part in 10^8. Its Zeeman effect is essentially the same as the Breit–Rabi diagram of Fig. 11.4 except for this energy scaling factor. The muon lifetime of about 2.2 μs limits the full linewidth of such precision measurements to about 150 kHz. The first excited state has been observed in the same way as in Ps, namely by a second-order transition near twice the Lyman-α wavelength $\lambda = 244$ nm (see Sec. 4.3.1).

24.3.3 Muonic hydrogen

Among the hadronic atoms, perhaps the simplest is muonic hydrogen, having a proton and a muon in place of the electron. It is written as $p^+\mu^-$. Because the muon is ~207 times as massive as the electron, its orbital radius is ~207 times smaller than a_0, and the energy-level structure is therefore much more sensitive to the properties of the proton, which has a hadronic structure, unlike that of the leptons in positronium or muonium. The transition energies among the Bohr levels are in the few keV range (X-ray), and the energy of the 3P → 1S transition has recently been measured [348]. The splitting between the 2P and 2S states is much larger than that between the $2P_{1/2}$ and $2P_{3/2}$, just the opposite of the case of H, as shown in Fig. 24.1 [349].

The experimental measurement of the fine structure [349] is almost the same as that performed by Lamb in the 1940s [70, 72]. A few of the $p^+\mu^-$ atoms formed in highly excited states as the muons collide with H_2 in the target chamber cascade down to the metastable 2S state, and transitions to the 2P states are induced by laser light at $\lambda \sim 6\,\mu m$ (in H the transition wavelength is ~30 cm). The decay of the 2P-states to the 1S ground state emits 2 keV X-rays (Lyman-α radiation at energy ~207 times that of H) whose dependence on the laser frequency is measured to determine the splitting. This measured "Lamb shift" is compared with calculations of the energy interval, which include a contribution from the proton

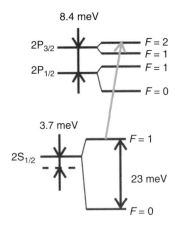

Figure 24.1 Fine structure of the $n = 2$ state of muonic hydrogen ($p^+\mu^-$). The green arrow shows the laser-induced transition of energy 206 meV at $\lambda \sim 6\,\mu$m (not to scale). (Figure from Ref. [349].)

form factor, and the results provide the most accurately known measure of the proton size [322].

24.3.4 Pionic hydrogen

Pionic hydrogen is the bound state of a negative pi meson and a proton, and is written $p^+\pi^-$. Its spectroscopy can provide a considerable amount of new information about the strong interaction since the binding energy is very small compared with the rest masses. Thus measurements are equivalent to scattering experiments at nearly zero relative energy. Very recent X-ray experiments on the hydrogen-like spectroscopy of $p^+\pi^-$ are reported in Ref. [350]. The authors observed several transitions in the keV energy range corresponding to the decay of the nP states with n = 2–4, as well as some of the intermediate cascade transitions. Moreover, some pionic deuterium transitions were also observed, and since both particles are hadrons, the proton–deuteron differences are quite significant.

24.3.5 Anti-hydrogen

The exotic atoms discussed up to here have all been constituted from ordinary particles with substitution of one exotic species. Perhaps the most perfect exotic atom of all is anti-hydrogen, symbol $\bar{\text{H}}$, made purely from antimatter. It has an antiproton ($\bar{p}$) bound to a positron. At first thought it seems natural to assume that its structure and spectrum should match that of ordinary hydrogen, but those assumptions are exactly what need precise testing. It is still not known whether the effect of gravity on antimatter is the same as on matter, and such a test of the equivalence principle is also planned.

There are two independent experimental groups working in the same hall at CERN in Geneva [351, 352]. This lab is chosen because it is by far the best source of low-energy antiprotons, but low energy has to be understood from the perspective of the world's largest accelerator facility. The $\bar{p}$s arrive with kinetic energy of 5.3 MeV and are slowed by a standard accelerator run in reverse. Then they pass through some foils and then an electron cooling region to get their energy low enough for an ion trap.

Elsewhere in the apparatus, copious positrons from a ^{22}Na source are also slowed and steered into another trap region where they are accumulated until one of the $\bar{p}$ bunches arrives from the accelerator. Then they are mixed with the $\bar{p}$s to form $\bar{H}$, and these are confined in a magnetic trap. The experiment is extremely complicated, the timing of the pulses and trap switching is elaborate, and the detection method is indirect. However, there is little doubt that the first measurements will be presented soon.

24.4 Bell inequalities

The story of the famous Bell inequalities began in the first decade of quantum mechanics with the famous paper of Einstein, Podolsky, and Rosen (EPR) [17] and concurrently Schrödinger's expression of his sentiment that entanglement is not merely a feature of quantum mechanics, but is the heart of it [353]. Various views and discussions of the paradox were discussed in the subsequent decades [354], notably by Bohm, but a young theorist at CERN named John Bell was able to sort out some of the complexity and derive some testable relations [355, 356]. These are known as the Bell inequalities, and there is a very large set of them. Many of them are suitable to address the EPR paradox, and one is described below.

Consider a completely classical experiment with only two possible outcomes such as drawing a black or white ball from a sack with one of each, flipping a coin, choosing the male or female of a pair, etc. The Bell inequalities refer to such classical considerations: there is nothing quantum mechanical about them (see Ref. [357]).

Consider three such sacks, each with a black and white ball, and let the experimenter choose one ball from one sack, then choose another ball from another sack, and record the results. The outcome will be two known colors and one unknown, and will be written as the triad (b, w, x) for the case of black chosen from sack 1, white from sack 2, and nothing from sack 3, etc. Multiple trials will result in a distribution of the eight possible outcomes with different number of occurrences n, and one of the Bell inequalities states that

$$n(x, b, w) \leq n(b, x, w) + n(w, b, x). \tag{24.1}$$

The proof of Eq. (24.1) is straightforward using

$$n(b, x, w) = n(b, b, w) + n(b, w, w) \tag{24.2a}$$

since the only possible choices for the untouched second sack are b or w. Similarly,

$$n(w, b, x) = n(w, b, w) + n(w, b, b), \tag{24.2b}$$

and adding Eqs. (24.2a) and (24.2b) gives

$$n(b, x, w) + n(w, b, x) = n(b, b, w) + n(b, w, w) + n(w, b, w) + n(w, b, b) \tag{24.2c}$$
$$= n(x, b, w) + n(b, w, w) + n(w, b, b) \leq n(x, b, w).$$

The sum of the first and third terms of Eq. (24.2c) is the left-hand side of Eq. (24.1), again since the only two possibilities for sack 1 are b or w. Since the other two terms are each ≥ 0, the proof of Eq. (24.1) is done. This Bell inequality is a consequence of simple counting, not even probability, and has nothing to do with quantum mechanics.

In spite of its obvious "truth" as supported by Eqs. (24.2), Eq. (24.1) is violated by quantum mechanical systems. Quantum mechanics posits that the left-hand side of Eq. (24.1) cannot be described properly by summing the first and third terms of Eq. (24.2c) because the contents of the unobserved first sack are not either black or white, but necessarily form a superposition state of the two possibilities that has no classical analog.

A quantum mechanical test of a Bell inequality can be made by making a pair of measurements of some quantity that has only two possible outcomes, in analogy to the black or white balls. Examples are the direction of a spin along some chosen axis or the path taken by a light beam through a polarizing beam splitter, e.g. a calcite crystal. Of course, such a pair of measurements has no meaning unless the measurements are made on a single quantum system with two correlated parts, such as the gamma rays emitted by electron–positron annihilation (EPR example) or the light from a two-step atomic decay cascade through an intermediate level. As long as it is known that two emitted partners from such a process have some property that is assured to be different for each partner, such as spin direction or polarization, then the black or white criterion is satisfied. The analog of the three different sacks derives from arbitrary and possibly different choices of the orientation of the measured spin axis or of the calcite crystal, as long as there are only three alternatives to choose from.

Many such tests have been made on many different experimental parameters, but the most decisive one to date is based on the selection rules of atomic physics, hence the inclusion of this topic in this book. In Ref. [358] the authors made measurements that violated a Bell inequality by more than 45 standard deviations, compelling evidence indeed that quantum mechanics provides an accurate description of physical reality. The philosophical implications of the result have been interpreted that the x of Eqs. (24.1) and (24.2) cannot represent a "hidden variable" that is carried by the components of the system but cannot be observed. Instead the entanglement of the system components generated at their production dictates the possible outcomes of measurements.

The experiment measures the correlations between the orientation of the linear polarization of light emitted in the cascade decay of the $(4p^2)^1S_0$ state of Ca through the $(4s4p)^1P_1$ level at $\lambda = 551$ nm to the ground $(4s^2)^1S_0$ state at $\lambda = 423$ nm. This doubly excited $(4p^2)^1S_0$ state is an example of the kind of state described in Sec. 15.5 whose energy lies below the first ionization threshold so that it decays only by spontaneous emission. The authors were extraordinarily careful with the optics and electronics, compromising their counting rates to reduce the effects of a variety of systematic errors. They chose polarizer angles that maximized the difference between the Bell inequalities and the quantum mechanical predictions of 22.5° and 67.5°.

In a later experiment [359] the authors used modulators that could switch the orientation of the polarizers in a time shorter than the flight time of light from the source to the detectors. Such a "delayed choice" measurement was able to close certain loopholes in the interpretation of previous experiments that also violated Bell's inequalities in favor of quantum mechanics by as much as 13 standard deviations.

24.5 Parity violation and the anapole moment

Soon after the discovery of parity violation in the 1950s there were discussions of atomic examples, but the subject first appeared in the forefront of atomic physics in the 1974 papers of Marie-Anne and Claude Bouchiat. These and many others are cited in an excellent review in Ref. [360]. Mirror symmetry (parity conservation) is a requirement of quantum electrodynamics, and its violation in atoms is allowed only because of the unification of the weak force with electrodynamics, since the weak force does not conserve parity. However, the effects are very small, of the order of a few parts in 10^6 at best.

Parity non-conservation (PNC) can most easily be described by considering that some interactions behave differently when the coordinates are reversed, and hence the total wavefunction of the system is not invariant (except for a sign change) under reflection. A consequence is that neither the energy differences nor the transition rates between a pair of states and their mirror images are equal.

In their 1974 papers the Bouchiats noted two very important features that could enable observation of PNC interactions in atoms. First, the effects under consideration scaled with atomic number as Z^3, suggesting that these would be amplified considerably in heavy atoms. Second, one need not look for a tiny effect that would be characterized by the square of a PNC Hamiltonian matrix element, but instead could choose a system where such a PNC term could interfere with the PNC-mimicking effect, in particular the Stark effect. Thus there would be a linear PNC term multiplied by the odd-parity Stark term that would be detected by reversing the sign of the latter. The essence of the experiment would be to look for some variation that depended on the direction of an applied electric field that could mix states of different parity.

Although the internal electric fields of molecules seem like an attractive option for such experiments, there are other problems related to these. Thus atomic experiments have dominated the field, and Cs has emerged as a better choice than the earlier experiments in Tl, Pb, Bi, and Dy. One problem with the molecules and other atoms arises because interpretation of the measurements requires detailed knowledge of the atomic or molecular wavefunctions which are much more easily obtained for the "one-electron" Cs (see Chap. 10) than for the multi-electron atoms.

The basic idea of the Cs experiment is to drive the dipole-forbidden transition $6^2S_{1/2} \to 7^2S_{1/2}$ at $\lambda = 540$ nm ($\Delta \ell = 0$). It is weakly allowed as an M1 transition but is strongly suppressed by the radial matrix elements so that the M1 component is negligible. The $7^2S_{1/2}$ state is excited in an applied electric field of a few hundred volts/cm that mixes in some P-state character, and the $7^2S_{1/2}$ state spontaneously decays to the $6^2P_{1/2}$ by the emission of infrared light at $\lambda = 1.36$ μm. The dependence of the IR radiation flux on field reversal, polarization reversal, propagation direction reversal, choice of hfs states, and a host of others is carefully measured.

In a different experiment the $6^2S_{1/2} \to 7^2S_{1/2}$ is detected by excitation of a previously emptied hfs sublevel of the $6^2S_{1/2}$ that can be populated by the $7^2S_{1/2}$ decay cascade, providing a better signal, but the essence of the experiment is the same [361]. The result for the PNC transition rate is parametrized in terms of an electric field, and the two values for two different hfs transitions are ~1.6 mV/cm with uncertainty of ~8 μV/cm, about 0.5%.

An important feature of the Ref. [361] result is ~80 μV/cm difference between the results for the two hfs states. Most of this difference arises from the nuclear anapole moment, a completely classical effect that was first described in the 1950s as another consequence of parity violation. It arises from a toroidal current distribution within the nucleus and is a point interaction so it can only be observed with atomic wavefunctions that overlap the nucleus, namely S states. The origin of the anapole moment is somewhat counterintuitive so its properties are rarely studied or discussed, but it can be measured by these very precise PNC experiments in atoms.

The results of Ref. [361] suggest that atomic physics measurements offer a new probe of the hadronic weak interaction, and call for further study. One such step that is being considered is the use of the highly unstable alkali francium (Fr) which has been produced, laser cooled, and trapped [362]. The PNC term in Fr is nearly 20 times larger than that of Cs, and the experiment would involve E1-allowed hfs transitions in the ground state that are enabled by the PNC terms, including the anapole moment. Experiments could be done on a wide range of isotopes, from neutron-rich to neutron-poor. Similar experiments over a range of isotopes are being planned for Dy. Finally, there are proposals for improved detection schemes such as stimulated emission from the states populated by a PNC transition [363, 364]

Parity violation studies in atomic physics have developed into a small industry that overlaps elementary particle physics. The measurements up to now have had a profound impact on the field, and the future ones may be even more important.

24.6 Measuring zero

There is a large class of experiments designed to measure small deviations from conventionally accepted "laws", for example the famous Eötvös experiment to measure the equivalence of gravitational and inertial mass in the late nineteenth century that has been recently pushed to a sensitivity of 10^{-13} [365], and the Coulomb law written as $1/r^{2+q}$ that reached $q \sim 10^{-16}$ [366]. Many other such "fundamental" investigations have been carried out, including tests of the gravitational law at sub-mm distances and measurement of the electrical neutrality of matter. Among the ones that still have current interest and use the techniques of atomic physics are the pursuit of a non-zero permanent electric dipole moment (EDM) of fundamental particles such as the electron, discussed below, and of the neutron [367, 368]. An excellent review and summary of EDM research is in Ref. [369].

In the case of the electron, a non-zero EDM would imply a "shape" other than that of a point particle. Detection of a non-zero result near the current experimental limitations of atomic measurements ($\sim 10^{-28}$ e·cm) already has implications for various supersymmetry theories of physics beyond the standard model and consequences for the fundamental symmetries of CPT. Such small-scale, atomic physics laboratory experiments are often described as having comparable impact to those done at major high-energy facilities. The atomic physics experiments are complicated and often tedious, and this section presents an abbreviated discussion.

To begin, any measurement of an EDM requires an applied electric field that would accelerate an electron out of a measuring region too fast for sensitive experiments. Thus it needs to be done in a neutral system such as an atom or molecule. That being the case, important studies initiated in the 1960s by one of the authors of Ref. [369] (P. Sandars) showed that, in spite of anticipated partial cancelations that could impede the experiments, relativistic effects in heavy atoms and molecules enhanced the effect of an EDM by orders of magnitude (see citations in Ref. [369]). For this reason there have been recent experiments in atoms of Tl, Hg, Cs, Fr, and molecules of PbO, TlF, ThO, YbF, and others.

The basic idea of any of these experiments exploits the possible tiny energy difference between the allowed orientations of an EDM in an applied electric field $\vec{\mathcal{E}}$. Usually there is also an applied magnetic field $\vec{\mathcal{B}}$ so that the frequency corresponding to the energy-level splitting between the two states of a spin-$1/2$ particle is simply $\omega_Z = g\vec{\mu}_B \cdot \vec{\mathcal{B}}/\hbar$. This is much larger than the electric field induced splitting because both electrons and neutrons have an intrinsic permanent magnetic dipole moment, which is what is measured. If there is also an electric dipole moment $\vec{\mu}$, there is an additional energy $-\vec{\mu}\cdot\vec{\mathcal{E}}$, always much smaller than $\hbar\omega_Z$ so that reversing $\vec{\mathcal{E}}$ produces small shifts of $\omega'_Z \equiv \omega_Z \pm \vec{\mu} \cdot \vec{\mathcal{E}}/\hbar$.

The protocol of the typical measurements is to reverse $\vec{\mathcal{E}}$ and seek small synchronous variations of ω'_Z. However, the shifts are minuscule because of the tiny size of the presumed EDM and can be mimicked by many other effects. For example, $\vec{\mathcal{E}} \sim 10$ kV/cm typically produces a frequency shift of 1 mHz. Such systematic effects are checked by reversing other parameters such as $\vec{\mathcal{B}}$, the polarization or sweep direction of the measuring field at frequency ω'_Z, the phase difference between the two pulses of the measuring field (done in the Ramsey configuration), and others. In the experiment described in Ref. [370] there are nine such pairs resulting in 512 independent sets of measurements for each data point.

There are also non-reversing systematic effects deriving from stray fields, leakage currents, thermal drifts, relativistic shifts as atoms move through the static applied fields, etc. All these effects have to be calculated and considered when interpreting the measurements. The current best upper limit of the electron EDM is $\sim 10^{-28}$ e·cm [370, 371].

Part IV
Appendix

Appendix A
Notation and definitions

Table A.1: Notation and definitions

Parameter	Definition	Description	Section	Page
$\vec{\mathcal{A}}(\vec{r},t)$		Vector potential electromagnetic field	3.B.1	52
$\hat{\vec{\mathcal{A}}}_k$		Quantized vector potential of electromagnetic field	5.B	90
$\mathcal{A}^q_{\ell'm',\ell m}$	$\langle \ell'm'\|Y_{1q}\|\ell m\rangle$	Angular part, dipole moment	3.5.2 / 7.6.1	48 / 119
A		Einstein A-coefficient	5.2.1	82
A	$\langle \xi(r)\rangle$	Spin–orbit constant	8.3.4	136
$\hat{a}^\dagger, \hat{a}$		Creation, annihilation operator	5.A	87
α	$e^2/4\pi\varepsilon_0 \hbar c$	Fine-structure constant	8.1	131
a		Hyperfine constant	9.3.1	154
a_0	$\dfrac{4\pi\varepsilon_0 \hbar^2}{me^2} = \dfrac{\hbar}{m\alpha c}$	Bohr radius	1.4.2	9
a_μ	ma_0/μ	Scaled Bohr radius	7.4.2	114
$\vec{\mathcal{B}}(\vec{r},t)$		Magnetic component of electromagnetic field	3.B.1	52
$\hat{\vec{\mathcal{B}}}_k$		Quantized magnetic component of electromagnetic field	5.B	90

continued on next page

continued from previous page

Parameter	Definition	Description	Section	Page
B_{ge}		Einstein B-coefficient	5.2.1	82
B	$\hbar^2/2I$	Rotational constant molecule	16.3.3	281
D_{eq}		Depth molecular potential	16.2	274
δ	$\omega - \omega_{eg}$	Laser detuning	2.5	18
δ_ℓ		Quantum defect	10.1	165
$\vec{\mathcal{E}}(\vec{r},t)$		Electric component of electro-magnetic field	3.B.1	52
$\hat{\vec{\mathcal{E}}}_k$		Quantized electric component of electromagnetic field	5.B	90
E_N	$\alpha^2 R_\infty m/M$	Hyperfine shift	9.3.1	153
$\hat{\varepsilon}$		Polarization of the light field	3.5.2	47
F	$J + I$	Total angular momentum, atom	9.3.1	154
f_{kj}	$2m\omega_{kj}|\hat{\varepsilon}\cdot\vec{r}_{kj}|^2/\hbar$	Oscillator strength	3.4	46
g_e	≈ 2	Landé g-factor electron	8.3.1	134
g_p	$+5.58$	Landé g-factor proton	9.3	152
g_n	-3.82	Landé g-factor neutron	9.3	152
g_j		Landé g-factor, fine structure	11.3.2	184
g_F		Landé g-factor, hyperfine structure	11.3.2	185
γ	$e^2\omega_0^2/6\pi\varepsilon_0 mc^3$	Classical decay rate	1.2.1	4
γ	$\omega^3\mu^2/3\pi\varepsilon_0\hbar c^3$	Spontaneous decay rate	5.4	86
γ'	$\gamma\sqrt{1+s_0}$	Linewidth, power-broadened	6.4	99
γ_p	$\gamma\rho_{ee}$	Scattering rate	6.4	99
$\mathcal{H}'_{jk}$	$\langle\phi_j|\mathcal{H}'|\phi_k\rangle$	Coupling matrix element between states j and k	3.2.1	40
I		Nuclear spin	9.3	152

continued on next page

Notation and definitions 469

continued from previous page

Parameter	Definition	Description	Section	Page
I	$\tfrac{1}{2}\varepsilon_0 c \mathcal{E}_0^2$	Light intensity	3.B.1	53
I_s	$\pi h c / 3 \lambda^3 \tau$	Saturation intensity	6.4	99
I	$\mu_N R_{eq}^2$	Rotational moment molecule	16.3.3	281
j	$\ell + s$	Total angular momentum, electron	8.3.4	136
J	$L + S$	Total angular mom., electrons	15.4	251
L		Orbital angular mom., electrons	15.4	251
ℓ		Orbital angular mom., electron	7.3	109
λ	$2\pi c / \omega$	Optical wavelength	3.B.1	52
μ	$\dfrac{m M_N}{m + M_N}$	Reduced mass electron	7.A	122
μ_N	$\mu_B (m/M)$	Nuclear magneton	9.3	152
μ_N	$\dfrac{M_A M_B}{M_A + M_B}$	Reduced mass molecule	16.2	274
μ_B	$e\hbar/2m$	Bohr magneton	8.1	131
$\vec{\mu}_e$	$-g_e \mu_B \vec{s}/\hbar$	Magnetic moment electron	8.8	134
$\vec{\mu}_I$	$g_I \mu_N \vec{I}/\hbar$	Magnetic moment nucleus	9.3	152
$\vec{\mu}_{eg}$	$e \langle e|\vec{r}|g \rangle$	Dipole moment	7.6.1	118
n		Principal quantum number (hydrogen)	7.4.1	113
n^*		Principal quantum number (alkali-metal)	10.1	165
ω_{eg}	$\omega_e - \omega_g$	Atomic resonance frequency	2.2	17
ω		Laser frequency	1.3.1	6
ω_N	$\sqrt{k_e/\mu_N}$	Rotational frequency molecule	16.3.3	281
ω_L	eB/m	Larmor frequency	1.B	11
Ω	$eE_0 \langle e|r|g \rangle / \hbar$	Rabi frequency	3.9	42
$\vec{\Omega}$	$(-\Omega_r, \Omega_i, -\delta)$	Rabi frequency vector	2.5	23

continued on next page

continued from previous page

Parameter	Definition	Description	Section	Page
Ω'	$\sqrt{\|\Omega\|^2 + \delta^2}$	Generalized Rabi frequency	2.3.3	19
P_{12}		Exchange operator	14.2.1	228
$\sigma_{x,y,z}$		Pauli matrices	2.A	26
π		Linearly polarized light	3.5.2	48
$\vec{R}$	(u, v, w)	Bloch vector	2.5	23
R_{eq}		Equilibrium distance molecule	16.2	274
ρ	$\|\Psi\rangle\langle\Psi\|$	Density operator	6.2.1	94
ρ_{ij}	$\langle\phi_i\|\rho\|\phi_j\rangle$	Density matrix element	6.2.1	94
$\mathcal{R}_{n'\ell',n\ell}$	$\langle n'\ell'\|\|r\|\|n\ell\rangle$	Radial part, dipole moment	7.6.2 / 10.5.1	119 / 172
R_∞	$Ze^2/8\pi\varepsilon_0 a_0 = \frac{1}{2}mc^2\alpha^2$	Rydberg constant	1.4.2 / 7.4.1	8 / 113
s		Spin angular momentum, electron	8.3.1	134
S		Spin angular momentum, electrons	15.4	251
$\vec{S}$	$\vec{\mathcal{E}} \times \vec{\mathcal{H}}$	Poynting vector	3.B.1	52
s	$s_0/(1 + (2\delta/\gamma)^2)$	Saturation parameter, off-resonance	6.4	99
s_0	$2\|\Omega\|^2/\gamma^2$	Saturation parameter, on-resonance	6.4	99
σ_{eg}	$3\lambda^2/2\pi$	Atom–light cross-section	1.3.2 / 6.4	6 / 100
$\sigma^\pm$		Left- (−) or right-handed (+) circularly polarized light	3.5.2	48
τ	$1/\gamma$	Lifetime of excited state	5.4	87
T_D	$\hbar\gamma/2k_B$	Doppler temperature	19.5	356
u, v, w		Bloch vector elements	2.5	23
v_r	$\hbar k/M$	Recoil velocity	19.7.3	364

Appendix B
Units and notation

B.1 Atomic units

In this book most formulas are given in SI units, because this is the most frequently used system of units in physics, and laboratory quantities are the most useful for applications. However, in atomic physics, use is very often made of so-called atomic units for two reasons. First, the formulas become considerably simplified because many constants can be dropped. Second, atomic units set the scale for all properties of atoms (see Sec. 7.5) and provide a natural sense of their values. Atomic units are based on the choices of the fundamental constants e, $\hbar$, m, $4\pi\varepsilon_0$, and αc all being set to unity. All other units as shown in Tab. B.1 are derived using this condition.

B.2 Spectroscopic notation

There is a long history of notation to label the various atomic energy levels that is summarized by the statement

$$n^{2S+1}L_j^\sigma. \tag{B.1}$$

Here n represents the principal quantum number, the total spin is S and may be different from $1/2$ if an atom has more than one electron, and j is the magnitude of the total atomic angular momentum. The capital L refers to the orbital angular momentum, and is replaced by S for $\ell = 0$, P for $\ell = 1$, D for $\ell = 2$, F for $\ell = 3$, and G, H, I for succeeding values of ℓ. Thus the entries in the top row of the Tab. B.1.

For a single electron, the Bohr level $n = 2$ can have $\ell = 0$ or 1, but $\ell = 1$ has two j-values $j = \ell \pm 1/2$. However, the $\ell = 0$ state has only $j = 1/2$ and so $n = 2$ has three states. Similarly, for the Bohr level $n = 3$ there are five states. Now an energy-level diagram for hydrogen as shown in Fig. 8.1 can be constructed.

There is no traditional spectroscopic notation for $\vec{F}$, and so the notation of Eq. (B.1) is simply supplemented by a parenthetical ($F = fvalue$). For example, for hydrogen in the ground state, $\ell = 0$, $s = 1/2$, so $j = 1/2$. The spin of the proton is $I = 1/2$, so the two possible values of F are 0 and 1. Then the states are written as $1^2S_{1/2}$ ($F = 0$) or $1^2S_{1/2}$ ($F = 1$).

472 · Units and notation

Quantity	Unit	Property	Value
Mass	m	Electron mass	9.110×10^{-31} kg
Charge	e	Electron charge	1.602×10^{-19} C
Angular momentum	$\hbar$	Plank's constant	1.055×10^{-34} J s
Length	$a_0 = \dfrac{4\pi\varepsilon_0 \hbar^2}{me^2}$	Bohr radius	5.292×10^{-11} m
Velocity	αc	Velocity in Bohr orbit	2.19×10^{6} m/s
Time	$\dfrac{a_0}{\alpha c}$	Time for one Bohr orbit	2.419×10^{-17} s
Energy	$2R_\infty = \dfrac{e^2}{4\pi\varepsilon_0 a_0} = mc^2\alpha^2$	Twice the ionization potential of hydrogen	4.360×10^{-18} J = 27.2116 eV
Wavenumber	$\dfrac{2\hbar c R_\infty}{}$	Twice the Rydberg constant	2.195×10^{7}/m
Electric field	$\dfrac{2R_\infty}{ea_0} = \dfrac{e}{4\pi\varepsilon_0 a_0^2}$	Electric field nucleus at Bohr radius	5.1×10^{9} V/m
Magnetic field	$\dfrac{R_\infty}{\mu_B} = \dfrac{\hbar}{ea_0^2}$	Corresponding magnetic field	2.35×10^{5} T
Electric dipole moment	ea_0	Electron charge separated by Bohr radius	8.478×10^{-30} C m
Electric quadrupole moment	ea_0^2		4.486×10^{-40} C m^2
Magnetic dipole moment	$\hbar e/m$		1.854×10^{-23} J/T

Table B.1 Atomic units. The fine-structure constant $\alpha = e^2/(4\pi\varepsilon_0 \hbar c)$ is $\approx 1/137$.

Appendix C
Angular momentum in quantum mechanics

The fundamental conservation laws of classical mechanics for energy, momentum, and angular momentum also hold in quantum mechanics. The first two are clearly embodied in the nature of the Schrödinger equation and the operators that can be used on its solutions. Angular momentum is of equally fundamental importance, but its role is not so obvious from the Schrödinger equation. For this reason there are many entire books entitled "Angular Momentum in Quantum Mechanics" or something similar, and the reader is referred to these for a more complete discussion than given here [58, 100, 101].

C.1 Orbital angular momentum

The components of the orbital angular momentum operator are given by

$$\ell_x = i\hbar \left(\sin\phi \frac{\partial}{\partial\theta} + \cot\theta \cos\phi \frac{\partial}{\partial\phi} \right)$$
$$\ell_y = i\hbar \left(-\cos\phi \frac{\partial}{\partial\theta} + \cot\theta \sin\phi \frac{\partial}{\partial\phi} \right)$$
$$\ell_z = -i\hbar \frac{\partial}{\partial\phi}, \tag{C.1}$$

and the operator $\vec{\ell}^{\,2}$ is

$$\vec{\ell}^{\,2} = \frac{-\hbar^2}{\sin\theta} \left[\frac{\partial}{\partial\theta}\left(\sin\theta \frac{\partial}{\partial\theta}\right) + \frac{1}{\sin\theta}\frac{\partial^2}{\partial\phi^2} \right] \tag{C.2}$$

Note that $\vec{\ell}^{\,2}$ plays a special role in atomic physics because the ∇^2 operator in the spherical coordinates appropriate for the Coulomb field of an atomic nucleus is essentially given in Eq. (C.2). The eigenfunctions are the spherical harmonics given in Tab. 7.1, where the eigenvalue is $\ell(\ell+1)\hbar^2$ for $\vec{\ell}^{\,2}$ and $m_\ell\hbar$ for ℓ_z. It is possible to show that the operators of Eq. (C.1) obey simple commutation rules and that they can be combined into "ladder" operators that raise and lower the m_ℓ-values as presented in Eq. (C.6) (see problems in Chap. 7).

C.2 The operators

The properties of the angular momentum operators are closely related to topics in group theory, and there are many discussions of the subject that are purely mathematical. Angular momentum algebra is a very convenient calculus for many problems in atomic physics, and some of its consequences are summarized here.

Because of the very frequent appearance of total angular momenta in many areas of quantum mechanics, the simple relationships among the operators are usually given in terms of $\vec{j}$. Description of these relationships begins with the obvious vector relation

$$\vec{j}^2 = j_x^2 + j_y^2 + j_z^2. \tag{C.3}$$

The commutation relations are

$$\left[j_x, j_y\right] = i\hbar j_z \qquad \left[j_y, j_z\right] = i\hbar j_x \qquad \left[j_z, j_x\right] = i\hbar j_y \tag{C.4}$$

from which

$$\left[\vec{j}^2, j_i\right] = 0 \qquad i = x, y, z. \tag{C.5}$$

The components of an angular momentum operator can be combined to make the ladder operators

$$j_\pm = j_x \pm i j_y. \tag{C.6}$$

It is also useful to consider the eigenfunctions of these operators in a general way, and these are written in terms of their quantum numbers only, such as $|j, m_j\rangle$. Then it can be shown that

$$\vec{j}^2 |jm\rangle = j(j+1)\hbar^2 |jm\rangle \qquad j_z |jm\rangle = m\hbar |jm\rangle \tag{C.7}$$

and also that

$$j_\pm |jm\rangle = \sqrt{j(j+1) - m(m \pm 1)}\,\hbar\, |j\, m \pm 1\rangle. \tag{C.8}$$

C.3 Addition of angular momenta

Because angular momentum is a vector, two or more angular momenta can be added according to the rules of vector algebra as in $\vec{j} = \vec{j}_1 + \vec{j}_2$ (see Fig. C.1). This leads to a set of discrete possible values for j depending on the orientation of $\vec{j}_1$ and $\vec{j}_2$, in the range $|j_1 - j_2| \leq j \leq j_1 + j_2$, where j changes in steps of unity. Thus, there are $2j_<$ possible values for j, where $j_< = min(j_1, j_2)$. The z-component of $\vec{j}$ is just the sum of the individual z-components, or $m = m_1 + m_2$. The coefficients of this summation are written in terms of the quantum numbers that determine them as $\langle j_1 m_1\, j_2 m_2 | jm \rangle$ and are called Clebsch–Gordan coefficients. Thus the state that represents the sum of $j_1 + j_2$ is written as

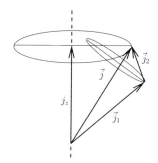

Figure C.1 Vectorial picture of the coupling of $\vec{j}_1$ and $\vec{j}_2$ to make $\vec{j}$.

$$|(j_1 j_2) jm\rangle = \sum_{m_1=-j_1}^{j_1} \sum_{m_2=-j_2}^{j_2} |j_1 m_1 \, j_2 m_2\rangle \langle j_1 m_1 \, j_2 m_2 | jm\rangle. \tag{C.9}$$

A common alternative to the Clebsch–Gordan coefficients are called $3j$-symbols, which are defined in a more symmetric way, and are written as

$$\langle j_1 m_1 \, j_2 m_2 | jm\rangle = (-1)^{j_1-j_2+m} \sqrt{2j+1} \begin{pmatrix} j_1 & j_2 & j \\ m_1 & m_2 & -m \end{pmatrix}. \tag{C.10}$$

Tabulated values of the $3j$-symbols and/or the Clebsch–Gordan coefficients are readily found in many books and on the web. A few of these are given in Fig. C.2, where the format of the tables is as follows:

$j_1 \times j_2$		j		
		m	j	j'
m_1	m_2	v	m'	m'
m'_1	m'_2	v'	v'	v''
m''_1	m''_2	v''	v'''	v''''

Here j_1 and j_2 are coupled to j with $|j1 - j2| \le j \le j1 + j2$. For each value of m_1 and m_2 with $|m_i| \le j_i$ ($i = 1, 2$) the Clebsch–Gordan coefficient v is given, where a square root is understood for each entry. If a minus sign is present, it goes outside the radical.

The Clebsch–Gordan coefficients satisfy certain sum rules called closure, given by

$$\sum_{m_1=-j_1}^{j_1} \sum_{m_2=-j_2}^{j_2} \langle jm|j_1 m_1 \, j_2 m_2\rangle \langle j_1 m_1 \, j_2 m_2 | j'm'\rangle = \delta_{jj'}\delta_{mm'} \tag{C.11}$$

and

$$\sum_{j=|j_1-j_2|}^{j_1+j_2} \sum_{m=-j}^{j} \langle j_1 m_1 \, j_2 m_2 | jm\rangle \langle jm|j_1 m'_1 \, j_2 m'_2\rangle = \delta_{m_1 m'_1}\delta_{m_2 m'_2} \tag{C.12}$$

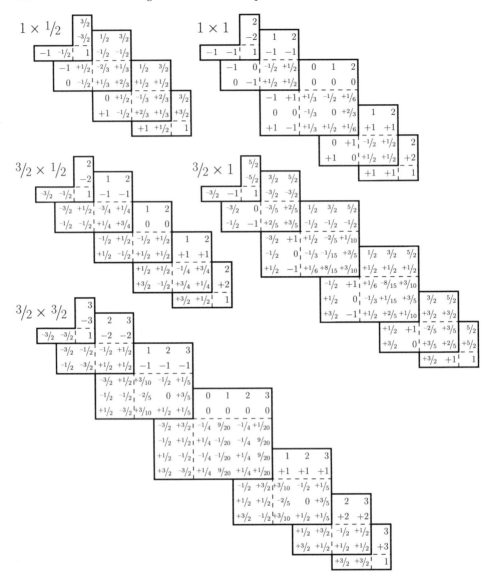

Figure C.2 Clebsch–Gordan coefficients for values of $j_1 = 1/2$, 1, and $3/2$. Equation (C.9) shows that two angular momenta may couple to form a third one, and since they are vectors that precess about one another, their coupling is not straightforward. Of course, their z-component is preserved, so $m_1 + m_2 = m$. Outside the outlined boxes are the two initial j-values, and the top two rows inside the outlined boxes give the allowed values of the total (j, m), one over the other. The side two columns of the outlined boxes give the values (m_1, m_2) of the initial states, one beside the other. The values inside the square part of the outlined boxes are the squares of the Clebsch–Gordan coefficients that are related to the $3j$-symbols by Eq. (C.10). If a minus sign is present, it goes outside the radical. (Figure adapted from [372].)

C.4 Matrix elements

The last topic to be discussed here relates to the matrix elements, including expectation values, of combinations of angular momentum operators. A very common need is for the dot product of two operators, such as $(\vec{j}_1 \cdot \vec{j}_2)$. There are two commonly used ways to find these dot products, and the choice depends on the basis set chosen for evaluation as in the different cases cited here. For the coupled basis, $\vec{j} = \vec{j}_1 + \vec{j}_2$, the good quantum numbers are j, j_1, j_2 and m, and squaring this relation leads to

$$\vec{j}_1 \cdot \vec{j}_2 = (\vec{j}^{\,2} - \vec{j}_1^{\,2} - \vec{j}_2^{\,2})/2. \tag{C.13}$$

On the other hand, in the uncoupled basis, where j_1, j_2, m_1 and m_2 are good quantum numbers, the ladder operators can be combined to give

$$\vec{j}_1 \cdot \vec{j}_2 = j_{1z} j_{2z} + (j_{1+} j_{2-} + j_{1-} j_{2+})/2. \tag{C.14}$$

Each of these two forms has its own special uses as illustrated in many places throughout the text.

In a more general case, where the expectation value becomes a more complicated function of $\vec{j}_1$ and $\vec{j}_2$ than their dot product, one can use the conservation of the total angular momentum $\vec{j}$. Thus, the component of a vector $\vec{V}$ in the direction of $\vec{j}$ is also conserved, and this can be simply written as

$$\langle \alpha' j' m' | \vec{V} | \alpha j m \rangle = C \langle \alpha' j' m' | \vec{j} | \alpha j m \rangle \tag{C.15}$$

where α represents all the other quantum numbers (e.g. radial parts of an atomic wavefunction) and C is a constant. Using the identity operator $\sum_{m'} = |\alpha j m'\rangle \langle \alpha j m'|$ the constant C can be obtained by using

$$\langle \alpha j m | \vec{V} \cdot \vec{j} | \alpha j m \rangle = \sum_{m'} \langle \alpha j m | \vec{V} | \alpha j m' \rangle \cdot \langle \alpha j m' | \vec{j} | \alpha j m \rangle$$

$$= C \sum_{m'} \langle \alpha j m | \vec{j} | \alpha j m' \rangle \cdot \langle \alpha j m' | \vec{j} | \alpha j m \rangle$$

$$= C \langle \alpha j m | \vec{j}^{\,2} | \alpha j m \rangle$$

$$= C j(j+1) \hbar^2 \tag{C.16}$$

where the term on the left-hand side can be evaluated easily depending on the specific form of $\vec{V}$.

It is also particularly useful to know that for tensor operators T_{kq} of all orders, there is a theorem that allows separation of the angular part from the rest of any matrix element of $T_{k,q}$. This Wigner–Eckart theorem gives

$$\langle \alpha' j' m' | T_{kq} | \alpha j m \rangle = \frac{1}{\sqrt{2j'+1}} \langle jm\, kq | j' m' \rangle \langle \alpha' j' \| T_k \| \alpha j \rangle, \tag{C.17}$$

where the double-bar symbol $\langle \alpha' j' \| T_k \| \alpha j \rangle$ is called a reduced matrix element and their evaluation depends on the radial part of the wavefunctions involved.

C.5 Spherical harmonics

The spherical harmonics $Y_{\ell m}(\theta, \phi)$ are the eigenfunctions of the orbital angular momentum $\vec{\ell}^2$ and its projection ℓ_z on the z-axis:

$$\vec{\ell}^2 Y_{\ell m}(\theta, \phi) = \ell(\ell + 1)\hbar^2 Y_{\ell m}(\theta, \phi) \qquad \ell_z Y_{\ell m}(\theta, \phi) = m\hbar Y_{\ell m}(\theta, \phi) \qquad \text{(C.18)}$$

where $\ell < n$ and $-\ell \leq m \leq \ell$. The lowest orders of the spherical harmonics are given in Tab. 7.1. The functions satisfy the orthonormality relation

$$\int_0^{2\pi} d\phi \int_0^{\pi} \sin\theta d\theta \, Y^*_{\ell m}(\theta, \phi) \, Y_{\ell' m'}(\theta, \phi) = \delta_{\ell\ell'} \delta_{mm'}. \qquad \text{(C.19)}$$

It can be shown that a product of two spherical harmonics can be expressed as a series of spherical harmonics:

$$Y_{\ell_1 m_1}(\theta, \phi) Y_{\ell_2 m_2}(\theta, \phi) = \sum_{\ell=|\ell_1-\ell_2|}^{\ell_1+\ell_2} \sum_{m=-\ell}^{\ell} \left(\frac{(2\ell_1 + 1)(2\ell_2 + 1)}{4\pi(2\ell + 1)} \right)^{1/2}$$
$$\times \langle \ell_1 0 \, \ell_2 0 | \ell 0 \rangle \langle \ell_1 m_1 \, \ell_2 m_2 | \ell m \rangle \, Y_{\ell m}(\theta, \phi), \qquad \text{(C.20)}$$

where the elements indicated with triangular brackets are defined in App. C. Special important examples of this relation are

$$\sqrt{\frac{8\pi}{3}} Y_{1,\pm 1}(\theta, \phi) Y_{l,m\pm 1}(\theta, \phi) = \sqrt{\frac{(l \pm m)(l + 1 \pm m)}{(2l + 1)(2l + 3)}} Y_{l+1,m}(\theta, \phi)$$
$$- \sqrt{\frac{(l \mp m)(l + 1 \mp m)}{(2l + 1)(2l + 3)}} Y_{l-1,m}(\theta, \phi) \qquad \text{(C.21)}$$

and

$$\sqrt{\frac{4\pi}{3}} Y_{1,0}(\theta, \phi) Y_{l,m}(\theta, \phi) = \sqrt{\frac{(l + 1)^2 - m^2}{(2l + 1)(2l + 3)}} Y_{l+1,m}(\theta, \phi)$$
$$+ \sqrt{\frac{l^2 - m^2}{(2l + 1)(2l + 3)}} Y_{l-1,m}(\theta, \phi). \qquad \text{(C.22)}$$

The relation of Eq. (C.20) in combination with the normalization of Eq. (C.19) can be used to evaluate the product of three spherical harmonics:

$$\int \sin\theta d\theta d\phi \, Y_{\ell_1 m_1}(\theta, \phi) \, Y_{\ell_2 m_2}(\theta, \phi) \, Y_{\ell_3 m_3}(\theta, \phi) \qquad \text{(C.23)}$$
$$= (-1)^{m_3} \left(\frac{(2\ell_1 + 1)(2\ell_2 + 1)}{4\pi(2\ell_3 + 1)} \right)^{1/2} \langle \ell_1 0 \, \ell_2 0 | \ell_3 0 \rangle \langle \ell_1 m_1 \, \ell_2 m_2 | \ell_3 -m_3 \rangle,$$

where the last two factors are Clebsch–Gordan coefficients, which are defined in App. C.3.

Appendix D
Transition strengths

The following diagrams show the transition strength for alkali-metal atoms for optical transitions from the ground state $n^2S_{1/2}$ to the first excited states $n^2P_{1/2,3/2}$. Since most alkali-metal atoms have a half-integer nuclear spin I, the diagrams are in the order $I = 1/2, 3/2, 5/2$ and $7/2$. The diagrams can be used for:

I	Element				
$1/2$	^{1}H				
1	^{6}Li				
$3/2$	^{7}Li	^{23}Na	^{39}K	^{41}K	^{87}Rb
$5/2$	^{85}Rb				
$7/2$	^{133}Cs				

For each value of I, diagrams are shown for the D_1-line (left page) and D_2-line (right page). The diagram at the top of each page is for π-polarization (pol), whereas the diagram on the bottom is for σ^+-polarization. The transition strength for σ^--polarization can be found by using the diagram for σ^+-polarization and replacing each M by $-M$. The transition strength is normalized for each line so that the strength of the weakest allowed transition becomes an integer. The strength is calculated by using the square of μ_{eg} in Eq. 10.23. In order to compare the strength of the D_1 line with the D_2 line, the numbers for the D_1 line have to be multiplied by a factor 2 ($I = 1/2$), 10 ($I = 1$), 5 ($I = 3/2$), 140 ($I = 5/2$), or 105 ($I = 7/2$). Figures are from Ref. [1].

Transition strengths

D_1-line: $^2S_{1/2}$–$^2P_{1/2}$

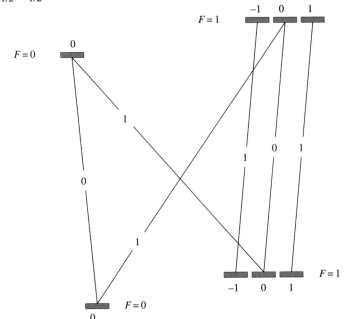

$I = 1/2$ π-pol

D_1-line: $^2S_{1/2}$–$^2P_{1/2}$

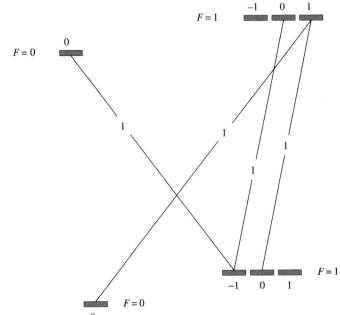

$I = 1/2$ σ^+-pol

Transition strengths

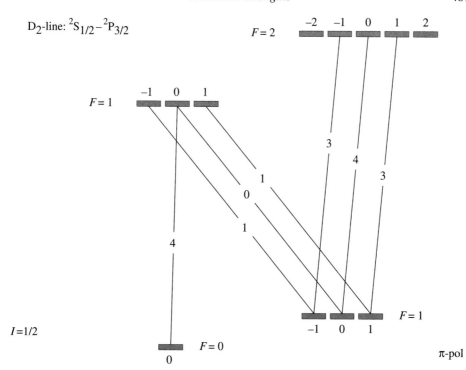

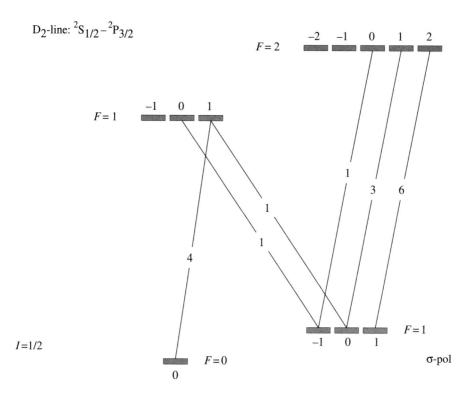

482 Transition strengths

D$_1$-line: $^2S_{1/2} - ^2P_{1/2}$

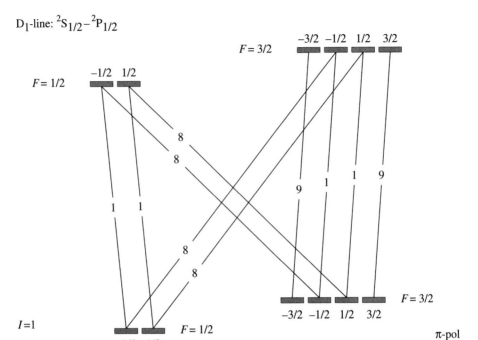

π-pol

D$_1$-line: $^2S_{1/2} - ^2P_{1/2}$

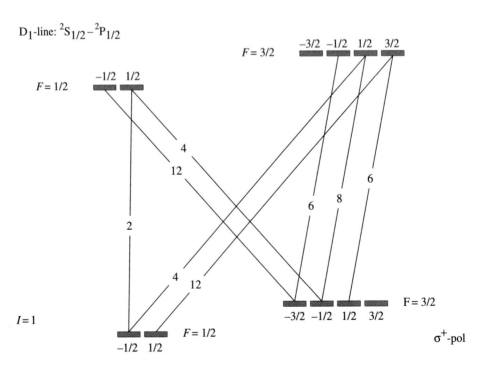

σ$^+$-pol

Transition strengths

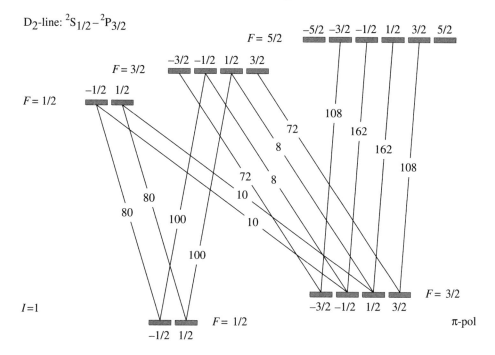

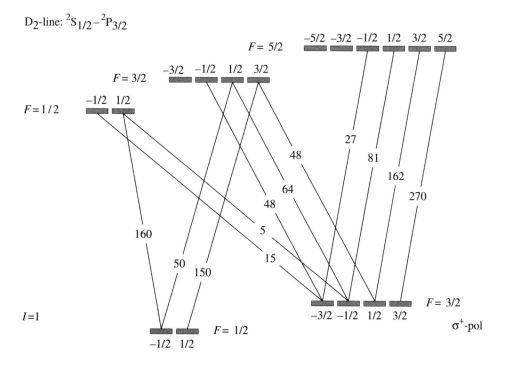

D$_1$-line: ^{2}S$_{1/2}$–^{2}P$_{1/2}$

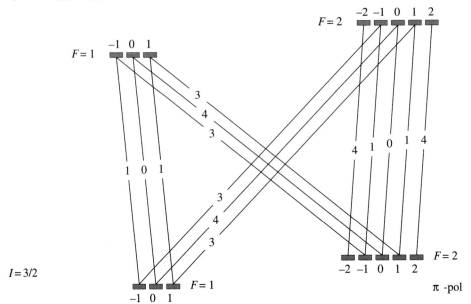

$I = 3/2$, π-pol

D$_1$-line: ^{2}S$_{1/2}$–^{2}P$_{1/2}$

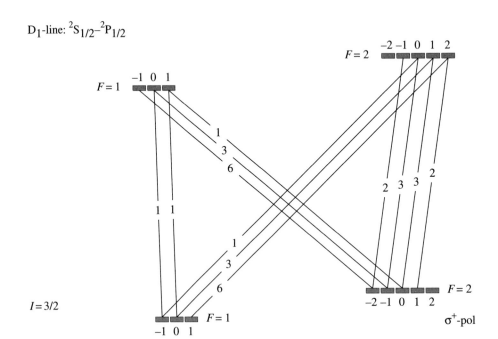

$I = 3/2$, σ$^+$-pol

Transition strengths

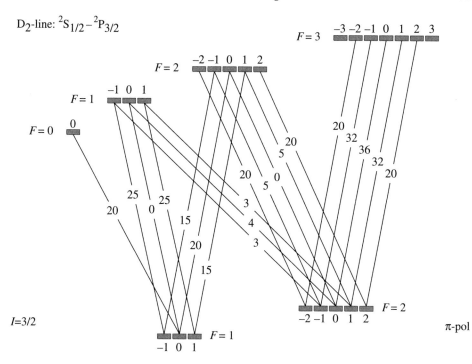

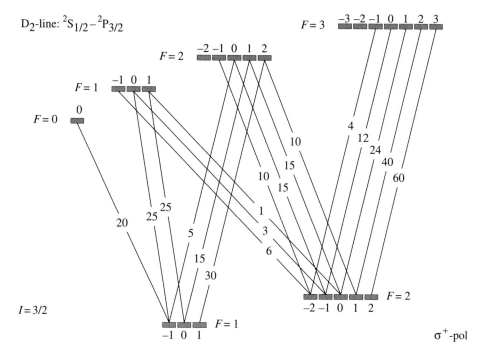

D$_1$-line: $^2S_{1/2} - ^2P_{1/2}$

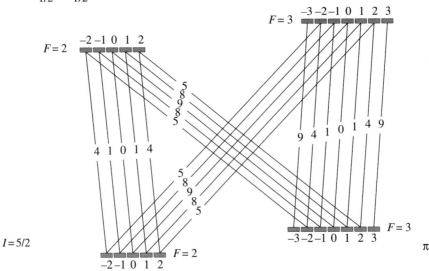

π-pol

D$_1$-line: $^2S_{1/2} - ^2P_{1/2}$

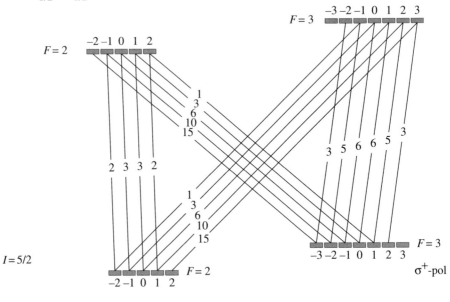

σ$^+$-pol

Transition strengths   487

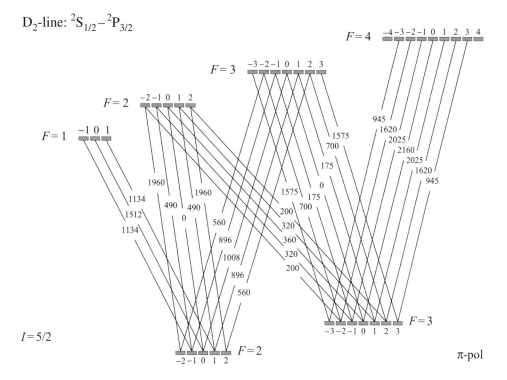

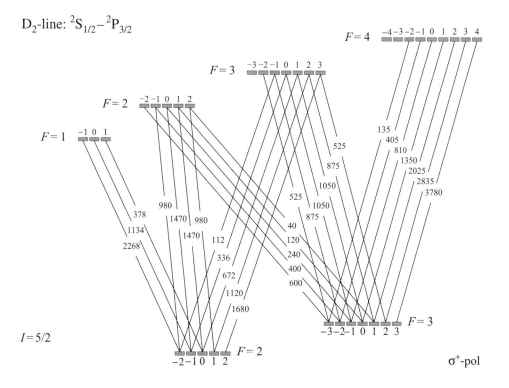

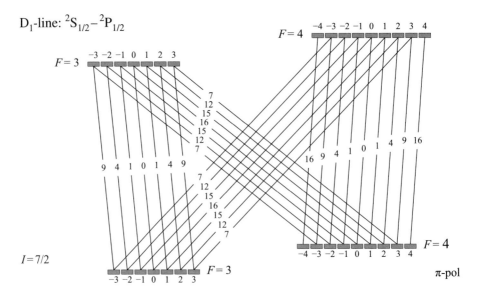

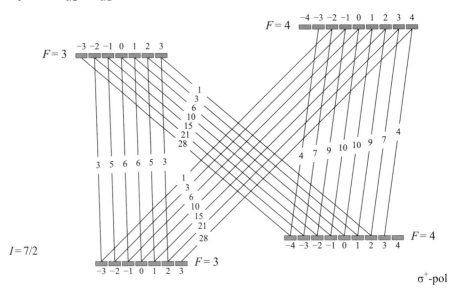

Transition strengths

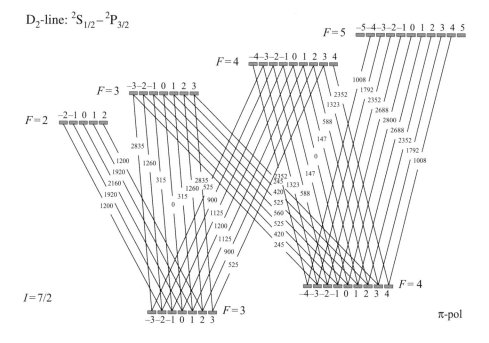

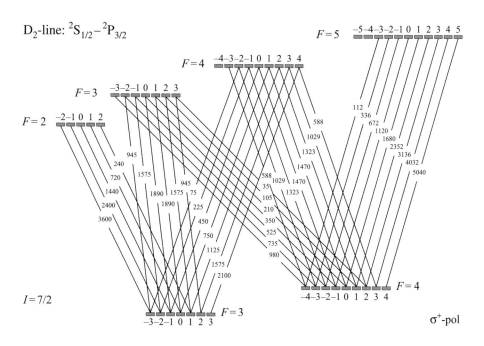

References

[1] H. Metcalf and P. van der Straten. *Laser Cooling and Trapping*. New York: Springer (1999).
[2] N. Bohr. On the Constitution of Atoms and Molecules. *Philos. Mag.* **26**, 1–25 (1913).
[3] N. Bohr. The Structure of the Atom. In *Nobel Lectures, Physics 1922–1941*. Elsevier (1965). Also available from Nobelprize.org.
[4] Louis de Broglie. *Recherches sur la Théorie des Quanta*. Ph.D. thesis, Faculté des Sciences de Paris (1924).
[5] John David Jackson. *Classical Electrodynamics*. New York: John Wiley and Sons (1975).
[6] L. Allen and J. H. Eberly. *Optical Resonance and Two-Level Atoms*. New York: Dover (1975).
[7] C. Cohen-Tannoudji, B. Diu, and F. Laloë. *Quantum Mechanics*. New York: Wiley (1977).
[8] B. R. Mollow. Power Spectrum of Light Scattered by Two-Level Systems. *Phys. Rev.* **188**, 1969–1975 (1969).
[9] E. T. Jaynes and F. W. Cummings. Comparison of Quantum and Semiclassical Radiation Theories with Application to the Beam Maser. *Proc. IEEE* **51**, 89–109 (1963).
[10] R. Feynman, F. Vernon, and R. Hellwarth. Geometrical Representation of the Schrödinger Equation for Solving Maser Problems. *J. Appl. Phys.* **28**, 49 (1957).
[11] Norman F. Ramsey. A Molecular Beam Resonance Method with Separated Oscillating Fields. *Phys. Rev.* **78**, 695–699 (1950).
[12] R. Wynands and S. Weyers. Atomic Fountain Clocks. *Metrologia* **42**, S64 (2005).
[13] Steven Chu. Cold Atoms and Quantum Control. *Nature* **416**, 206–210 (2002).
[14] I. D. Abella, N. A. Kurnit, and S. R. Hartmann. Photon Echoes. *Phys. Rev.* **141**, 391–406 (1966).
[15] E. L. Hahn. Spin Echoes. *Phys. Rev.* **80**, 580–594 (1950).
[16] S. Guérin, R. G. Unanyan, L. P. Yatsenko, and H. R. Jauslin. Adiabatic Creation of Entangled States by a Bichromatic Field Designed from the Topology of the Dressed Eigenenergies. *Phys. Rev. A* **66**, 032311 (2002).
[17] A. Einstein, B. Podolsky, and N. Rosen. Can Quantum-Mechanical Description of Physical Reality Be Considered Complete? *Phys. Rev.* **47**, 777–780 (1935).
[18] N. F. Ramsey. *Molecular Beams*. Oxford: Clarendon Press (1956).
[19] Leonard Mandel and Emil Wolf. *Optical Coherence and Quantum Optics*. Cambridge: Cambridge University Press (1995).

[20] A. Einstein. On the Quantum Theory of Radiation. *Phys. Z.* **18**, 121 (1917).
[21] D. J. Griffiths. *Introduction to Quantum Mechanics*. New Jersey: Prentice Hall (1994).
[22] Peter W. Milonni and Joseph H. Eberly. *Lasers*. New York: Wiley & Sons (1988).
[23] C. Cohen-Tannoudji, J. Dupont-Roc, and G. Grynberg. *Photons and Atoms, Introduction to Quantum Electrodynamics*. New York: Wiley & Sons (1989).
[24] P. A. Franken. Interference Effects in the Resonance Fluorescence of "Crossed" Excited Atomic States. *Phys. Rev.* **121**, 508–512 (1961).
[25] M. E. Rose and R. L. Carovillano. Coherence Effects in Resonance Fluorescence. *Phys. Rev.* **122**, 1185–1194 (1961).
[26] Peter Schenck, Robert C. Hilborn, and Harold Metcalf. Time-Resolved Fluorescence from Ba and Ca Excited by a Pulsed Tunable Dye Laser. *Phys. Rev. Lett.* **31**, 189–192 (1973).
[27] J. C. Baird, John Brandenberger, Ken-Ichiro Gondaira, and Harold Metcalf. Determination of the Atomic-Hydrogen Fine Structure by Level Crossing in the $2P$ States of Hydrogen, a Measurement of the Fine-Structure Constant. *Phys. Rev. A* **5**, 564–587 (1972).
[28] W. C. Martin and W. L. Wiese. Atomic Spectroscopy. In G. W. F. Drake, editor, *Atomic, Molecular, and Optical Physics Handbook*, chapter 10. Woodbury, NY: AIP Press (1996).
[29] A. Gold. *Proc. Fermi Summer School XLII*. New York: Academic Press (1969).
[30] Maria Göppert-Mayer. Über Elementarakte mit Zwei Quantensprüngen. *Ann. Phys.* **9**, 273 (1930).
[31] G. Breit and E. Teller. Metastability of Hydrogen and Helium Levels. *Astrophys. J.* **91**, 215 (1940).
[32] F. Biraben, B. Cagnac, and G. Grynberg. Experimental Evidence of Two-Photon Transition without Doppler Broadening. *Phys. Rev. Lett.* **32**, 643–645 (1974).
[33] M. D. Levenson and N. Bloembergen. Observation of Two-Photon Absorption without Doppler Broadening on the 3S–5S Transition in Sodium Vapor. *Phys. Rev. Lett.* **32**, 645–648 (1974).
[34] P. A. Franken, A. E. Hill, C. W. Peters, and G. Weinreich. Generation of Optical Harmonics. *Phys. Rev. Lett.* **7**, 118–119 (1961).
[35] D. Pritchard, J. Apt, and T. W. Ducas. Fine Structure of Na $4d^2D_2$ Using High-Resolution Two-Photon Spectroscopy. *Phys. Rev. Lett.* **32**, 641–642 (1974).
[36] Y. R. Shen. *The Principles of Non-Linear Optics*. Hoboken NJ: Wiley Classics (2002).
[37] Robert W. Boyd. *Nonlinear Optics*, Third Edition. Amsterdam: Academic Press (Elsevier) (2008).
[38] Marc D. Levenson. *Introduction to Nonlinear Laser Spectroscopy*. Amsterdam: Elsevier (1982).
[39] Shaul Mukamel. *Principles of Nonlinear Optical Spectroscopy*. Oxford: Oxford University Press (1999).
[40] V. Weisskopf and E. Wigner. Berechnung der natürlichen Linienbreite auf Grund der Diracschen Lichttheorie. *Z. Phys.* **63**, 54 (1930).
[41] E. M. Purcell. Spontaneous Emission Probabilities at Radio Frequencies. *Phys. Rev.* **69**, 681 (1946).
[42] Daniel Kleppner. Inhibited Spontaneous Emission. *Phys. Rev. Lett.* **47**, 233–236 (1981).
[43] Serge Haroche and Jean-Michel Raimond. *Exploring the Quantum: Atoms, Cavities, and Photons*. Oxford: Oxford University Press (2006).
[44] R. Loudon. *The Quantum Theory of Light*. Oxford: Clarendon Press (1973).

[45] Gilbert Grynberg, Alain Aspect, and Claude Fabre. *Introduction to Quantum Optics: From the Semi-classical Approach to Quantized Light.* Cambridge: Cambridge University Press (2010).
[46] Paul Berman and Vladimir Malinovsky. *Principles of Laser Spectroscopy and Quantum Optics.* Princeton: Princeton University Press (2011).
[47] Marlan O. Scully and M. Suhail Zubairy. *Quantum Optics.* Cambridge: Cambridge University Press (1997).
[48] Ulf Leonhardt. *Essential Quantum Optics: From Quantum Measurements to Black Holes.* Cambridge: Cambridge University Press (2010).
[49] Pierre Meystre and Murray Sargent. *Elements of Quantum Optics.* Springer (1990).
[50] Timothy H. Boyer. Random Electrodynamics: The Theory of Classical Electrodynamics With Classical Electromagnetic Zero-Point Radiation. *Phys. Rev. D* **11**, 790–808 (1975).
[51] M. D. Crisp and E. T. Jaynes. Radiative Effects in Semiclassical Theory. *Phys. Rev.* **179**, 1253–1261 (1969).
[52] C. R. Stroud and E. T. Jaynes. Long-Term Solutions in Semiclassical Radiation Theory. *Phys. Rev. A* **1**, 106–121 (1970).
[53] D. Bouwmeester, R. J. C. Spreeuw, G. Nienhuis, and J. P. Woerdman. Neoclassical Radiation Theory as an Integral Part of the Monte Carlo Wave-Function Method. *Phys. Rev. A* **49**, 4170–4175 (1994).
[54] D. ter Haar. Theory and Applications of the Density Matrix. *Rep. Prog. Phys.* **24**, 304 (1961).
[55] U. Fano. Description of States in Quantum Mechanics by Density Matrix and Operator Techniques. *Rev. Mod. Phys.* **29**, 74–93 (1957).
[56] C. Cohen-Tannoudji. Atoms in Strong Resonant Fields. In R. Balian *et al.*, editor, *Proceedings of Les Houches XXVII*, page 3 (1977). North Holland: Amsterdam.
[57] S. Stenholm. *Foundations of Laser Spectroscopy.* New York: Wiley & Sons (1984).
[58] A. R. Edmonds. *Angular Momentum in Quantum Mechanics.* Princeton: Princeton University Press (1957).
[59] H. A. Bethe and E. E. Salpeter. *Quantum Mechanics of One- and Two-Electron Atoms.* Berlin: Springer-Verlag (1957). Also published by Plenum, New York, 1977 (paperback).
[60] B. H. Bransden and C. J. Joachain. *Introduction to Quantum Mechanics.* New York: Longman (1989).
[61] E. Merzbacher. *Quantum Mechanics.* New York: Wiley & Sons (1961).
[62] D. J. Griffiths. *Introduction to Quantum Mechanics.* New Jersey: Prentice Hall (1995).
[63] Charles Burkhardt and Jacob Leventhal. *Foundations of Quantum Physics.* New York: Springer (2008).
[64] H. Hellmann. *Einfhürüng in die Quantumchemie.* Leipzig: Deuticke (1937).
[65] R. P. Feynman. Forces in Molecules. *Phys. Rev.* **56**, 340–343 (1939).
[66] A. Einstein and W. J. de Haas. Experimenteller Nachweis der Ampereschen Molekularstörme. *Deutsche Physikalische Gesellschaft, Verhandlungen* **17**, 152–170 (1915).
[67] Abraham Pais. *Subtle Is the Lord: The Science and the Life of Albert Einstein.* Oxford: Oxford University Press (1982).
[68] L. H. Thomas. I. The Kinematics of an Electron with an Axis. *Philos. Mag. Series 7* **3**, 1–22 (1927).

[69] B. H. Bransden and C. J. Joachain. *Physics of Atoms and Molecules*. New York: Wiley & Sons (1983).

[70] Willis E. Lamb and Robert C. Retherford. Fine Structure of the Hydrogen Atom by a Microwave Method. *Phys. Rev.* **72**, 241–243 (1947).

[71] H. A. Bethe. The Electromagnetic Shift of Energy Levels. *Phys. Rev.* **72**, 339–341 (1947).

[72] Sol Triebwasser, Edward S. Dayhoff, and Willis E. Lamb. Fine Structure of the Hydrogen Atom. V. *Phys. Rev.* **89**, 98 (1953).

[73] E. E. Salpeter. The Lamb Shift for Hydrogen and Deuterium. *Phys. Rev.* **89**, 92–97 (1953).

[74] Theodore A. Welton. Some Observable Effects of the Quantum-Mechanical Fluctuations of the Electromagnetic Field. *Phys. Rev.* **74**, 1157–1167 (1948).

[75] H. Bethe and E. Salpeter. *Quantum Mechanics of One and Two Electron Atoms*. New York: Academic Press (1957). Also published by Dover, New York, 2008 (paperback).

[76] Boris M. Smirnov. *Reference Data on Atomic Physics and Atomic Processes*, volume 51 of *Springer Series on Atomic, Optical, and Plasma Physics*. Berlin, Heidelberg: Springer (2008).

[77] A. E. Kramida. A Critical Compilation of Experimental Data on Spectral Lines and Energy Levels of Hydrogen, Deuterium, and Tritium. *At. Data Nucl. Data Tables* **96**, 586–644 (2010).

[78] Robley C. Williams. The Fine Structures of Hα and Dα Under Varying Discharge Conditions. *Phys. Rev.* **54**, 558–567 (1938).

[79] G. W. Series, editor. *The Spectrum of Atomic Hydrogen: Advances*. Singapore: World Scientific (1988).

[80] T. W. Hänsch, M. H. Nayfeh, S. A. Lee, S. M. Curry, and I. S. Shahin. Precision Measurement of the Rydberg Constant by Laser Saturation Spectroscopy of the Balmer α Line in Hydrogen and Deuterium. *Phys. Rev. Lett.* **32**, 1336–1340 (1974).

[81] C. Wieman and T. W. Hänsch. Precision Measurement of the $1S$ Lamb Shift and of the $1S$–$2S$ Isotope Shift of Hydrogen and Deuterium. *Phys. Rev. A* **22**, 192–205 (1980).

[82] Arthur Matveev, Christian G. Parthey, Katharina Predehl, Janis Alnis, Axel Beyer, Ronald Holzwarth, Thomas Udem, Tobias Wilken, Nikolai Kolachevsky, Michel Abgrall, Daniele Rovera, Christophe Salomon, Philippe Laurent, Gesine Grosche, Osama Terra, Thomas Legero, Harald Schnatz, Stefan Weyers, Brett Altschul, and Theodor W. Hänsch. Precision Measurement of the Hydrogen $1S$–$2S$ Frequency via a 920-km Fiber Link. *Phys. Rev. Lett.* **110**, 230801 (2013).

[83] T. van der Veldt, W. Vassen, and W. Hogervorst. Mass-Polarization Effects in the 1s2s 1S and 3S States of Helium. *Phys. Rev. A* **41**, 4099–4101 (1990).

[84] H. De Witte, A. N. Andreyev, N. Barre, M. Bender, T. E. Cocolios, S. Dean, D. Fedorov, V. N. Fedoseyev, L. M. Fraile, S. Franchoo, V. Hellemans, P. H. Heenen, K. Heyde, G. Huber, M. Huyse, H. Jeppessen, U. Koster, P. Kunz, S. R. Lesher, B. A. Marsh, I. Mukha, B. Roussiere, J. Sauvage, M. Seliverstov, I. Stefanescu, E. Tengborn, K. Van de Vel, J. Van de Walle, P. Van Duppen, and Yu. Volkov. Nuclear Charge Radii of Neutron-Deficient Lead Isotopes Beyond $N = 104$ Midshell Investigated by In-Source Laser Spectroscopy. *Phys. Rev. Lett.* **98**, 112502 (2007).

[85] C. Townes and A. Schawlow. *Microwave Spectroscopy*. New York: McGraw Hill, NY (1955).

[86] L.-B. Wang, P. Mueller, K. Bailey, G. W. F. Drake, J. P. Greene, D. Henderson, R. J. Holt, R. V. F. Janssens, C. L. Jiang, Z.-T. Lu, T. P. O'Connor, R. C. Pardo, K. E.

Rehm, J. P. Schiffer, and X. D. Tang. Laser Spectroscopic Determination of the ^6He Nuclear Charge Radius. *Phys. Rev. Lett.* **93**, 142501 (2004).

[87] L. Essen, R. W. Donaldson, M. J. Bangham, and E. G. Hope. Frequency of the Hydrogen Maser. *Nature* **229**, 110–111 (1971).

[88] N. Ramsey. *Nuclear Moments*. Wiley and Sons (1953).

[89] E. Arimondo, M. Inguscio, and P. Violino. Experimental Determinations of the Hyperfine Structure in the Alkali Atoms. *Rev. Mod. Phys.* **49**, 31 (1977).

[90] David J. Griffiths. Hyperfine Splitting in the Ground State of Hydrogen. *Am. J. Phys.* **50**, 698–703 (1982).

[91] H. M. Goldenberg, D. Kleppner, and N. F. Ramsey. Atomic Hydrogen Maser. *Phys. Rev. Lett.* **5**, 361–362 (1960).

[92] Yu. Ralchenko, A. E. Kramida, J. Reader, and NIST ASD Team. NIST Atomic Spectra Database (2014).

[93] R. Freeman and D. Kleppner. Core Polarization and Quantum Defects in High-Angular-Momentum States of Alkali Atoms. *Phys. Rev. A* **14**, 1614 (1976).

[94] C. Bottcher. An Iterative Perturbation Solution of the Inverse Potential Problem. *J. Phys. B* **4**, 1140 (1971).

[95] Jon C. Weisheit. Photoabsorption by Ground-State Alkali-Metal Atoms. *Phys. Rev. A* **5**, 1621–1630 (1972).

[96] C. Bottcher. An Iterative Perturbation Solution of the Inverse Potential Problem. *J. Phys. B* **4**, 1140 (1971).

[97] M. Marinescu, H. R. Sadeghpour, and A. Dalgarno. Dispersion Coefficients for Alkali-Metal Dimers. *Phys. Rev. A* **49**, 982–988 (1994).

[98] Enrico Clementi and Carla Roetti. Roothaan–Hartree–Fock Atomic Wavefunctions: Basis Functions and their Coefficients for Ground and Certain Excited States of Neutral and Ionized Atoms, $Z = 54$. *At. Data Nucl. Data Tables* **14**, 177–478 (1974).

[99] D. R. Bates and A. Damgaard. The Calculation of the Absolute Strenghts of Spectral Lines. *Phil. Trans. R. Soc. A* **242**, 101 (1949).

[100] R. N. Zare. *Angular Momentum*. New York: Wiley (1988).

[101] M. E. Rose. *Elementary Theory of Angular Momentum*. New York: Wiley and Sons (1957).

[102] M. Rotenberg, N. Metropolis, R. Birins, and J. Wooten Jr. *The 3j and 6j Symbols*. Cambridge: Technology Press (1959).

[103] E. Fermi. *Notes on Quantum Mechanics*. Chicago: University of Chicago Press (1961).

[104] R. Parsons and V. Weisskopf. The Spectrum of the Alkali Atoms. *Z. Phys.* **202**, 492 (1967).

[105] Steven Chu, Allen P. Mills, and John L. Hall. Measurement of the Positronium $1S13 - 2S13$ Interval by Doppler-Free Two-Photon Spectroscopy. *Phys. Rev. Lett.* **52**, 1689–1692 (1984).

[106] Willis E. Lamb. Fine Structure of the Hydrogen Atom. III. *Phys. Rev.* **85**, 259–276 (1952).

[107] J. von Neumann and E. Wigner. Über das Verhalten von Eigenwerten bei adiabatischen Prozessen. *Phys. Z.* **30**, 467–470 (1929).

[108] R. S. Knox and A. Gold. *Symmetry in the Solid State*. New York: W. A. Benjamin NY (1964).

[109] Jan R. Rubbmark, Michael M. Kash, Michael G. Littman, and Daniel Kleppner. Dynamical Effects at Avoided Level Crossings: A Study of the Landau–Zener Effect Using Rydberg Atoms. *Phys. Rev. A* **23**, 3107 (1981).

[110] T. Lu, X. Miao, and H. Metcalf. Nonadiabatic Transitions in Finite-Time Adiabatic Rapid Passage. *Phys. Rev. A* **75**, 063422 (2007).

[111] Tianshi Lu. Population Inversion by Chirped Pulses. *Phys. Rev. A* **84**, 033411 (2011).

[112] Matteo Leone, Alessandro Paoletti, and Nadia Robotti. A Simultaneous Discovery: The Case of Johannes Stark and Antonino Lo Surdo. *Phys. Persp.* **6**, 271 (2004).

[113] J. Stark. Observation of the Separation of Spectral Lines by an Electric Field. *Nature* **92**, 401 (1913).

[114] N. Bohr. On the Effect of Electric and Magnetic Fields on Spectral Lines. *Philos. Mag.* **27**, 506 (1914).

[115] Myron L. Zimmerman, Michael G. Littman, Michael M. Kash, and Daniel Kleppner. Stark Structure of the Rydberg states of Alkali-Metal Atoms. *Phys. Rev. A* **20**, 2251–2275 (1979).

[116] E. Luc-Koenig and A. Bachelier. Systematic Theoretical Study of the Stark Spectrum of Atomic Hydrogen I. *J. Phys. B* **13**, 1743 (1980).

[117] E. Luc-Koenig and A. Bachelier. Systematic Theoretical Study of the Stark Spectrum of Atomic Hydrogen II. *J. Phys. B* **13**, 1769 (1980).

[118] David A. Harmin. Theory of the Stark Effect. *Phys. Rev. A* **26**, 2656–2681 (1982).

[119] Harris J. Silverstone. Perturbation Theory of the Stark Effect in Hydrogen to Arbitrarily High Order. *Phys. Rev. A* **18**, 1853–1864 (1978).

[120] Thomas Gallagher. *Rydberg Atoms*. Cambridge: Cambridge University Press (1994).

[121] Serge Haroche. Nobel Lecture: Controlling Photons in a Box and Exploring the Quantum to Classical Boundary. *Rev. Mod. Phys.* **85**, 1083–1102 (2013).

[122] Harold Metcalf. Highly Excited Atoms. *Nature* **284**, 127 (1980).

[123] C. Fabre, S. Haroche, and P. Goy. Millimeter Spectroscopy in Sodium Rydberg States: Quantum-Defect, Fine-Structure, and Polarizability Measurements. *Phys. Rev. A* **18**, 229–237 (1978).

[124] E. A. Hessels, W. G. Sturrus, and S. R. Lundeen. Microwave Spectroscopy of High-L Helium Rydberg States: $10H–10I$, $10I–10K$, and $10K–10L$ Intervals. *Phys. Rev. A* **38**, 4574–4584 (1988).

[125] C. Deutsch. Rydberg States of HeI using the Polarization Model. *Phys. Rev. A* **13**, 2311–2313 (1976).

[126] K. T. Lu and U. Fano. Graphic Analysis of Perturbed Rydberg Series. *Phys. Rev. A* **2**, 81–86 (1970).

[127] R. H. Garstang. Atoms in High Magnetic Fields (White Dwarfs). *Rep. Prog. Phys.* **40**, 105 (1977).

[128] Myron L. Zimmerman, Jarbas C. Castro, and Daniel Kleppner. Diamagnetic Structure of Na Rydberg States. *Phys. Rev. Lett.* **40**, 1083–1086 (1978).

[129] R. J. Fonck, F. L. Roesler, D. H. Tracy, K. T. Lu, F. S. Tomkins, and W. R. S. Garton. Atomic Diamagnetism and Diamagnetically Induced Configuration Mixing in Laser-Excited Barium. *Phys. Rev. Lett.* **39**, 1513–1516 (1977).

[130] K. T. Lu, F. S. Tomkins, H. M. Crosswhite, and H. Crosswhite. Absorption Spectrum of Atomic Lithium in High Magnetic Fields. *Phys. Rev. Lett.* **41**, 1034–1036 (1978).

[131] Michael G. Littman, Myron L. Zimmerman, Theodore W. Ducas, Richard R. Freeman, and Daniel Kleppner. Structure of Sodium Rydberg States in Weak to Strong Electric Fields. *Phys. Rev. Lett.* **36**, 788–791 (1976).

[132] Michael G. Littman, Michael M. Kash, and Daniel Kleppner. Field-Ionization Processes in Excited Atoms. *Phys. Rev. Lett.* **41**, 103–107 (1978).

[133] Joel Gersten and Marvin H. Mittleman. Atomic Transitions in Ultrastrong Laser Fields. *Phys. Rev. A* **10**, 74–80 (1974).

[134] T. F. Gallagher and W. E. Cooke. Fine-structure Intervals and Polarizabilities of Highly Excited D States of K. *Phys. Rev. A* **18**, 2510–2516 (1978).

[135] Keith B. MacAdam and William H. Wing. Fine Structure of Rydberg States. III. New Measurements in D, F, and G States of ^{4}He. *Phys. Rev. A* **15**, 678–688 (1977).

[136] E. A. Hessels, F. J. Deck, P. W. Arcuni, and S. R. Lundeen. Precision Spectroscopy of High-L, $n=10$ Rydberg Helium: An Improved Test of Relativistic, Radiative, and Retardation Effects. *Phys. Rev. Lett.* **65**, 2765–2768 (1990).

[137] P. L. Jacobson, R. D. Labelle, W. G. Sturrus, R. F. Ward, and S. R. Lundeen. Optical Spectroscopy of High-L, $n=10$ Rydberg States of Nitrogen. *Phys. Rev. A* **54**, 314–322 (1996).

[138] R. F. Ward, W. G. Sturrus, and S. R. Lundeen. Microwave Spectroscopy of High-L Rydberg States of Neon. *Phys. Rev. A* **53**, 113–121 (1996).

[139] P. L. Jacobson, D. S. Fisher, C. W. Fehrenbach, W. G. Sturrus, and S. R. Lundeen. Determination of the Dipole Polarizabilities of $H_2^+(0,0)$ and $D_2^+(0,0)$ by Microwave Spectroscopy of High-L Rydberg States of H_2 and D_2. *Phys. Rev. A* **56**, R4361–R4364 (1997).

[140] Julie A. Keele, S. R. Lundeen, and C. W. Fehrenbach. Polarizabilities of Rn-like Th^{4+} from RF Spectroscopy of Th^{3+} Rydberg Levels. *Phys. Rev. A* **83**, 062509 (2011).

[141] G. D. Stevens, C.-H. Iu, T. H. Bergeman, H. J. Metcalf, I. Seipp, K. Taylor, and D. Delande. Precision Measurements of Lithium Atoms in an Electric Field Compared with R-Matrix and Other Stark Theories. *Phys. Rev A* **53**, 1349 (1996).

[142] P. Nussenzveig, F. Bernardot, M. Brune, J. Hare, J. M. Raimond, S. Haroche, and W. Gawlik. Preparation of High-Principal-Quantum-Number "Circular" States of Rubidium. *Phys. Rev. A* **48**, 3991–3994 (1993).

[143] M. Brune, F. Schmidt-Kaler, A. Maali, J. Dreyer, E. Hagley, J. M. Raimond, and S. Haroche. Quantum Rabi Oscillation: A Direct Test of Field Quantization in a Cavity. *Phys. Rev. Lett.* **76**, 1800–1803 (1996).

[144] M. Brune, P. Nussenzveig, F. Schmidt-Kaler, F. Bernardot, A. Maali, J. M. Raimond, and S. Haroche. From Lamb Shift to Light Shifts: Vacuum and Subphoton Cavity Fields Measured by Atomic Phase Sensitive Detection. *Phys. Rev. Lett.* **72**, 3339–3342 (1994).

[145] M. Saffman, T. G. Walker, and K. Mølmer. Quantum Information with Rydberg Atoms. *Rev. Mod. Phys.* **82**, 2313–2363 (2010).

[146] T. Wilk, A. Gaëtan, C. Evellin, J. Wolters, Y. Miroshnychenko, P. Grangier, and A. Browaeys. Entanglement of Two Individual Neutral Atoms Using Rydberg Blockade. *Phys. Rev. Lett.* **104**, 010502 (2010).

[147] L. Isenhower, E. Urban, X. L. Zhang, A. T. Gill, T. Henage, T. A. Johnson, T. G. Walker, and M. Saffman. Demonstration of a Neutral Atom Controlled-NOT Quantum Gate. *Phys. Rev. Lett.* **104**, 010503 (2010).

[148] L. I. Schiff. *Quantum Mechanics*. New York: McGraw-Hill (1968).

[149] P. van der Straten and R. Morgenstern. Shapes of Excited Atoms and Charge Motions Induced by Ion–Atom Collisions. *Comments At. Mol. Phys.* **17**, 243–260 (1986).

[150] D. R. Hartree. The Wave Mechanics of an Atom with a Non-Coulomb Central Field. Part II. Some Results and Discussion. *Math. Proc. Cambr. Phil. Soc.* **24**, 111–132 (1928).

[151] B. H. Bransden. *Atomic Collision Theory*. Reading: Benjamin/Cummings, (1983).

[152] V. Fock. Näherungsmethode zur Lösung des Quantenmechanischen Mehrkörperproblems. *Z. Phys.* **61**, 126–148 (1930).

[153] C. E. Moore and P. W. Merrill. *Partial Grotrian Diagrams of Astrophysical Interest*, volume 23. Gaithersburg, MD: NSRDS N.B.S. (1968).

[154] P. Langevin. Sur la Théorie du Magnétisme. *J. Phys. (Paris)* **4**, 678 (1905).

[155] P. Langevin. Magnetism et Theory des Electrons. *Ann. Chim. Phys.* **5**, 70 (1905).

[156] J. H. Weaver and H. P. R. Frederikse. Optical Properties of Selected Elements. In D. R. Lide, editor, *CRC Handbook of Chemistry and Physics*, pages 12.133–12.156. Boca Raton, FL CRC Press Inc. (2002).

[157] J. P. Desclaux. Relativistic Dirac–Fock Expectation Values for Atoms with $Z = 1$ to $Z = 120$. *At. Data Nucl. Data Tables* **12**, 311–406 (1973).

[158] Herzberg. *Spectra of Diatomic Molecules*. New York: D. van Nostrand Company, Inc. (1950).

[159] N. Bohr. On the Constitution of Atoms and Molecules. Part II Systems Containing Only a Single Nucleus. *Philos. Mag.* **26**, 476–502 (1913).

[160] N. Bohr. On the Constitution of Atoms and Molecules. Part III Systems Containing Several Nuclei. *Philos. Mag.* **26**, 857–875 (1913).

[161] Krzysztof Pachucki. Born–Oppenheimer Potential for H_2. *Phys. Rev. A* **82**, 032509 (2010).

[162] Anatoly A. Svidzinsky, Marlan O. Scully, and Dudley R. Herschbach. Bohr's 1913 Molecular Model Revisited. *Proc. Nat. Acad. Sci. U.S.A.* **102**, 11985–11988 (2005).

[163] Anatoly A. Svidzinsky, Marlan O. Scully, and Dudley R. Herschbach. Simple and Surprisingly Accurate Approach to the Chemical Bond Obtained from Dimensional Scaling. *Phys. Rev. Lett.* **95**, 080401 (2005).

[164] Anatoly Svidzinsky, Marlan Scully, and Dudley Herschbach. Bohr's Molecular Model, a Century Later. *Phys. Today* **67**, 33–39 (2014).

[165] John M. Brown and Alan Carrington. *Rotational Spectroscopy of Diatomic Molecules*. Cambridge University Press (2003).

[166] Philip M. Morse. Diatomic Molecules According to the Wave Mechanics. II. Vibrational Levels. *Phys. Rev.* **34**, 57–64 (1929).

[167] F. London. Zur Theorie und Systematik der Molekularkräfte. *Z. Phys.* **63**, 245–279 (1930).

[168] M. Karplus and R. N. Porter. *Atoms and Molecules*. Benjamin/Cummings (1970).

[169] Edmund S. Rittner. Binding Energy and Dipole Moment of Alkali Halide Molecules. *J. Chem. Phys.* **19**, 1030–1035 (1951).

[170] P. Atkins and R. Friedman. *Molecular Quantum Mechanics*. Oxford: Oxford University Press (2005).

[171] F. Hund. Zur Deutung einiger Erscheinungen in den Molekelspektren. *Z. Phys.* **36**, 657–674 (1926).

[172] J. Franck and E. G. Dymond. Elementary Processes of Photochemical Reactions. *Trans. Faraday Soc.* **21**, 536–542 (1926).

[173] Edward Condon. A Theory of Intensity Distribution in Band Systems. *Phys. Rev.* **28**, 1182–1201 (1926).

[174] W. T. Hill and C. H. Lee. *Light–Matter Interaction*. Weinheim: Wiley (2007).

[175] W. Heitler and F. London. Wechselwirkung neutraler Atome und homöopolare Bindung nach der Quantenmechanik. *Z. Phys.* **44**, 455–472 (1927).

[176] Y. Sugiura. Über die Eigenschaften des Wasserstoffmoleküls im Grundzustande. *Z. Phys. A* **45**, 484–492 (1927).

[177] Clarence Zener and Victor Guillemin. The B-State of the Hydrogen Molecule. *Phys. Rev.* **34**, 999 (1929).

[178] N. Rosen. Calculation of Interaction between Atoms with s-Electrons. *Phys. Rev.* **38**, 255–276 (1931).
[179] N. Rosen. The Normal State of the Hydrogen Molecule. *Phys. Rev.* **38**, 2099–2114 (1931).
[180] Hubert M. James and Albert Sprague Coolidge. The Ground State of the Hydrogen Molecule. *J. Chem. Phys.* **1**, 825 (1933).
[181] J. O. Hirschfelder and J. W. Linnett. The Energy of Interaction between Two Hydrogen Atoms. *J. Chem. Phys.* **18**, 130 (1950).
[182] W. Kolos and L. Wolniewicz. Potential-Energy Curves for the X $^1\Sigma_g^+$, b $^3\Sigma_u^+$, and C $^1\Pi_u$ States of the Hydrogen Molecule. *J. Chem. Phys.* **43**, 2429 (1965).
[183] W. Kolos and L. Wolniewicz. Potential-Energy Curve for the B $^1\Sigma_u^+$ State of the Hydrogen Molecule. *J. Chem. Phys.* **45**, 509 (1966).
[184] W. Kolos and L. Wolniewicz. Improved Theoretical Ground-State Energy of the Hydrogen Molecule. *J. Chem. Phys.* **49**, 404 (1968).
[185] Krzysztof Pachucki. Two-Center Two-Electron Integrals with Exponential Functions. *Phys. Rev. A* **80**, 032520 (2009).
[186] J. C. Slater. The Virial and Molecular Structure. *J. Chem. Phys.* **1**, 687–691 (1933).
[187] M. Marinescu and A. Dalgarno. Dispersion Forces and Long-Range Electronic Transition Dipole Moments of Alkali-Metal Dimer Excited States. *Phys. Rev. A* **52**, 311–328 (1995).
[188] M. Marinescu. Dispersion Coefficients for the nP-nP Asymptote of Homonuclear Alkali-Metal Dimers. *Phys. Rev. A* **56**, 4764–4773 (1997).
[189] M. E. Rose. The Electrostatic Interaction of Two Arbitrary Charge Distributions. *J. Math. Phys.* **37**, 215 (1958).
[190] A. Amelink and P. van der Straten. Photoassociation of Ultra-Cold Atoms. *Phys. Scr.* **68**, C82-89 (2003).
[191] K. M. Jones, P. S. Julienne, P. D. Lett, W. D. Phillips, E. Tiesinga, and C. J. Williams. Measurement of the Atomic Na(3P) Lifetime and of Retardation in the Interaction between Two Atoms Bound in a Molecule. *Europhys. Lett.* **35**, 85 (1996).
[192] E. Tiesinga, C. J. Williams, P. S. Julienne, K. M. Jones, P. D. Lett, and W. D. Phillips. A Spectroscopic Determination of Scattering Lengths for Sodium Atom Collisions. *J. Res. Nat. Inst. Stand. Technol.* **101**, 505 (1996).
[193] W. C. Stwalley, Y. Uang, and G. Pichler. Pure Long-Range Molecules. *Phys. Rev. Lett.* **41**, 1164 (1978).
[194] Robert J. LeRoy and Richard B. Bernstein. Dissociation Energy and Long-Range Potential of Diatomic Molecules from Vibrational Spacings of Higher Levels. *J. Chem. Phys.* **52**, 3869–3879 (1970).
[195] P. Molenaar. *Photoassociative Reactions of Laser-Cooled Sodium*. Ph.D. thesis, Utrecht University (1995).
[196] Thorsten Köhler, Krzysztof Góral, and Paul S. Julienne. Production of Cold Molecules via Magnetically Tunable Feshbach Resonances. *Rev. Mod. Phys.* **78**, 1311–1361 (2006).
[197] Paul S. Julienne. Cold Binary Atomic Collisions in a Light Field. *J. Res. Nat. Inst. Stand. Technol.* **101**, 487 (1996).
[198] Cheng Chin, Rudolf Grimm, Paul Julienne, and Eite Tiesinga. Feshbach Resonances in Ultracold Gases. *Rev. Mod. Phys.* **82**, 1225–1286 (2010).
[199] P. Ehrenfest. Bemerkung über die angenäherte Gültigkeit der klassischen Mechanik innerhalb der Quantummechanik. *Z. Phys.* **45**, 455 (1927).
[200] W. Phillips and H. Metcalf. Laser Deceleration of an Atomic Beam. *Phys. Rev. Lett.* **48**, 596 (1982).

[201] J. Prodan, W. Phillips, and H. Metcalf. Laser Production of a Very Slow Monoenergetic Atomic Beam. *Phys. Rev. Lett.* **49**, 1149 (1982).

[202] J. Gordon and A. Ashkin. Motion of Atoms in a Radiation Trap. *Phys. Rev. A* **21**, 1606 (1980).

[203] J. Dalibard and W. Phillips. Stability and Damping of Radiation Pressure Traps. *Bull. Am. Phys. Soc.* **30**, 748 (1985).

[204] S. Chu, L. Hollberg, J. Bjorkholm, A. Cable, and A. Ashkin. Three-Dimensional Viscous Confinement and Cooling of Atoms by Resonance Radiation Pressure. *Phys. Rev. Lett.* **55**, 48 (1985).

[205] S. Chu and C. Wieman (Eds.). Laser Cooling and Trapping of Atoms. *J. Opt. Soc. Am. B* **6**, 1961–2288 (1989).

[206] P. Gould, P. Lett, and W. D. Phillips. New Measurement with Optical Molasses. In W. Persson and S. Svanberg, editors, *Laser Spectroscopy VIII*, page 64, Berlin: Springer (1987).

[207] T. Hodapp, C. Gerz, C. Westbrook, C. Furtlehner, and W. Phillips. Diffusion in Optical Molasses. *Bull. Am. Phys. Soc.* **37**, 1139 (1992).

[208] C. Cohen-Tannoudji and W. D. Phillips. New Mechanisms for Laser Cooling. *Phys. Today* **43**, October, 33–40 (1990).

[209] P. Lett, R. Watts, C. Westbrook, W. Phillips, P. Gould, and H. Metcalf. Observation of Atoms Laser Cooled below the Doppler Limit. *Phys. Rev. Lett.* **61**, 169 (1988).

[210] P. D. Lett, R. N. Watts, C. E. Tanner, S. L. Rolston, W.D. Phillips, and C.I. Westbrook Optical Molasses. *J. Opt. Soc. Am. B* **6**, 2084–2107 (1989).

[211] J. Dalibard and C. Cohen-Tannoudji. Laser Cooling Below the Doppler Limit by Polarization Gradients — Simple Theoretical-Models. *J. Opt. Soc. Am. B* **6**, 2023–2045 (1989).

[212] P. J. Ungar, D. S. Weiss, S. Chu, and E. Riis. Optical Molasses and Multilevel Atoms — Theory. *J. Opt. Soc. Am. B* **6**, 2058–2071 (1989).

[213] B. Sheehy, S. Q. Shang, P. van der Straten, S. Hatamian, and H. Metcalf. Magnetic-Field-Induced Laser Cooling Below the Doppler Limit. *Phys. Rev. Lett.* **64**, 858–861 (1990).

[214] A. Ashkin. Acceleration and Trapping of Particles by Radiation Pressure. *Phys. Rev. Lett.* **24**, 156 (1970).

[215] S. Chu, J. Bjorkholm, A. Ashkin, and A. Cable. Experimental Observation of Optically Trapped Atoms. *Phys. Rev. Lett.* **57**, 314 (1986).

[216] A. Ashkin. Application of Laser Radiation Pressure. *Science* **210**, 1081–1088 (1980).

[217] A. Ashkin and J. M. Dziedzic. Observation of Radiation-Pressure Trapping of Particles by Alternating Light Beams. *Phys. Rev. Lett.* **54**, 1245 (1985).

[218] A. Ashkin and J. M. Dziedzic. Optical Trapping and Manipulation of Viruses and Bacteria. *Science* **235**, 1517 (1987).

[219] C. S. Adams, H. J. Lee, N. Davidson, M. Kasevich, and S. Chu. Evaporative Cooling in a Crossed Dipole Trap. *Phys. Rev. Lett.* **74**, 3577–3580 (1995).

[220] T. Takekoshi and R. J. Knize. CO_2-Laser Trap for Cesium Atoms. *Opt. Lett.* **21**, 77–79 (1996).

[221] H. Metcalf and W. Phillips. Electromagnetic Trapping of Neutral Atoms. *Metrologia* **22**, 271 (1986).

[222] N. Davidson, H. J. Lee, C. S. Adams, M. Kasevich, and S. Chu. Long Atomic Coherence Times in an Optical Dipole Trap. *Phys. Rev. Lett.* **74**, 1311–1314 (1995).

[223] A. Siegman. *Lasers*. Mill Valley: University Sciences (1986).

[224] N. Simpson, K. Dholakia, L. Allen, and M. Padgett. The Mechanical Equivalence of Spin and Orbital Angular Momentum of Light: an Optical Spanner. *Opt. Lett.* **22**, 52 (1997).

[225] D. McGloin, N. Simpson, and M. Padgett. Transfer of Orbital Angular Momentum from a Stressed Fiber Optic Waveguide to a Light Beam. *Appl. Opt.* **37**, 469 (1998).

[226] M. Beijersbergen. *Phase Singularities in Optical Beams*. Ph.D. thesis, University Leiden (1996).

[227] D. Wineland, W. Itano, J. Bergquist, and J. Bollinger. *Trapped Ions and Laser Cooling*. Technical Report 1086, NIST (1985).

[228] A. Migdall, J. Prodan, W. Phillips, T. Bergeman, and H. Metcalf. First Observation of Magnetically Trapped Neutral Atoms. *Phys. Rev. Lett.* **54**, 2596 (1985).

[229] T. Bergeman, G. Erez, and H. Metcalf. Magnetostatic Trapping Fields for Neutral Atoms. *Phys. Rev. A* **35**, 1535 (1987).

[230] T. H. Bergeman, N. L. Balazs, H. Metcalf, P. Mcnicholl, and J. Kycia. Quantized Motion of Atoms in a Quadrupole Magnetostatic Trap. *J. Opt. Soc. Am. B* **6**, 2249–2256 (1989).

[231] E. Raab, M. Prentiss, A. Cable, S. Chu, and D. Pritchard. Trapping of Neutral-Sodium Atoms with Radiation Pressure. *Phys. Rev. Lett.* **59**, 2631 (1987).

[232] H. Metcalf. Magneto-Optical Trapping and Its Application to Helium Metastables. *J. Opt. Soc. Am. B* **6**, 2206–2210 (1989).

[233] V. S. Lethokov. Narrowing of the Doppler Width in a Standing Light Wave. *JETP Lett.* **7**, 272 (1968).

[234] C. Salomon, J. Dalibard, A. Aspect, H. Metcalf, and C. Cohen-Tannoudji. Channeling Atoms in a Laser Standing Wave. *Phys. Rev. Lett.* **59**, 1659 (1987).

[235] Y. Castin and J. Dalibard. Quantization of Atomic Motion in Optical Molasses. *Europhys. Lett.* **14**, 761–766 (1991).

[236] P. Verkerk, B. Lounis, C. Salomon, C. Cohen-Tannoudji, J. Y. Courtois, and G. Grynberg. Dynamics and Spatial Order of Cold Cesium Atoms in a Periodic Optical-Potential. *Phys. Rev. Lett.* **68**, 3861–3864 (1992).

[237] P. S. Jessen, C. Gerz, P. D. Lett, W. D. Phillips, S. L. Rolston, R. J. C. Spreeuw, and C. I. Westbrook. Observation of Quantized Motion of Rb Atoms in an Optical-Field. *Phys. Rev. Lett.* **69**, 49–52 (1992).

[238] B. Lounis, P. Verkerk, J. Y. Courtois, C. Salomon, and G. Grynberg. Quantized Atomic Motion in 1D Cesium Molasses with Magnetic-Field. *Europhys. Lett.* **21**, 13–17 (1993).

[239] R. Gupta, S. Padua, C. Xie, H. Batelaan, T. Bergeman, and H. Metcalf. Motional Quantization of Laser Cooled Atoms. *Bull. Am. Phys. Soc.* **37**, 1139 (1992).

[240] R. Gupta, S. Padua, T. Bergeman, and H. Metcalf. Search for Motional Quantization of Laser-Cooled Atoms. In E. Arimondo, W. Phillips, and F. Strumia, editors, *Laser Manipulation of Atoms and Ions, Proceedings of Fermi School CXVIII, Varenna*. Amsterdam: North Holland (1993).

[241] B. P. Anderson and M. A. Kasevich. Macroscopic Quantum Interference from Atomic Tunnel Arrays. *Nature* **282**, 1686–1689 (1998).

[242] M. Greiner, O. Mandel, T. Esslinger, T. W. Hänsch, and I. Bloch. Quantum Phase Transition from a Superfluid to a Mott Insulator in a Gas of Ultracold Atoms. *Nature* **415**, 39–44 (2002).

[243] G. Birkl, M. Gatzke, I. H. Deutsch, S. L. Rolston, and W. D. Phillips. Bragg Scattering from Atoms in Optical Lattices. *Phys. Rev. Lett.* **75**, 2823–2826 (1995).

[244] C. I. Westbrook, R. N. Watts, C. E. Tanner, S. L. Rolston, W. D. Phillips, P. D. Lett, and P. L. Gould. Localization of Atoms in a 3-Dimensional Standing Wave. *Phys. Rev. Lett.* **65**, 33–36 (1990).

[245] M. Bendahan, E. Peik, J. Reichel, Y. Castin, and C. Salomon. Bloch Oscillations of Atoms in an Optical-Potential. *Phys. Rev. Lett.* **76**, 4508–4511 (1996).
[246] Charles E. Hecht. The Possible Superfluid Behaviour of Hydrogen Atom Gases and Liquids. *Physica* **25**, 1159–1161 (1959).
[247] Willian C. Stwalley and L. H. Nosanow. Possible "New" Quantum Systems. *Phys. Rev. Lett.* **36**, 910–913 (1976).
[248] Isaac F. Silvera and J. T. M. Walraven. Stabilization of Atomic Hydrogen at Low Temperature. *Phys. Rev. Lett.* **44**, 164–168 (1980).
[249] Harald F. Hess, Greg P. Kochanski, John M. Doyle, Naoto Masuhara, Daniel Kleppner, and Thomas J. Greytak. Magnetic Trapping of Spin-Polarized Atomic Hydrogen. *Phys. Rev. Lett.* **59**, 672–675 (1987).
[250] Naoto Masuhara, John M. Doyle, Jon C. Sandberg, Daniel Kleppner, Thomas J. Greytak, Harald F. Hess, and Greg P. Kochanski. Evaporative Cooling of Spin-Polarized Atomic Hydrogen. *Phys. Rev. Lett.* **61**, 935–938 (1988).
[251] Yu Kagan, I. A. Vartanyantz, and G. V. Shlyapnikov. Kinetics of Decay of Metastable Gas Phase of Polarized Atomic Hydrogen at Low Temperatures. *Sov. Phys. JETP* **54**, 590 (1980).
[252] D. Fried, T. Killian, L. Willmann, D. Landhuis, S. Moss, D. Kleppner, and T. Greytak. Bose Einstein Condensation of Atomic Hydrogen. *Phys. Rev. Lett.* **81**, 3811 (1998).
[253] Lincoln D. Carr, David DeMille, Roman V. Krems, and Jun Ye. Cold and Ultracold Molecules: Science, Technology and Applications. *New J. Phys.* **11**, 055049 (2009).
[254] Dallin Durfee and Wolfgang Ketterle. Experimental Studies of Bose–Einstein Condensation. *Opt. Express* **2**, 299–313 (1998).
[255] M. R. Andrews, C. G. Townsend, H.-J. Miesner, D. S. Durfee, D. M. Kurn, and W. Ketterle. Observation of Interference between Two Bose Condensates. *Science* **275**, 637–641 (1997).
[256] Yvan Castin and Jean Dalibard. Relative Phase of Two Bose–Einstein Condensates. *Phys. Rev. A* **55**, 4330–4337 (1997).
[257] L. D. Landau. The Theory of Superfluidity of Helium II. *J.Phys.* **V**, 71–90 (1941).
[258] Laszlo Tisza. The Theory of Liquid Helium. *Phys. Rev.* **72**, 838–854 (1947).
[259] D. M. Stamper-Kurn, H.-J. Miesner, S. Inouye, M. R. Andrews, and W. Ketterle. Collisionless and Hydrodynamic Excitations of a Bose–Einstein Condensate. *Phys. Rev. Lett.* **81**, 500–503 (1998).
[260] R. Meppelink S. B. Koller, and P. van Straten: Sound Propagation in a Bose–Einstein Condensate at Finite Temperatures. *Phys. Rev. A* **80**, 043605 (2009).
[261] Russell J. Donnelly. The Two-Fluid Theory and Second Sound in Liquid Helium. *Phys. Today* **62**, 34–39 (2009).
[262] K. I. Petsas, A. B. Coates, and G. Grynberg. Crystallography of Optical Lattices. *Phys. Rev. A* **50**, 5173–5189 (1994).
[263] P. Nozières and D. Pines. *The Theory of Quantum Liquids*, volume II. Redwood City: Addison-Wesley (1989).
[264] E. Taylor, H. Hu, X.-J. Liu, L. P. Pitaevskii, A. Griffin, and S. Stringari. First and second sound in a strongly interacting Fermi gas. *Phys. Rev. A* **80**, 053601 (2009).
[265] D. Jaksch, C. Bruder, J. I. Cirac, C. W. Gardiner, and P. Zoller. Cold Bosonic Atoms in Optical Lattices. *Phys. Rev. Lett.* **81**, 3108–3111 (1998).
[266] Matthew P. A. Fisher, Peter B. Weichman, G. Grinstein, and Daniel S. Fisher. Boson Localization and the Superfluid–Insulator Transition. *Phys. Rev. B* **40**, 546–570 (1989).

[267] Henk T. C. Stoof. Bose–Einstein Condensation: Breaking up a Superfluid. *Nature* **415**, 25–26 (2002).
[268] D. van Oosten, P. van der Straten, and H. T. C. Stoof. Quantum Phases in an Optical Lattice. *Phys. Rev. A* **63**, 053601 (2001).
[269] Markus Greiner, Olaf Mandel, Tilman Esslinger, Theodor W. Hansch, and Immanuel Bloch. Quantum Phase Transition from a Superfluid to a Mott Insulator in a Gas of Ultracold Atoms. *Nature* **415**, 39–44 (2002).
[270] L. E. Reichl. *A Modern Course in Statistical Physics*. Edward Arnold (1980).
[271] William C. Stwalley, Roman V. Krems, and Bretislav Friedrich, editors. *Cold Molecules, Theory, Experiment, Applications*. Boca Raton: CRC Press (2009).
[272] E. S. Shuman, J. F. Barry, and D. DeMille. Laser Cooling of a Diatomic Molecule. *Nature* **467**, 820–823 (2010).
[273] M. D. Di Rosa. Laser-cooling Molecules. *Eur. Phys. J. D* **31**, 395–402 (2004).
[274] M. A. Baranov, M. Dalmonte, G. Pupillo, and P. Zoller. Condensed Matter Theory of Dipolar Quantum Gases. *Chem. Rev.* **112**, 5012–5061 (2012).
[275] T. Breeden and H. Metcalf. Stark Acceleration of Rydberg Atoms in Inhomogeneous Electric Fields. *Phys. Rev. Lett.* **47**, 1726 (1981).
[276] Hendrick L. Bethlem, Giel Berden, and Gerard Meijer. Decelerating Neutral Dipolar Molecules. *Phys. Rev. Lett.* **83**, 1558–1561 (1999).
[277] Hendrick L. Bethlem, Giel Berden, Floris M. H. Crompvoets, Rienk T. Jongma, Andre J. A. van Roij, and Gerard Meijer. Electrostatic Trapping of Ammonia Molecules. *Nature* **406**, 491–494 (2000).
[278] W. Wing. On Neutral Particle Trapping in Quasistatic Electromagnetic Fields. *Prog. Quant. Elect.* **8**, 181 (1984).
[279] Wesley C. Campbell and John M. Doyle. Cooling, Trap Loading, and Beam Production Using a Cryogenic Helium Buffer Gas. In William C. Stwalley, Roman V. Krems, and Bretislav Friedrich, editors, *Cold Molecules: Theory, Experiment, Applications*, chapter 13. Boca Raton: CRC Press (2009).
[280] Jonathan D. Weinstein, Robert deCarvalho, Thierry Guillet, Bretislav Friedrich, and John M. Doyle. Magnetic Trapping of Calcium Monohydride Molecules at Millikelvin Temperatures. *Nature* **395**, 148–150 (1998).
[281] Nicholas R. Hutzler, Hsin-I Lu, and John M. Doyle. The Buffer Gas Beam: An Intense, Cold, and Slow Source for Atoms and Molecules. *Chem. Rev.* **112**, 4803–4827 (2012).
[282] Kevin M. Jones, Eite Tiesinga, Paul D. Lett, and Paul S. Julienne. Ultracold Photoassociation Spectroscopy: Long-Range Molecules and Atomic Scattering. *Rev. Mod. Phys.* **78**, 483–535 (2006).
[283] C. M. Dion, C. Drag, O. Dulieu, B. Laburthe Tolra, F. Masnou-Seeuws, and P. Pillet. Resonant Coupling in the Formation of Ultracold Ground State Molecules via Photoassociation. *Phys. Rev. Lett.* **86**, 2253–2256 (2001).
[284] Andrew J. Kerman, Jeremy M. Sage, Sunil Sainis, Thomas Bergeman, and David DeMille. Production of Ultracold, Polar RbCs* Molecules via Photoassociation. *Phys. Rev. Lett.* **92**, 033004 (2004).
[285] Jeremy M. Sage, Sunil Sainis, Thomas Bergeman, and David DeMille. Optical Production of Ultracold Polar Molecules. *Phys. Rev. Lett.* **94**, 203001 (2005).
[286] J. Deiglmayr, A. Grochola, M. Repp, K. Mörtlbauer, C. Glück, J. Lange, O. Dulieu, R. Wester, and M. Weidemüller. Formation of Ultracold Polar Molecules in the Rovibrational Ground State. *Phys. Rev. Lett.* **101**, 133004 (2008).
[287] Matthieu Viteau, Amodsen Chotia, Maria Allegrini, Nadia Bouloufa, Olivier Dulieu, Daniel Comparat, and Pierre Pillet. Optical Pumping and Vibrational Cooling of Molecules. *Science* **321**, 232–234 (2008).

[288] Juris Ulmanis, Johannes Deiglmayr, Marc Repp, Roland Wester, and Matthias Weidemüller. Ultracold Molecules Formed by Photoassociation: Heteronuclear Dimers, Inelastic Collisions, and Interactions with Ultrashort Laser Pulses. *Chem. Rev.* **112**, 4890–4927 (2012).

[289] Jens Herbig, Tobias Kraemer, Michael Mark, Tino Weber, Cheng Chin, Hanns-Christoph Nägerl, and Rudolf Grimm. Preparation of a Pure Molecular Quantum Gas. *Science* **301**, 1510–1513 (2003).

[290] S. Inouye, M. R. Andrews, J. Stenger, H.-J. Miesner, D. M. Stamper-Kurn, and W. Ketterle. Observation of Feshbach Resonances in a Bose–Einstein Condensate. *Nature* **392**, 151–154 (1998).

[291] Cindy A. Regal, Christopher Ticknor, John L. Bohn, and Deborah S. Jin. Creation of Ultracold Molecules from a Fermi Gas of Atoms. *Nature* **424**, 47–50 (2003).

[292] S. Jochim, M. Bartenstein, A. Altmeyer, G. Hendl, C. Chin, J. Hecker Denschlag, and R. Grimm. Pure Gas of Optically Trapped Molecules Created from Fermionic Atoms. *Phys. Rev. Lett.* **91**, 240402 (2003).

[293] M. W. Zwierlein, C. A. Stan, C. H. Schunck, S. M. F. Raupach, S. Gupta, Z. Hadzibabic, and W. Ketterle. Observation of Bose–Einstein Condensation of Molecules. *Phys. Rev. Lett.* **91**, 250401 (2003).

[294] Markus Greiner, Cindy A. Regal, and Deborah S. Jin. Emergence of a Molecular Bose–Einstein Condensate from a Fermi Gas. *Nature* **426**, 537–540 (2003).

[295] K. Winkler, F. Lang, G. Thalhammer, P. v. d. Straten, R. Grimm, and J. Hecker Denschlag. Coherent Optical Transfer of Feshbach Molecules to a Lower Vibrational State. *Phys. Rev. Lett.* **98**, 043201 (2007).

[296] Johann G. Danzl, Elmar Haller, Mattias Gustavsson, Manfred J. Mark, Russell Hart, Nadia Bouloufa, Olivier Dulieu, Helmut Ritsch, and Hanns-Christoph Nägerl. Quantum Gas of Deeply Bound Ground State Molecules. *Science* **321**, 1062–1066 (2008).

[297] K.-K. Ni, S. Ospelkaus, M. H. G. de Miranda, A. Pe'er, B. Neyenhuis, J. J. Zirbel, S. Kotochigova, P. S. Julienne, D. S. Jin, and J. Ye. A High Phase-Space-Density Gas of Polar Molecules. *Science* **322**, 231–235 (2008).

[298] F. Lang, K. Winkler, C. Strauss, R. Grimm, and J. Hecker Denschlag. Ultracold Triplet Molecules in the Rovibrational Ground State. *Phys. Rev. Lett.* **101**, 133005 (2008).

[299] D. J. Heinzen. Collisions of Ultracold Atoms in Optical Fields. In, D. Wineland, C. Wieman, and S. Smith, editors. *Atomic Physics 14*, page 369. New York: AIP Press (1995).

[300] P. D. Lett, P. S. Julienne, and W. D. Phillips. Photoassociative Spectroscopy of Laser Cooled Atoms. *Annu. Rev. Phys. Chem.* **46**, 423 (1995).

[301] P. A. Molenaar, P. van der Straten, and H. G. M. Heideman. Long-Range Predissociation in Two-Color Photoassociation of Ultracold Na Atoms. *Phys. Rev. Lett.* **77**, 1460 (1996).

[302] P. D. Lett, K. Helmerson, W. D. Phillips, L. P. Ratliff, S. L. Rolston, and M. E. Wagshul. Spectroscopy of Na_2 by Photoassociation of Laser-Cooled Na. *Phys. Rev. Lett.* **71**, 2200 (1993).

[303] L. P. Ratliff, M. E. Wagshul, P. D. Lett, S. L. Rolston, and W. D. Phillips. Photoassociative Spectroscopy of 1_g, 0_u^+ and 0_g^- States of Na_2. *J. Chem. Phys.* **101**, 2638 (1994).

[304] J. Weiner, F. Masnou-Seeuws, and A. Giusti-Suzor. Associative Ionization: Experiments, Potentials and Dynamics. *Adv. At. Mol. Phys.* **26**, 209 (1989).

[305] O. Dulieu, S. Magnier, and F. Masnou-Seeuws. Doubly-excited States for the Na$_2$ Molecule: Application to the Dynamics of the Associative Ionization Reaction. *Z. Phys. D* **32**, 229–240 (1994).

[306] Boichanh Huynh, Olivier Dulieu, and Françoise Masnou-Seeuws. Associative Ionization between Two Laser-Excited Sodium Atoms: Theory Compared To Experiment. *Phys. Rev. A* **57**, 958–975 (1998).

[307] A. Amelink, K. M. Jones, P. D. Lett, P. van der Straten, and H. G. M. Heideman. Spectroscopy of Autoionizing Doubly Excited States in Ultracold Na$_2$ molecules produced by photoassociation. *Phys. Rev. A* **61**, 042707 (2000).

[308] Bruce Shore. *The Theory of Coherent Atomic Excitation, volume 2*. New York: Wiley and Sons (1990).

[309] Bruce W. Shore. *Manipulating Quantum Structures Using Laser Pulses*. Cambridge: Cambridge University Press (2011).

[310] G. Alzetta, A. Gozzini, L. Moi, and G. Orriols. An Experimental Method for the Observation of R. F. Transitions and Laser Beat Resonances in Oriented Na Vapour. *Il Nuovo Cimento B Series 11* **36**, 5–20 (1976).

[311] E. Arimondo and G. Orriols. Nonabsorbing Atomic Coherences by Coherent Two-Photon Transitions in a Three-Level Optical Pumping. *Nuovo Cimento Lett.* **17**, 333–338 (1976).

[312] William A. Davis, Harold J. Metcalf, and William D. Phillips. Vanishing Electric Dipole Transition Moment. *Phys. Rev. A* **19**, 700–703 (1979).

[313] Petr M. Anisimov, Jonathan P. Dowling, and Barry C. Sanders. Objectively Discerning Autler–Townes Splitting from Electromagnetically Induced Transparency. *Phys. Rev. Lett.* **107**, 163604 (2011).

[314] U. Fano. Effects of Configuration Interaction on Intensities and Phase Shifts. *Phys. Rev.* **124**, 1866–1878 (1961).

[315] U. Gaubatz, P. Rudecki, M. Becker, S. Schiemann, M. Kulz, and K. Bergmann. Population Switching between Vibrational Levels in Molecular Beams. *Chem. Phys. Lett.* **149**, 463 (1988).

[316] J. R. Kuklinski, U. Gaubatz, F. T. Hioe, and K. Bergmann. Adiabatic Population Transfer in a Three-Level System Driven by Delayed Laser Pulses. *Phys. Rev. A* **40**, 6741 (1989).

[317] K. Bergmann, H. Theuer, and B. W. Shore. Coherent Population Transfer Among Quantum States of Atoms and Molecules. *Rev. Mod. Phys.* **70**, 1003–1025 (1998).

[318] Yuan Sun and Harold Metcalf. Nonadiabaticity in Stimulated Raman Adiabatic Passage. *Phys. Rev. A* **90**, 033408 (2014).

[319] M. P. Fewell, B. W. Shore, and K. Bergmann. Coherent Population Transfer among Three States: Full Algebraic Solutions and the Relevance of Non Adiabatic Processes to Transfer by Delayed Pulses. *Aust. J. Phys.* **50**, 281–308 (1997).

[320] A. Aspect, E. Arimondo, R. Kaiser, N. Vansteenkiste, and C. Cohen-Tannoudji. Laser Cooling Below the One-Photon Recoil Energy by Velocity-Selective Coherent Population Trapping. *Phys. Rev. Lett.* **61**, 826 (1988).

[321] J. Hack, L. Liu, M. Olshanii, and H. Metcalf. Velocity-Selective Coherent Population Trapping of Two-Level Atoms. *Phys. Rev. A* **62**, 013405 (2000).

[322] Stanley J. Brodsky, Carl E. Carlson, John R. Hiller, and Dae Sung Hwang. Constraints on Proton Structure from Precision Atomic-Physics Measurements. *Phys. Rev. Lett.* **94**, 022001 (2005).

[323] John E. Nafe and Edward B. Nelson. The Hyperfine Structure of Hydrogen and Deuterium. *Phys. Rev.* **73**, 718–728 (1948).

[324] H. M. Foley and P. Kusch. On the Intrinsic Moment of the Electron. *Phys. Rev.* **73**, 412–412 (1948).

[325] D. T. Wilkinson and H. R. Crane. Precision Measurement of the g Factor of the Free Electron. *Phys. Rev.* **130**, 852–863 (1963).

[326] D. Hanneke, S. Fogwell, and G. Gabrielse. New Measurement of the Electron Magnetic Moment and the Fine Structure Constant. *Phys. Rev. Lett.* **100**, 120801 (2008).

[327] Michael Lombardi, Thomas Heavner, and Steven R. Jefferts. NIST Primary Frequency Standards and the Realization of the SI Second. *NCSL Intern. Meas.* **2**, 74 (2007).

[328] Norman Ramsey. History of Atomic Clocks. *J. Res. Nat. Bur. Stand.* **88**, 301 (1983).

[329] J. R. Zacharias. Precision Measurements with Molecular Beams. *Phys. Rev.* **94**, 751 (1954).

[330] R. Weiss. Contribution to "Festschrift for Jerrold R. Zacharias". Private communication.

[331] C. W. Chou, D. B. Hume, J. C. J. Koelemeij, D. J. Wineland, and T. Rosenband. Frequency Comparison of Two High-Accuracy Al$^+$ Optical Clocks. *Phys. Rev. Lett.* **104**, 070802 (2010).

[332] Jeremy Darling. Methods for Constraining Fine Structure Constant Evolution with OH Microwave Transitions. *Phys. Rev. Lett.* **91**, 011301 (2003).

[333] Eric R. Hudson, H. J. Lewandowski, Brian C. Sawyer, and Jun Ye. Cold Molecule Spectroscopy for Constraining the Evolution of the Fine Structure Constant. *Phys. Rev. Lett.* **96**, 143004 (2006).

[334] D. DeMille, S. Sainis, J. Sage, T. Bergeman, S. Kotochigova, and E. Tiesinga. Enhanced Sensitivity to Variation of m_e/m_p in Molecular Spectra. *Phys. Rev. Lett.* **100** 043202 (2008).

[335] A. Cingöz, A. Lapierre, A.-T. Nguyen, N. Leefer, D. Budker, S. K. Lamoreaux, and J. R. Torgerson. Limit on the Temporal Variation of the Fine-Structure Constant Using Atomic Dysprosium. *Phys. Rev. Lett.* **98**, 040801 (2007).

[336] N. Leefer, C. T. M. Weber, A. Cingöz, J. R. Torgerson, and D. Budker. New Limits on Variation of the Fine-Structure Constant Using Atomic Dysprosium. *Phys. Rev. Lett.* **111**, 060801 (2013).

[337] T. Rosenband, D. Hume, P. O. Schmidt, C. W. Chow, L. Lorini, W. H. Oksay, R. Drullinger, T. Fortier, E. Stalneker, S. Diddams, W. Swann, N. R. Newbury, D. Wineland, and J. Bergquist. Frequency Ratio of Al$^+$ and Hg$^+$ Single-Ion Optical Clocks; Metrology at the 17th Decimal Place. *Science* **319**, 1808 (2008).

[338] L. Lorini, N. Ashby, A. Brusch, S. Diddams, R. Drullinger, E. Eason, T. Fortier, P. Hastings, T. Heavner, D. Hume, W. Itano, S. Jefferts, N. Newbury, T. Parker, T. Rosenband, J. Stalnaker, W. Swann, D. Wineland, and J. Bergquist. Recent Atomic Clock Comparisons at NIST. *Eur. Phys. J. Special Topics* **163**, 19 (2008).

[339] R. M. Godun, P. B. R. Nisbet-Jones, J. M. Jones, S. A. King, L. A. M. Johnson, H. S. Margolis, K. Szymaniec, S. N. Lea, K. Bongs, and P. Gill. Frequency Ratio of Two Optical Clock Transitions in ^{171}Yb$^+$ and Constraints on the Time Variation of Fundamental Constants. *Phys. Rev. Lett.* **113**, 210801 (2014).

[340] N. Huntemann, B. Lipphardt, Chr. Tamm, V. Gerginov, S. Weyers, and E. Peik. Improved Limit on a Temporal Variation of m_p/m_e from Comparisons of Yb$^+$ and Cs Atomic Clocks. *Phys. Rev. Lett.* **113**, 210802 (2014).

[341] Martin Deutsch. Evidence for the Formation of Positronium in Gases. *Phys. Rev.* **82**, 455–456 (1951).

[342] K. F. Canter, A. P. Mills, and S. Berko. Observations of Positronium Lyman-α Radiation. *Phys. Rev. Lett.* **34**, 177–180 (1975).

[343] D. Hagena, R. Ley, D. Weil, G. Werth, W. Arnold, and H. Schneider. Precise Measurement of $n = 2$ Positronium Fine-Structure Intervals. *Phys. Rev. Lett.* **71**, 2887–2890 (1993).
[344] M. S. Fee, A. P. Mills, S. Chu, E. D. Shaw, K. Danzmann, R. J. Chichester, and D. M. Zuckerman. Measurement of the Positronium 1^3S_1–2^3S_1 Interval by Continuous-Wave Two-Photon Excitation. *Phys. Rev. Lett.* **70**, 1397–1400 (1993).
[345] D. B. Cassidy and A. P. Mills. Jr. The Production of Molecular Positronium. *Nature* **449**, 195–197 (2007).
[346] M. W. Ritter, P. O. Egan, V. W. Hughes, and K. A. Woodle. Precision Determination of the Hyperfine-Structure Interval in the Ground State of Positronium. V. *Phys. Rev. A* **30**, 1331–1338 (1984).
[347] V. Hughes. Muonium. *Phys. Today* **20**, 29–40 (1967).
[348] D. S. Covita, D. F. Anagnostopoulos, H. Gorke, D. Gotta, A. Gruber, A. Hirtl, T. Ishiwatari, P. Indelicato, E.-O. Le Bigot, M. Nekipelov, J. M. F. dos Santos, Ph. Schmid, L. M. Simons, M. Trassinelli, J. F. C. A. Veloso, and J. Zmeskal. Line Shape of the μH (3p–1s) Hyperfine Transitions. *Phys. Rev. Lett.* **102**, 023401 (2009).
[349] Randolf Pohl, Aldo Antognini, François Nez, Fernando D. Amaro, François Biraben, João M. R. Cardoso, Daniel S. Covita, Andreas Dax, Satish Dhawan, Luis M. P. Fernandes, Adolf Giesen, Thomas Graf, Theodor W. Hänsch, Paul Indelicato, Lucile Julien, Cheng-Yang Kao, Paul Knowles, Eric-Olivier Le Bigot, Yi-Wei Liu, José A. M. Lopes, Livia Ludhova, Cristina M. B. Monteiro, Françoise Mulhauser, Tobias Nebel, Paul Rabinowitz, Joaquim M. F. dos Santos, Lukas A. Schaller, Karsten Schuhmann, Catherine Schwob, David Taqqu, João F. C. A. Veloso, and Franz Kottmann. The Size of the Proton. *Nature* **466**, 213–216 (2010).
[350] Detlev Gotta, F. Amaro, D. Anagnostopoulos, P. Bühler, H. Gorke, D. Covita, H. Fuhrmann, A. Gruber, M. Hennebach, A. Hirtl, T. Ishiwatari, P. Indelicato, E.-O. Le Bigot, J. Marton, M. Nekipelov, J. dos Santos, S. Schlesser, Ph. Schmid, L. Simons, Th. Strauch, M. Trassinelli, J. Veloso, and J. Zmeskal. Muonium. *Hyp. Int.* **209**, 57–62 (2012).
[351] G. Gabrielse, R. Kalra, W. S. Kolthammer, R. McConnell, P. Richerme, D. Grzonka, W. Oelert, T. Sefzick, M. Zielinski, D. W. Fitzakerley, M. C. George, E. A. Hessels, C. H. Storry, M. Weel, A. Müllers, and J. Walz. Trapped Antihydrogen in its Ground State. *Phys. Rev. Lett.* **108**, 113002 (2012).
[352] Alpha Collaboration. A Source of Antihydrogen for In-flight Hyperfine Spectroscopy. *Nat. Commun.* **5** doi: 10.1038/ncomms4089 (2014).
[353] E. Schrödinger. Discussion of Probability Relations between Separated Systems. *Math. Proc. Cambr. Phil. Soc.* **31**, 555–563 (1935).
[354] John Clauser and Abner Shimony. Bell's Theorem : Experimental Tests and Implications. *Rep. Prog. Phys.* **38**, 1881–1927 (1978).
[355] J. S. Bell. On the Einstein–Podolsky–Rosen Paradox. *Physics* **1**, 195 (1964).
[356] John S. Bell. On the Problem of Hidden Variables in Quantum Mechanics. *Rev. Mod. Phys.* **38**, 447–452 (1966).
[357] J. S. Bell. Bertlmann's Socks and the Nature of Reality. *J. Phys. Coll.* **42**, C2-41–C2-62 (1981).
[358] Alain Aspect, Philippe Grangier, and Gérard Roger. Experimental Realization of Einstein–Podolsky–Rosen–Bohm *Gedankenexperiment*: A New Violation of Bell's Inequalities. *Phys. Rev. Lett.* **49**, 91–94 (1982).
[359] Alain Aspect, Jean Dalibard, and Gérard Roger. Experimental Test of Bell's Inequalities Using Time-Varying Analyzers. *Phys. Rev. Lett.* **49**, 1804–1807 (1982).

[360] Marie-Anne Bouchiat and Claude Bouchiat. Parity Violation in Atoms. *Rep. Prog. Phys.* **60**, 1351–1396 (1997).

[361] C. S. Wood, S. C. Bennett, D. Cho, B. P. Masterson, J. L. Roberts, C. E. Tanner, and C. E. Wieman. Measurement of Parity Nonconservation and an Anapole Moment in Cesium. *Science* **275**, 1759–1763 (1997).

[362] E. Gomez, S. Aubin, G. D. Sprouse, L. A. Orozco, and D. P. DeMille. Measurement Method for the Nuclear Anapole Moment of Laser-Trapped Alkali-Metal Atoms. *Phys. Rev. A* **75**, 033418 (2007).

[363] Jocelyne Guéna, Michel Lintz, and Marie-Anne Bouchiat. Proposal for High-Precision Atomic-Parity-Violation Measurements by Amplification of the Asymmetry by Stimulated Emission in a Transverse Electric and Magnetic Field Pump-Probe Experiment. *J. Opt. Soc. Am. B* **22**, 21–28 (2005).

[364] L. Bougas, G. E. Katsoprinakis, W. von Klitzing, J. Sapirstein, and T. P. Rakitzis. Cavity-Enhanced Parity-Nonconserving Optical Rotation in Metastable Xe and Hg. *Phys. Rev. Lett.* **108**, 210801 (2012).

[365] S. Schlamminger, K.-Y. Choi, T. A. Wagner, J. H. Gundlach, and E. G. Adelberger. Test of the Equivalence Principle Using a Rotating Torsion Balance. *Phys. Rev. Lett.* **100**, 041101 (2008).

[366] E. R. Williams, J. E. Faller, and H. A. Hill. New Experimental Test of Coulomb's Law: A Laboratory Upper Limit on the Photon Rest Mass. *Phys. Rev. Lett.* **26**, 721–724 (1971).

[367] J. H. Smith, E. M. Purcell, and N. F. Ramsey. Experimental Limit to the Electric Dipole Moment of the Neutron. *Phys. Rev.* **108**, 120–122 (1957).

[368] P. G. Harris, C. A. Baker, K. Green, P. Iaydjiev, S. Ivanov, D. J. R. May, J. M. Pendlebury, D. Shiers, K. F. Smith, M. van der Grinten, and P. Geltenbort. New Experimental Limit on the Electric Dipole Moment of the Neutron. *Phys. Rev. Lett.* **82**, 904–907 (1999).

[369] Norval Fortson, Patrick Sandars, and Stephen Barr. The Search for a Permanent Electric Dipole Moment. *Phys. Today* **56**, 33 (2003).

[370] J. J. Hudson, D. M. Kara, I. J. Smallman, B. E. Sauer, M. R. Tarbutt, and E. A. Hinds. Improved Measurement of the Shape of the Electron. *Nature* **473**, 493–496 (2011).

[371] J. Baron (The ACME Collaboration), W. C. Campbell, D. DeMille, J. M. Doyle, G. Gabrielse, Y. V. Gurevich, P. W. Hess, N. R. Hutzler, E. Kirilov, I. Kozyryev, B. R. O'Leary, C. D. Panda, M. F. Parsons, E. S. Petrik, B. Spaun, A. C. Vutha, and A. D. West. Order of Magnitude Smaller Limit on the Electric Dipole Moment of the Electron. *Science* **343**, 269–272 (2014).

[372] Clebsch–Gordan Coefficients, Spherical Harmonics, and D Functions. *Eur. Phys. J. C* **15**, 208 (2000).

Index

Bold page numbers indicate the location where the indexed item is the primary topic of discussion.

21 cm line, 67, 156, 189
3 j-symbol, 121, 174, 475, 476
6 j-symbol, 174

absorption, 43, 81, 82
accidental degeneracy, 114
adiabatic motion, 372
adiabatic rapid passage, 35, **34–36**, 38
alkali-metal atoms, 46, 123, 137, 157, 158, 175, 176, **164–180**, 183, 205, 211, 227, 236, 244, 249, 250, 256, 259, 265, 291, 385, 432, 444, 479
    dimers, 291, 322
alkaline earth atoms, 211, 259
Allan variance, 162, 452
anapole moment, **461–463**
angular momentum, 108, 117, **473–478**
    addition, **474–477**
    matrix element, **477**
    operator, 109, 123, 126, 200, **474**
    orbital, 8, 471, **473**, 478
anharmonicity, 282, 299
anti-bonding, 306, 312
anti-crossing, 62, 192, 193, 196, 216, 222
antimatter, **455–459**
astrophysics, 171
atom–light interaction, **40–63**, 347
atomic beam collimation, **352**
atomic beam slowing, **352**
atomic clock, 29, 30, 39, 64, 161, 448, **450–452**
atomic density, 378
atomic force microscopy, 117
atomic orbit, 372
atomic radius, 249, 250
atomic thermal wavelength, 385
atomic traps, **367–381**
atomic unit, 108, 117, 169, 181, 198, 213, 217, 304, **471**, 472
Aufbau principle, 245, 248, **248–249**, 269, 270, 306
Auger electron, 211

Autler–Townes, 77, **438–440**, 444, 445

bad collisions, 331
Balmer formula, 8, 107, 114, 448
Balmer-α line, 144–146, 162, 449
band head, 301, 302
band system, 291
Bates–Damgaard model, 171, 172
BEC, *see* Bose–Einstein condensation
Bell inequalities, **459–461**
binding energy, 284, 309
Biot–Savart law, 153, 162
black body radiation, 53, 89, 92, 209, 224
Bloch equation, 97
Bloch oscillation, 379, 380
Bloch sphere, 25, **23–25**, 28, 31, 33–35, 37, 39, 102
Bloch theorem, 377
Bloch vector, 25, **23–25**, 25, 35, 38, 103, 451
Bloch–Siegert shift, 43, 102
Bogoliubov approximation, 390
Bogoliubov dispersion relation, 390, 399
Bogoliubov speed of sound, 399, 400
Bohr formula, 113, 143, 210
Bohr magneton, 58, 131, 450, 457
Bohr model, **7–9**, 14, 82, 107, 129, 139, 217, 225, 259, 448
    molecular, 276
Bohr radius, 8, 41, 55, 117, 143, 385
Boltzmann factor, 262
bonding, 306, 312
Born–Oppenheimer approximation, 273, 274, **278**, 303, 373
Bose–Einstein condensation, 370, 372, 378, **382–412**, 432, 448
    road to, **384–385**
Bose–Einstein distribution, 409, 410
Bose–Hubbard model, 401
boson, 333, 383, 386, 393, 405, 410
Bragg diffraction, 378, 379

508

Index 509

Breit–Rabi diagram, 189, 457
brightness, 352
Brillouin zone, 379, 381
buffer gas cooling, 414, **418–420**

canonical momentum, 62, 78
capture range, 375
Cartesian coordinates, 316
center-of-mass frame, 122, 149, 231, 272, 277
center-of-mass motion, **122–123**, 149
central-field Hamiltonian, 245, 246, 249
centrifugal barrier, 332
chemical potential, 387, 389, 390, 395, 401, 403, 404, 408
circular state, 115, 225
classical radius of electron, 143
classically forbidden region, 329
Clebsch–Gordan coefficient, 120, 160, 174, 186, 325, 362, 475, 476, 478
closed channel, 338–340
coherent population trapping, **436–437**, 439, 444, 445, 447
coherent state, 393
coinage elements, 265
cold molecules, **413–432**
collisional physics, 331
commutation relation, 63, 87
commutator, 124
commuting operators, 108, **124**
Compton wavelength, 141
condensed matter physics, 383, 400, 401, 448
Condon point, 432
confocal elliptical coordinates, 304, 305, 314, **316–317**, 317, 318
conservation law, 473
cooling limit
  Doppler, 356, 359
    cooling below, **359–360**
  recoil, 364
core polarization, 211
Coulomb blockade, **224–225**
Coulomb gauge, 41, 52, 66, 181
Coulomb integral, 234, 240
coupling
  many-electron atoms, **249–254**
  two electrons, **253–254**
covalent bonding, **284–285**
covalent character, 313
covalent state, 309
cross-section
  absorption, 6, 46, 100, 118, 209
  collision, 118, 367
  differential, 333
  elastic, 432
  total, 333
Cs clock, 29
Curie constant, 262, 263
cylindrical coordinates, 128

D-line, 174–176, 368, 444
damping coefficient, 4
damping force, 354, 356, 364, **363–364**, 365
damping rate, 364, 375
dark state, 436–438, 444
Darwin term, 115, 132, 137, **138**, 138
de Broglie wave atom optics, 376
de Broglie wavelength, 179, 372, 377, 379, 386, 393
de Broglie waves, 9, **9–10**, 14
Debye–Waller factor, 379
decay rate, 87, 97, 121
degeneracy, 225, 242
degenerate Bose gas, 334
density matrix, 84, **93–103**
density of states, 387, 388, 410, 411, 412, **410–412**
density operator, 94
diamagnetism, 211, 214
diffusion coefficient, 356
dipolar interaction, 415
dipole force, 350, 368, 376
dipole matrix element
  angular part, **120–121**, 121
  geometrical part, 119
  radial part, **119–120**
dipole moment, 83, 118, 210, 287, 347, 432
  octopole, 210
  quadrupole, 210
dipole operator, 291, 292
  molecular, 292, 293
dipole–dipole interaction, 322, 324, 325
  second order, 324
dipole–dipole potential, 330, 331
Dirac energy, 148
Dirac equation, 131, 139, **138–140**, 140, 143, 145, 449
Dirac Hamiltonian, 132, 138
direct integral, 305
dissipative force, *see* radiation pressure force
dissociation energy, 275, 281, 284, 287, 299–301
dissociation limit, 321, 322, 326, 330
distribution functions, **405–410**
  bosons, **409–410**
  classical particles, **406–408**
  fermions, **408–409**
  photons, **410**
Doppler effect, 100
Doppler laser cooling, 360, 364, 415
Doppler shift, 144, 352, 353, 360, 375, 379
  second order, 147
Doppler width, 61, 146, 150
doublet, 131, 144, 156
dressed state, 22, **21–23**, 35, 36, 102, 195

E1, *see* electric dipole transition
E2, *see* electric quadrupole transition
echoes, **30–34**
effective potential, 112
egg-crate potential, 378, 379
Ehrenfest theorem, 108, 124, 348, 349

Einstein coefficient, 20, **80–83**, 92
Einstein–deHaas experiment, 134
Einstein–Podolski–Rosen paradox, 459
electric dipole
    approximation, **40–42**, 48, **55–56**, 62, 64, 65, 70, 75, 79, 291, 349
        higher-order, **78–79**
    interaction, 78, 90, 324
    moment, 56, 85, 202, 291, 293, 294, 300, 448, 453, **463–464**
        permanent, 291
    transition, 57, 65, 146, 208, 434, 454
electric moment
    nuclear, 156
electric quadrupole
    interaction, 78
    moment, 68, 78, 157
    transition, 64, **68–70**
electromagnetic fields, **50–54**
electromagnetically induced transparency, 77, **438–440**, 444, 445
electron affinity, 283
electron distribution, 306, 312, 315
electron magnetic moment
    anomalous, **449–450**
electron spin resonance, 66
electron–electron interaction, 231, 233, 242, 244, 256, 309
    molecular hydrogen, **318–319**
emission rate, 46
energy band structure, 377
entanglement, **36–37**, 224
equilibrium distance, 274, 275, 281, 284, 298, 299
Euler angles, 294
evaporative cooling, 64, 372, 384, 414, 420
exchange integral, 234, 240, 257, 305
exchange operator, 228, **228–230**, 232, 245, 247, 304
exchange symmetry, 325
exotic atoms, **455–459**

Fermi contact term, 155, 188
Fermi energy, 409
Fermi golden rule, 45
Fermi–Dirac distribution, 408
fermion, 383, 405
Feshbach coupling, 414
Feshbach molecules, **337–342**
Feshbach resonance, 322, 338–341, 384, 385, 424
Feynman–Hellmann theorem, 129
field ionization, **217–218**, 218, 219, 222
field quantization, **89–91**
fine structure, 46, 59, 143, **131–148**, 259, 322, 326, 448, 457
    measurement, **144–147**
fine-structure constant, 131, **142–144**, 454
fine-structure interaction, 173, 175, 195, 327
first sound, 397–399
Fock state, 393

forbidden transition, **64–79**, 300, 452, 462
force operator, 349
Fortrat diagram, 298, 302
fountain clock, 30, 31, 451, 452, 454
four-wave mixing, 77
Fourier theorem, 54
Fourier transform limit, 26, 162, 451
Franck–Condon factor, 295, 296, 300, 415, 422, 432
Franck–Condon principle, 293, 294, 428
fundamental physics, **448–464**
fundamental progression, 296

Gauss law, 163, 179
Gaussian laser beam, 51, 53, 367
gerade, 288, 304, 306, 307, 311, 312, 318
gold
    color of, **264–269**
good collisions, 331
Göppert–Mayer work, 72–74, 147, 433
Gross–Pitaevskii equation, 389, 391
Grotrian diagram, 259
group velocity, 443
gyromagnetic ratio, 134, 184, 185, 222

H, *see* hydrogen, 175
H I region, 208
halogens, 250, 259
Hamiltonian
    helium, **230–233**
    hydrogen, **108–109**
        angular part, **109–110**
        radial part, **110–112**
        radial solution, **114–117**
    Zeeman effect, **182–183**
Hanle effect, 12, **11–12**, 62
harmonic oscillator, 88, 89, 281
    damped, 5, **4–5**
    damped driven, **6–7**
    quantum mechanical, **87–88**
Hartree model, 257
Hartree–Fock calculations, 165, 166
Hartree–Fock equation of state, 398
Hartree–Fock model, 171, 258, **256–258**
heat capacity, 397, 408
Heisenberg equation of motion, 41, 96
Heisenberg picture, 55
Heitler–London approach, *see* valence bond approach
helium, 221, **227–243**, 244, 255, 256, 259
    dimer, 283, 336
    liquid, 395, 414, 419
    metastable, 242
    ortho-, para-, 236
    singly excited, 236
    singly ionized, 127
    variational method, **233–239**
        doubly excited state, **237–239**
        ground state, **233–235**
        singly excited state, **236–237**

Helmholtz equation, 52
Hönl–London factor, 295
Hund case, 289, 326, 327, 331
Hund's rules, 245, **254–256**, 261–264, 270
Hund–Mulliken approach, *see* molecular orbital approach
hydrogen, **107–130**, 164, 274
   anti-, **458–459**
   molecular ion, **303–306**, 308, 318
   molecule, 222, **303–320**
   muonic, **457–458**
   pionic, **458**
   radial wavefunction, 113, 115, 116, 119, 125, 169, 211
   spectrum, 7
   spin-polarized, 385
hyperbolic model, 194
hyperfine energies
   hydrogen, **155–156**
   other atoms, **156–157**
hyperfine interaction, 143, 157, 173–175, 228, 327
hyperfine structure, 46, 59, 155, **152–157**, 259, 322, 326, 449
   two spin-$1/2$ particles, **160–161**
hyperfine-structure, 143

independent particle model, **231**, 236, **245–246**, 252
indistinguishable, 228, 333, 408
intercombination transition, 260
interferometry, **30–34**
Ioffe trap, 371, 372
ionic bonding, **285–286**
ionic state, 309
ionization potential, 176, 249, 250, 283
iron group, 264
isotope, 149, 150, 152
isotope shift, 150

j–j coupling, 251, **253**
Jaynes–Cummings model, 22, 91
Josephson effect, 145

Kerr effect
   Raman induced, 77
Kronig–Penney model, 377

L–S coupling, *see* Russel–Saunders coupling
laboratory frame, 122, 294
Lagrange multiplier, 407, 408
Laguerre polynomial, 113–115, 246
Lamb shift, 92, 132, 138, **140–142**, 143–146, 202, 207, 224, 448, **449**, 456, 457
Λ configuration, 434
Λ-doubling, 288, 453
Landé g-factor, 58, 134, 182
Landé interval rule, 155, 157, 253
Landau–Zener transition, **194–195**, 196
lanthanides, 260, 262

Larmor precession, 373, 381
laser cooling, **347–366**, 384, 413, 414, 422, 427
laser-atomic beam spectroscopy, 150
LCAO method, *see* linear combination of atomic orbitals
Legendre polynomial, 110, 234, 240, 318
LeRoy–Bernstein method, 322, **328–330**
level crossing spectroscopy, 49, 145
lifetime, 4, 87, 121, 147, 202
   hydrogen, **121–122**
light, **50–54**
   circularly polarized, 48, 49, 62, 67, 121, 174, 182, 361
   laser, **51–53**
   linearly polarized, 49, 58, 61, 67, 121, 294, 361
light pressure force, *see* radiation pressure force
light shift, **20–21**, 23, 350, 351, **362**, 363, 368, 376, 378, 435
line-broadening mechanism, 100
linear combination of atomic orbitals method, 304
linewidth, 99
   power-broadened, 99, 102, 355
Liouville–von Neumann equation, **101**
long-range molecular potentials, **322–328**
long-range molecule, 428
long-wavelength approximation, 55
Lorentz field, 135
Lorentz force, 11
Lorentz model, 84
Lorentz triplet, 184
Lorentz-invariant, 131, 138
low saturation limit, 21
Lyman-α radiation, 207, 456, 457

M1, *see* magnetic dipole transition
magnetic dipole
   interaction, 78, 154, **157–159**
   moment, 66, 79, 159, 340
   transition, 64, **66–68**
magnetic moment, 195, 263, 265, 370
   electron, 131, 133–135, 140, 153, 154, 157, 159, 160, 448
   neutron, 152
   nuclear, 131, 149, 152–157, 159, 160
   proton, 152
   rotational, 289
magnetic resonance, 17, 64
magnetic resonance imaging, 448
magnetic trap, 185, 370, **370–373**
magneto-association, **424–425**
magneto-optical trap, 31, 367, **373–376**
   atom cooling, **374–375**
many-body physics, 382
maser
   ammonia, 161
   hydrogen, 156, 161, **161–162**, 189
mass polarization, 123, 150
matrix elements

radial wavefunctions, **125**
matter-wave interference, **391–394**
Maxwell equations, 51
Maxwell–Boltzmann distribution, 408, 409
mean-field theory, **388–391**
mixed state, **94–96**, 103
mixing angle, 21
model potential, **170–171**
molecular frame, 292, 294
molecular orbital approach, 303, **306–310**, 312, 313, 319
molecular spectra, 273
molecules, **272–302**
    bonding, **282–286**
    diatomic, 272, 275, 323
    electronic states, **286–290**
    heteronuclear, 291, 414, 421, 423, 432
    homonuclear, 288, 290, 291, 414, 422, 423
    nuclear eigenfunctions, **279–281**
    nuclear motion, **276–282**
    rovibrational energies, **281–282**
Mollow triplet, 21
moment of inertia, 281
Morse potential, **299–300**, 300, 315
MOT, *see* magneto-optical trap
Mott-insulating phase, 401–404
multiconfiguration Dirac–Fock calculations, 268
multiplet, 131, 252
    inverted, 253
    normal, 253
muonium, **456–457**

noble-gas atoms, 249, 259, 261
    metastable, 432
    pair, 284
node, 115
    angular, 115
    radial, 113, 115, 116, 127, 129
non-cloning theorem, 37
non-crossing theorem, **192–194**, 215
non-linear optics, **74**, 76, 77
non-linear Schrödinger equation, *see* Gross–Pitaevskii equation
non-penetrating orbit, **168–169**, 210
nuclear magnetic resonance, 66, 97, 222
nuclear magneton, 152
nuclear motion, **149–150**
nuclear shape, **151–152**
nuclear size, **150–151**, 163
Numerov method, 171, **176–178**

OBE, *see* optical Bloch equation
OM, *see* optical molasses
open channel, 337–340
operator, 87
    annihilation, 90, 92, 393, 401
    creation, 90, 92, 401
    ladder, 160

    lowering, 22, 162, 473
    raising, 22, 162, 473
optical Bloch equation, **96–98**, 101, 349, 353, 355
optical coherence, 94, 98, 349
optical force, **347–366**
optical lattice, 361, 367, **376–380**, 400, 402, 403, 405
optical molasses, 352, 354, **353–355**, 360, 367
    experiments, **356–359**
optical pumping, 49, 359, 360, 362–364
optical transitions
    alkali-metal atoms, **171–175**
        angular part, **173–175**
        radial part, **172–173**
    hydrogen, **118–122**
    molecules, **290–298**
        rotational effects, **297–298**
        transition strength, **291–296**
        vibrational effects, **296–297**
optical trap, 361
    dipole force, **367–370**
optical tweezers, 13, **12–14**, 368
orbital
    anti-bonding, 300
    bonding, 300
oscillator strength, 47, **46–47**, **50**, 63, 83, 129

P-branch, 297, 298
parabolic coordinates, 108, **123–124**, 199, 203, 205
paramagnetism, **261–264**
parity, 115, 251
parity operator, 199, 200
parity violation, **461–463**
parity-forbidden, 266
Parsons–Weisskopf model, 179
Paschen notation, 253, 261
Pauli matrices, **25–26**, 38, 91, 134, 241
Pauli symmetrization principle, 227, 228, 232, 242, 244, **247–248**, 253, 257–259, 263, 283, 304, 307, 326, 405, 409
penetrating orbit, 210
periodic potential, 377
periodic system, **244–271**
Periodic Table, 227, 245, **258–261**
    column, 259
    row, 259
permutation operator, 242
perturbation approximation, **42–43**, 64, 70
perturbation theory, 204
    degenerate, 225
phase shift, 167, 333, 334
phase-space density, 384, 385, 416, 417
photo-association, 321, 414, **422–424**
photo-association spectroscopy, 385, **426–432**
photon, 90, 91, 410
π-pulse, 27, 32, 33, 35
$\pi/2$-pulse, 20, 25, 27, 28, 30, 32–34, 37
Planck
    formula, 7, 8, 80, 82, 276

hypothesis, 80, 114
    radiation formula, 53, 410, 412
plasma physics, 171
polarizability, 170, 202, 283, 286
    alkali-metal atoms, 168
    dipole, 168–170
    quadrupole, 168–170, 211
polarization, 45, 118, 120, 292, **360–362**
    induced, 286
polarization gradient, 360, 361, 363, 364
    lin ⊥ lin, 361, 363
positronium, 146, 161, 190, 191, 195, 455, **455–456**
power broadening, 100, **98–101**
Poynting vector, 4, 10, 52, 54
pressure broadening, 100
probability density, 114, 125, 127, 128
proton size, 458
pure state, 94, **93–94**, 95–97, 102

Q-branch, 297, 298
QED, *see* quantum electrodynamics
quadrupole–quadrupole interaction, 325, 431
quality factor, 5, 14
quantum
    communication, 37
    computing, 37
    cryptography, 37
    teleportation, 37
quantum beat, 49, 58, 60
quantum controlled *not* gate, 224
quantum defect, 133, 137, 164, 165, 168, **166–168**,
    169, 170, **175–176**, 177–179, 208, 211, **209–211**,
    214, 226, 227, 244, 259
quantum electrodynamics, 131, 132, 138, 142, 143,
    145, 156, 190, 449, 450, 456, 457, 461
    cavity, 86, 209, 223
quantum hydrodynamics, 383, 395, **394–399**
quantum information, 209, 224
quantum jump, 72
quantum number
    effective, 164, 165
    orbital angular, 115
    principal, 115, 204, 208
    principle , 471
    vibrational, 274, 293, 330
quantum optics, 76, **89–91**
quantum phase transition, 402
quantum statistics, **385–388**
quantum Zeno effect, 224
quantum-field theory, 389
qubit, 224

R-branch, 297, 298
R-centroid approximation, 295
Rabi frequency, 18, 25, 42, 91, 102, 118, 224, 349,
    351, 353, 435, 436
    vacuum, 85
Rabi oscillation, 27, 91, 97, 102, 433, 442

Rabi two-level problem, 96
radial wavefunction, 112, 225
radiation pressure force, 348, **348–352**
Raman effect, 73
    coherent, 77
    inverse, 77
    spontaneous, **435–436**
    stimulated, 73, **435–436**
Raman spectroscopy, 444
    coherent anti-Stokes, 77
    opto-acoustic, 77
Ramsey method, **26–30**, 32, 451, 464
Ramsey–Bordé scheme, 33
Rayleigh length, 53
Rayleigh scattering, 75
recoil energy, 355
reduced mass, 108, 123, 150, 176
    molecular, 332
    nuclei, 274
reflection symmetry, 327
refractive index, 443
relativistic effect, 137, 139, 267–269
relativistic mass term, **132–133**
retardation, 221, 222, 426
rigid rotor, 281
Rittner model, 286
Ritz combination principle, 8
ro-vibrational transitions, 274, 291
Rosen–Zener model, 194
rotating frame transformation, 19, **18–20**, 434
rotating wave approximation, 19, **18–20**, 22, **43–44**,
    90, 102, 349, 351, 434
rotational barrier, 334
rotational constant, 281, 301
Russel–Saunders coupling, 251, **252–253**, 255
Rutherford model, 8
RWA, *see* rotating wave approximation
Rydberg atoms, 205, **208–226**
    external fields, **211–218**
Rydberg constant, 14, 113, 132, 176
Rydberg electron, 211, 217, 219
Rydberg formula, 107, 164
    modified, 165
Rydberg gas
    cold, 209
Rydberg spectroscopy, 211, 216, 218, 222, **219–225**
Rydberg state, 117, 167, 416
    circular, **223–224**

saddle-point, 217
saturation, **98–101**
saturation parameter, 99, 100, 102
    on-resonance, 99
scalar potential, 51, 52, 55
scale of atoms, **117–118**
scanning tunneling microscopy, 117
scattering force, 368

scattering length, 322, 334, 335, 337, **334–337**, 339–341, 388–390, 403, 426, 432
scattering rate, 99, 100, 350, 352, 365
scattering theory, **331–334**
Schrödinger equation
  angular part, 198
  radial, 198
screened charge, 233, 234
second sound, 383, 395, 397–399
selection rule, **47–50**, 67, 71, 78, 184, 260
  electric dipole, 47, 270
  molecular, 295
shell, 248, 258
SI unit, 107, 471
singlet state, 230, 232, 236, 260, 290, 291, 307
Sisyphus laser cooling, 363, 414
Slater determinant, 247, 257
slow light, 77, **442–444**
sodium, 47, 210, 216, 220, 258, 269
spaghetti region, 216
spectral energy density, 53, 54, 81
spectral intensity, 44, 45, 50, 53, 54, 71, 75
spectral lines of hydrogen, 107
spectral width, 44, 54, 71
spectroscopic notation, **471**
spectroscopic stability, 188
spherical coordinates, 108, 109, **123**, 198, 199, 203, 205, 318, 473
spherical harmonic, 110, 473, **478**
spherical unit vector, 118
spin, 134, 227, 247, 471
  electron, 133
spin–orbit interaction, 148, 182–185, 198, 228, 252, 257, 288
  nuclear, 268
  other atoms, **137**
spin–orbit term, **133–137**, 186
spontaneous emission, 23, 57, 81, 82, **80–92**, 95–97, 102, 209, 347–352, 362–364, 414, 422, 423, 428, 436, 437, 443, 461
spring constant, 274, 275
  vibrational, 284
square well, 210, 335, 339
standing wave, 21, 361
Stark effect, **198–207**, 462
  hydrogen, **202–203**
  linear, **203–204**
  non-linear, **200–202**
  parabolic coordinates, **202–205**
  quadratic, **204–205**
  Rydberg atoms, 214
  spherical coordinates, **199–202**
Stark shift, 212, 222
  linear, 221
  quadratic, 201
Stark slowing molecules, 413, **415–418**
state
  electronic, 290–292

rotational, 290–292
vibrational, 290–292
statistical mixture, 94–97
Stern–Gerlach experiment, 133, 134, 161, 370
stimulated emission, 44, 81, 82, 347, 350, 354
stimulated rapid adiabatic passage, 77, **440–442**, 444, 445
  molecules, **425–426**
STIRAP, see stimulated rapid adiabatic passage
stochastic electrodynamics, 91
Stokes transition, 77
sub-Doppler laser cooling, 360, 362, 364, 415
sub-Doppler spectroscopy, 146, 299
superfluid, 384, 394, 395, 397, 400–403
superfluid–Mott insulator transition, **400–405**
superposition, **36–37**, 126, 436, 437
sympathetic cooling, see buffer gas cooling
synchronous molecule, 416, 417

temperature, 358
  limits, **355–356**
term symbol, 252
Thomas correction, **135–136**
Thomas–Fermi limit, 389
Thomas–Fermi radius, 399
Thompson model, 7
three-body recombination, 425, 426
three-level systems, **433–447**
time-resolved fluorescence, **56–62**
transition strength, 172, 291, **479**
triplet state, 230, 236, 260, 290, 291, 307
two-fluid hydrodynamics, see quantum hydrodynamics
two-level atom, 81, 83, 93, 97, 102, 347, 348, 358, 359, 362, 436
two-photon transition, 74

ultra-cold atoms, 331
ultra-cold chemistry, **321–343**
ultra-cold collision, 328
uncertainty principle, 315, 364
ungerade, 288, 304, 306, 307, 312, 318

valence bond approach, 276, 303, **311–312**, 312, 313, 319
van der Waals interaction, 273, 274, **283–284**, 285, 324, 327, 415, 423
vapor cell MOT, 376
variation of constants, **452–455**
variational method, 235, **239**, 243
vector model, 136
vector potential, 41, 51, 55, 78, 181
vertical transition, 293
vibrational level, 291, 378
virial theorem, 314, 315
virtual states, 76

Welton calculation, 140, 141, 148

Wigner function, 224
Wigner–Eckart theorem, 119, 173, 477
Wigner–Weisskopf model, **84–87**
WKB condition, 329
WKB method, 167, 168, 205, 207, 328

X-ray diffraction, 379

Young's double slit experiment, 58

Zeeman effect, 49, 60, **181–197**, 215, 216, 340, 373, 375, 438, 448, 455–457
  anomalous, 133
  hydrogen
    ground state, **188–190**
  positronium, **190–192**
  quadratic, 182, 213
  Rydberg atoms, **212–214**
  spin–orbit interaction, **183–188**
zero-point energy, 84, 89, 91, 209, 281
zero-point field, 140, 141